W0041982

THE CLIMATE

OF THE

CONTINENT OF AFRICA

THE CLIMATE

OF THE

CONTINENT OF AFRICA

BY

ALEXANDER KNOX, B.A. (Cantab.), F.R.G.S.

MEMBER OF CONVOCATION OF THE UNIVERSITY OF THE
CAPE OF GOOD HOPE

AUTHOR OF *NOTES ON THE GEOLOGY OF AFRICA*
GLOSSARY OF GEOGRAPHICAL AND TOPOGRAPHICAL TERMS
(STANFORD'S COMPENDIUM), ETC.

Cambridge:

at the University Press

1911

CAMBRIDGE UNIVERSITY PRESS
Cambridge, New York, Melbourne, Madrid, Cape Town,
Singapore, São Paulo, Delhi, Tokyo, Mexico City

Cambridge University Press
The Edinburgh Building, Cambridge CB2 8RU, UK

Published in the United States of America by Cambridge University Press, New York

www.cambridge.org
Information on this title: www.cambridge.org/9781107600713

© Cambridge University Press 1911

First published 1911
First paperback edition 2011

A catalogue record for this publication is available from the British Library

ISBN 978-1-107-60071-3 Paperback

Additional resources for this publication at www.cambridge.org/9781107600713

Cambridge University Press has no responsibility for the persistence or
accuracy of URLs for external or third-party internet websites referred to in
this publication, and does not guarantee that any content on such websites is,
or will remain, accurate or appropriate.

PREFACE

MY thanks are due, in the first instance, to Dr W. N. Shaw, LL.D., Sc.D., F.R.S., Director of the Meteorological Office, who, with the assistance of Mr R. G. K. Lempfert, M.A., also of the Meteorological Office, very kindly read the MS. of this work, and pronounced upon it.

The book, however, could not have seen the light, but for the very practical interest taken in the matter by the Council of the Royal Geographical Society, who not only paid for the reproduction of the maps, but also supplied the greater part of the funds necessary before the printing of the text could be undertaken. Towards this end assistance was also received from

The Royal Colonial Institute,
The African Society,
and   The Royal Meteorological Society:
and I beg here gratefully to acknowledge the kindness of these four Societies.

This united action on their part constitutes a most interesting fact, rare, though perhaps not quite unique, in the history of the publication of geographical works, and augurs well for the future.

My thanks are due to the Controller of H.M. Stationery Office and to the Hydrographer for leave to make use of *The Anglo-Egyptian Sudan* by Colonel Count Gleichen, K.C.V.O., D.S.O., and of *The Africa Pilot* respectively; and to Captain H. G. Lyons, LL.D., F.R.S., who, while Surveyor-General of Egypt, gave me permission to make general use of the publications of his Department, as well as of his work *The Physiography of the Nile and its Basin.*

The Agents-General of the Cape of Good Hope and of Natal I have to thank for the loan of the annual Reports of the Meteorological Commission, and those of the Government Astronomer respectively; and the Commissioner of Lands, Transvaal, for leave to make extracts from his Report.

I take this opportunity to convey my thanks to the many learned societies and kind friends who have in various ways assisted me in the compilation of the work.

The Royal Geographical Society, through its Secretary, Dr J. Scott Keltie, l.L.D., granted me permission to make general use of the *Geographical Journal,* and of this I have freely availed myself.

The Royal Scottish Geographical Society allowed me to make use of passages which occur in the pages of the *Scottish Geographical Magazine,* and the Manchester Geographical Society has, similarly, permitted me to quote from its *Journal.*

The Italian Geographical Society, through its Secretary, Signor Ferd. Rodizza, not only allowed me to make use of matter contained in its *Bolitino* and in Captain Ferrandi's *Lugh,* but Signor Rodizza also pointed out where other additional valuable material could be found, relating to other localities.

To the Royal Meteorological Society and its Secretary, Mr Marriott, I am indebted for the use of the Library and for valuable assistance.

La Société Royale de Médecine et de Topographie Médicale, Brussels, very kindly sent me the exhaustive report prepared by the late Dr Lancaster and Dr Moulmein on Belgian Congo, and of the meteorological matter contained in this a general use has been made.

Le Comité de l'Afrique française I have to thank for permission to use certain matter contained in its interesting *Renseignements Coloniaux* (supplements to the *Bulletin*).

To the British South Africa Company I am indebted for valuable MS., and printed series of meteorological observations; and to the Nyasa and Mozambique Companies for their Reports.

Dr Barot and Dr Ribot, in the course of correspondence, communicated valuable material with regard to their experiences in West Africa, and M. Gautier with reference to his journey across the Sahara from Algeria to Timbuktu, especially in connection with the rainfall in the neighbourhood of the Niger bend.

Colonel Sir Charles Watson, K.C.M.G., C.B., placed at my disposal the temperature observations which he recorded while at Suakin, and his Diary while up the Nile with General Gordon; and Lt.-Col. C. F. Close, C.M.G., R.E., similarly lent me his notes, made while acting as Boundary Commissioner both on the Cross River, and on the Nyasa-Tanganyika plateau.

I have to thank Mr C. Atchley, C.M.G., I.S.O., the Librarian, Colonial Office, and Mr F. J. Hudleston, the Librarian, War Office, for their cordial and sympathetic help.

It was thought advisable that those who live in Europe, or intending emigrants, should be enabled to compare the climatic conditions of their country with those of their possessions in Africa, and, with this end in view, tables for the capitals of the countries, which have the greatest territorial interest in Africa, have been introduced as Appendix II.

M. Angot, Director of the *Bureau Central Météorologique*, most kindly constructed for me a meteorological table for Paris.

M. Vincent of the *Service Météorologique* of the Royal

Observatory, Brussels, gave information relative to a similar table for Brussels, and Dr Kreuser of the *Meteorologisches Institut*, Berlin, sent me a table for that capital.

Signor Luigi Palazzo, Director of the *Ufficio centrale di Meteorologia e di Geodinamica*, very kindly sent me four published works by Tracchini and Eredia on the climate, sunshine, rain, and rainy days at Rome, and also the proof sheets of a work, then unpublished, by the latter, on the temperature for 50 years; and from these I was able to construct the table for Rome; while Senhor A. A. Vidal, Director of the *Observatorio meteorologico e magnetico do Infante D. Luiz*, most courteously constructed the table for Lisbon.

My thanks also are due to a considerable number of authors and publishers for leave to quote from, or to summarise passages in, their works, all of which will be found acknowledged in the text.

Especially am I indebted to those kind friends who actually contributed to the text: to Major-General J. K. Trotter, C.M.G., C.B., for his paper on Sierra Leone; to Major C. W. Gwynn, C.M.G., R.E., for his contribution on Abyssinia; to Capt. E. A. Steel, R.A., and Mr F. F. Hopkins, for papers on Southern and Northern Nigeria respectively; to Mr L. H. L. Huddart, M.A., A.R.S.M., for his notes on Mozambique: and, last, but by no means least, to my daughter, Mary S. Knox, who not only assisted me with the tables and the proofs, but contributed the explanatory botanical glossary which appears as Appendix I.

A word as to the arrangement of the text. The maps are first dealt with, then the general climatic conditions are considered for each month, after which the various countries, colonies, and protectorates are treated in detail in separate sections, grouped into four great divisions—North, Tropical West, Tropical East, and South—the second being subdivided into a Northern and a Southern area.

A paragraph on productions will be found at the end of each section (or group of sections), followed by climatological tables, the rainfall at selected stations being given last.

Owing to the fact that the first eighty pages of the text were in print, and the type distributed, before the black foundation of the maps was finally approved by the Royal Geographical Society, the spelling of some five or six names is not identical in the text and on the maps. The Algiers of the maps appears in its French form Alger in the text; the Kilimane of the text appears in its Portuguese form Quilimane on the maps; again the Marocco of the text is the Morocco of the maps and so on. This, though regrettable, was unavoidable in the circumstances.

In conclusion, this book, which has cost much time and labour, does not pretend, in any way, to be a meteorological treatise, but merely represents an endeavour to place before those who require it such accurate information concerning the climate of Africa, as is available.

A. K.

August, 1911

CONTENTS

	PAGE
Preface	v
The Maps	1
The general climatic conditions month by month	13
North Africa	32
Marocco	33
Algeria	44
Tunis	52
Tripoli	58
Lower Egypt	64
Upper Egypt	72
The Sahara	78
Tropical Africa.—The tropical seasons	84
Western Tropical Africa: **North Section**:—	
West Coast	92
British West Africa :—	
Gambia	99
Sierra Leone	104
Gold Coast	115
Ashanti	126
Northern Territories of the Gold Coast	128
Southern Nigeria	132
Northern Nigeria	146
French West Africa :—	
Senegal	156
Upper Senegal	161
Kasamanse	165
French Guinea	166
French Niger Territory	171
Ivory Coast	177
Dahomey	182
German West Africa :—	
Togoland	187
Cameroons	193
Portuguese Guinea	203
Liberia	205

PAGE

Western Tropical Africa : **South Section**:—
Spanish Guinea 207
French Equatorial Africa :—
French Congo 208
Ubangi-Shari-Chad Region 213
Belgian Congo 222
Angola, with Kabinda 243

Eastern Tropical Africa :—
The Winds and Rains of East Africa 254
Anglo-Egyptian Sudan 258
Eritrea 274
Abyssinia 284
The Horn of Africa, or Somaliland 296
Uganda Protectorate 308
East Africa Protectorate 317
German East Africa 329
Northern Rhodesia :—
North-Eastern Rhodesia 353
North-Western Rhodesia 362
Nyasaland 369
Mozambique 379

South Africa 390
Southern Rhodesia 395
The Transvaal 418
Orange Free State 429
Natal 436
Cape Colony 445
Basutoland 480
Bechuanaland 483
German South-West Africa 489
Walfisch Bay 500
Gazaland 504

Appendices :
I. Explanatory Glossary of the principal vegetable productions
except Timber-trees 511
II. Meteorological Tables for Brussels, Lisbon, Rome, Berlin,
Paris, and London 519
III. Tables 525
IV. Explanation of Terms 530
Index 533

*These maps are now available for download from
www.cambridge.org/9781107600713

ILLUSTRATIONS

to face page

Plate 1. Map showing the Mean Annual Rainfall 10

,, 2. Map showing the Mean Annual Rainfall in January

,, 3. ,, ,, ,, ,, ,, ,, February

,, 4. ,, ,, ,, ,, ,, ,, March

,, 5. ,, ,, ,, ,, ,, ,, April

,, 6. ,, ,, ,, ,, ,, ,, May

,, 7. ,, ,, ,, ,, ,, ,, June

,, 8. ,, ,, ,, ,, ,, ,, July

,, 9. ,, ,, ,, ,, ,, ,, August

,, 10. ,, ,, ,, ,, ,, ,, September

,, 11. ,, ,, ,, ,, ,, ,, October

,, 12. ,, ,, ,, ,, ,, ,, November

,, 13. ,, ,, ,, ,, ,, ,, December

In pocket at end*

,, 14. Diagram to illustrate the Section on Tropical Seasons 90

ABBREVIATIONS

abs.	= absolute
A. F.	= Bulletin du Comité de l'Afrique française (Paris)
A. G.	= Annales de Géographie (Paris)
alt.	= altitude
B. S. G.	= Bulletin de la Société de Géographie (La Géographie) (Paris)
B. S. G. I.	= Bollettino della Società geografica italiana (Rome)
C.	= Centigrade
D. K.	= Deutsches Kolonialblatt (Berlin)
Fahr.	= Fahrenheit
G. J.	= The Geographical Journal
G. Z.	= Geographische Zeitschrift (Leipzig)
J. R. A. M.	= Journal of the Royal Army Medical Corps
m.	= metres
max.	= maximum
M. D. S.	= Mitteilungen aus den Deutschen Schutzgebieten (Berlin)
M. G.	= Mouvement Géographique (Brussels)
min.	= minimum
mm.	= millimetres
P. M. G.	= Petermann's Mitteilungen, Gotha
R. G.	= Revue de Géographie (Paris)

THE MAPS

The Rainfall maps, which accompany this book, are new compilations, except that, for the extreme south of the continent, by permission of Dr J. G. Bartholomew, Dr Buchan's maps, which appeared in his large Atlas *Meteorology*, have been used.

These maps, showing the rainfall in the Cape Colony and neighbouring regions, were originally compiled from a very large amount of material extending over a considerable period, and it was thought quite unnecessary to attempt to recompile for this area.

It would be both superfluous, and uninteresting, except to a very few, to print here all the data on which the rain maps are based, and it will suffice if some of the main points in connection with the compilation of the "Annual Rainfall" map be stated, similar considerations underlying and guiding the construction of the monthly maps. In making an ordinary topographical map, the first desideratum is a series of points, fixed by triangulation, on which to base the compilation. Now, if we have the mean of ten years' rainfall, it takes a very considerable divergence from this mean, in subsequent years, to make any appreciable difference in the mean for the resultant number of years. I have therefore taken such positions as have a record extending over ten years and upwards for the basis of the maps, to act, so to speak, as my points fixed by triangulation.

Overleaf is a list of these long-record stations, forming the basis of the whole. And now to go to the other extreme. The lack of reliable information forces us at times to have recourse to biological considerations, and we have to rely on our knowledge of the area over which the equatorial forest extends, with the interaction between forest growth

K. A.   1

and rainfall, to form, in some of these equatorial regions, an estimate of the annual rainfall, confirmed here and there by a few more reliable details in the shape of the number of rainy days at various stations, and now and again short

		in.	mm.			in.	mm.
Algeria.	Alger	21	538	Cameroons.	Debunja	412	10469
	Batna	16	399		Duala	156	3953
	Constantine	25	632	Angola.	Loanda	11	270
	Oran	18	446	German	Olukonda	19	494
	Bougie	41	1036	S.W. Africa.	Waterberg	21	521
	Orleansville	17	442		Okahanja	16	404
	Bu Saada	11	270		Windhoek	15	382
	Biskra	7	171		Schaaprivier	14	354
	Laghuat	7	188		Rehoboth	10	265
	Ghardaia	4	104		Walfisch Bay	0·3	7·6
Tunis.	Maktar	14	365		Omaruru	14	352
	Tunis	18	451	Eritrea.	Addi Ugri	21	538
Tripoli.	Tripoli	17	420	British East	Kismayu	16	398
	Benghazi	12	300	Africa.	Malindi	41	1052
Egypt.	Alexandria	9	221		Takaungu	43	1101
	Cairo (Abbasia)	1·3	33		Mombasa	48	1211
	Suez	0·8	20		Shimoni	57	1435
	Ismailia	1·9	48		Machakos	37	950
	Port Said	0·8	21		Kikuyu(Ft Smith)	45	1144
	Suakin	9	217		Mumia's	71	1809
	Wadi Halfa	0	0		Entebbe	58	1473
Marocco.	Cape Juby	5	115	German East	Dar-es-Salam	46	1179
Senegal.	St Louis	16	413	Africa.	Tanga	61	1543
	Goree	21	521	Nyasaland.	Zomba	54	1382
Gambia.	Bathurst	50	1265	Portuguese	Lourenço-		
Sierra Leone.	Freetown	166	4220	E. Africa.	Marques	28	717
Gold Coast.	Christiansborg	23	575		Mopea	42	1077
Togo.	Misahöhe	59	1494	S. Africa.	Johannesburg	30	771
	Kete Krachi	52	1311		Pretoria	27	684

Note. The millimeters over 50 have been converted to the nearest inch.

series of observations. The absence of vegetation, on the other hand, leads to a diametrically opposite conclusion. This class of information might almost be compared with "native information" of topographical compilers, though of really a more reliable nature than the usual "native information" inserted in the maps of little-known areas, because the deductions are scientifically drawn. Between these two, the triangulated points and the local information, the data are numerous, and may be divided into such as would, in a topographical map, be represented by astronomically fixed positions of varying accuracy, say positions with rainfall records for from five to ten years to correspond with fairly well fixed positions. Stations with only from two to four years' record would then have the relative value of stations fixed by compass traverse, also of varying accuracy. Stations with only one year's record, though useful in drawing general conclusions, should be used only with care, as possibly leading to very unreliable results. Many of the stations will be mentioned in the short discussion which follows, especially those which have a governing influence on the isohyets. Others will be found in the lists of selected stations at the end of each section.

Having plotted all the stations on the map, due regard being had to their relative value and the one-year results being viewed with a certain amount of suspicion, we may begin to draw the lines of equal rainfall, having determined beforehand what the magistral lines shall be. The Annual Rainfall map is divided into areas or layers the limits of which are the isohyets (in mm.) 0, 250, 500, 750, 1000, 1500, 2000, 3000, 4000, there being no higher limit, all stations having a rainfall of over 4000 mm. being ranked together. The corresponding figures in inches are 0, 10, 20, 30, 40, 60, 80, 120, 160. I believe this is the first occasion on which an isohyet of 750 mm. (30 inches) has been introduced on any English rainfall map of Africa, the data, hitherto, not having, apparently, been deemed sufficient to warrant such a course. Dr Fraunberger, however, constructed a map, which appeared in *Petermann's Mitteilungen*, Band LII, No. 4, 1906, with isohyets of 200, 400, 600, 800, 1000, 1300, 1600, 2000, 3000, 4000 mm., but the data seem scarcely to warrant so much differentiation in the layers throughout the whole of the continent, though for certain areas there may be sufficient information available for the purpose. The latest English rainfall maps of Africa, as a whole, are those of Dr Herbertson, which appeared in *The Distribution of Rainfall over the Land*, published by the Royal Geographical Society in 1901, and since then very many rain stations have been established, giving a correspondingly large number of results. The most noticeable additions to our knowledge arise from the establishment of a large number of stations in the Anglo-Egyptian Sudan, in British East Africa, in Nigeria and the Gold Coast, in Rhodesia, and in the German Colonies, especially in German South-West Africa, German East Africa and Togo, and from the inauguration of a *Service Météorologique* for French West Africa.

For the purpose of drawing the isohyets with as little liability to error as possible, the stations were originally plotted on a map on the scale of 1 : 7,500,000, as a matter of fact on tracings of Stieler's map on that scale, the sheets of which form part of his Hand Atlas; where the stations were very close, maps on larger scales were used, viz. 1 : 6,336,000, 1 : 4,000,000, and 1 : 3,421,440. These maps, when the isohyets had been drawn, were reduced to a scale of 1 : 28,512,000 or one inch to 450 miles, and a further reduction to 1 : 40,000,000 was made for publication.

In the extreme north is a number of small *enclaves*, some of greater rainfall, some of less, than that of the surrounding regions. That on the extreme west is governed by Cape Spartel and Tangier, the former having a mean annual rainfall of 32·59 inches (828 mm.) according to Dr Fischer[1], who quotes from Dr Hahn, and 27·99 inches (711 mm.) according to the table of meteorological observations given by the Hydrographer in the *Africa Pilot*. The mean of the two records is 30·28 inches (769 mm.); and the mean annual rainfall at Tangier is 32·16 inches (817 mm.); and therefore these two stations lie in the layer of between 30 and 40 inches (750 and 1000 mm.) of rainfall. Passing eastwards, the next *enclave* is one of lower precipitation than the adjacent neighbourhood, and rests on the following evidence, viz.

	in.	mm.
Nemours with a rainfall of	18·43	468
Sidi bel Abbes	15·67	398
Oran	17·56	446
Orleansville	17·40	442

Alger, with 21·18 inches (538 mm.), is responsible for the next *enclave*; and Bougie, with 40·8 inches (1036 mm.), and Fort National, with 44·14 inches (1121 mm.), for the next. When we get into Tunis, we know that Tabarka has a rainfall of over 40 inches (1000 mm.), and that the

---

[1] Dr Fischer has constructed a rain map for Marocco (*Zeitschrift der Gesellschaft für Erdkunde zu Berlin*, Band XXXV, 1900, p. 365) in which he links up this coastal *enclave* with the summits of the Atlas, showing no intermediate zone, which appears somewhat arbitrary.

mountains in the rear are similarly favoured. For the first of the two remaining *enclaves*, Tripoli, with 16·58 inches (420 mm.), is responsible, while the second is due to our knowledge that Benghazi has a rainfall of 12 inches (300 mm.), and that Tobruk has a fall equal to that of any place in Barka, and hence its well-known fertility. The remaining isohyets in the Algeria-Tunis area are governed by the following groups of stations :—

(1) Stations having a rainfall between 30 and 40 inches (750 and 1000 mm.).

		in.	mm.
Blida	... ...	36·30	922
Medea	... ...	29·70	754
Dellys	... ...	35·19	894
Tizi-Uzu ...	...	32·38	823
Philippeville	...	30·16	766
La Calle ...	...	33·85	860

(2) Stations having a rainfall between 20 and 30 inches (500 and 750 mm.).

		in.	mm.
Tlemsen ...	...	24·84	631
Tiaret	... ...	29·29	744
Teniet-el-Had	...	25·00	635
Constantine	...	24·88	632
Gelma	... ...	25·08	637
Maktar	... ...	22·24	565

(3) Stations having a rainfall between 10 and 20 inches (250 and 500 mm.).

		in.	mm.
Tunis	... ...	17·76	451
Susa	... ...	15·71	399
Batna	... ...	15·71	399
Setif	... ...	17·72	450

Sfax with 8·19 inches (208 mm.), Biskra with 6·97 inches (177 mm.), and Laghuat with 7·36 inches (187 mm.) settle the limit, on the south, beyond which the isohyet of 10 inches (250 mm.) must not pass. The position of Laghuat shows that not only the 10 inches (250 mm.) line, but also the 20 inches (500 mm.) line must pass between that station on the south and Tiaret and Teniet-el-Had on the north.

With regard to the Maroccan area, we have a fair amount of information so far as the coast is concerned, but for the interior there are next door to no instrumental data, and here we must be governed by biological considerations. The belt of country lying nearly adjacent to the Atlantic coast, and including the middle courses of the streams which flow from the Atlas, is the granary of the country. This belt includes all coastal areas up to an altitude of about 1600 feet; and to the east lies a steppe region with, certainly, less rain than the coastal lands. Now we know that, in about the same latitude, in Tunis, corn will grow anywhere where the precipitation is as much as, or a little more than, $17\frac{3}{4}$ inches (450 mm.) in the year. And therefore this region, up to an altitude of 1600 feet[1], including the town of Fez, in the north, but not reaching so far inland southwards, will be included in the hyetal zone between 10 and 20 inches (250 and 500 mm.), while the steppe region, including Márrakesh, for which we have some few particulars, will be placed in the zone between 0 and 10 inches (0 and 250 mm.).

The isohyets in the borderlands of the Sahara and Western Sudan depend mainly on the values for St Louis, Timbuktu and Lake Chad. The rainfall at the first is gradually growing less, as is pointed out in the section on Senegal; for the second we have a four-year mean; with regard to the third some explanation appears to be necessary. There is an old set of rainfall observations for a year taken by Corporal Church (one of Vogel's party) in 1854 at Kúkawa, the result of which is given in the *Journal of the Royal Geographical Society* (vol. xxv, 1855, p. 241). During the rainy season, from the beginning of July to the end of September, 19·5 inches of rain fell, and during the whole year no more than 21 inches. Thus the fall for the year is 534 mm., and for the rainy season 495 mm., and not 352 and 327 mm. respectively as miscalculated in the *Meteorologische Zeitschrift*[2], and copied by Dr Fraunberger[3] for the purposes of his map. Again, with regard to the number of rainy days, we have the following records[4]:

---

[1] *See* De Roquevaire's hypsometrical map of Marocco in *Annales de Géographie*, 1901, p. 330.

[2] Vol. xx, 1903, p. 476.

[3] *Petermann's Mitteilungen*, 1906, Band LII, p. 75.

[4] In a similar calculation in the *Meteorologische Zeitschrift*, Barth's record is omitted.

	July	August	September
Rohlfs (P. M. G. Ergänz. 34, 1872), 1866–7		11	
Nachtigal (Nach Sah. u. Sud. Vol. II), 1870		19	11
Barth (Vol. III), 1851		11	
Denham (Denham and Clapperton), 1825	17		
Mean	17	14	11 = 42

If the rainfall for the rainy season be distributed in the proportion of 17, 14 and 11, we obtain the following for the mean rainfall for July, August and September:

	in.	mm.
July	7·87	200
August	6·50	165
September ...	5·12	130

There appears, however, to be little doubt that, since 1854, the rainfall in this region has gradually diminished. Chevalier[1] reduces the season of rains to $2\frac{1}{2}$ months, and the number of falls to about 15, the season being much shorter than in Bagirmi, while in Kanem further north (N.E. of Lake Chad) the period during which it rains is reduced to two months; and at Bir Alali it rains on only 14 days in the year[2]. Chevalier gives the rainfall in Bagirmi as some 500 mm. and therefore that at Lake Chad must now be considerably less. The fall too cannot be reckoned on, for in 1907, the "rainy season," we are told by the Boundary Commissioners, consisted of one rainy day[3].

Again, in the region of the Niger bend, Dr Fraunberger sets great store by Sievers's *Florenkarte*, and takes Dr Herbertson to task for drawing a straight line through an area for which no data were available. This *Florenkarte* was originally published in 1891 (when there was little more information, in this region, for a botanical map than there was for a rain-map) and was reproduced, with little or no alteration, in this area, in 1901 and again in 1903. Dr Fraunberger makes his layer of 200—400 mm. of rain follow what he calls the Halfa zone of Sievers's map. It is quite true, of course, that the rainfall is reflected in the vegetation, and *vice versa*, but in this instance it appears to

---

[1] *Mission Chari—Lac Tchad*, par Aug. Chevalier, Challamel, Paris, 1908, pp. 397–735.

[2] Rouget, *L'Expansion Coloniale au Congo français*, Larose, Paris, 1906, p. 273.

[3] *Scottish Geographical Magazine*, 1908, p. 375.

be a case of the blind leading the blind. Dr Fraunberger also follows this zone in the Tripoli area. Now M. Gautier lately, on his way south to Gao, traversed a belt of mimosa bush (*Acacia arabica*) extending for some 400 or 500 kilometres north of the Niger bend, which lies in the so-called Halfa zone. I am informed by the Kew authorities that this bush would thrive on less than 10 inches (250 mm.) of rain, if there were water fairly near the surface; and, further, M. Gautier, who is well qualified to speak of both regions, tells me that the soil here is underlain by clay, and that water can be found at a short distance from the surface. He also says that nowhere in the Algerian Halfa region, which corresponds with the southern highlands, is there a rainfall of as much as $7\frac{3}{4}$ inches (200 mm.). It would thus appear that Dr Fraunberger has been doubly misled in following the *Florenkarte*.

The narrow bands of small precipitation to the East of Cape Three Points are governed by:

(1) Kwitta, with 19 inches or 482 mm.

(2) Accrá, to the west of Kwitta, with nearly 26 inches or 653 mm., and, on the east of the same place, Lome, with about 25 inches or 639 mm., and Kpeme, with about 25 inches or 624 mm.

(3) Sékondí, with 37 inches or 931 mm., and Cape Coast, with 34 inches or 862 mm. on the west, and Little Popo, with about 32 inches or 801 mm., on the east.

Aburi, with 40 inches or 1015 mm., shows that the three isohyets of 20, 30 and 40 inches or 500, 750 and 1000 mm. must pass to the south of this neighbourhood.

The isohyet of 40 inches, or 1000 mm., in crossing the Western Sudan, has to pass north of Gambaga (46 inches or 1165 mm.) and Sansanne Mangu, and north of Zúngeru ($44\frac{1}{2}$ inches or 1130 mm.), but dips down south of Yola (35 inches or 881 mm.), to rise again and pass north of Lai (44 inches or 1117 mm.), Fort Archambault (42 inches or 1059 mm.), Fort Crampel (50 inches or 1275 mm.). Further east it passes south of Wau (33 inches or 844 mm.), and its position is definitely determined by Mongalla and Gondókoro, between which it must lie, since the annual values for these two stations are 38 inches or 971 mm. and 42 inches or 1073 mm. respectively. The easternmost position of the isohyet of 60 inches or 1500 mm. is governed by the two stations Entebbe (60 inches or 1501 mm.) and Jinja at the Ripon Falls, with 46 inches or 1166 mm., the southerly trend of this line being determined by Mbarara (69 inches or 1762 mm.), while it passes near Kabambara ($58\frac{1}{2}$ inches or 1485 mm.), and to the south of both Lusambo (66 inches or 1677 mm.) and Luluabourg (61 inches or 1543 mm.).

In the Abyssinia-Sudan area Kássala, Gedáref, and Gállabat being in three different layers, and lying roughly N.—S., with the values $11\frac{1}{2}$, 24, and 33 inches or 293, 612, and 841 mm. respectively, play an important part in fixing the position of the isohyets. In the same layer as Kássala, we have Wad Médani, El Dueim, and El Obeid, with 16, 11, and 14 inches or 410, 282, and 354 mm. respectively; while with Gedáref we have Kodok (28·3 inches or 719 mm.), but Doleib Hilla and Nasser both lie in the layer of higher precipitation, thus determining the position of the isohyet of 30 inches or 750 mm., while Gaba Shamba (26 inches or 663 mm.) indicates that it must pass below that station. Again, Gambela on the Baro R., with 51 inches or 1305 mm., shows that the 40 inches or 1000 mm. isohyet must include this station within its area.

The small area near Massaua, of over 20 inches or 500 mm., is accounted for by Ginda with 24 inches or 610 mm., and Addi Ugri with 21 inches or 538 mm.

Along the Zanzibar coast, reaching at least from Malindi on the north to a little below Dar-es-Salam on the south, is an area of heavy rainfall, enclosing a smaller coastal belt of still higher precipitation, as is shown by the following figures for the total mean annual rainfall at the places mentioned (from north to south):

	in.	mm.
Malindi	41·43	1052
Takaungu	43·36	1101
Mombasa	47·69	1211
Shimoni	56·51	1435
Tanga	60·25	1530
Pangani	47·31	1203
Bagamoyo	40·56	1030
Dar-es-Salam ...	46·43	1179

Shortly below the last-named place this high-precipitation belt ends, and the figures for Lindi and Makindani are 32 and 36 inches or 813 and 918 mm. respectively.

This belt extends at least as far inland as Mrogoro, which has a mean annual fall of nearly 49 inches (1234 mm.). Inland there is a large area of comparatively small rainfall which is situated mainly in German East Africa, but extends northwards into the British sphere, and just touches Victoria Nyanza at Shirati. If we take a set of stations from east to west the existence of this area of small rainfall becomes apparent, and the gradual fall from the coastal area, mentioned above, through an intermediate zone of less precipitation is also shown, as well as the subsequent rise again on the west, the intermediate zone encircling the low precipitation area.

Stations from East to West.	Mean annual rainfall.	
	in.	mm.
Dar-es-Salam	46·43	1179
Mrogoro	48·59	1234
⎧Mpwapwa	29·21	742
⎩Kilossa	28·07	713
Kondoa-Irangi	24·02	611
Tabóra	26·50	675
Ujiji	32·75	832
Kabambara (W. of Tanganyika)	60·92	1547

In the north we have the line Fort Hall—Naivasha—Shirati, showing a similar fall (east to west).

	in.	mm.
Fort Hall	47·57	1208
Naivasha	31·02	788
Shirati	25·43	646

Again, if we take two lines running north and south we shall find that they intersect the low rainfall area, thus (south to north)

	in.	mm.
Manow	96·67	2455
⎧Tabóra	26·58	675
⎩Kondoa-Irangi ...	24·02	610
Muanza	42·61	1082

and again,

	in.	mm.
Mahenge	73·44	1865
Kilossa	28·07	713
Mpwapwa	29·21	742
Kondoa-Irangi	24·02	610
Ukamba Province : Nairobi	40·01	1016

This area of low precipitation pushes up the Rift Valley and reaches beyond Baringo (26 inches or 662 mm.), and

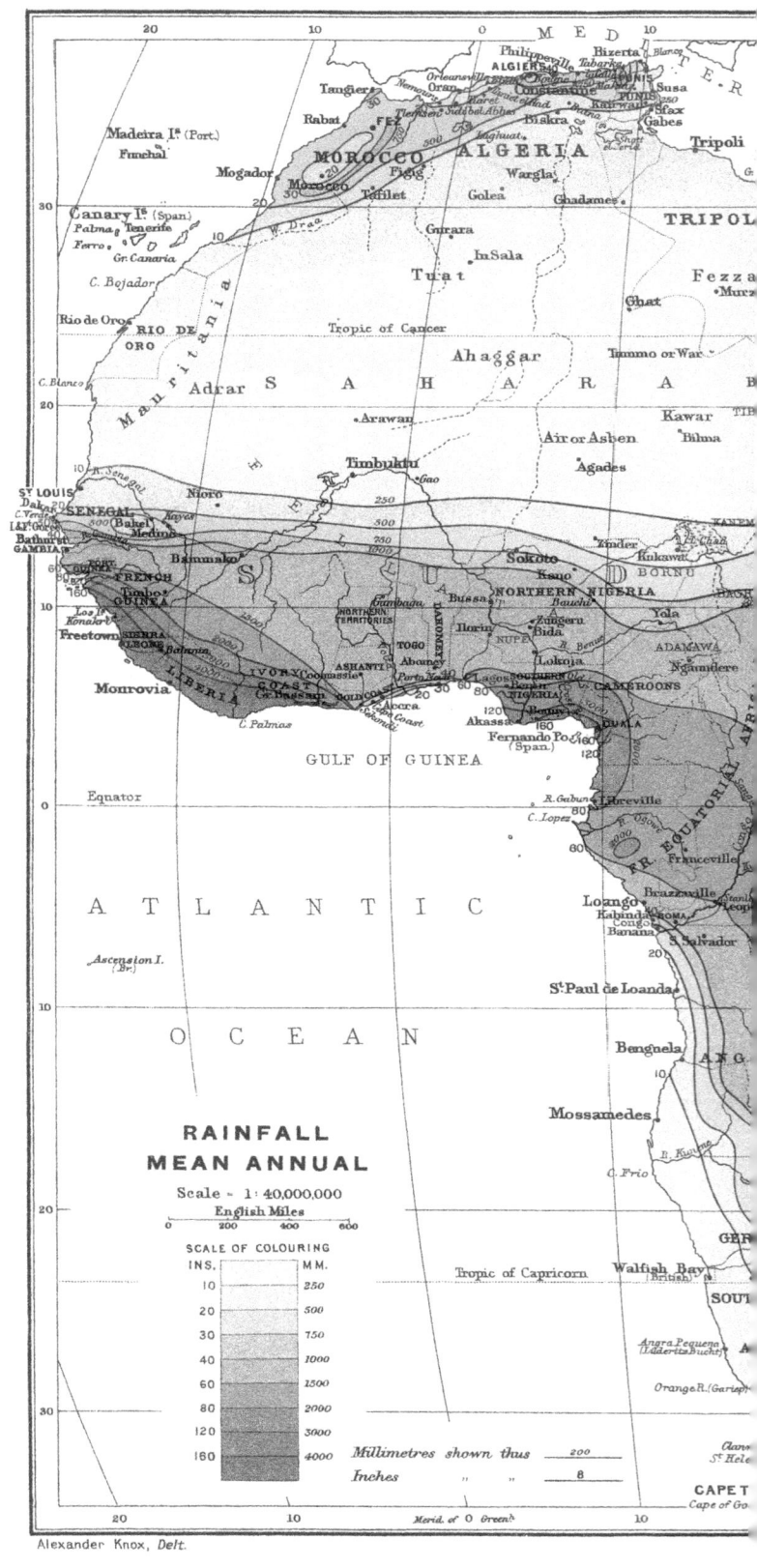

RAINFALL
MEAN ANNUAL

Scale ~ 1:40,000,000

English Miles

SCALE OF COLOURING

INS.		M.M.
10		250
20		500
30		750
40		1000
60		1500
80		2000
120		3000
180		4000

Millimetres shown thus _____ 200 _____

Inches   "   "   8 _____

Alexander Knox, Delt.

Bartholomew, Edin.

on the north-east extends towards the railway line in the vicinity of Makindu.

If we consider the mean annual rainfall along the railway, and to the north of it, at Machakos, Fort Hall, and Kitui, we find a drop from the high figures of the coastal belt to $37\frac{1}{2}$ inches (950 mm.) at Mackinnon Road, $36\frac{1}{2}$ inches (924 mm.) at Voi, 35 inches (889 mm.) at Makindu, $30\frac{3}{4}$ inches (781 mm.) at Athi River, and $37\frac{1}{2}$ inches (950 mm.) at Machakos; and this points to the fact that the intermediate belt (30 to 40 inches or 750 to 1000 mm.) surrounding the low precipitation area (20 to 30 inches or 500 to 750 mm.) passes northwards to join the similar area, which, beginning at the coast from Kismayu southwards, includes Lake Rudolf, and encircles the Abyssinian plateau. There is also a second junction between the latter and the southern surrounding intermediate area in the northward extension of the Rift Valley, as is clearly shown by the figures $33\frac{1}{2}$ inches (830 mm.) for the mean annual rainfall at Naivasha, and $31\frac{1}{4}$ inches (794 mm.) at Nakuru; while between the two northerly extensions of the low precipitation area, the one towards Makindu, the other towards Baringo, lies an area of high precipitation, in evidence of which we have the following figures for the mean annual rainfall at

	in.	mm.
Kitui	42·81	1087
Fort Hall ...	47·57	1208
Nairobi	40·13	1019
Kikuyu	45·05	1144

This area forms, as it were, an easterly outlier of the great equatorial high-precipitation belt which is continued from the west through Uganda into the East Africa Protectorate as shown by the following figures for the annual rainfall—

	in.	mm.
Entebbe	59·07	1500
Kisumu	51·59	1310
Port Florence... ...	57·65	1464
Nandi	79·67	2023
Molo	73·01	1754

And here, between Molo and Kikuyu, the break occurs, but, as we have seen, the great equatorial belt reappears again in the outlier above mentioned and in the neighbourhood of the east coast.

The small area of comparatively low rainfall on the north-east of Tanganyika is accounted for by Ujiji ($32\frac{3}{4}$ inches or 832 mm.) and Usambura ($32\frac{3}{4}$ inches or 832 mm.).

In the more southern isohyets Lealui plays an important part as a somewhat isolated station with an eight years' record giving 34·43 inches or 872 mm., and Mopea on the Zambezi delta with 42 inches or 1067 mm. is another governing station. The sharp curve of the 30 inches or 750 mm. isohyet north of Salisbury ($34\frac{1}{2}$ inches or 875 mm.) in Rhodesia, is caused by the values for Sipolilos and Mt Darwin, viz. 24·29 and 26·79 inches or 617 and 680 mm. respectively. The small area near Salisbury with over 40 inches or 1000 mm. is accounted for by Chishawasha with 41·09 inches or 1044 mm.

On entering German South-West Africa the direction of the isohyet of 10 inches or 250 mm. is governed by the situation of—

			in.	mm.
Franzfontein	...	...	8·94	227
Outyo	...	...	17·60	447
Omaruru	...	...	10·87	276
Okombahe	...	...	7·13	181
Karibib	...	...	6·02	153
Wilhelmstal	...	...	14·92	379
Heusis	...	...	11·30	287
Rehoboth	...	...	10·47	266
Hoakhanas	...	...	9·37	238
Gokhas	...	...	9·73	247

while the governing stations for the 20 inches or 500 mm. isohyet are—

			in.	mm.
Onjipa	...	...	22·87	581
Ondangui	...	...	22·44	570
Olukonda	...	...	22·68	576

which lie close to one another, and the isohyet must pass to seaward of them. Immediately to the south lies Okankulyo with 19·4 inches or 492 mm., and therefore the line must almost touch this station, and is governed further south by the pair of stations Ekotoweni with $16\frac{1}{2}$ inches or 421 mm., and Naidaus with $23\frac{1}{3}$ inches or 592 mm., so that the isohyet must pass between them. Continuing southwards we have Otyikango with $18\frac{3}{4}$ inches or 477 mm., and Omalako with 20 inches or 500 mm. The line then curves in an easterly direction and leaves the Protectorate.

THE GENERAL CLIMATIC CONDITIONS MONTH BY MONTH

JANUARY.

The sun is near its southern limit in January, and is in the zenith at noon between 23° S. and 17° 20′ S. approximately in any year and the thermal equator is far south of the geographical equator. The highest isotherm, so far as Africa is concerned[1], is 90° Fahr., including an area having that temperature and over, but under 95°, which lies roughly in the form of an ellipse, with its major axis extending from the Ruaha Valley, in German East Africa, to the junction of the Vaal River with the Orange River, and including the southern part of Tanganyika and Nyasa, the minor axis being about one-third the length of the major axis.

Corresponding with this area of high temperature is a lowest-pressure (cyclonic) area of approximately the same form, but narrower in the south and extending further north, its bounding line meeting the Congo River about Stanley Falls, the Nile at Wádelai and, curving round south of Lake Rudolf, almost reaching as far east as Mombasa. Speaking generally, the low-pressure areas are in the tropical regions.

Of the two isotherms of 80°, the northern enters Africa north of Sierra Leone, bends slightly north and extends across the continent to Zeila (Somaliland) and then dips south-eastward, cutting off the extremity of the Horn. The southern touches the Ivory Coast and the mouths of the Niger and then runs southwards, roughly parallel

---

[1] The remarks on temperature and pressure are based on the maps in Dr Bartholomew's Atlas *Meteorology*.

with the coast, but inland, cutting the Lower Congo, and, curving round the south of the $90°$ elliptical area, leaves the continent not far from Beira. These two $80°$ curves may be said to include within them the Tropical Province, which is a low-pressure area.

Two great high-pressure (anticyclonic) areas lie east and west of South Africa, the latter being the more formidable, and its centre closer to the coast, while the North-Atlantic high-pressure area includes Marocco, Algeria, and Tunis.

On the north-west, from Gibraltar to Sierra Leone, the general direction of the wind is from the north-east; and these winds are the "North-East Trades." If we now trace the winds from Gibraltar along the north coast, down the Red Sea, along the east coast, round the Cape and up to Sierra Leone again, we shall find that, with one or two local exceptions, the wind tends to follow the clock. Along the north coast (except at Benghazi, where the direction is N.E.) the general direction is from the north-west; down the Red Sea it is north, and (omitting the entrance to the Red Sea, where the winds are S. and S.W.) then turns slightly to the east, being N.E. over the coastal regions of Abyssinia (gradually turning, in the interior of Abyssinia and over the Egyptian Sudan down to about $5°$ N., to north, on account of the friction over the land) and N.E. all down the east coast, as far as, and beyond Cape Delgado, in consequence of the influence exerted by the inland cyclonic area. Here the winds turn S.E. to as far south as below Natal, but remain N.E. inland over Natal, the Transvaal and Southern Rhodesia. At Port Elizabeth the winds are south-west, south at Cape Town, south-west and west on the German South-West Africa coast, owing to the western high-pressure area (west at Loanda), and south-west along the coasts of the Gulf of Guinea as far as Liberia.

Within the high-temperature and low-pressure area we have the most rain, culminating with the highest rainfall for the month (8—12 inches) around Nyasa and westwards to the upper waters of the Zambezi, Kasai and Kunene, and extending northwards to include Lusambo and Kabambara. There is also rain almost all along a narrow coastal strip from Gibraltar to Egypt, the greatest amount (2—4 inches) being measured in Algèria and Tunis, on the northern slopes of the mountains. The West Coast rainfall is at its lowest, while, on the western coast south of the Congo, the influence of the cyclonic area over the South African plateau is shown in the weakening and westing of the winds (*e.g.* at Loanda) and lessening the rainfalls, which also reach a minimum at this season, to rise later and then fall again. This is one of the rainy months on the equatorial lake plateau, but in the region to the north, including the Bahr-el-Jebel, Bahr-el-Ghazal and Sobat areas, it is one of the dryest months of the year.

FEBRUARY.

In February the sun is in the zenith at noon in any year between $17°\ 3'$ S. and $7°\ 48'$ S. approximately. The climatic conditions in this month are very similar to those in January. The distribution of temperature is practically the same, except that a high-temperature area of 90° Fahr. and over has developed north of the equator, in the form of a triangle, with its apex between Lake Chad and the Ubangi bend, and including the lower Sobat River, Bahr-el-Jebel and Bahr-el-Ghazal regions. The pressure conditions are also practically unaltered, except that the extreme low-pressure area lies now in a N.W.—S.E. position, extending from the Niger-Benue district to Madagascar, and including the whole of the lake region, the Congo River as low as the Ubangi junction, and Cameroons. The winds are also practically the same as in January.

The rainfall decreases on the North African coast, the upper limit of the tropical rain area is moved a trifle farther north, and the West Coast coastal slopes receive a greater precipitation. On the east, the immediate Nyasa region receives even a slightly greater rainfall than in January. February is a relatively dry month in the region of the equatorial lakes and northwards. But the rain is steadily moving northwards and is increasing in the south of German East Africa.

MARCH.

In this month the sun is in the zenith at noon in any year between $7° 26'$ S. and $11° 19'$ N. approximately, and the thermic equator is pushed much farther north. The extreme high-temperature area of $90°$ and over, to the south of the equator, has diminished to a little ellipse lying farther west, almost midway between the east and west sides of the continent from Matabeleland to the latitude of Bangweulu. On the other hand the northern area has largely increased in dimensions and extends across the continent covering the Western Sudan, and including the upper White Nile and Bahr-el-Jebel and reaching to the Abyssinian highlands.

The high-pressure areas in the South Atlantic and Indian Ocean occupy about the same position, but have become more formidable in character, the central curves being marked 30·15 and 30·20 inches respectively in place of 30·10 and 30·15 inches in February, and at the same time the high-pressure system has approached, and, in the extreme south and west, covered South Africa, the 29·95 inch isobar entering the continent about Walfisch Bay, curving down to the Karoo, and re-entering the ocean south of Natal. The low-pressure area has expanded along its minor axis and turned somewhat more N.—S., and extends, along its major axis, from the north of Nyasa up to and including Tibesti, and, touching the tropic of Cancer, with an extreme depression of 29·75 inches in the centre, reaching from a little south of the equator to the south of Tibesti, having followed the sun northwards. The Marocco—Algeria—Tunis area still lies in the northern high-pressure zone. On the whole the low-pressure lines conform more to the shape of the continent, being deepest in the heart.

Up the Mozambique Channel the winds are S.E., S.W. and S.E. again, about Mombasa N.E., and S.E. on the South Benadir coast. On the north the winds are still N.W., except on the Western Tripoli coast, where they are N.E. There are areas of calm about the Niger delta, over Tanganyika and to the south of Victoria Nyanza, while the south-westerly winds still continue to blow parallel with the extreme southern coast of the continent. The prevailing winds on the Abyssinian plateau and in the Egyptian Sudan are practically the same as in January.

With regard to the rainfall, we must consider the effect of the low-pressure area on the winds and the regions from which these winds come. From dry regions on the north, extreme north-west, and north-east, dry winds rush in to the northern part of the low-pressure area and consequently bring no rain. The rains are less in the south of the Nyasa region and begin to be felt in German East Africa and the Congo districts to the west, though between 6° N. and 20° S. over 100 mm. (4 inches) fall. South of this there is less rain, except on the S.E. coast, where the trades are moist. In the centre is an area with 200 mm. (8 inches) of rainfall. In the north-west, the low-pressure area extends over the Sahara-Sudan region and the air rushes in from the Gulf of Guinea, moisture-laden, from over the ocean, and releases this moisture as it rises over the higher lands. The coastal strip south of Loanda is still a region of low precipitation, less than 50 mm. (2 inches) being precipitated.

APRIL.

In April the sun is in the zenith at noon in any year between 4° 42' N. and 14° 54' N. approximately. The southern area of 90° Fahr. and over has disappeared, and the larger area north of the equator, which had a south-easterly extension touching Albert Nyanza, has moved to a more general E.—W. position, and is more of a latitudinal belt in form. The northern isotherm of 80° Fahr. occupies about the same position on the west, entering the continent at the Gambia, and curving northwards, but, on the east, it is bent down along the south-western shore of the Red Sea, and runs along the coast of Somaliland and the Horn. The southern isotherm of 80° enters the continent north of the equator, on the Cameroons coast; its position lies slightly to the east of the March position, and it also does not reach quite so far south, leaving the continent in Gazaland.

The pressure has risen considerably over the southern hemisphere and the whole of South Africa lies within the great southern high-pressure area ; all, except the coastal fringe, of Marocco, with Algeria, has been removed from the great northern high-pressure area. The great central low-pressure area has reversed its position from N.W.— S.E. to one stretching from the Niger to the S.W. coast of Arabia, and lying wholly north of the equator, in sympathy with the great northern heat area.

In south-eastern Africa, in the summer rain area, the precipitation has very sensibly decreased, from about 4 inches to 3 inches, and, on the other hand, that of the Cape Peninsula has doubled, the March figure being about $1\frac{1}{2}$ inches and that for April 3 inches. On the Cameroons coast the region of the Cameroons Mountain has a precipitation of over 12 inches, and so also has the Libreville neighbourhood ; and the region receiving eight inches and over has moved in a north-westerly direction ; isolated areas with the same amount of rainfall occur on the coast of the north part of German East Africa and the extreme south of British East Africa and also about midway between the coast and Victoria Nyanza, as well as on the coast between Capes Palmas and Three Points. There are heavy rains in the north-eastern Congo Forest area. Slight rains fall throughout Somaliland and right across Africa, including a narrow strip touching the Bahr-el-Ghazal, and extending to the coast of French Guinea, and increase in intensity southwards over the Sudan.

The most noticeable change in the winds is the reversal of direction on the northern Benadir coast from N.E. to S.W. The wind in the eastern part of South-East Africa, over the land, has acquired a more easterly course in Natal, the Transvaal, and Orange Free State, with S.E. winds further south. Variable winds are experienced in the neighbourhood of Khartoum and to the north of Victoria Nyanza, and there are areas of calms on northern Tanganyika and to the east, and also on the Niger delta.

MAY.

In this month the sun is in the zenith at noon in any year between $15° 21'$ N. and $21° 59'$ N. approximately. This is the transition month between the northern spring and summer; and the thermal equator, the low-pressure area and the rain belt are well established in a northern position. High temperature prevails in all the tropical and sub-tropical regions. The region of highest temperature ($90°$ Fahr. and over) lies over the Sahara and the greater part of the Sudan in a form closely resembling the shape of the continent, bulging northwards towards Algeria and westwards towards Senegal and including, on the east, the Abyssinian highlands and the Ghazal region. This isotherm of $90°$ is followed on the north by those for $85°$, $80°$, $75°$, and $70°$, the last of which skirts the coast inland from Cape Blanco to the Nile delta. The southern isotherm of $80°$ touches the Niger delta and Cameroons, runs S.E. to the north of the Limpopo River bend, and leaves the continent at the Zambezi delta. The great southern high-pressure area extends northwards to above the tropic, and its northern isobaric border lies, roughly, immediately south of lat. $20°$ S.; and an anticyclonic area of $30·10$ inches pressure appears over the inland portion of S.E. Africa. The low-pressure area of $29·80$ inches has left the west coast, and moved eastwards, touching the Niger from Say to the Benue River junction, and including the east coast from about $25°$ N. to the middle of the Benadir coast. The extreme low-pressure area of $29·70$ inches lies in the form of an ellipse with its major axis extending from Berber to Lake Chad. From the Zambezi mouths northwards to the Horn the prevailing winds are south, with more and more westing, till S.W. is reached at Guardafui. Southerly winds are experienced from the south of Tanganyika to Gondókoro, easterly winds in the Orange Free State, westerly winds at East London, north-westerly at Port Elizabeth, south-westerly at Cape Town, and north-westerly over Little Namaqualand and Clanwilliam.

There is an area of calms in the centre of the southern anticyclonic area, and variable winds are still the rule at Khartoum.

2—2

The tropical rain area is bounded on the north by a line which, running from Portuguese Guinea to the south of what remains of Lake Chad and curving northwards, passes south of Wad Médani, through Gedáref and leaves Addi Ugri to the south ; and on the south by a line from the Congo mouth, which, curving first slightly southwards and then slightly northwards, cuts the south end of Tanganyika, and, turning S.E. and then sharply to the south, cuts the east coast midway between Mozambique and Kilimane. The precipitation increases gradually from these two lines towards the middle of the belt, except in German East Africa. The band 4—8 inches rainfall extends right across the continent, bounded on the north by a waving line from the north of French Guinea, touching Lake Rudolf on the south-east, and reaching the coast near Mogdishu ; on the south the bounding line begins a little north of the mouth of the Ogowe River, follows the general trend of the southern limit of the rain zone, described above, at approximately the same distance throughout, till it reaches the neighbourhood of Tanganyika, where it bends northwards, and then drops suddenly to cut the coast to the north of Lindi. The area receiving between 8 and 12 inches does not reach the east coast, but its bounding line starts at Dahomey, curves round to the west of the Lakes, and, cutting the Congo above Nouvelle Anvers, reaches the west coast again at the southern boundary of Cameroons. There are two areas which receive over 12 inches of rain, the one including the southern part of Sierra Leone, Liberia, and the Ivory Coast, with a small area of excessive rainfall along the coast from Cape Palmas to Cape Three Points ; the other curving round Benin and Cameroons Mountain region.

In the north the rain area includes Northern Marocco, Algeria and Tunis. In the south the rains visit the country west and south of Delagoa Bay, and the coastal belt thence to the Cape Peninsula, and an inland band thence to the Orange River. The Orange Free State itself is a rain area, and so also are Basutoland and Kaffraria, as well as some of the eastern and midland parts of Cape Colony. The greatest precipitation is experienced on the south coast. But the most noticeable feature is the increasing rainfall in the west, and the diminishing fall in the east, the Cape Peninsula receiving 4 inches, and the south coast 2 inches and over.

JUNE.

The approximate limits of the zenithal position of the sun at noon, in any year, during this month, are 22° 7′ N. at the beginning of the month and 23° 10′ N. at the end, but the position on the 20th and 21st is 23° 27′ N. The isotherms have now taken up their summer positions in the north. The isotherm of 95° Fahr., which in May extended round the region from westward of Lake Chad to that included by the White and Blue Nile, has now shifted its position so that its axis has a more N.W.—S.E. direction, the eastern limit remaining the same, but the whole included area being shifted farther north on the western side, so as to exclude Lake Chad and reach beyond the tropic, and almost as far west as the meridian of Greenwich. The isotherm of 90° has not altered much in general shape, but is pushed farther north in the Marocco-Algeria region and also in the Nile region, passing now through Abu Hammed. On the north and north-west the isotherms of 85°, 80°, and 75° follow closely the general direction of the 90° line, that for 75° passing through the Nile delta north of Cairo, skirting the coast and leaving the western shores of the continent in 20° N. The isotherm of 70° runs inland along the coast of Marocco from Tangier to Cape Juby.

The southern isotherm of 80° skirts the West Coast (inland) from the Ivory Coast to Lagos, passes north of the Niger delta, and then dips south to the northern boundary of Rhodesia, and bends northwards touching the north of Nyasa, and leaves the coast near the Anglo-German boundary in East Africa. With less and less southerly dips this line is followed by the isotherms of 70° and 60°, the former entering the continent at Benguela, touching the southern tropic and leaving the continent near Sofala; the latter entering at the Orange River mouth and leaving in Pondoland. The isotherm of 55° just touches the Cape Peninsula. This is the first month of the northern summer and southern winter.

The general distribution of pressure corresponds with that of temperature. North of the equator, roughly, low pressures prevail, and south of the equator high pressures, though the extreme north-west, including Tunis, Algeria and Marocco, is still in the northern high-pressure area. The extreme low-pressure isobar of 29·70 inches enters the Red Sea littoral abreast of Wadi Halfa, reaches as far west as 10° E. and leaves the coast at Jibuti, enclosing the highlands of Tibesti and passing north of Lake Chad. But the general low-pressure area extends as far as the Niger bend. The isobar of 29·95 inches just touches the southern shores of Victoria Nyanza and those of 30 and 30·10 inches pass through the continent from E.—W. generally parallel with the equator, the latter passing through Walfisch Bay and Sofala. The extreme high-pressure areas lying to the east and west of South Africa have moved farther from the coast and an isolated extreme pressure area lies over the south-eastern portion of the continent in the form of an ellipse, its major axis having a general N.E.—S.W. direction and reaching from the tropic in the Limpopo bend to the neighbourhood of Port Elizabeth. This is encircled by the isobar of 30·20 inches and has its centre (30·30) abreast of Zululand.

The rains during this month are almost confined to a belt bounded on the north by a line drawn from Cape Verde and reaching almost to Jibuti and curving round south of the sea-board to Cape Guardafui; and on the south by a line drawn from the south of Spanish Guinea, passing south of the junction of the Ubangi River with the Congo, just touching Tanganyika at Usambura, then curving slightly northwards, and bending south to leave the coast at Zanzibar. This rain band thus lies almost entirely north of the equator, and the intensity of the precipitation increases from the north and south limits inwards, and also from east to west; the two rainiest areas, with 16 inches and over, lie the one round the Niger delta and Cameroons mountains, and the other cut off by a line drawn from the south of French Guinea to Cape Three Points. Outside this north tropical rain band, rains (over 2 inches) are also experienced in the eastern Algerian Atlas in the north, and in the neighbourhood of the Cape Peninsula (over 4 inches) in the south, where the precipitation has increased; and there are over 2 inches in the southern coastal areas. Of this month, Capt. Lyons, in *The Rains of the Nile Basin and the Nile Flood of 1906*, says :—" By this time the low pressure centre which occupies the north-western portion of India from June to September has developed, and the gradient from the Nile Valley eastwards towards the Persian Gulf is well defined. In connection with this the southerly winds rapidly extend to lat. 16° N. and the monsoon rains set in over the Sudan plains, the Abyssinian tableland, and in Eritrea. The Blue Nile and the Sobat are rising steadily, fed by a rainfall of 150 to 200 mm."

The pressure conditions, it will be noticed, are almost the exact reverse of those in January, in a general sense, though the forms of the low-pressure areas are roughly at right angles to one another in the two months, and the effect of this on the winds is most noticeable on the east coast where, except in the extreme south, the direction of the winds is almost reversed, for while in January there was a general northerly tendency, in July there is a decided southing from abreast of Madagascar northwards.

JULY.

The approximate limits of the zenithal position of the sun in July, in any year, are 23° 6' N. and 18° 11½' N. This is the typical summer month in the north, and winter month in the south, the thermic equator lying north of the tropic of Cancer. The area enclosed by the isotherm of 95° has its greatest extent, and is shifted, in its western part, still farther to the N.W., so as to include the southern parts of Algeria, the Nile from Wadi Halfa to 13° N., and the plains of the middle and lower Atbara. The isotherm of 90° follows closely the shape of that of 95°, extending as far S.E. as Jibuti, and, skirting the Red Sea coast, touches Lake No, and passes slightly N. of the Niger Gorge at Tosaye, then running northwards into Algeria, and finally trending S.E. to the Nile in the neighbourhood of Aswan. The northern isotherm of 80° enters the continent on the west somewhat to the south of 20° N., runs N.E. to the north coast at the Marocco-Algerian frontier, and then skirts the whole north coast. The southern 80° line enters in French Guinea, runs to the north of the West Coast equatorial forest belt and, passing near Yola, bends S.E. to the Ubangi-Congo junction, passes through the northern part of Tanganyika, and, enclosing the greater part of the lake plateau, leaves the continent at Kismayu, at the mouth of the Jub River. The southern isotherm of 70° enters at the mouth of the Congo River, runs S.E. to where the tropic of Capricorn is cut by the upper Limpopo River, and then trends due east, the isotherms of 60° and 55° having precisely similar, though more southerly positions, the latter skirting the south-west portions of the Cape Colony. There is a small northerly elliptical area of 70° enclosing the southern coasts of Marocco.

Corresponding with the temperature, the low-pressure areas lie north of the equator, except the Maroccan and Algerian coasts, which lie on the outskirts of the North Atlantic anticyclonic area, and the high-pressure areas in the south. The increased heat in the north appears to result in the widening apart of the isobars right across the continent from the Red Sea to Sierra Leone. In between the two anticyclonic areas, which lie to the east and west of South Africa, over the Atlantic and Indian Oceans respectively, is a third such area encircling the Orange Free State and the adjacent portions of the Cape Colony. In fact, during this month, the regions of minimum pressure and maximum temperature, and those of maximum pressure and minimum temperature are practically coincident.

The equatorial rain-band has moved still farther north and, with the exception of the extreme east and a dip down below the Ubangi-Congo junction, lies wholly north of the equator, in sympathy with the high temperature and low pressure. The southern portion of the coast region of Portuguese Guinea, the western part of French Guinea and Sierra Leone, as well as Liberia, receive a precipitation of over 16 inches, and so too does the area round the mouths of the Niger and Cameroons Rivers. An area of over 12 inches appears on the Abyssinian plateau. Roughly speaking, the whole area included between the equator on the south, latitude 15° N. in the north, and limited on the east by the 40th meridian, receives a rainfall of over 4 inches, the coast east of Cape Three Points being the most noticeable exception. At the same time the northern rain limit has been made to include portions of Air and Tibesti, for though we have no absolute figures to warrant this, yet it seems highly probable from such information as is at our disposal, that the rains in these lofty regions are such as are indicated by the isohyets at this period. In the extreme south the rains are less than in June on both sides of the continent. In the north the whole Mediterranean area is very dry, as well as the Atlantic coast lands. Both to the north and south of Nyasa, as well as round Nkata Bay, there are small rain areas, the precipitation in the north amounting to over 4 inches.

AUGUST.

In August the sun is in the zenith at noon, approximately, in any year, between 17° 57′ N. and 8° 32′ N. The climatic conditions remain practically the same in character as in the previous month. There is, however, a slight tendency of the isotherms, south of the equator, to move farther south, especially in the east; while, in the north, the area included by the isotherm of 90°, lying over the Sahara, becomes contracted in a north-south direction, but the general appearance of the isotherms undergoes no change. With regard to the pressure, there is a slight diminution over almost the whole continent, marked, however, in the extreme south, by the dissipation of the high-pressure area, which lay over the Orange Free State and the adjacent parts of the Cape Colony. On the whole the isobars south of the equator move slightly southwards in sympathy with the isotherms.

With regard to the rainfall, in the great equatorial belt, the two western areas receiving a precipitation of over 8 inches coalesce, and form one great unbroken area, including almost the whole of the Western Sudan. Except in the vicinity of Victoria Nyanza, and on the Zanzibar coast, British East Africa has become almost entirely rainless, while the precipitation in the small rainy area to the north of Nyasa has diminished considerably. In the south, the rain area is broadened out, while the fall on the Cape Peninsula has decreased about one inch. The rains borne by the south-east winds are at a minimum, and those brought by the south-west winds are at a maximum. (*See* the section on "South Africa.")

SEPTEMBER.

The limits of the zenithal position of the sun at noon in this month are $8° 9'$ N. and $2° 57'$ S. approximately in any year, so that the sun crosses the equator in its journey south. In the north, the $95°$ isotherm has disappeared and the highest is $90°$, which has now, towards its western part, dropped south, so that the form is somewhat like an ellipse, broader in the west than in the east, whose major axis, nearly four times the length of the minor axis, extends from Jibuti to the intersection $20°$ N. with $5°$ E. The southern isotherm of $80°$ enters the continent at the Gambia, follows the trend of the coast to the bend of the Cross River, and then dives south, cutting the Congo River at the junction of the Kasai River, just touching the Limpopo River bend, and turns north again, with a general form resembling that of the continent, leaving the coast to the south of Zanzibar. The other southern isotherms follow similar courses, that of $60°$ running inland along the coast of German South-West Africa, cutting off the south-west of the Cape Colony and passing out at the Knysna. The northern isotherm of $80°$ enters a little to the north of St Louis and runs N.E., then passes along the southern portions of Algeria and Tunis, skirts, inland, the Tripolitan coast, and leaves the Nile delta to the north. A small area of $70°$ still lies over the southern Marocco coast.

Pressure has risen over the whole of the northern half of Africa, and the whole of Marocco, with Algeria and Tunis, and the western Tripolitan coast are now included in the North Atlantic anticyclonic area, whereas in August this position was only accorded to the Maroccan littoral between Capes Spartel and Verde. On the other hand, pressure has diminished in the south, though the extreme south-east of the Cape Colony is still included in the layer bounded by the isobars of 30·10 and 30·15 inches, whereas in August the whole of the land south of the tropic was so placed.

In its passage northwards the sun crossed the equator in March, and it might naturally be supposed that the rainfall in that month, at least in equatorial or inter-tropical regions, would bear some resemblance to the September map. But these two maps are utterly unlike. In fact the May map, especially as regards the area which receives 4 inches of rain and over, bears a much stronger likeness to that for September, and for this reason, that the sun, when in the north, is in the land hemisphere, as opposed to the water hemisphere, and has not the opportunity to collect so much vapour to carry with it on its southerly course, as it had when travelling north, and by May, on its northern course, it has lost a great portion of its force as a rain agent, and hence the rain conditions of May and September approximate more nearly than those of March and September.

In September the rainfall on the Abyssinian plateau, as well as in the Sudan plains, has fallen very considerably, the 12-inch (300 mm.) area having disappeared altogether, and the 8-inch area having become very much smaller. The isohyet of 4 inches bends south in the Congo area between the 20th and 25th meridians, so as to include Luluabourg and Lusambo, while in the Nyasa region the rainy areas to the north and round Nkata Bay have disappeared. In the West Coast region the area receiving a rainfall of 8 inches and more has become narrower. In the south the rains brought by the south-east winds gradually increase.

OCTOBER.

The limiting positions of the zenithal sun at noon during October in any year are approximately $3° 20'$ S. and $14° 15'$ S., so that the sun for the whole of the month is in the southern hemisphere. But both the temperature and low-pressure monthly equators are still north of the terrestrial equator. There is a slight rise in temperature from Darfur southwards, in the east of the continent, which, however, is more marked in the regions around Nyasa; and there is also a slight rise along the western coast-lands south of Marocco, but between these two areas is a band of slightly lower temperature, extending the whole length of the continent. In the north the temperature has fallen slightly, the fall being most appreciable in Algeria and Tunis and the hinterland. The highest isotherms in Africa are $90°$ in the north, enclosing an elliptical area lying east and west, from northern Abyssinia to Lake Chad and enclosing Darfur, and $90°$ in the south, enclosing a similar area, having a north-south position, reaching from the Victoria Nyanza to the Zambezi and embracing Tanganyika, Mweru, and Bangweulu, as well as the northern part of Nyasa.

In sympathy with the temperature, a low-pressure system becomes well developed in the north, the isobar of $29\cdot60$ inches encircling an area which covers Eritrea, northern Abyssinia, the Nile from Abu Hammed to the Sobat River, and extending westwards to Lake Chad. The effect of this is to divert the S.W. winds so that they take a south-easterly direction, blow over the ocean, and bring the rains to south-eastern Somaliland. The two great southern anticyclonic areas of the Atlantic and Indian oceans still join up and include Southern Africa.

The rains have almost entirely disappeared from the extreme south-west of the continent, in the area which receives its rainfall in the southern winter. In the north, the beginning of the rain area has, on the coast, moved south of St Louis, and the area of extreme precipitation, 16 inches and over, no longer includes Freetown, but has moved farther south. The coast to the east of Cape Three Points receives a smaller rainfall. On the lake plateau the rains gradually increase, and an area of over 8 inches fall occurs around Lake Edward. The isohyets have bent farther south in the south-central region, and there are slight rains at the Victoria Falls. The Nyasa rains increase in the south, and appear in the north. The isohyet of 4 inches (100 mm.) in its dip southwards is governed in its position by Lai, Fort Archambault, Fort Crampel, and Basoko at the confluence of the Aruwimi with the Congo. In the extreme east, the easterly bend of the isohyet of 1 inch or 25 mm. is due to the rains at Lugh, where in this month there are nine rainy days in all, three of them being recorded as heavy rains, with storms.

NOVEMBER.

The position of the sun, when in the zenith at mid-day, varies in November, in any year, between the limits $14°\ 34'$ S. and $21°\ 43\frac{1}{2}'$ S. approximately at the beginning and end of the month respectively. The isotherms have all moved slightly southwards, and the high-temperature area enclosed by the $90°$ line in the north has disappeared, while the southern high-temperature area of $90°$ has grown both in length and breadth, and now reaches as far south as the Limpopo bend. The whole of the West Coast, as far east as Lagos, has a higher temperature. In the south only a fringe of the coast from Cape Frio south-ward, and the southern part of Cape Colony, are without the $70°$ line.

In sympathy with the temperature, the low-pressure area in the north has vanished; the isobars have moved slightly southwards; and the two great high-pressure areas of the South Atlantic and Indian Ocean, which were formerly linked up so as to include Southern Africa, have now almost parted company, and only embrace the narrow coast-belt alluded to above. The result of these changes is a decided westing of the winds on the western side, south of the equator, and an easting on the eastern side, most noticeable near Zanzibar; while in the north there is an easting over the Sahara and near Lake Chad.

In the extreme north the rains have increased, especially in Algeria and Tunis, and now include a coastal belt of Egypt and Tripoli, while the neighbourhoods of Benghazi-Derna and Tripoli receive two inches and over. The great equatorial rain-belt has moved farther south, following the sun, and in these regions there is now practically no rain north of $10°$ N., except in the extreme west, where there are rains in French Guinea. On the West Coast the rain-fall has diminished greatly, the only part receiving over 16 inches being the immediate neighbourhood of Grand Bassam, but in Cameroons, French Congo, and the Congo regions, the area receiving 8 inches and over has a large development and extends along the coast from the Niger delta to Kabinda, reaching inland eastwards so as to include Lake Edward, Lake Kivu, and Bukoba on Victoria Nyanza. In the east, in British East Africa and on the Zanzibar-Mombasa coast the precipitation exceeds 4 inches, and, in the latter region, 8 inches; while in Southern Rhodesia is an area of 4 inches and over. In the Red Sea, both Suakin and an area immediately south of Massaua receive rains.

DECEMBER.

The zenithal position of the midday sun on the first of this month in any year is approximately $21° 53'$ S. and on the 31st $23° 4\frac{1}{2}'$ S., it having on the 20th, 21st and 22nd been so far south as $23° 27'$ S. The climatic conditions in December begin to bear a striking resemblance to those of January in all respects. The southern area enclosed by the isotherm of $90°$, in the east of the continent, moves southwards, leaves Victoria Nyanza to the north, only includes the extreme south of Tanganyika and extends southwards over the Transvaal; the southern isotherm of $80°$ enters the continent at the equator on the west, and that of $70°$ just skirts the south-western and southern coasts, while the northern $80°$ line enters at Sierra Leone, dips southwards to the Ubangi bend, and curves northwards to the northern coast of Eritrea; and the northern $70°$ line runs between the Tropic and $20°$ north. Pressure has diminished over the whole of the south of the continent south of the equator; the South Atlantic and Indian Ocean anticyclonic areas have parted company, the former merely touching the coasts, in a similar manner to the $70°$ isotherm; while a low-pressure area has developed in a north-south direction, corresponding roughly with the area encircled by the $90°$ isotherm, and extending from a little south of the equator to midway between $20°$ S. and the Tropic. In the north the Atlantic anticyclonic central area includes Marocco. The effect of the southern development on the winds in the east is to give a north-easterly direction to those south of the equator and in the Mozambique channel, which had a distinctly southerly direction in November.

The West Coast rains have still further diminished. The area receiving 8 inches of rain does not reach the coast on either side of the continent,—though there is a small enclave round Libreville which receives that amount— and is bent in a N.W.—S.E. direction so as to include, on the north-west, Vivi, Stanley Pool, Bolobo and Nouvelle Anvers, but to exclude Irebu, at the Ubangi-Congo junction, and in the south-east to include nearly all Nyasa, except near Fort Johnston, and dip down to Mopea, on the Zambezi delta, and also include, in German territory, Mahenge, Manow and Neu Langenberg, but exclude Fort Hill and Fife. In the north-east of German South-West Africa the rains have considerably increased. In the Red Sea, Suakin and the region south of Massaua still receive rain, and the latter area is continued along the northern Somali coast.

NORTH AFRICA

Taken as a whole, the rainfall in the extreme north of Africa diminishes from the coast inland. And if the highlands behind the northern sea-board of Tunis be taken as a starting-point, the rainfall, in the coastal regions, diminishes both westwards and eastwards, except that the Atlas of Marocco shows a marked increase, after the rainfall has sunk to a low figure over the outer plateau.

The northern parts of Marocco, Algeria and Tunis, the extreme north of the promontory of Barka and the Nile delta, being in the Mediterranean climatic province, are within the realm of winter rains, with dry or almost dry summer, though this does not hold good beyond a limited distance from the coast. The summer drought is particularly pronounced in Algeria, Tunis and farther east. In the Atlas regions, the largest amount of rain usually falls in January in Western Algeria, but in December further east; in the north of Marocco there is a small amount of rain even in July and August, which are the driest months, November being the rainiest; while further south the two former months are quite rainless. With increase in altitude, in an inland direction, the summer drought is moderated, while throughout the whole interior of this Atlas region the spring is the rainiest period.

PRODUCTIONS.

Any region in North Africa where the rainfall amounts to 450 mm. (17·72 inches), distributed over from sixty to eighty days or more, is suitable for the production of all British cereals, as well as the millets, maize and root crops. In fact the usual crops of Southern Europe may be cultivated in any locality which enjoys not less than the above-mentioned rainfall. In the Tell or "Hill" country, between the maritime zone and the high plateau zone, wheat does uncommonly well. In the northern region the autumn is both warm and dry, with sufficient rainfall in the remainder of the year, in the western parts, for the cultivation of the grape, for which the Tell is apparently best suited. The raisins of some districts have been celebrated for years. In these western parts, especially in Algeria and Tunis, olives flourish. The Egyptian varieties of cotton do well in Algeria, especially in the province of Oran. Good crops of flax and tobacco are obtained. In the south is the Alpha zone, which produces the *Alpha* grass, largely used in the manufacture of paper. The date palm gives the staple food of the southern desert regions. Many fruits, such as oranges, almonds, lemons, shaddocks, and figs, are produced in large quantities. The region of Marocco which is known as the steppe country, lying between the corn-producing zone and the Atlas, is especially noted for its fruit trees, while the *Acacia gummifera* is a native of Mogador. Carob beans are produced in Algeria and henna in Tunis. In the north-west of Tunis and the neighbouring Algerian territory the cork-oak abounds. Chick-peas and semsem are grown in Marocco, as well as hemp, coriander, cumin, and caraway.

Egypt regularly exports cotton, sugar, beans, lentils, wheat, and opium, which is also produced in Tunis; rice and gum are produced in large quantities. In Algeria lucerne, vetches and other fodder plants do well, and also indigo, madder, and saffron. Even on the western coastal strip of Egypt both barley and wheat are grown.

MAROCCO.

Speaking generally, the climate of Marocco is healthy. From March to September the sky is clear and serene, and, even during the rainy season, which extends from September to March, it is a rare occurrence for a whole day to pass without the sun being occasionally visible. In winter the rains are fairly abundant, and fall at regular periods; but should the rains be prolonged beyond March, there is danger of fever.

K. A.

3

The coast districts and the Tell enjoy a temperate equable climate. The mean temperature in the Tell is $66°\cdot9$ Fahr. ($19°\cdot4$ C.) with a maximum of $71°$ Fahr. ($21°\cdot8$ C.) in August and a minimum of $61°\cdot7$ Fahr. ($16°\cdot5$ C.) in February. At Mogador the thermometer seldom rises above $77°$ Fahr. ($25°$ C.) owing to the regularity of the "North-East Trades." The mean temperature at Márrakesh, which stands at an altitude of 530 metres (1733 feet), is about the same as at Tangier, namely $63°$ or $64°$ Fahr. ($18°$ C.). Throughout the country generally, the heat is moderated by the proximity of both sea and mountain. In the coast regions the thermometer never sinks below $41°$ Fahr. ($5°$ C.). At Fez, alt. 420 metres (1373 feet), the mean summer temperature is between $59°$ and $61°$ Fahr. ($15°$ and $16°$ C.). In the interior, however, both heat and cold are excessive. The cold rains begin about the middle of October, and December and January are the coldest months. The north and west winds are very strong in March[1].

De Roquevaire has constructed a hypsometrical map of Marocco[2], from a study of which we may divide the whole country, exclusive of the extreme south, into three main zones :—

1. From sea level to an altitude of 1600 feet, including the coastal area from the latitude of El Arish to Cape Ghir, that south of the Rif country, with Fez district, and the region which contains the lower courses of the coastal streams. This zone is the granary of Marocco.

2. From 1600 to 3250 feet : this zone lies to the east of the above, and has a general elliptic form, being shut in on the east by the foothills of the Atlas which occupy a comparatively narrow belt. This zone is a steppe region, characterised by trees and notably fruit trees.

3. The summit zone of the Rif and the Atlas.

From the Atlantic eastwards there is at first a gentle increase in the mean annual precipitation, followed by a rapid decrease. At Cape Spartel the fall, according to

---

[1] Canal, *Géographie générale du Maroc*, A. Challamel, Paris, 1902, pp. 55 *sqq.*
[2] *Annales de Géographie*, 1901, p. 330.

Dr Fischer[1], is 828 mm. (32·59 inches), while, according to the table of Meteorological Observations, compiled from five years' observations, and published by the Admiralty (*see* below), it is given as 28·02 inches. Combining these two divergent statements, however, the precipitation would appear to be not far from 770 mm. or a trifle over 30 inches. At Tangier, close by, the mean annual fall is 817 mm. or about 32 inches, but it diminishes rapidly towards the south, when we enter the first of the zones mentioned above. At Casablanca (Dar-el-Beida) the mean annual rainfall appears to be from 15 to 17 inches (380 to 432 mm.), at Safi it is said to be only a little over 13 inches (330 mm.), while inland at Márrakesh the total annual rainfall is about 9½ inches. Márrakesh may be taken as an example of places in the second zone, after the rapid decrease mentioned above, though not as a typical example, for, from the accounts of travellers it would appear that many parts of this tree-zone are not favoured with so copious a fall. At Mogador the fall is 407 mm. (16 inches) and in the extreme south at Cape Juby, in latitude 27° 57′ N., Saharan conditions prevail and the annual precipitation only amounts to 4·14 inches (105 mm.).

The Marocco rains are essentially winter rains, though at some stations the spring rains are as abundant as those of the winter months, as at Márrakesh, but the excessive dryness of the summer is more marked in the south, where, for instance, there is no rain at all at Safi and Mogador in June and July. It is notable, in connexion with the diminution of rainfall east and west of Tunis, on the coast, that the north coast of Marocco is less favoured than many parts of the Algerian Tell.

In the Shawía district the tempering influence of the ocean is experienced for quite 50 miles from the coast. At Casablanca the mean annual temperature is about 64°·4 Fahr. (18° C.), that of January is 55°·4 Fahr. (13° C.), and that of July 73°·4 Fahr. (23° C.). The thermometer never goes below 41° Fahr. (5° C.) even on the coolest nights of winter, and very seldom rises above 86° Fahr.

---

[1] *Zeitschrift der Gesellschaft für Erdkunde zu Berlin*, Band XXXV, 1900, p. 365.

(30° C.) on the hottest summer days. Inland, Weisgerber observed as high a temperature as 104° Fahr. (40° C.) in the Beni Meskin desert, and experienced white frosts at Alwa, but these extremes of temperature are not trying, owing to the dryness of the atmosphere on the elevated plains. On the coast, on the contrary, the relative humidity is high, and mists, which are a serious danger to navigation, are frequent, and dews abundant, growing less and less as the interior is approached. On the high plateaux dews are not experienced. The rainy season, broken by spells of beautiful days, lasts from October to April, the average number of rainy days at Casablanca being 67, and the total annual rainfall from 15 to 17 inches. Sometimes it rains in September and in May, but rarely in June, July and August. Inland, though the precipitation is less, the number of days on which it rains is greater. Snow is unknown, but there are, on occasions, hail-storms in the winter in the interior. The S.W. wind brings the rain; the prevailing winds are the N.E. trades. During the summer when the wind is in the latter direction, the sky is blue and the temperature agreeable, and the heat is only trying on the rare occasions when it is not blowing; and even then there is usually a sea-breeze in the middle of the day from 10 to 4 o'clock. Storms are frequent in winter. On the whole, Shawía enjoys an agreeable and healthy climate.

At Tangier the mean reading of the barometer oscillates between 30·180 inches in December and 30·021 in August. The month in which the barometer varies most is March, when there is a difference of no less than 1·132 inches between the extreme readings. It varies least in August, the corresponding difference being ·314 inch. The mean annual temperature is 63°·4 Fahr., derived from readings taken at 9 a.m. and 9 p.m. For the monthly temperature *see* the table below. By far the most prevalent wind is from the east, the next most frequent wind, which is only half as frequent, is from the south-west; and this is followed in order of frequency by winds from the west and south-east.

At Cape Juby the barometer varies between a maximum of 30·190 inches, which is the mean reading in January, and

30·069 in October. The extreme monthly range of the barometer is greatest in January, when there is a difference of ·927 inches between the highest and lowest readings. The barometer has the lowest range of oscillation in August, in the same month, that is, as at Tangier, and the range. is here ·251 inch, somewhat less than further north. The prevailing winds are from the north-east ("trades") and south. In fact the wind is almost wholly confined to these directions. "The climate of Cape Juby," says Dr Zaytoun, who lived there for five years, "is salubrious. On the north-west coast of Africa, the excessive heat of the Sahara is modified by the cool N.N.E. trade winds, which prevail during eight months out of the twelve. The air is pure, dry, and clear, and the weather is not subject to sudden changes. The soil being dry,......the diseases of hot and malarious districts, such as dysentery, intermittent and blackwater fevers, are unknown, and during the twelve years Cape Juby was occupied by Europeans no case occurred. The temperature rises above 80° Fahr. (26°·7 C.) only when the Harmattan is blowing from the desert. The Harmattan blows for two or three days at a time and makes the atmosphere hazy. The difference between the summer and winter temperature is small. There are heavy night dews, and at dawn the country is as wet as if it had been raining copiously. The south-west is the rainy quarter. The barometer falls for S.E. and S.W. winds, and rises for W., N., and N.W. winds. Fogs and mists are both rare and of short duration, and, when they do occur, make their appearance in the morning and vanish in the afternoon. Thunder is rare, but sheet lightning is frequently seen in the period from October to April."

Casablanca enjoys a good climate and is considered to be healthy, though the town is dirty.

Mogador, on the other hand, is surrounded by swamp, which is inundated at spring tides. The rain here is irregular and cannot be depended on from year to year, the mean annual rainfall (16·02 inches) being derived from such divergent figures as 20·67 inches in 1894–5 and 0·11 inches in 1895–6.

At Márrakesh the tempering influence of the ocean is

not felt, and consequently the climate is almost of the continental type, marked by extremes of temperature. In summer the heat is great, and even in winter, when the thermometer nears the freezing-point at night, it will sometimes rise to 78° or 80° in the day-time.

(For Productions *see* under North Africa.)

CAPE SPARTEL. Lat. 35° 47′ N., Long. 5° 55′ W.

Compiled from 5 years' observations.

	Jan.	Feb.	March	April	May	June	July	August	Sept.	Oct.	Nov.	Dec.	Year
Mean temperature (Fahr.)[1]	54°·1	55°·9	57°·2	59°·4	63°·0	69°·5	73°·3	73°·7	70°·6	66°·7	61°·0	56°·5	63°·4
Mean daily range[2]	9°·0	9°·2	10°·0	10°·1	10°·9	12°·3	14°·8	13°·2	10°·8	10°·7	9°·2	8°·3	10°·7
Absolute maximum during the five years	68°·1	71°·5	81°·3	79°·5	82°·6	92°·2	99°·2	96°·5	89°·0	90°·4	79°·1	72°·8	99°·2 (highest)
Absolute minimum	30°·0	43°·4	43°·4	45°·7	50°·4	54°·0	46°·7	55°·0	57°·9	41°·7	45°·4	44°·0	30°·0 (lowest)
Mean relative humidity[3] (per cent.)	87·6	86·2	83·9	85·7	83·2	77·8	77·0	78·0	82·1	81·5	84·1	84·7	82·7
Amount of cloud (0—10) ...	5·0	4·0	4·9	4·2	3·6	2·9	1·8	1·9	3·6	4·5	4·6	4·8	3·8
No. of rainy days	12·8	8·0	11·4	8·4	8·0	3·2	0	·4	3·4	7·4	9·4	10·4	82·8
Mean rainfall (in inches) ...	4·45	2·82	4·17	1·95	1·44	·52	0	·01	·50	3·18	4·56	4·42	28·02

From the *Africa Pilot*. It is not there stated' what the specific 5 years were. The observations were taken at 9 a.m. and 9 p.m.

[1] The mean temperature is derived from $\frac{9a+9p}{2}$.

[2] The daily range is max.—min.

[3] The relative humidity is derived from $\frac{9a+9p}{2}$.

MOGADOR. Lat. 31° 30' N, Long. 9° 42' W.

Compiled from 8 years' observations[1].

	Jan.	Feb.	March	April	May	June	July	Aug.	Sept.	Oct.	Nov.	Dec.	Year
Mean temperature[2] (Fahr.)	61·2	62·1	64·0	66·4	68·9	70·3	71·1	71·1	71·2	68·9	64·8	61·3	66·6
Mean maximum temperature ...	68·0	70·0	70·5	74·3	75·6	77·7	77·7	77·9	78·4	75·0	70·7	67·8	73·6
Mean minimum temperature ...	55·8	56·5	58·5	61·3	63·9	66·2	67·8	67·6	67·1	64·2	61·0	55·2	62·0
Absolute monthly maxima during the 8-year period	78·8	77·0	75·2	78·8	82·4	84·2	86·0	87·8	86·0	82·4	75·4	71·6	87·8 (highest)
Absolute monthly minima during the 8-year period	51·8	53·6	55·4	57·2	60·8	62·6	64·4	65·8	65·1	59·0	55·4	50·7	51·8 (lowest)
Mean daily range[3] ...	12·2	13·5	12·0	13·0	11·7	11·5	9·9	10·3	11·3	10·8	9·7	12·6	11·6

[1] 1867—1874. A table for 6 years (April 1894—June 1900) will be found in *Étude sur l'Hygiène au Maroc*, by Dr L. Raynaud, Alger, 1902.

[2] The mean temperature is derived from $\dfrac{8a + 2p + 10p}{3}$.

[3] The daily range is max.—min.

CAPE JUBY. Lat. 27° 57′ N., Long. 12° 56′ W.

Compiled from 20 years' observations.

	Jan.	Feb.	March	April	May	June	July	Aug.	Sept.	Oct.	Nov.	Dec.	Year
Mean temperature (mean of 9 a.m. and 9 p.m.) Fahr.	60°·6	61°·3	62°·5	63°·8	65°·0	66°·9	67°·6	68°·4	68°·9	67°·5	65°·4	62°·4	65°·0
Mean daily range[1]	10°·0	10°·2	8°·3	7°·0	6°·4	6°·0	5°·1	5°·4	6°·3	7°·6	8°·8	10°·4	7°·6
Absolute maximum	86°·0	103°·6	90°·4	91°·2	86°·0	78°·0	77°·7	81°·8	83°·6	94°·0	89°·0	81°·4	103°·6 (highest)
Absolute minimum	46°·0	45°·0	50°·0	50°·5	52°·4	57°·7	62°·1	62°·2	58°·7	55°·2	50°·2	44°·0	44°·0 (lowest)
Mean relative humidity[2] (per cent.)	81·5	82·6	84·9	84·0	84·4	86·6	90·5	91·8	91·1	89·2	86·1	84·0	86·4
Mean amount of cloud (0—10)	3·7	3·7	4·5	4·7	5·4	5·9	6·2	5·6	3·8	4·0	4·1	4·1	4·6
Mean rainfall (in inches) ...	·467	·503	·464	·172	·053	·001	·008	·004	·244	·457	·684	1·082	4·139
Mean No. of days' rain ...	4·8	4·2	5·7	2·9	0·5	0·1	0·6	0·3	2·0	5·3	7·4	6·2	40·0

The prevailing wind is from the north-east and north.
From the *Africa Pilot*. The specific years forming the 20-year period are not stated. The observations were taken at 9 a.m. and 9 p.m.

[1] The daily range is max.—min.

[2] The mean relative humidity is derived from $\dfrac{9a+9p}{2}$.

MÁRRAKESH (MAROCCO CITY). Lat. 31° 35′ N., Long. 12° 17′ W. Alt. 442 metres.

Compiled from 2 years' observations [1].

	Jan.	Feb.	March	April	May	June	July	Aug.	Sept.	Oct.	Nov.	Dec.	Year
Mean temperature [2] (Fahr.) ...	50°·7	54°·9	56°·3	64°·4	67°·3	74°·8	79°·9	81°·5	71°·6	64°·9	58°·5	**54°·5**	64°·9
Mean maximum temperature ...	65°·1	68°·0	70°·9	82°·4	84°·9	93°·9	101°·8	101°·7	88°·9	80°·2	71°·4	69°·1	81°·5
Mean minimum temperature ...	41°·2	46°·9	48°·0	53°·4	56°·3	63°·1	66°·0	68°·2	62°·2	55°·9	49°·5	43°·0	54°·5
Mean daily range [3]	23°·9	21°·1	22°·9	29°·0	28°·6	30°·8	35°·8	33°·5	26°·7	24°·3	21°·9	26°·1	27°·0
Absolute maxima during the period	77°·4	84°·6	87°·6	100°·2	101°·3	110°·8	114°·6	114°·3	103°·6	95°·4	84°·9	81°·5	114°·6
Absolute minima during the period	35°·2	35°·6	38°·3	40°·5	48°·6	54°·1	59°·5	56°·8	54°·9	43°·0	38°·8	34°·7	34°·7
Mean relative humidity [2] (per cent.)	70	75	71	60	60	56	50	44	59	64	65	63	61

[1] 1900, 1901.

[2] The means are derived from $\dfrac{6a + 1p + 10p}{3}$

[3] The daily range is max.—min.

MEAN ANNUAL RAINFALL IN INCHES.

	Jan.	Feb.	March	April	May	June	July	Aug.	Sept.	Oct.	Nov.	Dec.	Year
*Cape Spartel (according to Dr Fischer)	4·37	2·95	4·06	1·97	1·69	0·55	0·05	0·03	3·98	3·86	4·92	4·49	32·9
Cape Spartel (according to Admiralty) 5 years	4·45	2·80	4·17	1·96	1·46	0·51	0	0	0·51	3·19	4·57	4·41	28·0
Tangier[1] (5 years)	4·65	3·54	5·04	4·69	2·48	0·28	0·12	0·36	0·39	3·35	2·95	4·33	32·2
Casablanca[2] (4 years)	2·21	2·05	4·69	0·75	0·95	0·14	0	0	0·30	1·38	3·74	1·81	18·0
Mogador[3] (6 years)	3·82	2·60	2·68	0·55	0·39	0	0	0	0·34	1·61	2·56	1·50	16·0
Cape Juby[4] (20 years)	0·47	0·50	0·46	0·54	0·05	0	0	0	0·24	0·45	0·68	1·06	4·4
Márrakesh[5] (21½ years)... ...	1·61	1·33	1·46	1·07	0·53	0·53	0·06	0	0·26	0·62	1·53	0·41	9·42

* It will be noticed that the chief difference between the two records occurs in September, the results for the remaining months being fairly similar. The Admiralty *Pilot* does not give the precise years. For Dr Fischer's figures *see Zeitschrift der Gesellschaft für Erdkunde*, 1900, p. 391.

[1] 1880—1885. *Meteorolog. Zeitschrift*, 1887, p. 27.
[2] Sept. 1896—May 1900.
[3] Sept. 1894—May 1900.
[4] The specific years are not stated in the Admiralty *Pilot*.
[5] Jan, Feb, March, Sept., Oct., Nov., Dec. 1886, Jan, Feb, March 1887, 1900, 1901.

ALGERIA.

The climate of Algeria is very varied ; in fact nearly all French writers differentiate between five areas, each with its own peculiar climate :—

1. The Sahel or maritime zone.
2. The Tell or "hill" country.
3. The high-lying plateaux.
4. The more elevated parts of the Department of Constantine, whose climate partakes partly of that of the Tell, and partly of that of the highlands.
5. The Saharan regions.

In the Sahel the winter, which lasts from October to April, is beautifully mild. At Alger and Oran the mean temperature for January is about $53°\cdot6$ Fahr. $(12°$ C.) ; April and May are transition months, the first with $61°\cdot7$ Fahr. $(16°\cdot5$ C.) as a mean, and the latter with $64°\cdot4$ or $66°\cdot2$ Fahr. $(18°$ or $19°$ C.). The hot season begins in June with a mean of $69°\cdot8$ to $73°\cdot4$ Fahr. $(21°$ to $23°$ C.), which gradually works up to from $75°\cdot2$ to $78°\cdot8$ Fahr. $(24°$ to $26°$ C.) in July and August.

On the high plateaux extremes are experienced, the thermometer not infrequently registering $14°$ Fahr. $(-10°$ C.) in the depth of winter, and as much as $104°$ Fahr. $(40°$ C.) in summer, when the Sirocco is blowing. Bu Saada has a mean for August of $82°\cdot4$ to $87°\cdot8$ Fahr. $(28°$ to $31°$ C.).

The Tell is a kind of mean between the Sahel and the plateaux, and is characterised by winter temperatures lower than those of the former, but not so rigorous as those of the latter. Tlemsen, in January, has a mean temperature of $44°\cdot6$ to $50°$ Fahr. $(7°$ to $10°$ C.), Medea $37°\cdot4$ to $42°\cdot8$ Fahr. $(3°$ to $6°$ C.), and Constantine, which some writers regard as a plateaux situation, and others as belonging to the Tell, $37°\cdot4$ to $41°$ Fahr. $(3°$ to $5°$ C.). During the cold season, December to April, snow covers the mountains. In summer it is as hot as on the plains, with a mean temperature of $75°\cdot2$ to $84°\cdot2$ Fahr. $(24°$ to $29°$ C.).

In the Saharan zone the summer is torrid, Biskra having a July—August mean of $91°$ Fahr. $(33°$ C.), with a maximum of $115°$ and $120°$ Fahr. $(45°$ and $50°$ C.) at Wad Rhir

(Ghir) after sunset under the influence of nocturnal radiation. In South Oran (Figig, Duveyrier, etc.) during the summer the air is almost fresh in the very early morning. Towards 8 or 9 a.m. the heat increases, and reaches its maximum between noon and 3 p.m.

The west winds, which do so much for Europe, have not the same beneficial effect in Algeria. In the first place between 30° and 36° N. lat. they are not fully developed, and have neither the force nor regularity which they possess further north; and secondly the mountains of Marocco deprive them of their moisture before they reach Algeria, and accordingly they bring but little rain.

The N.W. and N. winds, though less frequent and shorter-lived, are stronger. In winter and spring they are cloud-laden, bringing with them torrential rains and leaving the mountain tops snow-capped. But on the Jurjura alone is there any lasting snow. These winds are of rare occurrence in summer, but when they do come, they have a refreshing effect.

The wind seldom blows directly from the east, but in summer the most frequent direction is from the north-east, and these winds are sometimes very strong, and the comparatively small amount of moisture, which they collect in their passage over the Mediterranean, is more generally apparent in mists and fogs, which sometimes last for whole days, than in the form of rain.

The winds from the Sahara come from the S.E. and S.W. as well as from the South. When they are strong they are known as the Sirocco, excessively hot currents, drinking in the moisture and parching the ground, equally distressing to animal and vegetable life. The Sirocco may occur at any season of the year, but is experienced most commonly in winter. It comes in its most offensive form in summer, though happily then of short duration.

There are however very much lighter, almost constant, winds from the Sahara, which are more in the nature of airs than winds, and are exceedingly debilitating in their effect. In summer and autumn there are regular sea breezes, which, to a great extent, serve to counteract the bad effects of the Sahara airs, in the region[1].

[1] Thévenet, *Essai de climatologie algérienne.*

In south Oran, at Duveyrier and thereabouts, the Sirocco is frequently experienced, generally speaking, simply as a warm breeze, very dry and enervating. But occasionally the force of the wind increases, its direction changes, and whirlwinds are formed, which generally have a zigzag course. These are the precursors of hurricanes, and the sand soon obscures the sky[1].

Mists are very frequent, not only on the littoral, where they are brought by the east winds, but further inland, where they are formed at night, and disperse at sunrise, though still, during the early hours of the day, clinging to the low-lying plains and valleys.

The following list gives the mean annual rainfall at stations

   (1)   On the littoral.
   (2)   In the Tell.
   (3)   On the plateau.
   (4)   In the Saharan region.

(1) On the littoral.	in.	mm.	(2) In the Tell.	in.	mm.
Nemours	18·43	468	Sidi-bel-Abbes	15·67	398
Oran	17·56	446	Tlemsen	24·84	631
Alger	21·18	538	Bufarile	30·47	774
Dellys	35·19	894	Orleansville	17·40	442
Bougie	40·80	1036	Blida	35·77	922
Philippeville	30·16	766	Tizi-Uzu	32·40	823
La Calle	33·85	860	Medea	29·69	754
			Fort National	44·14	1121
			Gelma	25·08	637

(3) On the plateau.	in.	mm.	(4) Saharan region.	in.	mm.
Afiu	14·53	369	Biskra	6·97	177
Tiaret	29·29	744	Laghuat	7·36	187
Teniet-el-Had	25·00	635	Ghardaia	4·02	102
Constantine	24·88	632	Wargla	3·58	91
Batna	15·71	399	El Golea	2·80	71
Setif	17·84	453			

In the whole Atlas region the precipitation diminishes from north to south, and this is very marked in eastern Algeria. The Tell is the most favoured region, though the fall is very uneven throughout, and diminishes from east to west; it is 41·5 inches (1094 mm.) at Tabarka (Tunis),

---

[1] De Pimodan, *Oran, Tlemcen, Sud Oranais.*

19·13 inches (486 mm.) at Oran and 18·43 inches (468 mm.) at Nemours. The variation in the breadth of the Mediterranean is probably one of the chief causes of the diminution from east to west; opposite Bougie it measures about 440 miles, but only about 125 miles in the meridian of Oran ; towards the extreme west, however, this is modified by the proximity of the Atlantic, and the decrease in the height of the mountains.

In the Atlas region, to the south of Orleansville, the greatest precipitation has been recorded at Tiaret, viz. 41·3 inches (1086 mm.), the mean being 29·9 inches (774 mm.), and it then diminishes rapidly as the steppe region is reached, *e.g.* 10·3 inches (270 mm.) at El Arisha ; but rises again as the Saharan Atlas region is neared, *e.g.* Géryville, 18·9 inches (498 mm.), and Aflu, 14·53 inches (369 mm.), both at considerable altitudes ; but this increase is not experienced in lower areas.

The rains of northern Algeria are essentially winter rains. In summer the winds from the Mediterranean are dry, and only give rise to mists and dews, with perhaps rain in the most elevated situations ; and on the west only 5 per cent. of the total rainfall is measured in summer. In the east the maximum is reached in December and in the west in January. As the altitude increases the maximum comes later, *e.g.* at El Arisha it occurs in April.

The rains begin, generally, about September, with occasional showers, and the season becomes fairly fixed by November ; they increase as December wears on, being more abundant by night than by day. The rain gradually diminishes from December to May, on the 1st of which month the dry season is said to begin, and this lasts till about the end of August. There are occasional showers, however, in May, June, July and August. Throughout, the rain is by no means constant in character, but varies considerably from year to year, and there have been occasions when it has not begun until nearly the end of winter. Being brought mainly by winds from the north-west, on the regularity of these winds depends the rainfall. They prevail, however, as a rule from the end of October up to May. In the dry season the prevalent wind is from the east or north-east, and these are cold winds.

In the Sahara region rains are rare; in the table below is given the percentage of their occurrence during the course of the year.

	Géryville (5 years)	Laghuat (8 years)	Biskra (14 years)
January	5	7	6
February	3	6	12
March	11	20	13
April	5	10	15
May	11	8	16
June	5	5	3
July	1	3	0
August	3	7	1
September	16	11	11
October	27	16	9
November	5	3	5
December	9	4	8

(For Productions *see* under North Africa.)

ALGER. Lat. 36° 47′ N., Long. 3° 4′ E. Alt. 125 feet.

Compiled from 6 years' observations[1], mean pressure and temperature from 15 years'[2].

	Jan.	Feb.	March	April	May	June	July	Aug.	Sept.	Oct.	Nov.	Dec.	Year
Mean barometric pressure (inches)	30·08	30·04	29·94	29·88	29·91	29·94	29·94	29·91	29·95	29·95	29·95	29·99	29·95
Mean temperature (Fahr.°) ...	54°·0	55°·0	56°·1	61°·2	64°·8	70°·7	76°·6	78°·1	73°·4	67°·6	59°·5	54°·7	64°·2
Mean maximum temperature ...	61°·3	64°·0	62°·7	70°·5	73°·2	80°·6	84°·6	86°·9	82°·6	76°·5	68°·9	62°·4	72°·9
Mean minimum temperature ...	49°·3	50°·0	52°·3	55°·9	59°·4	65°·8	70°·3	71°·5	68°·9	63°·0	56°·5	51°·3	59°·5
Mean daily range[3] ...	12°·0	14°·0	10°·4	14°·6	13°·8	14°·8	14°·3	15°·4	13°·7	13°·5	12°·4	11°·1	13°·4

[1] 1901—1906.
[2] The 15-year figures are from Dr Buchan's *Challenger* Report, 1870—1884.
[3] The daily range is max.—min.

SIDI-BEL-ABBES (in the Tell). Lat. 35° 2′ N., Long. 0° 39′ W. Alt. 1700 feet.

Compiled from 5 years' observations[1], mean temperature from 15 years'[2].

	Jan.	Feb.	March	April	May	June	July	Aug.	Sept.	Oct.	Nov.	Dec.	Year
Mean temperature (Fahr.°) ...	45°·1	48°·4	51°·3	55°·6	62°·8	70°·2	78°·8	76°·8	69°·3	60°·6	52°·2	46°·0	59°·7
Mean maximum temperature ...	57°·4	60°·6	66°·7	73°·9	79°·2	86°·4	98°·8	100°·0	88°·0	78°·1	65°·8	58°·8	76°·1
Mean minimum temperature ...	36°·0	36°·3	40°·1	45°·1	47°·5	54°·7	61°·2	62°·2	55°·2	48°·7	42°·9	38°·1	47°·3
Mean daily range[3] ...	21°·4	24°·3	26°·6	28°·8	31°·7	31°·7	37°·6	37°·8	32°·8	29°·4	22°·9	20°·7	28°·8
Mean rainfall (inches)	2·44	2·09	1·77	2·01	0·87	0·43	0·20	0·03	0·39	1·18	2·21	1·89	15·5

[1] 1902—1906.
[2] The 15-year figures are from Dr Buchan's *Challenger* Report, 1870—1884.
[3] The daily range is max.—min.

The following table, giving the Mean Temperature in Fahr. degrees for the 15 years 1870—1884, shows the difference in temperature between stations in the Sahel, Tell, High Plateaux, and Sahara. (Copied from Dr Buchan's *Challenger* Report.)

Station		Lat. N.	Long.	Alt. (ft.)	Jan.	Feb.	March	April	May	June	July	Aug.	Sept.	Oct.	Nov.	Dec.	Year
Alger	...	36° 47′	3° 4′ E.	72	53·96	55·04	56·12	61·16	64·76	70·70	76·64	78·08	73·40	67·64	59·54	54·68	64·2
Oran	...	35° 42′	0° 39′ W.	190	53·60	54·86	57·02	61·70	66·74	72·32	76·64	77·00	73·58	66·56	59·90	53·96	64·4
Bougie	...	36° 47′	5° 5′ E.	240	55·22	55·04	57·56	61·16	66·74	71·32	78·98	79·34	74·66	68·00	61·52	56·84	65·5
Orleansville	...	36° 10′	1° 21′ E.	422	49·82	51·62	55·40	60·62	69·26	77·36	86·00	84·56	76·64	66·92	56·48	49·64	65·4
Constantine	...	36° 22′	6° 36′ E.	2368	45·50	46·04	49·82	55·04	62·06	71·42	80·42	78·98	72·14	60·80	51·80	45·86	59·9
Tlemsen	...	34° 53′	1° 18′ W.	2955	47·66	48·74	50·36	56·22	61·70	68·54	77·36	77·00	70·52	60·98	55·22	46·94	60·3
Tebessa	...	35° 24′	8° 6′ E.	3175	44·06	45·32	49·46	55·22	63·50	73·58	79·52	78·44	70·70	60·62	50·54	44·24	59·5
Biskra (Sahara)		34° 5′	5° 40′ E.	446	51·98	54·32	60·00	67·64	74·66	84·02	89·96	89·24	81·32	70·70	59·18	52·70	69·6

MEAN RAINFALL IN INCHES FOR LONG PERIODS.

	Jan.	Feb.	March	April	May	June	July	Aug.	Sept.	Oct.	Nov.	Dec.	Year
Oran[1] ...	2·01	2·32	2·09	2·01	1·93	0·39	0·04	0·08	0·59	1·69	2·05	3·78	18·98
Alger[2] ...	3·82	3·31	3·11	2·01	1·30	0·59	0·08	0·28	1·02	2·87	3·90	4·61	26·90
Bougie[3] ...	4·72	4·96	5·67	3·23	1·93	2·21	0·24	0·47	1·89	4·16	5·12	6·50	41·10
Orleansville[4] ...	1·34	2·91	2·32	2·72	1·54	0·59	0·08	0·08	0·79	1·89	2·40	2·40	19·06
Constantine[5] ...	2·52	2·32	2·68	2·60	1·46	1·22	0·32	0·47	0·98	2·17	1·97	3·39	22·10

[1] 21 years, 1865—1885; the greatest yearly rainfall for any one year was 25·12 inches in 1870, and the least 9·69 inches in 1867.

[2] 54 years, 1843—1896.

[3], [4], [5] Taken from Dr Supan's list in *Petermann's Mitteilungen*, Ergänz, 124, 1898. The first is for 20 years, the second for 15 and the third for 14 years, but the specific years are not stated.

ALGERIAN SAHARA. Rainfall in inches.

	No. of years	Jan.	Feb.	March	April	May	June	July	Aug.	Sept.	Oct.	Nov.	Dec.	Year
Ayata[1] ...	5	0·87	0·55	1·06	0·32	0·28	0·04	0	0	0·16	0·32	0·12	1·14	4·86
Ghardaia[2] ...	6	0·75	0·28	0·59	0·32	0·67	0·04	0	0·08	0·16	0·12	0·39	1·10	4·50
Batna[3] ...	19	1·97	1·34	1·38	2·01	1·65	1·14	0·20	0·71	1·14	1·48	1·24	1·50	15·76
Biskra[4] ...	19	0·43	0·95	1·02	1·18	1·22	0·24	0·04	0·12	0·83	0·75	0·39	0·67	7·94

[1] 1888—1892.   [2] 1887—1892.

[3], [4] 19 years, 1860—1868, and 1875—1884. The greatest yearly rainfall for any one year was 16·30 inches at Biskra in 1884, and the least 2·44 inches in 1878.

4—2

TUNIS.

Tunis is bounded, both north and east, by the Mediterranean, and its mountain system is less accentuated; consequently the temperature is more regular than in Algeria; the air is softer and not so dry. The rains, too, are more abundant than in Algeria, and the average number of rainy days is about 90. The rainy season, speaking generally of the whole country, extends over the period October—March, but rains do occur in other months, sometimes even in summer. Sea breezes are the rule and not the exception, and these are charged with moisture, and refreshing. The average annual precipitation is 38·69 inches (985 mm.) on the elevated plateaux, which falls on some 99 days in the year. In the less elevated mountain regions it is only 18·23 inches (463 mm.) on 61 days, while in the Sahel zone it is 16·30 inches (414 mm.) on 65 days, and, in the region of the oases, there are only 30 rainy days, and the total fall only amounts to 9·09 inches (231 mm.).

Speaking generally, there are, in the Tunis region, four climatic zones :—

1. The littoral or maritime zone, the *Sahel*, which includes such towns as Bizerta, Tunis, Susa, Sfax and Gabes. Here the winter begins with the rains about the end of October. January and February are the coldest months, with a mean temperature of from $50°$ to $53°·6$ Fahr. ($10°$ to $12°$ C.). March, April and May have a mean temperature of about $60°·8$ Fahr. ($16°$ C.), with occasional rains. The summer heats begin in June ($71°·6$ to $73°·4$ Fahr. or $22°$ to $23°$ C.), and increase till the end of August ($77°$ to $80°·6$ Fahr. or $25°$ to $27°$ C.). When the wind blows from the north-east or north it has a moderating effect and the heat is not so much felt, but the south wind or Sirocco is always oppressive, and has been known to raise the temperature to $118°·4$ Fahr. ($48°$ C.), but this of course is exceptional, though the thermometer is always high when the wind is in this direction.

2. The Tell region, with such centres as Sok-el-Arba, Zaghuan, Kairuan, etc., which being, as the name implies, hill country, and lying farther from the sea than the Sahel,

experiences colder winters, the mean for January being $48^\circ\cdot2$ to 50° Fahr. (9° to 10° C.), and for the hottest month, August, $82^\circ\cdot4$ to 86° Fahr. (28° to 30° C.).

3. The elevated plateau-lands, such towns as Ain Draham, Sok-el-Jemaa, and Maktar being included in the area, have a type of climate more resembling the continental, extremes of both heat and cold being experienced. A temperature of 23° Fahr. (-5° C.) has been registered at Kef in winter, and $111^\circ\cdot2$ Fahr. (44° C.) in summer, but these figures are of course not usual. However, the mean temperature for the Tell for January is $39^\circ\cdot2$ to $44^\circ\cdot6$ Fahr. (4° to 7° C.), and for August, the warmest month, $73^\circ\cdot4$ to $80^\circ\cdot6$ Fahr. (23° to 27° C.).

4. The Saharan region or the region of the oases (Tuzur, Duz, Medenin) has a desert climate, with extreme variations of temperature, but as a rule is extremely healthy, except in the vicinity of the oases themselves. At Gafsa it is recorded that the thermometer has reached $118^\circ\cdot4$ Fahr. (48° C.), while in the winter of 1890 a temperature of $24^\circ\cdot8$ Fahr. (-4° C.) was registered. The mean temperature of this region in January is $46^\circ\cdot4$ to 50° Fahr. (8° to 10° C.), and in August 86° to $89^\circ\cdot6$ Fahr. (30° to 32° C.).

The climate of Tunis is very suitable for Europeans. The army doctor, M. Bertholon, has pointed out that the admissions to hospital, between August and March, were less than in any province in Algeria, and even less than in France; and that the death rate (calculated since 1882) for French residents was only between 24 and 25 per 1000, less than at Marseilles, or Rheims, or Toulouse, approximately equal to that of Paris, and less than the rate for the whole of France[1]. Cases of fever are less frequent in Tunis than in Algeria.

In the whole of the Atlas region the precipitation diminishes from north to south, and this is very marked in the area of the Tunisian Atlas. An isohyet of 20 inches would include a great part of the Tunisian Atlas, though on the actual coast and in the north-east the rainfall is less, and still less further south, at Susa $15\cdot7$ inches (399 mm.) and at Sfax only $8\cdot2$ inches (208 mm.).

The rains of the Atlas region are essentially winter

---

[1] Leroy-Beaulieu, *Algérie et Tunisie*, Paris, 1897.

rains. In summer the winds from the Mediterranean are dry and only give rise to mists and dews, with, perhaps, rain in the most elevated situations. On the coast only 5 per cent. of the total annual rainfall is measured in summer, while the maximum precipitation is generally reached in December.

In the vicinity of Tunis (town) the north-west wind is essentially the wet wind, and the south, the Sirocco or Shehili, the dry wind. The average annual rainfall for the 19 years 1885–1903 was 17·76 inches (451 mm.), the heaviest being 25·95 inches (659 mm.) in 1892 and the least 12·95 inches (329 mm.) in 1893. These figures are the means for an area of about 20 or 25 miles round Tunis. More rain is experienced in Jebel Ahmar and towards Jedeida, and less in the neighbourhood of Sejumi. The north-west winds are robbed of a good deal of their moisture by the Mejerda. The rains occur in the autumn and spring, the summer being practically rainless. The winter is never very severe at Tunis, the extreme minimum temperature ($25°·2$ Fahr. or $-3°·8$ C.) was experienced in the winter of 1890–1, which was quite an abnormal season. The following figures show two different types of winter at Tunis.

	1893—1894		1897—1898	
Month	Absolute minimum	Mean of minima	Absolute minimum	Mean of minima
December ...	$32°·90$ Fahr.	$41°·54$ Fahr.	$39°·20$ Fahr.	$47°·66$ Fahr.
January ...	$31°·10$	$36°·68$	$41°·00$	$45°·68$
February ...	$34°·70$	$37°·04$	$35°·60$	$43°·52$

Snow only falls at long intervals, the latest recorded instances being in 1854, 1891, and 1895.

The summer temperature never falls lower than that indicated below for the year 1895, and usually rises to some such height as that given for the year 1896.

When the N.W. winds are late the summer heat is not infrequently prolonged into October, and the mean of the maxima for that month seldom falls below 77° Fahr. (25° C.), and is usually about 86° Fahr. (30° C.).

Month	1895		1896	
	Absolute maximum	Mean of maxima	Absolute maximum	Mean of maxima
June	102°·20 Fahr.	74°·12 Fahr.	102°·20 Fahr.	84°·38 Fahr.
July	102°·20	86°·00	111°·20	95°·36
August ...	98°·24	89°·96	122°·00*	103°·46
September ...	89°·60	76°·64	104°·00	93°·20

* This is the highest ever recorded.

During the second half of the hot season north-easterly and easterly winds are frequent. The sea appears to have very little tempering effect on the climate, which is almost of the continental type[1].

In the region of Maktar, central Tunis, the prevalent wind is from the north. In 1899 it blew from this direction on no less than 227 days. Sometimes, however, during a certain period of the year, *i.e.* the first four months, there is a considerable number of days when the wind is from the south; in one year during this period there were 56 days on which it blew from the south against 48 from the north, on the remainder the wind was either E. or W. East and south-east winds are more frequent at Berbéron and on the Hamáda Kessera than at Maktar.

The south winds coming from a hotter to a cooler region frequently produce rain, but it is the N.W. wind which brings the greater quantity. The rainfall at Maktar is so variable that calculated means give only fair results, as will be seen from the following table, from which the only really reliable result that can be deduced is that the annual rainfall does not fall below 16 inches and that there is very little rain in the summer months.

It is worth noting that heavy falls of rain and snow are almost invariably synchronous with similar falls at Thala, Tebessa, and Constantine in Algeria.

[1] *Annales de Géographie*, 1904, p. 150.

Winter consists of the months December, January, and February, though it is sometimes prolonged into March, but the temperature in this month (mean minimum) does not sink below 41° Fahr. (+ 5° C.).

RAINFALL AT MAKTAR IN INCHES.

	1889	1890	1891	1892	1893	1894	1898	1899	1900
Jan.	5·04	3·11	3·39	2·60	4·33	2·72	—	1·10	3·56
Feb.	2·45	4·10	2·80	0·81	2·24	3·50	—	2·19	0·16
March	2·36	9·02	2·80	1·52	1·58	4·72	0·47	2·46	0·93
April	0·40	4·45	2·21	6·18	1·61	3·90	5·92	0·14	1·85
May	1·12	1·03	5·75	5·93	0·98	1·54	1·18	2·01	1·63
June	1·10	0·38	1·42	0·75	0·87	1·26	0·95	0·73	2·18
July	0·38	0·43	1·14	0·91	1·34	—	0·63	0·39	0·83
Aug.	0·11	0·58	0·67	1·77	0·51	—	0·73	2·19	0·55
Sept.	3·01	1·48	3·19	0·67	1·02	—	1·06	1·04	1·18
Oct.	0·57	2·40	2·28	0·55	1·61	—	1·55	0·75	0·75
Nov.	0·81	0·69	2·52	1·69	2·21	—	0·18	3·76	2·40
Dec.	3·38	2·30	2·36	1·42	4·72	—	2·09	0·63	0·62
Totals	20·73	29·97	30·53	24·80	23·02	—	—	17·39	16·64

The following are the figures for the temperature of the warmest and coldest recorded winters:

	1893—94 coldest winter	Number of days at or below 32°	1889—90 warmest winter	Number of days at or below 32°
December	30°·20 Fahr.	24	38°·3 Fahr.	2
January	28°·40	20	41°·90	none
February	30°·20	11	39°·20	none

The summer consists of June, July, and August, of which the last named is the hottest. The mean maximum temperature in that month was 93°·2 Fahr. (34° C.) in 1889, 90°·7 Fahr. (32°·6 C.) in 1898, and 83°·3 Fahr. (28°·5 C.) in 1899.

Towards the end of August and in September there are usually strong N.W. storms. With the change of the wind a rapid change of temperature is experienced[1].

(For Productions *see* under North Africa.)

---

[1] *Annales de Géographie*, 1901, p. 346.

BIZERTA. Lat. 37° 16′ N., Long. 9° 54′ E. (approx.).

Compiled from 5 years' observations[1].

	Jan.	Feb.	March	April	May	June	July	Aug.	Sept.	Oct.	Nov.	Dec.	Year
Mean temperature[2] (Fahr.°) ...	52°·4	54°·2	56°·5	59°·9	65°·7	73°·5	77°·9	78°·4	76°·6	69°·2	60°·4	54°·9	65°·0
Mean maximum temperature ...	57°·4	59°·4	62°·2	65°·7	72°·3	80°·2	85°·1	84°·0	81°·9	75°·4	66°·0	59°·7	70°·8
Mean minimum temperature ...	47°·1	49°·1	50°·7	54°·1	59°·0	66°·7	70°·7	72°·7	71°·2	63°·0	54°·7	49°·8	59°·2
Mean daily range[3]	10°·3	10°·3	11°·5	11°·6	13°·3	13°·5	14°·4	11°·3	10°·7	12°·4	11°·3	9°·9	11°·6

[1] 1901—1906.

[2] The mean temperature is derived from $\dfrac{\text{max.} + \text{min.}}{2}$.

[3] The daily range is max.—min.

MEAN RAINFALL IN INCHES.

	Jan.	Feb.	March	April	May	June	July	Aug.	Sept.	Oct.	Nov.	Dec.	Year
Gabes $(3\frac{1}{2})$[1]	0·59	0·71	0·32	0·43	0·12	0·04	0	0	1·14	1·18	1·38	1·46	7·37
Tunis (5)[2]	2·91	1·93	1·61	2·28	1·06	1·14	0·55	0·08	1·02	3·15	1·65	1·73	19·11
Ain Draham (5)[3] ...	10·95	9·53	10·16	8·62	2·84	0·87	0·28	2·13	1·65	6·22	7·72	8·98	69·95
Maktar $(8\frac{5}{12})$[4]	3·23	2·27	2·87	3·01	2·34	1·06	0·75	0·89	1·58	1·30	1·79	2·19	23·28

[1] 1886—1889.
[2] 1880—1884. A 19-year mean gives as a yearly total 17·76 inches (1885—1903).
[3] 1885—1890.
[4] 1889—1893, January—June 1894, March—November 1898, 1899, 1900.

The figures in brackets refer to the number of years for which the observations are calculated.

TRIPOLI.

The region generally included under the name of Tripoli consists of Tripoli proper, lying to the east of the southern territories of Tunis, on the south of the Gulfs of Gabes and Sidra, the northern promontory of Barka, reaching out into the Mediterranean, and the country known as Fezzan, lying to the south of Tripoli proper. This large region may be divided, so far as the purpose in hand is concerned, into two very unequal portions, (1) the seaboard, including the whole of the north of the Barka protrusion with the town of Benghazi, and (2) the Tripolitan Sahara, the line between the Climatic Provinces of the Mediterranean and Sahara just skirting the Tripoli coast, and cutting off the Barka portion mentioned above. The proximity of Jebels Nefus and Gharian have a marked effect on the climate of Tripoli, for while the gentle north-east winds of winter, after their passage over the Mediterranean pass freely over the eastern and lower part of the country, these heights act as condensers and arrest the moisture, with very beneficial results to the neighbouring region.

Tripoli town and neighbourhood being situated on the fringe of the Mediterranean Climatic Province enjoys such rain as it is favoured with during the winter months ; and the year is divided into two seasons, the hot dry season and the colder wet season, though it is seldom really very cold, the absolute minimum temperature recorded being $36°{\cdot}3$ Fahr. ($+2°{\cdot}4$ C.) on the 6th of January, 1895. Ayra, in his work *Tripoli e il suo Clima*, states that during the four-year period, 1893–6, there were, on an average, in each year, only 6 days when the thermometer went below $41°$ Fahr. ($5°$ C.), 55 when the lowest temperature was between $41°$ and $50°$ Fahr. ($5°$ and $10°$ C.), and 91 days when the lowest register was between $50°$ and $59°$ Fahr. ($10°$ and $15°$ C.); that is, in all, only 152 days when the thermometer fell below $59°$ Fahr. ($15°$ C.). The absolute maxima registered during the same period were $108°$ Fahr. ($42°{\cdot}3$ C.) on the 1st June, 1893, $106°$ Fahr. ($41°{\cdot}2$ C.) on the 20th June, 1895, and $109°$ Fahr. ($42°{\cdot}8$ C.) on the 27th June, 1896. The hottest time of the year is, therefore, apparently, the month of June, or, at any rate, the highest maxima of temperature generally occur in that month. In this period of four years there were, on an average, in each year, 10 days on which the maximum ranged between 95° and 104° Fahr. (35° and 40° C.), 40 days when it registered between 86° and 95° Fahr. (30° and 35° C.), and 112 days between 77° and 86° Fahr. (25° and 30° C.).

The mean daily range resulting from the four years' observations is 14° Fahr. (7°·8 C.), which is small, when the proximity of the desert is taken into consideration, for one of the marked characteristics of desert regions is the large range of temperature. On one morning in June, the hottest month, however, Martino, on his journey from Derna to Benghazi, found that his thermometer only recorded 53°·6 Fahr. (12° C.).

The greatest barometric pressure recorded was, during a period of over three years, 30·47 inches (773·9 mm.) on 31st January, 1896, and the least 29·35 inches (745·5 mm.) on 2nd April, 1894.

Except on rare occasions, the barometer registers from 29·53 to 30·32 inches (750 to 770 mm.).

The following list gives the percentage of the occurrence of wind, from the quarters mentioned, registered during a four-year period :

N.E.	22·0 per cent.		N.	10·3 per cent.
N.W.	19·0	,,	W.	9·8 ,,
S.W.	13·5	,,	E.	8·6 ,,
S.E.	11·3	,,	S.	5·5 ,,

The N.E. winds, which are the most frequent, are the "N.E. trades." The greatest heat is always experienced when the Ghebli, or south wind, blows, as it comes directly from the desert regions, which are scorched and parched by the sun.

With regard to the rainfall, for which there are particulars from April 1892 to June 1896, the following are the totals :

1892 (from April)	9·03 inches	or	229·3 mm.	
1893	21·56	,,	,, 547·6	,,
1894	26·08	,,	,, 662·8	,,
1895	8·82	,,	,, 224·0	,,
1896 (first six months)	12·58	,,	,, 319·7	,,

The mean for the three complete years is 18·82 inches or 478 mm. For the mean for the whole period *see below.*

It is obvious, however, that a mean derived from these divergent figures cannot give any very satisfactory result as a practical guide. The number of days on which rain, which is usually brought by the N.W. wind, was measured, was 67 in 1893, 64 in 1894, and only 36 in 1895.

The following list gives the number of days, in the period 1893–5, which were recorded as fine, mixed, and overcast.

	No. of fine days	No. of mixed days	No. of overcast days
1893	232	77	56
1894	208	127	30
1895	276	74	15

The result showing a splendid record of fine weather.

Observations[1] at Benghazi result in the following table showing the percentage of the direction of the wind, by seasons.

	N.	N.E.	E.	S.E.	S.	S.W.	W.	N.W.
Winter	12·1	1·1	4·4	6·4	37·4	9·0	21·5	7·4
Spring	32·7	2·7	2·9	3·5	24·5	5·2	13·9	14·6
Summer	70·5	4·2	4·7	1·0	5·7	1·4	4·8	7·7
Autumn	29·2	3·8	4·3	7·6	34·4	3·8	10·1	6·8

The following table gives the mean monthly temperature (Fahr.) :—

	1891	1892	1893	1894	1905	1906	1907	1908	1909	Mean
	°	°	°	°	°	°	°	°	°	°
Jan.	52·70	58·46	—	52·52	52·16	—	57·20	55·04	57·38	55·04
Feb.	51·62	59·90	—	56·12	57·74	—	58·64	56·84	59·90	57·20
March	63·68	63·14	—	62·60	61·88	—	61·52	62·96	62·78	62·60
April	65·66	68·54	62·60	64·40	69·98	—	66·38	68·18	64·92	66·38
May	72·68	69·08	72·68	71·24	74·66	—	70·70	70·70	72·86	71·78
June	76·82	76·10	75·56	75·02	—	—	72·86	74·12	75·56	75·20
July	80·06	79·70	78·26	78·26	—	—	78·80	77·90	75·20	78·26
Aug.	80·02	—	77·90	79·52	—	—	78·98	78·62	78·08	78·98
Sept.	78·62	—	79·16	77·90	—	—	78·44	78·62	78·26	78·44
Oct.	74·66	—	75·02	78·08	—	76·64	71·60	75·92	71·60	74·84
Nov.	66·20	64·22	68·72	65·30	—	68·72	61·88	69·08	65·30	66·20
Dec.	60·62	—	60·08	57·74	—	60·44	57·56	60·26	59·36	59·36

The mean temperature given above is derived from $\dfrac{max. + min.}{2}$.

[1] Hildebrand, *Cyrenaïka*, Carl Georgi, Bonn, 1904, pp. 204, 225; also *Meteorolog. Zeitschrift*, 1903, p. 222.

It will be noticed that the above tables show that there is a marked uniformity in the mean monthly temperature from year to year, except, perhaps, in the winter months. Another point which the temperature table shows is that, at Benghazi, the autumn is very nearly as warm as the summer, and considerably warmer than the spring. This can be accounted for by a consideration of the wind table. In the summer time the Sahara, the hottest place in the world, becomes surcharged with heat, the air over this heated region rises and the north-east trades are drawn in, in the form of a north wind, from over the Mediterranean. During the autumn, or at any rate the two earlier months of that period, the Sahara is still warm, and the south winds are, as in summer, when they are of infrequent occurrence, the hot winds, and Benghazi receives 34·4 per cent. of its wind from that quarter, as compared with only 5·7 per cent. in summer; in fact the dominant wind is from the south during the autumn. In the winter the circumstances are quite different. The Sahara is relatively cold, and the south winds, coming to Benghazi, which, being on the sea, has its climate tempered by its moderating effect, are felt as cold winds, and the mean temperature is in January as low as 54·5° Fahr. The rainfall cannot be depended on from year to year, nor from month to month; the only dependable circumstances connected with it are its total absence, as at Tripoli, during the months June, July, and August, and a very moderate total for the year. The mean annual rainfall given below, viz. 11·94 inches, is derived from such divergent figures as 24·25 inches in 1894, and 6·81 inches in 1897, 21·69 in 1893 and 7·09 in 1899. Tobruk has a rainfall equal to any in Barka and hence its well-known fertility.

Of the climate of the Tripolitan Sahara there is little to say. It is of the ordinary Saharan type, characterised by great heat, but a very large range of temperature, bright sunny days with clear blue skies, and an almost entire absence of moisture. The great curse of this region is the *Simoon*, a scorching southerly wind, which carries with it great pillars of overheated dust, which shroud the region traversed with a pall, worse and darker than a London "pea-soup" fog. Fezzan is considerably warmer than

Tripoli, but in December and January the thermometer frequently sinks to 43° and 41° Fahr. (6° and 5° C.), and rain is rare. Murzuk lies in a depression and is unhealthy and malarious. Ghadames is excessively hot in summer, but like other parts of this area, is relatively cold in winter.

(For Productions *see* under North Africa.)

TRIPOLI. Lat. 32° 48′ N., Long. 13° 3′ E. (approx.).

Compiled from 4 years' observations[1].

	Jan.	Feb.	March	April	May	June	July	Aug.	Sept.	Oct.	Nov.	Dec.	Year
Mean maximum temperature (Fahr.°)	61°·4	65°·5	67°·6	74°·5	73°·6	81°·7	84°·4	85°·6	85°·8	82°·8	73°·8	63°·7	75°·1
Mean minimum temperature ...	46°·4	52°·3	51°·4	57°·4	62°·1	67°·8	74°·1	72°·5	69°·6	68°·5	60°·4	50°·9	61°·1
Mean temperature[2] ...	53°·9	58°·9	59°·5	65°·9	67°·8	74°·8	79°·3	79°·1	77°·7	75°·7	67°·1	57°·3	68°·1
Mean daily range[3]	15°·0	13°·2	16°·2	17°·1	11°·5	13°·9	10°·3	13°·1	16°·2	14°·3	13°·4	12°·8	14°·0
Mean of monthly absolute maxima[4] ...	70°·9	75°·2	82°·8	94°·8	93°·7	101°·5	97°·2	96°·4	99°·9	94°·6	82°·6	70°·2	—
„ „ minima[4] ...	38°·5	42°·8	41°·4	48°·0	53°·2	61°·9	64°·9	66°·7	63°·4	57°·4	52°·7	42°·3	—
Mean relative humidity (per cent.) ...	61	66	62	65	68	71	69	67	61	59	63	63	64·6

[1] April 1892—Dec. 1895.

[2] The mean temperature is derived from $\dfrac{\text{max.} + \text{min.}}{2}$.

[3] The daily range is max.—min.

[4] The absolute maximum recorded was 108°·14 in June 1893, and the absolute minimum 36°·3 in January 1895.

MEAN RAINFALL IN INCHES.

	Jan.	Feb.	March	April	May	June	July	Aug.	Sept.	Oct.	Nov.	Dec.	Year
Tripoli (31½ years)	...	4·20	0·79	0·28	0·22	0	0	0	0·22	0·77	3·09	7·57	18·58
Benghazi (9 years)	...	3·66	1·81	0·67	0·08	0·08	0	0	0·08	0·28	2·21	3·07	11·94

The Benghazi results are for the years 1891—1899. The results for Sept., Oct., and Nov. 1892 are wanting. The Tripoli results are for April 1892—Dec. 1895. Dr Supan, in *Petermann's Mitteilungen*, Ergänz. 124, 1898, gives the means for 8½ years, included in the period 1879—March 1892, thus:

Jan.	Feb.	March	April	May	June	July	Aug.	Sept.	Oct.	Nov.	Dec.	Year
2·60	1·97	1·02	0·39	0·32	0	0	0·08	0·36	1·81	2·28	3·11	13·94

EGYPT

LOWER EGYPT.

In Lower Egypt two types of climate are experienced, the Mediterranean type and the Saharan type. The area embraced by the Mediterranean type includes the whole of the delta during the winter months, but, in the summer, only a very narrow coast-belt is embraced in this Climatic Province. It follows, therefore, that, in the winter, the delta lies in the region of "winter rains."

November, February, March and part of April are delightful months in Lower Egypt, while May and October, though hot, are not unpleasant. Up to December there are frequent heavy dews, and morning fogs hang over the delta, due to exhalations from the damp soil lately left uncovered by the receding flood, and the morning humidity is high. The air of Egypt is always fresh, light and invigorating, and the humidity is very low, except in the immediate vicinity of cultivated lands. In these cultivated areas the relative humidity is high, the temperature drops suddenly at sunset and the air is very moist.

The most frequent winds are from the north and north-west; in the winter, however, south-west winds, which are sometimes very cold, are not infrequently experienced. The northerly winds are sometimes replaced by winds from the south, when cyclonic disturbances are passing over the Mediterranean. In the late spring—usually in April—the Khamsin—meaning 50, because this hot wind is experienced during 50 days—blows from the south-west in fits of about two days a week, sometimes with great violence, intensely hot, and accompanied by clouds of sand. The Khamsin is usually a most unpleasant experience.

In December and January it is cold in the delta, and as these are also the rainiest months, the middle of winter may be said to be unpleasant. It will be noted from the tables below that the humidity in the delta begins to increase about June.

At Cairo the healthiest time is the end of February and March. During these months the mean temperature is about 59° or 60° Fahr. (15° C.) for February, and 62°·5 Fahr. (17° C.) for March. Though in the early afternoon 70° is measured in the shade in the first month, and 75° in the second, the nights are always cool, but new arrivals feel the air a little oppressive and long for the desert, where the humidity is less. December and January are cold and not unhealthy, the mean temperatures being 58° Fahr. (14°·5 C.) and 54°·5 Fahr. (12°·5 C.) respectively, and the mean minima 48° Fahr. (8°·8 C.) and 44°·5 Fahr. (6°·9 C.). Most of the houses are now being fitted with fireplaces, but cold combined with rain is not pleasant anywhere. The thermometer however seldom goes below 40° Fahr. The hot weather begins in April, and, after that up to October, the thermometer may mount up to 100° Fahr. and over any day, but the heat is not oppressive in the early part of this period, and the nights are cool, the daily range in April and July being nearly 30° Fahr., and in May and June nearly 31° Fahr. July is the hottest month.

The northern towns of the delta are like the usual healthy Mediterranean sea-side places, only a little warmer, on account of their southerly position. Alexandria has a drier atmosphere than Cairo and it is warmer there at night, for, being on the sea-coast, the daily range of temperature is not so large. The mean annual temperature is 69°·5 Fahr. (20°·9 C.), a trifle less than at Cairo. July and August are the hottest months, with mean temperatures of 79° and 80°·5 Fahr. (26°·2 and 26°·9 C.), but the mean maximum temperature is less by 11°·7 Fahr. (6°·5 C.) in July at Alexandria than at Cairo. January and February are the coolest and rainiest months, the mean temperature of the former being 57°·8 Fahr. (14°·3 C.) and of the latter 59°·9 Fahr. (15°·5 C.); the lowest mean minimum for any month occurs in January, viz. 51° Fahr. (10°·7 C.). At Alexandria the prevailing winds are from the north and north-west, the former being the most frequent; north-easterly winds are frequently experienced in the periods October—December and March— May; and during December, January and February there

K. A. 5

are also winds from other quarters. The mean annual rainfall is about $8\frac{1}{2}$ inches, but westward of Alexandria the precipitation is somewhat heavier.

The general conditions at Rosetta and Damietta are very similar to those which prevail at Alexandria.

The Tables for Ismailia and Suez will be of interest to the very large number of people who travel by the canal route.

With reference to the Tables below the following passages from the Meteorological Report for 1904 are of interest: "A discussion of the best hours for the daily observations showed that so far as Egypt is concerned 8 a.m., 2 p.m. and 8 p.m. give the best results, but owing to the want of a station in the Sudan furnished with continuously recording instruments it is impossible to say how far these hours meet the case there, though the matter is receiving consideration and will be discussed as soon as data, sufficient to give daily variations of the different elements, are available.

"It may not be out of place to refer to some of the principal difficulties in the way of securing accurate meteorological observations in Egypt and the Sudan. In both countries for a great part of the year (in places throughout the whole year) the air is excessively dry and the prevalent fresh winds raise quantities of dust, both of which circumstances react unfavourably on the hygroscopic observations. Despite all precautions it is impossible sometimes to prevent the reservoir of the wet-bulb thermometer drying up, and the cotton wick becoming clogged with dust, and it is only by scrupulous watchfulness that one can hope to get accurate observations with this instrument.

"Another difficulty of a different character arises from the hygroscopic tables in use, which are those of Jelinek. It is extremely doubtful how far these tables are applicable to the humidities that obtain in Upper Egypt and the Sudan, and it appears desirable that an investigation of this point should be undertaken.

"Another instrumental difficulty arises in connection with the minimum thermometer, which at most stations has alcohol for the thermometric substance. The high temperatures frequently encountered in Egypt and the Sudan seem to render alcohol unfit for use in these thermometers. Evaporation and recondensation in the upper part of the tube is frequent, and there have been several cases where apparently the meniscus has been broken down so that the index has protruded from the alcohol....This difficulty acquires increased importance from the want of scientific training, at least in handling instruments, on the part of most of the observers....Consequently a breakdown of the minimum thermometer, which is the chief offender, generally means delay until the instrument can be repaired by one of the trained staff of the Department."

5—2

ALEXANDRIA¹. Lat. 31° 11′ 39″ N., Long. 29° 53′ 30″ E. Alt. 105 feet.

	Jan.	Feb.	March	April	May	June	July	Aug.	Sept.	Oct.	Nov.	Dec.	Year
Mean maximum temperature (Fahr.°)	64°4	66°9	70°0	74°5	79°1	82°4	85°3	86°5	85°1	81°9	74°7	68°2	76°6
Mean minimum temperature	51°3	52°9	55°0	58°8	63°7	69°1	73°2	74°3	72°1	68°9	60°9	54°1	62°8
Mean temperature²	57°9	59°9	62°4	66°7	71°4	75°7	79°3	80°4	78°6	78°9	67°8	61°2	69°8
Mean daily range³	13°1	14°0	15°0	15°7	15°4	13°3	12°1	12°2	13°0	13°0	13°8	14°1	13°8
Mean relative humidity⁴ morning⁵	64	62	61	60	60	64	67	64	63	64	61	66	63
" " 2 or 3 p.m.	54	52	51	53	56	60	61	58	56	58	55	58	56

¹ Extracted from *The Physiography of the Nile and its basin*. The means are for 21 years, but it is not stated what the specific years were.

² The mean temperature is derived from $\dfrac{\text{max.}+\text{min.}}{2}$.   ³ The daily range is max.—min.

⁴ per cent.   ⁵ 8 a.m. in some years, 9 a.m. in others.

ISMAILIA¹.

	Jan.	Feb.	March	April	May	June	July	Aug.	Sept.	Oct.	Nov.	Dec.	Year
Mean maximum temperature (Fahr.°)	65°1	69°8	74°5	81°9	88°1	92°8	95°9	94°8	90°1	86°2	76°1	68°9	82°0
Mean minimum temperature	46°2	48°9	52°3	57°0	61°7	66°4	70°7	71°7	67°8	63°7	55°4	49°5	59°2
Mean temperature	55°8	59°4	63°5	69°4	74°8	79°7	83°3	82°9	79°0	74°8	65°8	59°2	70°6
Mean daily range	18°9	20°9	22°2	24°9	26°4	26°4	25°2	23°1	22°3	22°5	20°7	19°4	22°8
Mean relative humidity² morning	84	80	75	70	71	74	77	80	80	82	83	84	78
" " 2 or 3 p.m.	49	33	38	32	30	30	32	34	40	42	47	51	39

² per cent.

¹ The notes above for Alexandria apply to Ismailia.

SUEZ[1]. Lat. 29° 56′ 3″ N., Long. 32° 33′ 15″ E. Alt. 10 feet.

	Jan.	Feb.	March	April	May	June	July	Aug.	Sept.	Oct.	Nov.	Dec.	Year
Mean maximum temperature (Fahr.°)	66°·4	70°·7	75°·7	83°·8	91°·4	95°·5	98°·2	96°·9	92°·8	88°·5	77°·0	69°·9	83°·9
Mean minimum temperature	46°·6	49°·1	53°·1	58°·3	63°·7	68°·0	71°·6	72°·0	68°·4	64°·8	55°·8	49°·6	60°·0
Mean temperature[2]	56°·5	59°·9	64°·4	71°·1	77°·5	81°·9	84°·9	84°·6	79°·7	76°·6	66°·4	59°·9	72°·0
Mean daily range[3]	19°·8	21°·6	22°·6	25°·5	27°·7	27°·5	26°·6	24°·9	24°·4	23°·7	21°·2	20°·3	23°·9
Mean relative humidity[4] morning[5] ...	76	74	70	65	64	66	70	74	75	77	78	76	72
,, ,, 2 or 3 p.m. ...	40	35	32	27	25	24	27	28	30	34	38	41	32

[1] Extracted from *The Physiography of the Nile and its basin*, but it is not there stated in what years the observations were recorded.

[2] The mean temperature is derived from $\dfrac{\text{max.} + \text{min.}}{2}$.

[3] The daily range is max.—min.

[4] per cent.

[5] 8 a.m. or 9 a.m.

CAIRO[1]. Lat. 30° 4′ 36″ N., Long. 31° 17′ 15″ E. Alt. 98 feet.

Compiled from observations recorded during the 35 years 1869—1903.

	Jan.	Feb.	March	April	May	June	July	Aug.	Sept.	Oct.	Nov.	Dec.	Year
Mean maximum temperature (Fahr.°)	69°·9	70°·0	75°·6	83°·5	90°·7	95°·2	97°·0	94°·8	90°·0	86°·2	75°·7	68°·4	82°·7
Mean minimum temperature	44°·4	46°·8	49°·8	55°·0	60°·6	65°·3	69°·4	69°·4	66°·0	62°·8	54°·1	47°·8	57°·6
Mean temperature[2]	54°·1	56°·8	62°·4	70°·2	76°·6	82°·2	83°·5	82°·6	78°·1	74°·5	66°·0	58°·6	70°·5
Mean daily range[3]	20°·5	23°·2	25°·8	28°·5	30°·1	29°·9	27°·6	25°·4	24°·0	23°·4	21°·6	20°·6	25°·1
Mean relative humidity[4] morning[5] ...	72	70	61	54	50	53	61	67	68	72	72	74	64
,, ,, 2 or 3 p.m....	48	43	34	30	27	27	29	32	39	41	44	49	36

[1] Abbassia Observatory. "Its position on the edge of the cultivated lands of the Delta with the desert to the east and south probably has affected the results slightly, giving higher temperatures in the months when easterly winds prevail, and lower ones during westerly winds than would have been recorded at stations wholly in the cultivation or in the desert respectively."

[2] Extracted from *The Physiography of the Nile and its basin*, but it is not stated how the mean was obtained.

[3] The daily range is max.—min.

[4] per cent.   [5] 8 a.m. or 9 a.m.

MEAN ANNUAL RAINFALL IN INCHES.

	No. of Years	Jan.	Feb.	March	April	May	June	July	Aug.	Sept.	Oct.	Nov.	Dec.	Year
Alexandria [1]	29	2·21	1·34	0·75	0·12	0·04	0	0	0	0·08	0·28	1·58	2·13	8·53
Port Said [2]	20	0·83	0·36	0·39	0·20	0·04	0	0	0	0	0·08	0·51	0·91	3·32
Ismailia [3] ... .· ...	19	0·39	0·24	0·20	0·12	0·12	0	0	0	0	0·12	0·24	0·43	1·86
Suez [4]	19	0·28	0·12	0·04	0·04	0·12	0	0	0	0	0	0	0·16	0·76
Cairo [5]	20	0·29	0·19	0·15	0·08	0·06	0	0	0	0	0·16	0·17	0·25	1·35

[1] 1868—1896. [2] 1886—1905. [3] 1886—1904.

[4] The Monthly means for April varied between 0 and 1·02 inches, and for January between 0 and 1·18 inches. In May 1888 the total precipitation was 2·64 inches. The 19 years are from 1887 to 1905.

[5] In January 1890 the fall was 0·93 inches and in March 0·83 inches; in November 1895 there was a total fall of 1·06 inches, and in 1898 of 0·90 inches. The 20 years are from 1886 to 1905.

UPPER EGYPT.

Upper Egypt lies wholly in the Saharan Climatic Province and is subject to desert conditions. There is, in these circumstances, very little rain, but recently there have been showers in winter, the climate evidently undergoing a certain amount of change, owing to the barrages now erected. The heavy dews referred to above (under Lower Egypt) extend as far south as Assiut, and in a less degree sometimes as far as Aswan, but never beyond. In Upper Egypt it is always coldest a little before sunrise, and the hottest period of the day is about 3 p.m. When the crops are growing, two, and even three, degrees of frost are sometimes recorded in the fields, but not in the desert. The winter climate of Upper Egypt is almost unsurpassed, wonderfully salubrious, warm, dry, and sunny. Indeed Dr Sandwith asserts that for one absolutely cloudless winter day in England, even in the sunniest parts of the south coast, there are ten or a dozen in Upper Egypt. In the summer, of course, it is very hot. The mean temperature for June and July at Aswan is no less than 91° or 92° Fahr. (33° C.) and the mean maximum 107°·4 Fahr. (41°·8 C.); at Wadi Halfa it is almost as high; but at Assiut the mean temperature is 85° or 86° Fahr. (30° C.) and the mean maximum 100° Fahr. (37°·8 C.), and at Beni Suef the corresponding figures are 82°·5 Fahr. (28° C.) and 94°·8 Fahr. (34°·8 C.) in July. A glance at the figures for relative humidity, given in the tables below, will show at once the wonderful dryness of the atmosphere, and will also show that, as in the delta, the humidity begins to increase about June. The daily range of temperature, however, is very considerable, amounting to from 27° to 30° Fahr. (15° to 16° C.).

(For Productions *see* under North Africa.)

With reference to the tables below, *see* note under Lower Egypt at p. 66.

BENI SUEF. Lat. 29° 4′ 11″ N., Long. 31° 6′ 14″ E. Alt. 95 feet.

Compiled from observations recorded from 1893 to 1904.

	Jan.	Feb.	March	April	May	June	July	Aug.	Sept.	Oct.	Nov.	Dec.	Year
Mean maximum temperature (Fahr.°)	67°·1	69°·6	75°·6	82°·9	90°·5	93°·7	94°·8	92°·1	87°·4	84°·4	76°·6	70°·7	82°·1
Mean minimum temperature[1] ...	42·4	45·0	49·1	55·8	63·1	67·3	70·3	69·8	67·5	63·1	54·0	47·3	57·9
Mean temperature[2]	55·0	57·4	62·4	69·3	76·8	80·4	82·6	81·0	77·5	73·8	65·3	60·8	70·2
Mean daily range[3]	24·7	24·6	26·5	27·1	27·4	26·4	24·5	22·3	19·9	21·3	22·6	23·4	24·2

[1] Capt. Lyons considers that the winter minimum is not low enough, the instrument having been fixed under a verandah.

[2] The mean temperature is derived from $\dfrac{\text{max.}+\text{min.}}{2}$.

[3] The daily range is max.—min.

ASSIUT. Lat. 27° 11′ N., Long. 31° 12′ 36″ E. Alt. 180 feet.

	Jan.	Feb.	March	April	May	June	July	Aug.	Sept.	Oct.	Nov.	Dec.	Year
Mean maximum temperature (Fahr.°)	68°·4	76°·5	83°·1	90°·5	97°·9	99°·9	100°·2	100°·0	93°·2	88°·3	81°·9	72°·5	87°·7
Mean minimum temperature ...	40·5	44·1	48·9	57·2	65·3	70·2	72·7	73·2	69·3	63·3	51·6	44·6	58·4
Mean temperature[1] ...	54·5	60·3	66·0	73·9	81·5	85·1	86·4	86·5	81·3	75·7	66·7	58·6	73·0
Mean daily range[2]	27·9	32·4	34·2	33·3	32·6	29·7	27·5	26·8	23·9	25·0	30·3	27·9	29·3
Mean relative humidity[3] morning ...	76	68	59	45	36	40	45	48	60	67	72	74	58
,, ,, 2 or 3 p.m.	34	44	24	21	16	17	22	22	31	38	48	46	30

A Climatological station was established here in 1900, but the observations were irregularly recorded in 1901. The above table was compiled from observations extending to Dec. 1904.

[1] The mean temperature is derived from $\dfrac{\text{max.}+\text{min.}}{2}$.   [3] per cent.

[2] The daily range is max.—min.

ASWAN. Lat. 24° 2′ 25″ N., Long. 32° 52′ 40″ E. Alt. 325 feet.

Compiled from observations recorded from Jan. 1901 to Dec. 1904.

	Jan.	Feb.	March	April	May	June	July	Aug.	Sept.	Oct.	Nov.	Dec.	Year
Mean maximum temperature (Fahr.°)	74°·1	79°·0	86°·0	94°·8	102°·6	107°·4	107°·4	105°·0	101°·5	99°·1	87°·3	79°·0	93°·5
Mean minimum temperature ...	49°·1	52°·0	56°·1	65°·3	71°·1	75°·6	77°·2	76°·8	73°·8	69°·6	61°·5	52°·5	65°·1
Mean temperature[1]	61°·7	65°·5	71°·1	80°·1	86°·7	91°·6	92°·3	90°·9	87°·6	84°·4	74°·5	65°·8	79°·3
Mean daily range[2]	25°·0	27°·0	29°·9	29°·5	31°·5	31°·8	30°·2	28°·2	27°·7	29°·5	25°·8	26°·5	28°·4
Mean relative humidity[3] morning ...	58	48	38	34	29	28	27	28	37	42	47	54	39
″ ″ 2 or 3 p.m.	30	22	17	17	15	15	13	13	18	22	25	30	20

The temperatures are the means of the figures for the Military Hospital and of those for a point on the east
bank of the river a little more than 200 yards from the river, immediately below the reservoir.

[1] The mean is derived from $\dfrac{8a + 2p + 8p + min.}{4}$.

[2] The mean daily range is max.—min.   [3] per cent.

In the Libyan Desert Von Grünau found that, in the first half of December, the temperature was 46° Fahr. (7°·8 C.) at 6 o'clock in the morning, rose to 67° Fahr. (19°·5 C.) at noon, and reached its highest point, 70° Fahr. (21°·1 C.) at 2 p.m., sinking gradually to 53°·6 Fahr. (12° C.) by 8 o'clock in the evening. This gives a mean day temperature of 57°·9 Fahr. (14°·4 C.), with a daily range of temperature of 23°·9 Fahr. (13°·3 C.). The corresponding figures for January were 38°·8 Fahr. (3°·8 C.) at 6 a.m., 63°·7 Fahr. (17°·6 C.) at noon, 65°·6 Fahr. (18°·7 C.) at 2 p.m., and 47°·5 Fahr. (8°·6 C.) in the evening, showing a mean of 52° Fahr. (11°·2 C.), and a daily range of 26°·8 Fahr. (14°·9 C.). From the above figures it appears that the mean day temperature for December and January in the Libyan Desert is 55° Fahr. (12°·8 C.).

In Siwa Oasis the mean temperature for the latter part of December and the first few days of January was 56° Fahr. (13°·4 C.). The temperature in the early morning was the same as that in the Libyan Desert in the first half of December, viz. 46° Fahr. (7°·8 C.); the noon temperature was practically the same as that in the desert in January, viz. 63°·5 Fahr. (17°·5 C.); at 2 p.m. the thermometer showed 64°·4 Fahr. (18° C.), while in the evening the mean of the readings was 53°·2 Fahr. (11°·8 C.). In the oasis the daily range was less by 5°·4 Fahr. (3° C.) than in the desert. The relative humidity was much less variable in the oasis than in the desert, the range in the former being 28 per cent., and in the latter 45 per cent.[1]

Mr Silva White describes the day before he reached Siwa (2nd April) as unbearable, owing to the scorching heat of the wind and sun. From May to October, which is the period of the *Khamsin* at Siwa, malaria is very prevalent, and attacks all the inhabitants. But in the winter the climate is healthy. The sky is always clear, but in January and February there may be occasional showers of rain.

---

[1] *P.M.G.* 1907, p. 165.

In the Eastern Desert of Egypt[1] the average evening (about 7 p.m.) temperature was

Oct. 29—Nov. 11, 1897	... 65°·7 Fahr. or	18°·7 C.
Nov. 12—Nov. 30 ...	... 57°·4 ,,	14°·1
Dec.	... 55°·4 ,,	13°·0
Divided into { 1st half ...	58°·5 ,,	14°·7
{ 2nd half ...	52°·3 ,,	11°·3
1st week in Jan. ...	... 49°·5 ,,	9°·7

The average for February was 64°·4 Fahr. (18° C.), the maximum being 75°·2 Fahr. (24° C.), and the minimum 52° Fahr. (11°·1 C.); the evening average for March was 65°·7 Fahr. (18°·7 C.) with many changes, a maximum of 78°·8 Fahr. (26° C.), and a minimum of 53°·6 Fahr. (12° C.). In April the evening average was 70°·7 Fahr. (21°·5 C.), the maximum being 78°·8 Fahr. (26° C.), and the minimum 59° Fahr. (15° C.). During the spring months a sudden rise of temperature generally foreshadows a sandstorm, and is followed by a fall and clearing of the atmosphere.

The south-eastern portion of the Peninsula of Sinai consists of four climatic zones[2], viz.

1.　The western plains with high temperatures. On the coasts of this zone the prevailing winds are N.W.

2.　The central plateau with low temperatures in winter and sudden storms; temperate summer, with clear sky and calms.

3.　South Sinai with calms interrupted by sudden gusts, and gentle afternoon breezes; and with milder winter in the valleys.

4.　The Ákaba coast with violent winds, mainly N.E., and with lower temperature than the western plains.

For rainfall this region is entirely dependent on local storms, and the precipitation is very considerable, but, of course, cannot be relied upon. Snow occurs on the mountains, and hail has been experienced in February. The maxima and minima of temperature recorded were

	Oct.	Nov.	Dec.	Jan.	Feb.	March	April	May
Maxima (Fahr.°)	100°·6	80°·6	81°·7	73°·9	75°·9	95°·0	103°·1	97°·5
Minima ...	57°·9	41°·9	36°·5	35°·6	41°·0	45°·0	64°·0	63°·0

[1] Barron and Hume, *Geology of the Eastern Desert*, Cairo, 1902.
[2] Hume, *Topography and Geology of the Peninsula of Sinai* (South-eastern Portion). Cairo, 1906.

Mr Hume was occupied in surveying the western portion of Sinai during five and a half months, from the middle of October, 1898, to the end of March, 1899, and states that, in the middle of winter, the weather often hindered his work, on account of rain storms and, on one occasion, a snowstorm. In the spring, on the coastal plain, sandstorms and winds were of frequent occurrence, the former often causing the stoppage of work for a day, while the wind was often too strong to allow of the plane-table being set up on a hill-top, and the compass was so affected that it would not settle. "Up to the 6th November there was a decreasing temperature, then a rise with a sudden fall, followed again by a rise, which continued to the end of the month. During December there was a steadily falling temperature, which culminated in the first week in January, 1899, in a severe frost for about a week, the lowest temperature being measured on the 4th. From this date there was a gradual rise, a maximum being reached on the 4th February, followed by a rapid fall on the 7th, this being followed by another rise and corresponding drop to freezing point on the 21st. From this date to the 20th March the temperature was equable, a sudden fall taking place on that day, followed by a steady rise to the end of the month[1]."

During the period under consideration there was one fall of snow, while it rained on 20 days, on eight of which there were heavy downpours. In October, November and March there were frequent thunderstorms. Hoar-frost was seen on six days. The sky was usually clear, though on several days it was overcast for a time, and there were 35 really cloudy days. The winds on this coast have been alluded to above; these were specially noticeable in December, and occurred in the form of gales and storms towards the end of February and the early part of March. During the early part of the year very hot, strong, and unpleasant south-east winds were experienced, but the direction of the prevailing winds is from the north or north-west. On the coast a south-west wind is the precursor of rain.

[1] *The Topography and Geology of the Peninsula of Sinai* (Western Portion), by T. Barron. Cairo, 1907, p. 90.

THE SAHARA

The Sahara is by no means the absolute desert that most people suppose it to be. It is, no doubt, a very dry region, but rain is by no means unknown. In the north-central region September is the month when rains generally occur, but Foureau also records rain on five days in November and five at least in December; thus on 4/11 rain fell in drops on three occasions, on 7/11 there was a fall of rain, on 26/11 a pretty copious short shower, on 27/11 drops of rain, on 28/11 a series of short showers; on 1/12 drops of rain up to 9 a.m. and violent showers at night, and on 12/12 a slight fall of rain. This was six weeks out from Wargla, and the rain must have been fairly copious, for Foureau records that the caravan was stopped by it. Again he speaks of constantly meeting "numerous streams," and of a cascade with a beautiful pool of blue water at its foot, on one occasion. Further, in his account, there is no very long stretch of country without trees and herbaceous plants being noticed here and there[1]. Lapperine speaks of the abundant rains at Hassi Gernan, south of In Zize, and the magnificent pasturage, in the middle of April.

The climate of the northern Saharan oases is most trying and, needless to say, very hot. From observations made in July and August, 1900, it appears that at 3 p.m. the shade temperature was always over 104° Fahr. (40° C.), sometimes reaching 120°·2 Fahr. (49° C.). On the other hand in January, 1901, for several consecutive days, the readings at 5 a.m. were 23° Fahr. (−5° C.) and at noon 77° Fahr. (25° C.). In these oases it rains very seldom; the sky is sometimes clouded and a few drops fall, but good rains are rare. Storms of rain, however, are not unknown, and are greatly dreaded, because of the enormous damage they occasion when they do occur. Violent winds are almost constant, but change in direction sometimes several times in one day[2].

In the Zusfana and Saura region, in the month of February, 1901, the thermometer in the forenoon stood at

---

[1] Foureau, *Mission Saharienne*. Masson et Cie. Paris, 1905.
[2] *Mouvement Géographique*, 1902, p. 366.

104° Fahr. (40° C.), and at the end of March 122° Fahr. (50° C.) was the not unusual record, while on 9th April it stood at 132° Fahr. (56° C.). The nights and also the early mornings and the evenings are cold. In March the thermometer sinks below freezing point. In April the nights are not so cold. The winds are strong and seem to blow from every point of the compass within the space of a few hours.

In the Central Sahara, on the high plateaux, the climate belies its exceedingly bad character. The winter is fairly cool, but, of course, the summer is very hot, and Duveyrier observed the thermometer at 112°·3 Fahr. (44°·6 C.). The almost complete absence of all flora and fauna is due to the enormous range of temperature, and the same traveller says that on occasions the cold was so great that, even with two thicknesses of clothes, he did not feel sufficiently protected in the early morning, while on the same day the midday heat would be overpowering.

The climate of Tidikelt is characterised by an extreme dryness, the result of incessant wind. In spite of the dirt and want of carefulness of the inhabitants, illness is rare. Dead animals become, so to speak, embalmed, completely dried up skins enveloping the bones of dead oxen in a very short period after death. As in all other districts of the Sahara the range of temperature is very large, and not uncommonly the thermometer registers 32° Fahr. (0° C.) or lower in the morning and 77° to 82° Fahr. (25° to 28° C.) at midday. In summer, from the middle of June to the middle of September, the midday heat varies between 118° and 122° Fahr. (48° and 50° C.) and that at night between 86° and 95° Fahr. (30° and 35° C.).

In the Muidir district, in April, 1903, the minimum temperature ranged between 50° and 59° Fahr. (10° and 15° C.), and the maximum between 86° and 98° or 100° Fahr. (30° and 37° or 38° C.). The mean temperature for May was 107°·6 Fahr. (42° C.) and for June 113° Fahr. (45° C.). During April, May, and June the prevailing winds were from the west and south-west, sometimes violent, and always charged with fine sand. There were only two days on which there was any rain, viz. the 21st April, when there were a few drops in the middle of the day and then a fine rain, which continued during the night; and the 31st May, when there was intermittent rain during the afternoon. These light and purely local rains merely moistened the surface of the ground[1].

In the Hoggar Highlands, which rise to 9000 and 9500 feet, it only rains once in about three or four years, and the variations of temperature are excessive. In November on one occasion the thermometer registered 35°·6 Fahr. (2° C.) and 111° Fahr. (44° C.) within 24 hours.

In the Air region it rains every year and the storms reach there generally after the end of June and between that and October. There is no reliable rainy season. In some years there is a fair amount of rain, in others it rains very little and at detached places. In the latitude of Agades there are no regular rains, but the precipitation is greater than in the northern Sahara. The Telwa, which runs past Agades was in a flowing condition no less than seventeen times in 1905. The mean number of times, per year, that this happens is stated to be five or six, but in some years it only flows a couple of times.

In Air, Barth mentions tropical rain storms in the second half of August. Foureau records frequent rains, *e.g.* 20th May a little rain, 7th June rain, 14th June drops of rain, and also on the 15th, 20th, and 27th; good rains on 27th July, and still more in the mountainous parts; on 2nd August rain, on 3rd August the plains converted into a sea; 4th August some rain, 5th August pouring rain in the evening, 10th August drops of rain and also on 15th and 16th; on 18th August terrific rain, similar to that of 3rd August. And so on, with slight rains up to the middle of September[2].

In the Adghar mountain region of the Sahara, south of Tenezruft, roughly south of 22° N., there is quite a different vegetation from that found further north, the more northerly Saharan flora gradually disappearing, to be replaced by a more southerly vegetation in consequence of a change in climate. Rains occur with tornadoes, normally about three or four times a year, which generally begin about the first fortnight in May. In 1907 the first tornado occurred on the 15th May, and was felt as far north as Tesalit, north of

---

[1] *La Belgique Coloniale*, 1904, p. 213.
[2] Foureau, *Mission Saharienne*. Masson et Cie. Paris, 1905.

$20°$ N., where some drops fell. Up to the 15th August rains were frequent and the sky covered with clouds. The temperature was high and sometimes even moist. From 15th August to 15th May it is especially dry, and this is the period of the high winds.

The temperature gradually gets lower till it reaches a few degrees above freezing point during the December nights, and also in January and February. In April and May the great heat begins again. The amount of rain in Adghar is really peculiar for the Sahara, where some regions are six, seven, and eight years without a drop. Adghar is the most northerly district to receive regular rain. In consequence of the regularity of the rain, there is a regular vegetation[1].

In the southern Sahara, to the north and north-east of the Niger Bend, in the Tilemsi-Niger plain, calms are frequent in July and August, and when the wind blows it is from the west and south-west. In the winter the wind is generally from the north-east, and is sometimes very strong. All the tornadoes come from this direction, and bring the summer rain to the plain. Between the 27th July and 22nd August, Gautier saw ten tornadoes, only two of which, however, touched the neighbourhood in which he was. They were very violent and brought deluges of rain, reducing the height of the thermometer from $82°{\cdot}4$ Fahr. ($28°$ C.) at five in the morning to $75°$ Fahr. ($24°$ C.) at midday. At other times in summer the air is very dry and the midday heat about $95°$ Fahr. ($35°$ C.), the difference between the dry and wet bulbs on the Niger, in this region, being $21°{\cdot}6$ Fahr. ($12°$ C.). Both Timbuktu[2] and Gao are healthy and agreeable places to stay at in the summer after the tornado season has once begun.

In conclusion, the climate of the Sahara, were it not so inimical to the existence of flora and fauna alike, would render it an almost ideal abode for man, from a health point of view—perfectly dry air, the humidity being almost *nil*, bright blue skies, and brilliant sunshine being the rule.

From tables published in *Statistique générale de l'Algérie* it appears that, in the north-western Sahara, from observa-

---

[1] *Bulletin de la Société de Géographie* (La Géographie) Paris, 1908, Vol. XVII. p. 271.

[2] See under French Niger territory.

K. A. 6

tions recorded at Timimum, Adrar and In Sala, the prevailing wind throughout the year is between north and east, chiefly from N.N.E., followed in order of frequency by N.E. These were the predominant winds in 41 out of 44 months at Timimum, in 28 out of 36 at Adrar, and in 30 out of 42 at In Sala. In May and June the prevailing wind is sometimes between south and west, most frequently from the S.W. Only in one month, June, at In Sala, was the predominant wind from the N.W., while the north as the prevalent wind was only recorded at Timimum in July, 1906; at Adrar only in October, November and December, 1905, and January and February, 1906; and at In Sala only in March and October, 1904, and in December, 1905. This record quite upsets all preconceived ideas, and disagrees almost entirely with the maps of the *Challenger* Report, as well as with the theory elaborated by Schirmer in his *Sahara*[1].

PRODUCTIONS.

The main stay of the Saharan oases is, of course, the date-palm, but wheat and barley and millet are produced in considerable quantities where circumstances are favourable, and tobacco and henna are grown. Camels and asses are the beasts of burden, and poor sheep and goats provide meat, which is always eaten salted and dried. The cereals deteriorate very quickly, and the seeds become smaller and smaller each year, so that fresh seed is required every three or four years. The tobacco is only for home consumption as it is not of a quality that would find purchasers on the market.

[1] Schirmer, *La Sahara*, Paris, 1893.

Of the three following stations, Timimum is situated, approximately, in 0° 15′ E., 29° 15′ N., Adrar in 0° 15′ W., 27° 50′ N., and In Sala in 2° 27′ E., 27° 11′ N. The mean relative humidity for Adrar is for January to May, and September to December, 1903, and for the whole of 1904, that for In Sala is for the year 1904. The mean daily range is from January, 1903, to August, 1906, for Timimum, except February, March and June, 1905; from January, 1903, to July, 1906, for Adrar, except that the last four months are for only one year; from January, 1903, to February, 1906, for In Sala.

MEAN RELATIVE HUMIDITY (per cent.).

	Jan.	Feb.	March	April	May	June	July	Aug.	Sept.	Oct.	Nov.	Dec.
Adrar ...	56·7	48·8	43·7	46·1	54·5	68·0	63·0	62·0	53·4	55·4	50·5	51·2
In Sala ...	49·0	31·0	45·0	54·0	53·0	58·0	29·0	19·0	19·0	31·0	47·0	45·0

MEAN DAILY RANGE (Fahr.).

	Jan.	Feb.	March	April	May	June	July	Aug.	Sept.	Oct.	Nov.	Dec.
Timimum ...	28°·8	32°·1	34°·2	35°·8	36°·2	35°·1	36°·2	36°·9	37°·8	33°·3	28°·8	26°·1
Adrar ...	28°·8	37°·8	33°·3	33°·3	32°·0	37°·0	35°·7	41°·1	38°·4	35°·8	30°·2	28°·8
In Sala ...	29°·9	33°·0	34°·0	31°·5	35°·0	35°·4	33°·2	33°·3	33°·3	32°·4	29°·7	28°·8

TROPICAL AFRICA

THE TROPICAL SEASONS.

In temperate regions the seasons are differentiated by temperature, but not so in the torrid zone. In tropical regions it is the rain which determines the seasons. Theoretically all places near the equator should have four seasons, whilst the four seasons merge and form only two near the tropics. The sun and the thermic equator, the belt of calms, where the trade winds meet, and the resultant cloud-ring, all show a marked preference for the northern hemisphere, and the result is that the belt of land which has four seasons lies more in the northern hemisphere than the southern, the mean position of the cloud-ring being somewhere between 8° north latitude and 2° south latitude, and it is between these parallels, roughly, that we find the four-season area. As the sun moves northwards it is followed by the cloud-ring, and similarly the sun on its passage to the southern tropic is accompanied by this phenomenon, the extreme limits being 18° N. lat. and 10° S. lat. The climatic phenomena vary with the position of the sun, followed by the cloud-ring, the heat centre and the rain reservoir, but it must be remembered that the greatest heat and the greatest amount of rain do not always coincide with the period when the sun is in the zenith, but usually drag on somewhat later than the sun, just as our experience tells us that noon is not the hottest time of day.

Now of this preference of the determining factors for the northern hemisphere, a little explanation appears to be necessary. And first with regard to the sun :—at the winter solstice the earth is nearer to the sun than at the summer solstice, and, therefore, according to the law of central forces, which requires equal areas to be described in equal times, the earth must move faster at the former than at the latter, and it follows that the sun is detained for a longer period in the northern hemisphere than in the southern. The actual length of time is about eight days.

Again, one of the functions of the trade winds is to assist in the work of evaporation, that is, in the collection of moisture. Where the trade winds converge and ascend they are separated by a belt, known as the zone of equatorial calms, where, through their powerful evaporating agency, is formed a peculiar cloud phenomenon, constantly renewed, and so rendered permanent, known, as mentioned above, as the cloud-ring. Now in the northern hemisphere the north-east trades, unlike the south-east trades, pass mainly over land (which largely preponderates in the northern hemisphere, and which, it should be remembered, gives up its heat more quickly than water), and are thus deprived of the opportunity of exercising to the full their powers as evaporating agents. It is a well-known physical fact that for the reception of moisture, or in causing evaporation, the air necessarily gives up a portion of its heat, and, therefore, it follows that the temperature of the north-east trades is not so fully diminished as that of the south-east trades, which pass mainly over water, by the exhaustion of the heat, which would be given up for the work of evaporation. Consequently on nearing the equator the north-east trades are hotter and ready to ascend at a higher latitude than the south-east trades, which thus force their way across the equator before they attain to a temperature corresponding with that of the north-east trades, and in their turn rise into the higher realms of the atmosphere. The effect of these causes is that the belt of calms lies in the northern hemisphere, the cloud-ring shows a strong preference for that hemisphere, the thermic equator is pushed far north of the terrestrial equator. As a matter of fact the heat equator enters Africa at about the latitude of Sierra Leone, passes northwards beyond the tropic of Cancer due north of the Niger Bend, and curves down gradually till it leaves the continent at about the latitude of Jibuti, in French Somaliland.

The following table, giving the mean annual rainfall at places in various latitudes in Eastern Africa, shows clearly how the rain follows the sun in its passage from the southern to the northern tropic. The rainy season at Zomba, in Nyasaland, is fully developed by January, and it will be seen that the maxima of precipitation steadily advance northwards with the sun till it reaches the northern tropic in June, as is shown by the black figures. The return journey of the sun is not quite so well marked, but is indicated by the italics. The stations selected are all situated in Eastern Africa, where there is the largest supply of data to work upon, and where, also, the land is comparatively uniform, consisting of a huge plateau with a general meridional direction, flanked on either side by lower land. The figures in brackets after the place-names denote the number of years' observations from which the means are calculated. It should be borne in mind that this table does not deal with the relative amount of rain at various places, but with the time of year at which each place has most rain.

In more central regions the available data are more scanty, but the same tendency is visible in the table on p. 89, the stations selected being Lealui on the Upper Zambezi, Luluabourg and Basoko on the Lulua and Congo rivers respectively, in Belgian Congo, Wau in the Bahr-el-Ghazal region, and El Obeid in Kordofan.

The western section of tropical Africa has not the same uniformity of structure as the eastern, consisting, as it does, almost of two independent members, northern Guinea, with its coast running east and west, and southern Guinea, with its coast running at right angles to this, and its climate, especially in the southern portion, is subject to various influences, which are not found on the eastern side, such, for instance, as the cold Benguela current. Yet if a series of stations be taken from the equator northwards, such as those below, the influence of the sun is still perceptible. The stations chosen are Libreville at the mouth of the Gabun, Yaunde in Cameroons, Axim on the Gold Coast, Gribingi in French Congo, and Gambaga in the Northern Territories of the Gold Coast. These stations are widely

MEAN RAINFALL IN INCHES.

Latitude	Place	Jan.	Feb.	March	April	May	June	July	Aug.	Sept.	Oct.	Nov.	Dec.
9° 2′ N.	Addis Ábbaba (8)	0·32	1·30	3·50	3·15	1·73	6·14	11·58	10·67	6·73	0·55	0·43	0·16
5° 11′ N.	Mongalla (3)	0·08	0·47	2·21	4·41	3·74	5·55	2·84	7·21	4·41	5·59	1·73	0·20
2° 45′ N.	Wádelai (7)	1·10	0·75	4·37	3·82	4·88	3·43	3·90	4·61	4·02	6·61	4·49	0·87
0° 3′ S.	Entebbe (9)	3·03	3·39	7·21	8·70	6·77	5·08	2·84	2·60	3·30	3·54	6·77	5·87
5° 15′ S.	Tabóra (8)	4·53	4·17	4·69	3·15	0·79	0·16	0	0	0·28	0·59	3·15	5·68
9° 20′ S.	Fife (3)	5·64	8·02	6·29	2·94	0·35	0·03	0	0	0·02	0·53	3·59	6·99
15° 4′ S.	Zomba (12)	11·14	11·50	8·11	3·66	0·71	0·47	0·32	0·12	0·32	1·73	3·39	10·95

separated towards the north, and some of them are on the coast, where many influences are at work, but still, as the sun moves northwards, it will be seen here, too, that the rains follow in its track.

It may be concluded, then, that the rains follow in the track of the sun, in tropical regions. And now to proceed to the discussion of the seasons. In the diagrams on Plate 14, each bar represents, as near as may be, about 50 mm. of rain, and the stations are arranged from north to south. A rainfall of less than one inch (25 mm.) is omitted and the space under that month is left blank—from 25 to 74 mm. is reckoned as 50 mm., from 75 to 124 as 100 mm., from 124 to 174 as 150 mm., and so on. Taking the eastern diagram as perhaps the most typical, a brief examination shows that, at the equator and near it, it rains almost every day, and each month has its rainfall, as is evidenced by the stations Natete and Entebbe in the diagram. It will be seen that there are two seasons of greatest rainfall, the first occurring, in the case of both stations, in April, and the second in October at the northern of the two and in November/December in the southern. It will also be noticed that these two seasons of maximum rainfall at each place are separated by periods of less rain ; in the case of Natete these seasons extend over June, July and August, and over December and January ; in the case of Entebbe over July and August on the one hand and January on the other.

At Butiaba there are the two maxima of rainfall, as before, separated in the summer-time (northern summer) by a period of less rain, while in the winter-time (northern winter) there occurs a dry season, extending over January and February. At Wádelai there is the early maximum of rainfall followed by the season of less rain, after which comes the second maximum, and this is followed by a dry season, longer than that at Butiaba, further south, and extending over December, January and February. Still farther north, at Gondókoro, there are the two rainy seasons as before, separated by the season of less rain, and the dry season as at Wádelai, but it will be noticed that the two rainy seasons are approaching one another in the direction

MEAN RAINFALL IN INCHES.

Latitude	Place	Jan.	Feb.	March	April	May	June	July	Aug.	Sept.	Oct.	Nov.	Dec.
13° 11' N.	El Obeid (3½)	0	0	0·24	0	0·06	0·95	4·25	**4·33**	1·89	0·71	0	0
7° 42' N.	Wau (3)	0·16	0·98	0·59	0·79	4·02	**6·77**	3·78	7·28	4·57	2·84	1·46	0
1° 14' N.	Basoko (17 mths.)	2·17	3·66	6·06	**6·10**	3·47	7·36	6·30	4·45	6·73	3·35	6·02	4·37
5° 56' S.	Luluabourg (29 mths.)	5·75	5·83	**6·46**	5·04	3·78	0·24	0·12	2·44	6·22	6·58	9·09	7·09
15° 13' S.	Lealui (4)	7·01	**9·13**	5·47	0·91	0	0	0	0·04	0·02	0·83	3·15	7·87

MEAN RAINFALL IN INCHES.

Latitude	Place	Jan.	Feb.	March	April	May	June	July	Aug.	Sept.	Oct.	Nov.	Dec.
10° 32' N.	Gambaga (7) ...	0	0·05	0·40	2·34	5·51	5·01	7·77	**11·73**	8·95	3·71	0·39	0
7° 0' N.	Gribingi (1) ...	0·2	0·14	2·28	3·67	5·63	8·50	**10·79**	**10·08**	4·02	5·08	0	0
4° 47' N.	Axim (6) ...	1·79	2·72	3·67	6·67	**15·09**	**24·90**	5·57	1·87	1·09	6·34	7·13	5·39
3° 49' N.	Yaunde (2) ...	1·02	2·80	5·67	**8·58**	8·07	4·61	2·24	2·76	8·27	8·94	5·83	1·97
0° 3' N.	Libreville (6 1/12)	6·14	8·86	**13·78**	**14·21**	5·04	0·28	0·16	0·83	3·78	14·92	17·60	8·19

of the northern summer, until at Gaba Shamba and Nasser they meet and form one wet season, with a dry season extending over six months, including the winter, the summer being the rainy season.

Starting at the equator again and proceeding southwards, it will be seen that the break for the dry season occurs in the middle of the year, the southern winter, while the two rainy seasons approach one another and gradually form one wet season towards December/January, the southern summer, which again is the one wet season, the dry season occurring in the winter months. Similar diagrams are given for a central series of stations and for a western series, where the same general laws may be seen to occur, though the western series does not bring into prominence the fact that it rains all the year round near the equator, there being a break for a dry season at Libreville in June, July and August.

Comparing the four diagrams, it appears that the four-season belt does not extend north of $7°$ N. either on the east or on the west, but in the central group of stations, at Wau, in $7° 42'$ N. there is still a trace of a season of less rain between the two maxima of precipitation. As to the southern limit of the belt, the eastern group shows that it extends at least as far south as Muanza ($2° 31'$ S.) but does not reach to Tabóra ($5° 1'$ S.); in the central series of stations it does not extend so far south as Luluabourg ($5° 56'$ S.), but on the west coast it reaches as far south as Banana ($6° 0'$ S.), and on the east coast as far south as Mikindani ($10° 17'$ S.).

But there is abundance of evidence to show that the general southern limit may be safely placed at about $3°$—$4°$ S. in all parts, except in the neighbourhood of the coast, both east and west. At Kondoa-Irangi in $4° 55'$ S. there are only two seasons, and only two at Ujiji in $4° 52'$ S. These two stations are both in German East Africa. Again, in Belgian Congo, there are only two seasons at Kabambara in latitude $4° 37'$ S.; and at Lusambo, though it rains practically all the year round, there is only one maximum of precipitation, and only one season of less rain, namely, in December and January, and similar conditions prevail at Nyangwe in latitude $4° 19'$ S.

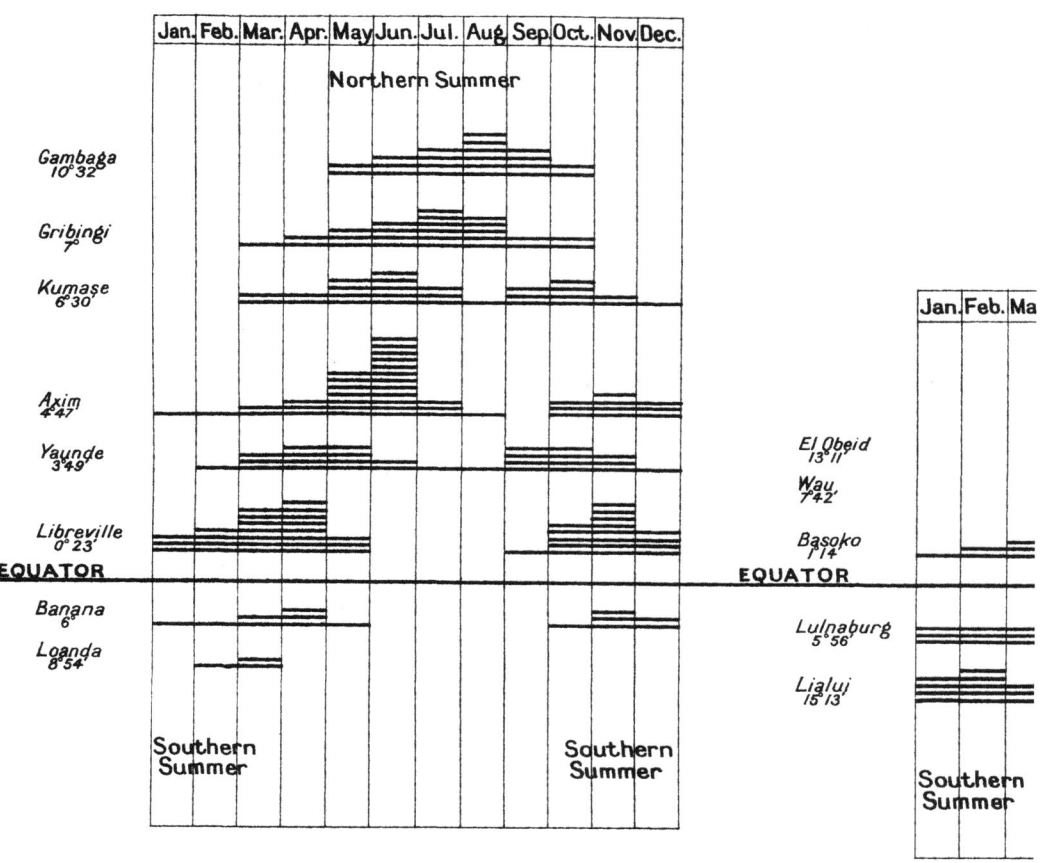

WEST

| | Jan. | Feb. | Mar. | Apr. | May | Jun. | Jul. | Aug. | Sep. | Oct. | Nov. | Dec. |

DIAGRAM TO ILLUSTRATE THE SECTION ON TROPICAL SEASONS

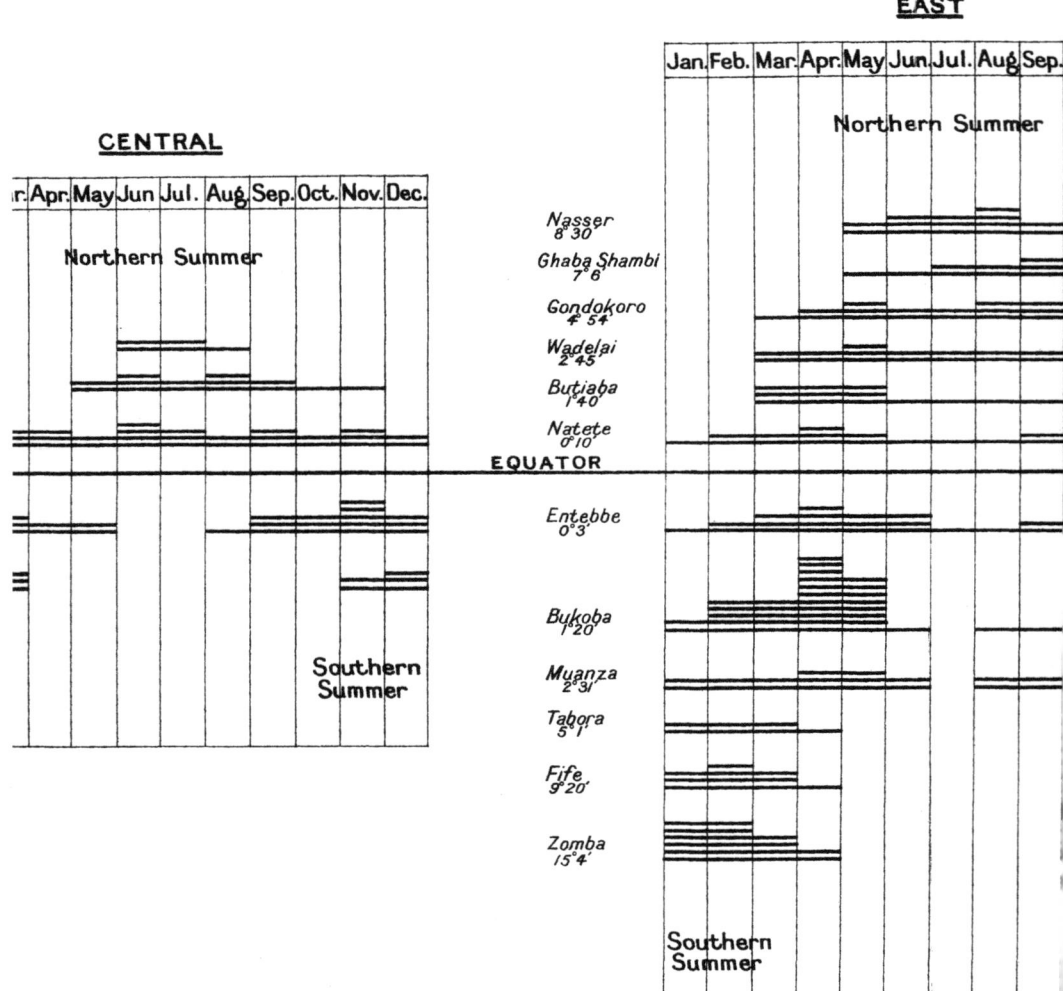

Bartholomew, Edin.

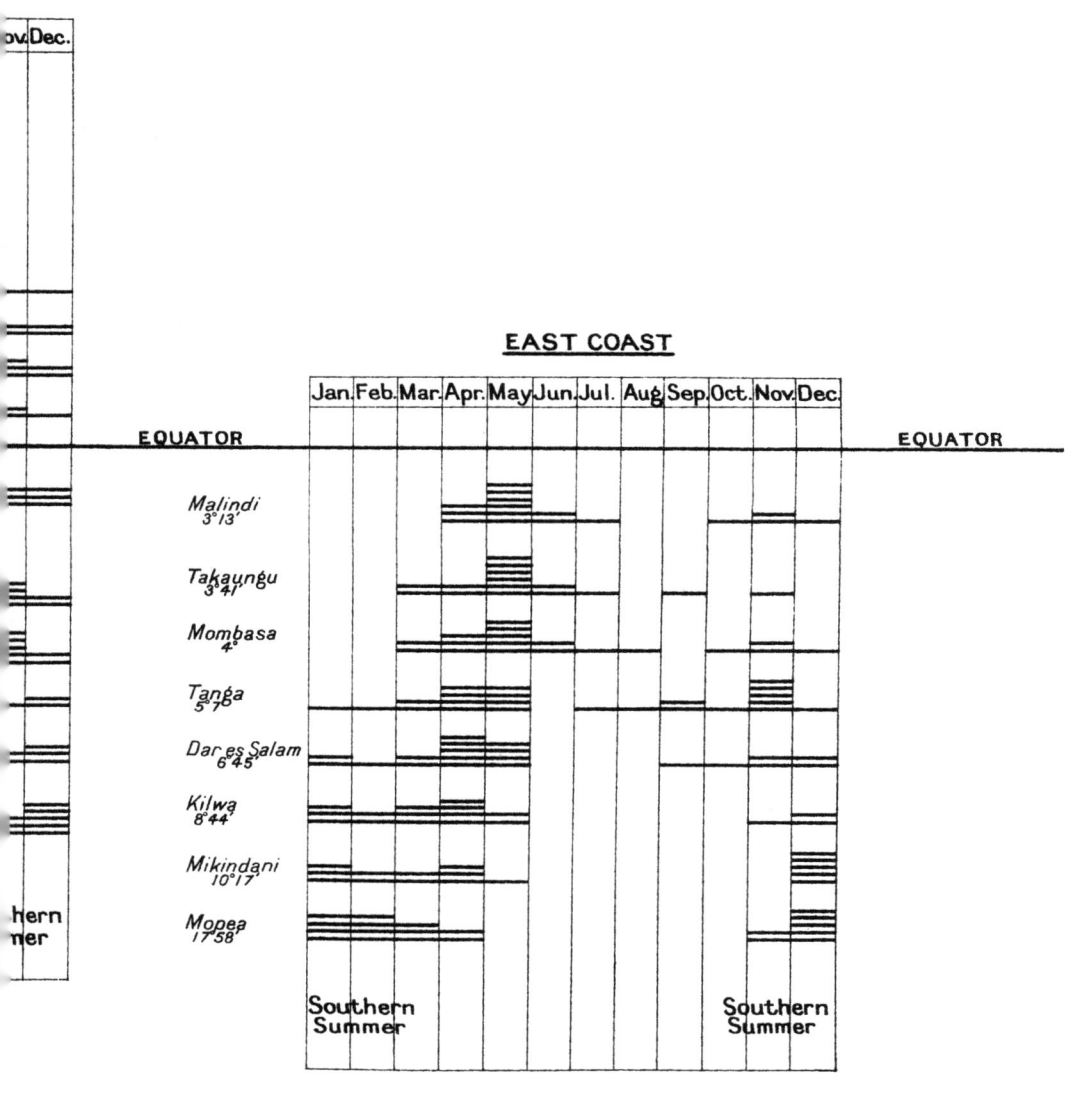

EAST COAST

One other fact may be gleaned from the diagrams, namely, that the rains are both of longer duration and greater intensity, especially in the southern hemisphere, as the sun travels from the southern tropic to the northern, than on its passage in the opposite direction. This is only natural, for the predominance of land in the northern hemisphere allows the sun comparatively little moisture to carry back on its return journey to the south.

WESTERN TROPICAL AFRICA

North Section.

"WEST COAST."

The "West Coast," as it is generally called, lies at the northern limit of the "South-East Trades" during the northern summer. The heated atmosphere of the Sahara ascends, and the air is drawn in from the ocean, and thus the "S.E. Trades" are diverted from their course, and blow from the south and south-west under the name of the "South-West Monsoon" of the West Coast. These winds bear moisture-laden clouds from the equatorial seas during the northern summer. In the northern winter the limit of the S.E. Trades moves southwards, and the belt of calms, which separates the northern "trades" from the southern, lies along the West Coast, with the result that the wind, blowing landwards, has very little strength, and contains less moisture ; and hence the chief part of the rainfall is measured rather in the summer than in the winter months.

The Harmattan is experienced as a wind which blows, especially in the months of December, January and February, from the N.E., and is a hot wind in some localities and a cold wind in others, according to circumstances. During this season, in the actual coast districts, it alternates with the land breezes, which blow from midnight till about sunrise, chiefly from the N.W. As a general rule the Harmattan begins early in the morning and falls off towards mid-day. Dried by its passage over the warm sands of the Sahara, it brings sandstorms in its course. Under its influence the lagoons of the coast region rapidly give up their water, while stagnant waters disappear altogether, as, except, perhaps, for an occasional storm, it does not rain during the Harmattan period. Further, in the process of this evaporation, it gives up a large amount of heat, and is thus rendered a comparatively cool wind. The Harmattan generally gives place to sea breezes from the S.W., while land and sea breezes alternate during the remainder of the year.

If the region of the Niger Bend, or a trifle north of this, be taken as a starting-point, the temperature decreases as the equator is approached; and the climatic zones are to a certain extent governed by the existence of three series of heights, running somewhat in a circular form, with the above-mentioned region as a common centre, and only feebly represented in the east by such heights as the Bauchi and Zaria Hills, and corresponding roughly with the limits of the zones of vegetation. And thus this tract of the "West Coast" and its hinterland may be divided into three approximately parallel climatic zones.

1. An equatorial zone corresponding roughly with the area occupied by the great equatorial forest belt.

2. A tropical zone corresponding with the more lightly wooded Sudan area.

3. A Saharan or desert zone.

In the equatorial zone there are two rainy seasons and two comparatively dry seasons in the year. The general direction of the winds is from north to south, while the tornadoes or storms follow a N.E.—S.W. direction. The rainfall amounts to nearly 60 inches, and in some cases, especially in the extreme S.W., considerably more, in the year. The sky is always of a greyish hue, and the atmosphere, saturated with water-vapour, veils the sun's rays.

In the tropical zone, or, at any rate, all except the extreme south, there are only two seasons in the year, one rainy season and one dry season. In the rainy season the general direction of the prevailing winds is from west to east, and in the dry season from east to west. It is during this latter period that the Harmattan is experienced as a hot wind, sometimes raising the temperature to 104° Fahr. (40° C.). The tornadoes, in this zone, have a general N.W.—S.E. direction. The rainfall is variable, being very slight in some of the northern parts and as much as 40 inches (1000 mm.) in the southern portion. The sky is usually clear, and the humidity is low, even in the rainy season.

In the Saharan zone rain is rare, there is scarcely any vegetation, and the prevailing winds have a general N.E.—S.W. direction, while the sky is always clear.

The following table[1] gives a general view of the climatic conditions prevailing in the West Coast and its hinterland. The dates, of course, must be taken only as approximate, for they vary slightly according to locality.

The predominant topographical feature of all this Upper Guinea hinterland is, of course, the great Niger river, one of the four great rivers of Africa, whose course, for ages, puzzled geographers and travellers alike. Lenfant[2], the gallant Frenchman, who ascended the river, in order to solve the problem of conveying supplies to the French territory to the north, has compiled a useful description of the climate of the riverain regions, which, by permission, I have condensed as follows :—

The seasons in the Niger Valley :

In the delta region rains are less frequent from November to March ; the first rains of the rainy season occurring in April gradually increase and then, becoming less frequent, die off in October. The weather is less oppressive from the end of November to the end of January. The heat is almost constant, but May, June and July are the hottest months.

On the Lower Niger, from Bussa to Lokoja, the dry season extends from November to mid February. The first rains occur in February and grow more frequent towards the end of March, and the last rains occur in November. December and January form the cooler season. February is a hot damp month and the greatest heat occurs in April.

In the region extending from Yelwa to Say the dry season extends from the end of November to April, and the first days of May, the first rains occurring in the latter month, and the last towards the 1st of November. December and January are the coolest months, the first great heat occurring in February and reaching its maximum in April.

In the Sorbo-Kabara region the dry season extends from the middle of October to the middle or end of May. The first rains occur towards the end of June, and the last

---

[1] Barot, *L'Afrique Occidentale.* E. Flammarion, Paris.
[2] Lenfant, *Le Niger.* Hachette et Cie., Paris.

RAINY SEASONS.

Zone	Date	Temperature		Sky	Humidity	Pressure inches
		Maximum	Minimum			
Saharan ...	Rains rare and very irregular					
Tropical ...	June—Mid October	100° Fahr.	60° Fahr.	Cloudy-tornadoes	Variable	29·72
Equatorial I.	Mid September—Mid December	90°	61°	Covered	Saturation	29·33
„ 2.	Mid March—Mid July	97°	64°	Covered	Saturation	29·33

DRY SEASONS.

Zone	Date	Temperature		Sky	Humidity	Pressure inches
		Maximum	Minimum			
Saharan ...	...	120° Fahr.	21° Fahr.	Clear	Nil	20·53
Tropical ...	Mid October—May	108°	34°	Clear	Very little	29·13
Equatorial I.	Early December—Mid March	100°	54°	Covered	Moderate	29·13
„ 2.	Mid July—Mid September	97°	60°	Covered	Moderate	

in the first days of October, the interval being marked by some 16 or 18 violent storms every year, but no continuous rains. The cooler season extends from the end of October to early March. The extreme heats begin in April and reach their height in June and July.

In the lake region the dry season covers the period from November to the end of April. The first rains come in May and the last occur in October. The cool season is from mid November to mid February. The first days of March bring the great heats, which reach their maximum at the end of April and May.

In the Bani delta region November to the end of April constitutes the dry season. The rains begin at the end of April, and leave off in mid October. The cool season includes December, January, and the first days of February. The heats begin in March and are at their highest in April.

In the Segu region November to May is the dry season. At the end of the latter month the rains begin, and leave off at the end of October. The cool season is from mid November to mid February. At the end of February the heats begin and are at their height from the end of March to May.

In the Sigiri region the dry season extends from the end of November to the end of March. The rains begin in April, and the rainy season extends from the end of April to the end of October, the last rains occurring in the early days of November. The cool season extends from mid November to the end of January. The hot season begins in mid February, and the greatest heat is experienced in the middle of March.

In the Upper Niger region the dry season extends from mid November to the end of March. The first rains occur in mid April, and the last towards the end of October. The cool season extends from mid November to the end of January. The extreme heat begins in early March, and reaches its maximum at the end of March or the beginning of April.

Throughout the whole of this Niger region, with the exception of the extreme north or Saharan section, there is a lesser rainy season, near the coast in February, from the

Sahara district onwards in January and the extreme upper Niger region in February again.

The general conclusion to be arrived at from a consideration of all the facts to hand is that in the Timbuktu region the Sudan is encroaching on the Sahara, while the reverse appears to be the case further east in the Chad-Zinder-Sókotó area.

PRODUCTIONS.

Throughout the whole West Coast coffee is indigenous, though but little is found north of Sierra Leone, except, perhaps, in the Rio Nunez valley in French Guinea. Indeed Sierra Leone is at the parting of the ways. To the north of this, as well as in the colony itself, ground nuts, whether under that name or as earth nuts, *amendoim, mandobim,* or *mancarra,* flourish, but to the east, though they grow on the Gold Coast, for instance, and in both Southern and Northern Nigeria, they are of inferior quality, though better results are now being obtained. Again the oil palm, though it appears in French Guinea and on the Gambia, is not really at home except to the east of Sierra Leone. The same may be said of the cocoa-nut palm and also of cacao or cocoa, except that the coastal strip east of Cape Three Points is not suited to this last, because it requires a very damp, in addition to a very warm, climate. North of Sierra Leone we find the *Acacia arabica* which produces gum. Cotton can be grown successfully throughout the whole of the West Coast. Rice does well in Sierra Leone and northwards, and also in Northern Nigeria and the Northern Territories of the Gold Coast. Guinea corn or dhurra, one of the millets, and Guinea wheat or maize are grown in most, if not all, of the colonies. Kola trees appear in French Guinea and eastwards. *Manioc, mandioc* or *manihot,* the product of which is called cassava, from which tapioca is made, grows from Portuguese Guinea eastwards. Tobacco does well, especially in French and Portuguese Guinea and in Northern Nigeria, and should do well in the Northern Territories of the Gold Coast. Rubber thrives everywhere in the forest from Kasamanse

K. A. 7

eastwards. Wheat will grow in the hinterland, and yams and potatoes ; Southern Nigeria, Sierra Leone, Liberia, and French Guinea produce ginger, and in the two first regions sesame or semsem seed is also grown. Sweet potatoes and chillies thrive in Northern Nigeria. Sugar has been tried successfully in Portuguese Guinea and in Nigeria, and should do well in the Northern Territories of the Gold Coast, whence eastwards through Nigeria the shea tree thrives.

It should be noticed that cereals, with the exception of maize and millet, do not reach maturity, except in the hinterland. Thus we find that wheat will flourish in Gurma, in French territory, and also in Northern Nigeria and in north Cameroons, but on the coast European cereals do not thrive. On the northern coast the banana is grown successfully, and St Louis boasts of its peas and beans ; while in the hinterland of the more eastern colonies too some European vegetables, such as beans, will grow. The acacia should do well in all the northern parts, and alpha grass in the extreme north. Other fibre plants might be tried with every prospect of success. Jute will grow in all the colonies, and bowstring-hemp is indigenous on the West Coast.

BRITISH WEST AFRICA

GAMBIA.

The climate of the Gambia differs in many respects from that of the other West African Colonies, and for six months in the year, at any rate, it is much more pleasant. For seven consecutive months, from November to May, inclusive, practically no rain falls. During these months the weather is often pleasantly cool, the thermometer sometimes being as low as $57°$ Fahr. ($14°·0$ C.) at 7 a.m. But the variations are somewhat trying, often ranging from $30°$ to $40°$ Fahr. ($16°$ to $22°$ C.) between 7 a.m. and 3 p.m. The daily range of temperature in the Protectorate is far less than in Bathurst, but it is impossible to employ white labour on the Gambia. From July to October is regarded as the unhealthy season, and during this period all business is at a standstill, and as many Europeans as possible leave for Europe.

The rainfall for the year 1902—June to October—was $57·13$ inches (1451 mm.). This is about 7 inches above the average. Of this total rainfall August was responsible for $35·87$ inches (911 mm.), of which 16 inches ($406·4$ mm.) fell on the 16th and 17th of that month. With one exception, when it was a trifle over 36 inches ($914·4$ mm.), this is the highest recorded rainfall for any one month for the past 20 years.

The table for 1903 (on p. 100) gives some idea of the monthly temperature, but a table of means, compiled from 6 years' observations, will be found below (on p. 102).

The following is from the official Handbook by F. B. Archer[1]:

"The climate of the Gambia during the dry season is very pleasant and healthy, which is more than can be said for any other of our possessions on the West Coast. But during the wet season the conditions are much the same

---

[1] *The Gambia Colony and Protectorate.* St Bride's Press. Mr Archer is the Treasurer of the Gambia and has lived there for many years.

7—2

as elsewhere, though the change from excessive dryness to the damp atmosphere of a miasmatical nature, so prevalent on the Gulf of Guinea, is probably more trying to the inhabitants.

		Shade max.	Shade min.	range	mean	Rain inches
		°	°	°	°	
January	(Fahr.)	93	56	37	71·9	—
February	...	100	60	40	75·3	—
March	...	100	61	39	76·5	—
April	...	93	64	29	75·2	—
May	...	88	67	21	75·4	—
June	...	89	69	20	79·2	5·91
July	...	89	68	21	79·0	7·13
August	...	89	70	19	79·3	35·87
September	...	89	67	22	79·4	4·15
October	...	91	67	24	79·7	4·07
November	...	92	63	29	79·0	—
December	...	89	57	32	73·1	—

" The lowest reading of the thermometer during the last dry season was, at Bathurst, 56° Fahr. (13°·3 C.), the highest recorded temperature being 100° Fahr. (38° C.); but up-country—at Boraba Kunda, on the south bank of the upper river district—the lowest temperature experienced by the Governor on a trip in February, 1904, was 40° Fahr. (4°·4 C.) the highest on the same day being 99° Fahr. (37°·2 C.). In the wet season of 1903 the lowest temperature recorded at head-quarters was 67° Fahr. (19°·4 C.) and the highest 91° Fahr. (32°·7 C.).

" The Harmattan wind usually sets in early in December, and is welcomed after the oppressive weather of October and November. It is an exceedingly dry wind, and, though cool and bracing in the morning, becomes hot under the influence of the sun. The highest temperature of the day is usually between two and three o'clock. The Harmattan is an intermittent wind, blowing for a few days and then being succeeded by a refreshing sea breeze, which, in turn, gives way again to the Harmattan (which usually blows for about a fortnight at a stretch). This state of things continues well into April, the month of February being the time when the Harmattan is most prevalent."

The period from July to October is considered the most unpleasant and unhealthy in Bathurst. In July the sea breeze has abated, and the atmosphere becomes very hot and oppressive, the days previous to the rains being most stifling, and generally closing with tornadoes and heavy rain. The first rains generally appear about the middle of June, but the wet season does not finally establish itself until well into July, that month, August and September seeing the principal rainfall of the season. The following are the records of rain for the decade 1895–1904.

Year	Highest occurring in one month in inches		Total for year in inches
1895	August	36·63	66·86
1896	August	17·30	51·18
1897	September	11·84	33·61
1898	August	19·20	48·65
1899	August	14·26	56·17
1900	July	20·53	43·38
1901	August	19·90	45·31
1902	August	14·36	29·42
1903	August	35·87	57·87
1904	August	17·27	38·20
Mean		20·72	47·06

As a rule a heavy rainy season is considered healthy, especially if tornadoes are frequent, as the air becomes unusually pure and clear.

The public health of the colony has of late years been good and the cleanliness of the town (Bathurst) exceptional. There has recently been no epidemic in the Gambia. The sanitary conditions of the town have been considerably improved.

(For Productions *see* under West Coast.)

BATHURST[1]. Lat 13° 27' 16" N., Long. 16° 34' 19" W.

(Means compiled from 6 years' observations, 1900—1905.)

	Jan.	Feb.	March	April	May	June	July	Aug.	Sept.	Oct.	Nov.	Dec.	Year
	°	°	°	°	°	°	°	°	°	°	°	°	°
Mean highest shade temperature at 3 p.m. (Fahr.°)	91·0	98·1	99·0	98·2	95·3	90·8	90·3	89·8	90·8	92·7	92·5	90·3	99·0
Mean lowest shade temperature at 3 p.m.	76·3	76·3	75·2	77·7	77·2	81·3	80·8	79·8	82·3	82·0	81·5	78·3	75·2
Mean of the average shade temperature at 3 p.m.	84·5	87·8	83·8	83·9	83·4	85·8	85·0	85·5	86·9	88·2	87·6	85·1	85·7
Mean highest shade temperature at 7 a.m.	66·2	68·2	70·0	70·2	72·2	75·3	76·3	77·0	76·7	76·5	75·0	69·0	77·0
Mean lowest shade temperature at 7 a.m.	54·7	59·3	60·2	60·7	62·7	67·3	67·8	69·0	67·7	68·3	65·0	58·7	54·7
Mean of the average shade temperature at 7 a.m.	61·0	63·8	64·4	66·6	67·2	71·6	72·8	72·7	72·5	72·3	70·2	64·2	68·2
Absolute maximum temperature during the six years at 3 p.m.	93	100	101	102	104	95	96	92	92	94	96	95	104
Ditto minimum	72	71	73	75	75	79	79	78	78	78	75	76	73
Absolute maximum temperature during the six years at 7 a.m.	68	71	72	73	76	77	78	79	78	79	79	72	79
Ditto minimum	46	57	57	59	60	62	66	67	64	64	63	53	46
Dew point	52	56·8	59·6	62·5	67	71·2	74	74·3	74·5	73·2	64·7	54·4	65·6

[1] Bathurst lies on a sandy point on St Mary I. at the mouth of the Gambia river.

RAINFALL IN INCHES.

	Jan.	Feb.	March	April	May	June	July	Aug.	Sept.	Oct.	Nov.	Dec.	Year
Mean rainfall in inches ...	0	0·04	0·02	0·01	0·03	3·1	12·3	20·1	7·2	3·6	0·12	0	46·4

The following table gives the death rate.

Year	per 1000	Places
1900	28·7	Bathurst, British Kommbo, Ceded Mile, McCarthy Island, Tenda Ba, Bai, Kannsala, Briffet and Bajianna.
1901	30·0	Bathurst, British Kommbo, Ceded Mile.
1902	30·4	Ditto
1903	36·7	Ditto (for Bathurst alone 42·1).
1904	34·4	Bathurst
1905	33·9	Bathurst

SIERRA LEONE.

The following notes on the climate of Sierra Leone (pp. 104—7) have been kindly written, specially for this volume, by Major-General J. K. Trotter, C.B., C.M.G., who was on the Boundary Commission of 1896-7, has been in the interior during every month in the year, and was General Officer Commanding the Troops from 1906 to 1908 :—

Sierra Leone has long enjoyed the reputation of being a place peculiarly fatal to human beings and domestic animals. It is only during the present century that scientific investigation has made it clear that the diseases which have wrought such havoc in the past are not necessarily indigenous in Sierra Leone, but that, given the adoption of sanitary measures, they can all be removed. But when malaria, blackwater fever and the other ailments propagated by the mosquito and tsetse fly are no longer known on the West Coast, the climate will not cease to be a trying one for the European constitution. The heat, as recorded, is not great; the shade temperature, in fact, rarely exceeds 95° Fahr. (35° C.); but the excessive moisture makes it far less tolerable than a higher temperature when the air is relatively dry, and the effect of the constant and excessive action of the skin is exhausting.

Temperatures of more than 90° Fahr. (32°·2 C.) are only recorded in the low country. In the higher parts the thermometer rarely reaches that point. In the Hill Station, near Freetown, at a height of 880 feet, 85° Fahr. (29°·5 C.) is the highest temperature recorded. The hottest months of the year are January, February, and March, but the air being then relatively dry, the heat is less felt than at any other time except during the rains. May and October would be classed as the hottest months, and they are certainly the months when the heat is most felt, the air being very moist, but according to records they are cooler than the winter months. And yet the records show very little variation in the maximum, except during the rains, when the weather is perceptibly cooler, and the maximum is generally not more than 75° Fahr. (24° C.).

There is no cold weather in Sierra Leone. The lowest temperatures in the colony, of which $66°$ Fahr. ($19°$ C.) may be regarded as the extreme, occur during the rainy season and occasionally during the prevalence of the Harmattan in January, February and March. In the high country of the Protectorate, particularly in the hills north of Bumban, temperatures of $56°$ Fahr. ($13°·3$ C.) are common during the nights of the dry season, which are markedly cooler than in the country near the coast. But throughout the year the daily range of temperature is very small, and often does not exceed $4°$ or $5°$ Fahr. ($2°·22$ or $2°·78$ C.).

During the rainy season the wind blows from the south-west and west, and the rain comes from the same direction. This season begins about the middle of June and lasts till the latter part of September. The greater part of the annual rainfall—165 inches (4191 mm.)—falls in this period. When the true rains end, the wind changes to the east or north-east, and the tornado season begins. The tornadoes come generally at night and with the tide. Preceded by brilliant lightning, they strike each locality with remarkable sudden-ness, one moment the air being perfectly still, the next a violent wind, accompanied by heavy rain and thunder, raging. The tornadoes are soon over, and rarely last more than half an hour, but they occasionally return an hour or two later. They occur about every alternate night after the end of the rains. As the season progresses they become less frequent, till in November they are occasional, and in December cease. They return in April, becoming more frequent as the rainy season approaches. Out of the rainy season the rain that falls is always accompanied by tornadoes with thunder storms; during the true rainy season storms are rare. The rainfall in the higher parts of the colony is probably greater than that recorded at Freetown. In the Protectorate it is less, but no continuous records exist to establish the difference.

During the wet season, when heavy rain has not fallen for some time, and during the period immediately before and after the rains, mists are frequent, and heavy cloud masses travel at low elevations beneath the hill tops. In the Harmattan season or "smokes," as they are called locally, a haze so dense frequently prevails that it is impossible to see any distant object, and navigation is then very difficult. This haze is caused by particles of red laterite dust blown from the interior and suspended in the air.

The dewfall varies in intensity, but it is always heavy. The ground in the early morning, particularly at the beginning of the dry season, is as wet as if heavy rain had fallen, and it is impossible to follow a bush path without being soaked, although the range of temperature is very small. The air is, in fact, so completely saturated that the least fall of temperature produces a heavy dew.

The mornings and evenings are not invigorating or fresh. The air is too moist, and the fall of temperature at night too slight to produce any stimulating effect. The one thing which contributes to freshness is the breeze, which is rarely absent in the evening, especially near the sea. The early mornings are generally still and rather oppressive. In Harmattan season, however (January, February, and March), the air is relatively dry, and to most Europeans much more agreeable than at other times of the year. But natives and Europeans long resident on the coast find the dry wind unpleasant. It dries the skin and produces cracks, chills and other ailments.

The general health of Europeans in Sierra Leone has greatly improved since the discoveries concerning the propagation of malaria by the mosquito were made. It is now quite possible to do a tour of service in the country without suffering from that disease, and not many Europeans, who take ordinary precautions, are seriously affected by it. Blackwater fever is still prevalent, but is much more successfully treated than formerly. The nature and means of prevention of this disease still remain to be discovered. Where stations, such as the Hill Station, near Freetown, are established, from which native families are carefully excluded, the risk of suffering from malaria is very small. When sufficient force can be accumulated to overcome the inertia of West Africa, sufficient resolution to combat native prejudice, and sufficient means to cover the cost of carrying out the most ordinary and obvious sanitary measures, there can be no doubt that the diseases affecting both man and beast, which have been so fatal in Sierra Leone, will disappear absolutely and for ever. There will then remain the depressing effect of the climate which makes a change after a year's service on the coast essential to almost every European constitution. Those who remain longer than a year in Sierra Leone, though suffering from no disease, obviously lose both tone and vigour, and have a washed-out appearance.

Dress should be carefully attended to. Clothes must be light and always woollen. A flannel belt should be worn at all times, and a thin flannel shirt and flannel trousers or thin woollen drawers. Linen collars should not be worn except in the evening. Frequent changes of clothing are necessary owing to the excessive action of the skin when any exercise is taken. A good helmet is essential when the sun is up.

The man who does best in Sierra Leone is he who is temperate in diet, particularly temperate as regards the use of alcohol, and who is not afraid of outdoor work, and even of exposure to the sun. A prudent man abstains from the use of alcohol at least during the daytime. Numerous and deadly as are the diseases of Sierra Leone, it is at least questionable whether any of them can contend with the abuse of alcohol for deadly effect. On the other hand, the very moderate use of stimulants at proper times is probably beneficial.

It is very important that the European in West Africa should be well fed. At present the food supply is far below the requirements of the climate. The only fresh supplies of good quality consist of fish (on the coast), fruit, and to some extent, vegetables. The native beef, mutton, and poultry are all of very inferior quality, and the eggs are very small. English fresh meat, fish, butter, etc., can be obtained from mail boats fitted with refrigerators, but the quality leaves much to be desired. Tinned foods should be avoided, more especially those which are liable to go bad. The occurrence of ptomaine poisoning in West Africa is probably much more frequent than is generally supposed. Anyone about to proceed to Sierra Leone is advised to take with him every description of good food which can be transported otherwise than in tin.

Finally, the rule for the European in Sierra Leone should be: eat good food, drink little, take plenty of exercise in the open air, and go home at the end of a year.

The total rainfall in 1900 was 175·43 inches (4456 mm.). In that year the mean temperature in April and in December was 80°·8 Fahr. (27°·1 C.) and 80°·2 Fahr. (26°·7 C.) in March, while in August, the coolest month, the mean registered temperature was 76°·3 Fahr. (24°·6 C.). The heaviest rainfall was in June (41·36 inches or 1050 mm.), in July there were 37·97 inches (964 mm.), and in August 32·12 inches (816 mm.), while in January and February there was no rain. In 1903 the total rainfall for the year was 162·01 inches as compared with 183·31 inches in 1902 (4115 and 4656 mm.). The heaviest fall in 1903 was in August, when 53·75 inches (1365 mm.) were recorded, and the 12th of that month witnessed the greatest rainfall in any one day, namely, 9·55 inches (243 mm.). There was no rainfall in February and only 0·16 inch (4 mm.) was measured in December. In 1904 the total rainfall for the year was 158·16 inches (4017 mm.), which was 3·85 inches less than in the previous year, and is the lowest recorded for some years. The highest monthly rainfall was in July, in which month 42·67 inches (1084 mm.) were measured. The rainfall for the five months June to October amounted to 143·08 inches (3634 mm.). The highest shade temperature recorded was 95° Fahr. (35° C.) on the 7th February, and the lowest 66°·4 Fahr. (19°·1 C.) on the 15th January. In 1905 the rainfall was 176·94 inches (4494 mm.), or 18·78 inches in excess of the figures for the previous year. The highest monthly rainfall was 57·25 inches (1454 mm.), measured in the month of July, and the highest fall on one day was 8·09 inches (205 mm.) on the 31st of that month. The highest shade temperature recorded was 98°·9 Fahr. (37°·1 C.) in March, and the lowest 67° Fahr. (19°·4 C.) in January. The average mean temperature for the year was 81° Fahr. (27°·2 C.).

The death rate for the Colony in 1903 was 22·6 per 1000, showing a decrease of 83 deaths on the 1902 figures. In 1905 the death rate was 28·2. In Freetown the death rate in

$$
\begin{array}{llll}
1901 & \text{was} & 28\cdot9 & \text{per } 1000 \\
1902 & ,, & 24\cdot9 & ,, \\
1903 & ,, & 23\cdot9 & ,, \\
1904 & ,, & 26\cdot7 & ,, \\
1905 & ,, & 29\cdot6 & ,,
\end{array}
$$

The death rate among Europeans was 18·5 per 1000 in 1905. The Report for that year states, "In connection with the death rate, attention may be called to the trite but most misleading description, 'Sierra Leone, or the White Man's Grave.' With the exception of Gambia, Sierra Leone (especially the Protectorate) is probably much more healthy than any other British Colony on the West Coast, not excluding Northern Nigeria. The expression 'White Man's Grave' became well known in older days, when all the places on the West Coast were under the Sierra Leone Government." Muirhead[1] also bears out this contention.

During the year 1905 the total number of days on which European officials, resident in Freetown and at Clive Town, were under medical treatment amounted to 518, out of which 297 or 57 per cent. were for malaria. But Hill Station is justifying itself by considerably reducing the incidence of malaria amongst the officials.

Speaking generally there is no doubt that the climate of Sierra Leone cannot be classed as healthy, especially at the beginning and end of the rainy season, which commences in May and lasts till October, the remaining months forming the dry season. Tornadoes or violent thunder storms, which, however, do but little damage, though accompanied by strong winds, are frequent at the beginning and end of the rainy season, during which period the atmosphere is very damp. The Harmattan, a very dry easterly wind from the Sahara is experienced between the months of December and March.

The chief diseases are malaria and blackwater fever; enteric fever is unknown here, due possibly to the pure water supply.

The prevailing wind at Freetown is from the west, but easterly winds (Harmattan), as mentioned above, are frequent between December and March, and winds from the same direction from October to December. Storms are frequent during the rains, but not so frequent in the rainiest months (July and August).

At Freetown the variation in barometric pressure is

[1] *J.R.A.M.* 1907, p. 199.

given by the following table which shows the difference between the highest and lowest monthly means.

Year	Highest	Lowest	Difference
1900	July	January	0·081 inches
1902	July	January	0·084
1903	July	January, April	0·098
1904	August	January	0·096

The following gives the lowest and highest recorded temperature during the years 1900–5.

Year	Lowest	Date	Highest	Date
1900	67°·0 Fahr. (19°·4 C.)	26th January	92°·0 Fahr. (33°·3 C.)	19th December
1902	67°·0 ,, (19°·4 C.)	20th June	94°·4 ,, (34°·6 C.)	11th June
1903	67°·0 ,, (19°·4 C.)	25th September	98°·0 ,, (36°·7 C.)	17th March
1904	64°·4 ,, (18° C.)	15th January	95°·0 ,, (35°·0 C.)	7th February
1905	67°·0 ,, (19°·4 C.)	2nd January	98°·9 ,, (37°·1 C.)	7th March

At Falabá, and in the hills of the Koinadugú and the central districts, the thermometer falls as low as 55° (12°·8 C.) at night. March and April are here the hottest months, and in December, before the Harmattan blows, the weather is also said to be oppressive.

The number of thunderstorms for the five years previous to 1907 was on the average 127·6, while in that year there were 123. The following are the average numbers of storms in the months in which they usually occur:

May 19·8	September 12·6
June 19·2	October 20·4
July 10·4	November 16·2
August 3·8	

Since the above was written I have received from General Trotter the daily records (centigrade) of the maximum and minimum temperature at the Hill Station, which stands 880 feet above sea-level, for the period April, 1907, to March, 1908; and it is interesting to compare the temperature on the hill with that at Freetown below. From April to December the comparison is complete, the figures being those for 1907 in both cases, but unfortunately the Blue Book for 1908 is not yet published, and I have therefore compared the 1907 figures for Freetown with the 1908 figures for the Hill Station for the three months January, February and March. As will be seen at once from the figures below (p. 112) the Hill Station has an immense advantage, the mean maximum being there some

10 to 11 degrees Fahr. less in the hottest months of the year, and never less than 7°·5 Fahr. lower than at Freetown. The mean minimum figures do not show so marked an advantage for the Hill Station, though the thermometer there goes decidedly lower than at the Station below.

On pp. 113, 114 are given two tables, one compiled from observations taken during six years in the period 1900–6, the other from observations for 19 years previous to 1899, and it is interesting to note that the mean temperature, from some cause, appears to be less for the former period and that the relative humidity and amount of cloud are also less. The rainfall for June, July and August has considerably increased, but is less for the remaining months. The second table gives some additional information not contained in the first, namely, the number of rainy days. The absolute maxima and minima of temperature for late years are given above. The mean temperature in both tables is the mean of the mean maximum and mean minimum. In the second table the mean daily range is given, in the first it is easily calculated by subtracting the mean minimum from the mean maximum.

(For Productions *see* under West Coast.)

MEAN MAXIMUM TEMPERATURE (Fahr.).

	Jan.	Feb.	March	April	May	June	July	Aug.	Sept.	Oct.	Nov.	Dec.
Freetown... ...	89·8	90·7	91·8	90·6	89·9	87·7	84·7	83·4	85·4	87·6	89·0	89·1
Hill Station ...	80·06	80·42	81·32	79·70	80·06	78·26	76·10	75·74	77·72	79·16	80·24	80·24
Difference	9·74	10·28	10·48	10·90	9·84	9·44	8·60	7·66	7·68	8·44	8·76	8·86

MEAN MINIMUM TEMPERATURE (Fahr.).

	Jan.	Feb.	March	April	May	June	July	Aug.	Sept.	Oct.	Nov.	Dec.
Freetown... ...	73·1	74·3	75·0	74·6	73·4	72·4	71·8	72·3	72·1	71·8	73·3	73·5
Hill Station ...	69·80	69·62	68·90	71·60	72·50	71·06	68·90	67·64	68·54	68·90	70·70	70·52
Difference ...	3·30	4·68	6·10	3·00	0·90	1·34	2·90	4·66	3·56	2·90	2·60	2·98

The Admiralty publication, *Africa Pilot*, Part I, issued in 1899, gives the following information compiled from 19 years' observations at Sierra Leone (Freetown).

	Jan.	Feb.	March	April	May	June	July	Aug.	Sept.	Oct.	Nov.	Dec.	Year
Mean temperature (Fahr.°)	81°·6	82°·3	82°·6	83°·1	82°·3	80°·9	79°·6	78°·7	79°·3	80°·6	81°·4	81°·7	81°·2
Mean daily range...	17°·1	17°·1	16°·9	18°·8	17°·4	15°·9	14°·5	13°·4	13°·8	16°·4	17°·1	16°·8	16°·2
Absolute maximum temperature	96°·0	98°·0	99°·8	98°·0	99°·0	98°·0	97°·0	99°·0	96°·0	98°·4	98°·9	96°·0	99°·8
Absolute minimum temperature	61°·0	61°·0	63°·0	60°·0	58°·2	62°·0	62°·0	60°·0	62°·0	64°·0	63°·0	62°·4	58°·2
Mean relative humidity (per cent.)	77·1	74·5	74·2	75·1	78·9	81·2	83·5	84·4	84·6	81·7	79·2	78·7	79·4
Mean amount of cloud (0—10) ...	3·2	3·2	3·3	3·5	3·8	3·8	3·9	3·7	3·9	3·8	3·6	3·2	3·6
Mean rainfall (in inches)... ...	·618	·522	1·050	5·339	14·809	21·337	36·834	39·600	32·503	15·209	5·302	1·286	174·409
Mean number of rainy days ...	·9	·6	2·9	6·7	16·5	21·8	25·4	24·8	24·2	19·4	9·7	3·1	155

The humidity and cloud are the means of readings at 9 a.m. and 3 p.m.
The mean temperature is the mean of the mean maximum and mean minimum, the observations having been taken at 9 a.m. and 9 p.m.

FREETOWN. Lat. 8° 29′ 30″ N., Long. 13° 9′ 17″ W. Alt. 224 feet.

From 6 years' observations recorded in the Blue Books for 1900—1906.

	Jan.	Feb.	March	April	May	June	July	Aug.	Sept.	Oct.	Nov.	Dec.	Year
Mean temperature[1] (Fahr.°) ...	80°·9	82°·0	82°·2	82°·0	81°·5	79°·5	77°·6	77°·4	78°·5	80°·2	80°·5	81°·4	80°·3
Mean maximum temperature ...	88·4	89·7	89·6	89·2	88·6	86·4	83·5	83·0	85·0	87·0	88·3	89·0	86°·5
Mean minimum temperature ...	73°·4	74°·3	74°·8	74°·9	74·4	72·8	72·3	72·0	72·2	71·7	73·3	74°·0	73°·3
Mean relative humidity[2] (morning)[3] ...	69·5	76·5	64·5	69·6	71·2	77·2	83·6	86·2	83·0	78·8	75·4	73·0	75·7
Mean relative humidity[2] (afternoon)[4] ...	64·2	61·8	60·4	66·2	68·0	75·4	80·4	81·6	78·2	74·2	70·0	65·0	70·5
Mean rainfall (in inches)	·37	0	2·23	5·28	7·48	26·14	42·72	42·07	26·72	11·96	5·40	·94	171·31
Mean amount of cloud (0—10) at 9 a.m. ...	2·6	2·6	2·6	3	3	3	3	3	3	3	3	3	2·9
,, ,, 3 p.m. ...	2·4	2	2·2	2·4	3	3	3	3	3	3	3	3	2·7

[1] The mean temperature is derived from $\dfrac{\text{max.} + \text{min.}}{2}$.

[2] Per cent.   [3] At 9 a.m.   [4] At 3 p.m.

GOLD COAST.

The Gold Coast lies, like the rest of the West Coast, in the torrid zone, and naturally has a tropical climate. The heat is accompanied by a large amount of moisture, both near the coast and in the inland region, which lies in the Great Equatorial Forest zone, and thus renders the climate, speaking generally, unhealthy, hot, moist, and malarious, and especially trying and dangerous to Europeans, though the Colonial Report for 1903 states that Europeans of sound constitution, who practise habits of moderation, generally suffer little from the effects of a not too prolonged residence in the Colony. In the equatorial belt, generally, the year is divided into four seasons, viz. two rainy seasons, with two intervening "seasons of less rain," sometimes called dry seasons. The latter seasons, at Axim, are made up of January to beginning of March, and August and September; at Cape Coast the one dry season consists of December to February, and the other of August and September. The wet seasons are less supportable by Europeans. The Report states that, as a rule, March is the hottest month and July and August the coolest, though sometimes both June, in some years, and September, in others, are comparatively cool. The dry seasons, though more trying in respect of the greater heat, are less unhealthy, for then the humidity is not so great, and, in addition, the Harmattan is experienced. The rainiest places are Axim and the mining districts of Tarkwa, Wassaw and the like. Small-pox and dysentery are the scourges from which the country suffers most, and the latter frequently attacks Europeans.

The percentage of wind frequency is shown below, the compass being, so to speak, divided into eight half-quadrants, S.W. including all winds which are nearer to S.W. than to S. or W. and so on. The table is the result of several years' experience at Christiansborg, to the east of Accrá.

PERCENTAGE OF WIND-FREQUENCY.

	N.	N.E.	E.	S.E.	S.	S.W.	W.	N.W.
December—February	1	1	—	2	1	62	3	30
March—May	1	—	—	1	2	67	5	24
June—August	—	—	—	—	3	75	8	14
September—November	—	1	—	1	1	75	4	18

8—2

From this it appears that the wind is almost entirely in the two half-quadrants S.W. and N.W.   The N.W. wind is a dry land wind, but the S.W. wind blows from over the ocean, and has a wonderful effect on the rainfall.   From the western frontier as far east as Cape Three Points the coast receives the full effect of these naturally rain-laden winds. But thence eastwards the coast bends up northwards and the wind is parallel, or almost so, with the coast, which consequently does not receive the precipitation it otherwise would, if the wind were to strike the coast, instead of passing parallel with it.   The following figures show clearly how the rainfall suddenly diminishes, after Cape Three Points is rounded, and gradually grows less and less.

Mean annual rainfall
(From west to east)

Axim	...	...	82·23 inches	2088·6 mm.
Sékondí	...	...	36·65 ,,	930·9 ,,
Cape Coast		...	33·95 ,,	862·3 ,,
Accrá	...	...	25·72 ,,	653 ,,
Christiansborg		...	21·34 ,,	542 ,,
Kwitta	...	...	19·15 ,,	487 ,,

It must be understood that this applies only to the actual coast, and the application of the rule applies as far west as Wida, in Dahomey.   But inland the precipitation increases again, *e.g.* at Aburi it is 41·91 inches (1064·5 mm.) and at Coomassie 51·69 inches (1313 mm.).

Other explanations, besides the effect of the wind blowing parallel with the coast, have been given to account for this lack of rainfall, but they do not carry conviction.   It is noticeable that the mean temperature rises, immediately Cape Three Points is rounded, during the rainy period, the figures for Sékondí being higher than those for Axim. And this rise in temperature is still more marked as the rainfall decreases along this coastal strip towards the east, so that we find that at Kwitta the mean temperature is 5° Fahr. higher than at Axim in June and July.   On the other hand, at Aburi, which is a little inland and where the rainfall is larger, the mean temperature in the rainy months is less than at Axim.

The meteorological records available for the Gold Coast, with the exception of those for Accrá, which appear in the Annual Blue Books laid before Parliament, are hidden away in the *Gold Coast Gazette*, which is issued weekly. It is a pity that, in many of the returns, especially those of earlier date, errors had been allowed to creep in. Such as were obviously errors have been omitted in constructing the tables, which appear below. For instance, in returning 108 as the percentage of relative humidity, there must have been an error somewhere, and consequently this figure would be discarded. Some other humidity returns, too, must be looked on with suspicion; *e.g.* for Sékondí the mean annual relative humidity works out to only 66·08 per cent., while that of other coast towns, omitting Axim, is Cape Coast 83·03 per cent., Accrá 79·5 per cent., and Kwitta 75·15 per cent. And it will be noticed that, if we omit Sékondí, the relative humidity for the other stations diminishes from west to east, as might be expected, in harmony with the gradual decrease of rainfall.

The tables show that the mean annual temperature, on the coast, is approximately 79°—80° Fahr. (26°·1—26°·7 C.), that March is usually the hottest month, and the cooler period occurs in the months July—September, the first mentioned being, as a general rule, the coolest.

The following table gives the highest mean maximum daily temperature and the lowest mean minimum daily temperature for any month during the whole period for which the meteorological returns are available at the places mentioned.

Station	Highest mean	Date	Lowest mean	Date
Axim	91°·41 Fahr. (33°·0 C.)	Mar. 1903	66°·86 Fahr. (19°·3 C.)	June 1902.
Cape Coast	90°·48 Fahr. (32°·5 C.)	Mar. 1906	69°·96 Fahr. (21°·1 C.)	Aug. 1901.
Kwitta	92°·74 Fahr. (33°·7 C.)	Mar. 1906	71°·66 Fahr. (22°·0 C.)	June 1906.
Aburi	[1]89°·32 Fahr. (31°·7 C.)	Jan. 1906	62°·19 Fahr. (16°·8 C.)	July 1901.

The mean daily range is small throughout, as might be expected in regions so near to the ocean. At Coomassie, however, it is higher, and still more so at Gambaga, where in the early months of the year it reaches 23° Fahr. (13° C.). There are, however, occasional exceptions, even on the coast, for in February, 1903, the greatest difference between the highest day temperature, at Accrá, and the lowest night temperature was 22°·6 Fahr. (12°·6 C.).

---

[1] 114°·80 Fahr. (46° C.) was returned as the mean daily maximum temperature for November 1905, but this appears to be improbable.

Neither the total annual rainfall, nor the monthly fall, from year to year, can be depended on, and a few instances of its vagaries will show that this is the case. At Accrá the mean annual fall is derived from totals varying between 13·11 inches (333 mm.) in 1905 and 35·75 inches (908 mm.) in 1901. At Cape Coast no less than 24 inches (610 mm.) were measured in May, 1906, but only 0·92 inches (23 mm.) in the same month in 1903, while, in June, 1902, 22 inches (559 mm.) were recorded, against 1·71 inches (43 mm.) in 1906; in May, 1903, 7·11 inches (180·5 mm.) fell in one day. At Axim, in the month of June, over 41 inches (1041 mm.) fell in 1905, of which 6·12, 6·30 and 6 inches (155·5, 160 and 152·5 mm.) fell on three days, while, in 1904, 32·5 inches (825 mm.) fell, with 8·5 inches (216 mm.) on one day, and in 1902, 30 inches (762 mm.) was the total for the month, with 8·12 and 7·30 inches (206 and 185 mm.) on two consecutive days, but in 1906 the total precipitation for the whole month only reached 9 inches (228·5 mm.). Again, at Kwitta in 1902, 1905, and 1906 the records for September were 0·05, 0·15, and 0·02 inches (1·2, 3·7, and ·6 mm.) respectively, but in 1901 there was a comparative downpour of 9·45 inches (240 mm.).

In dealing with the death rate it is difficult, among such small numbers, where one death causes a very considerable difference in the rate, to arrive at any very satisfactory conclusions, especially as the figures available include death by misadventure, as well as by disease. However, taking the average for five-year periods, the results are as follows:

		Death rate per 1000 non-officials	Death rate per 1000 officials
1891—1895	...	49·39	43·06
1896—1900	...	48·23	44·28
1901—1905	...	24·24	38·21

It may be gathered, therefore, on the whole, that the conditions are improving.

(For Productions *see* under West Coast.)

With reference to the stations for which meteorological tables are given below—

Axim lies on the shore between Kebrada Point on the north and Papre Point on the south, and faces almost due west. There are hills about two-thirds of a mile away, in an easterly direction, some 200 feet high.

Sékondí faces almost due east, and lies to the north of Sékondí Point, nestling in a bay behind it.

Cape Coast Castle stands on a rock on a southerly projecting point of the line of coast. From the landing-place the ground rises gradually to the castle, and the town stands in rear (to north, east, and west); westwards of the town is a small lagoon with swamps.

Accrá lies on the coast immediately to the east of Korli Lagoon, and faces south and south-east. The southern extremity of the rocky Accrá Point, on which Jamestown Fort is built, stands 36 feet high.

Kwitta is surrounded by trees, and lies on the narrow sandy strip, which separates Kwitta Lagoon from the sea.

Aburi lies inland, the nearest point on the coast being some 20 miles distant. It stands on comparatively high ground at the source of the little Ajamenta Su.

COAST STATIONS, FROM WEST TO EAST.

AXIM. Lat. $4°\ 52'\ 18''$ N, Long. $2°\ 14'\ 45''$ W.

Compiled from the weekly numbers of the *Gold Coast Gazette*, 1901—1906.

	Jan.	Feb.	March	April	May	June	July	Aug.	Sept.	Oct.	Nov.	Dec.	Year
Mean maximum temperature (Fahr.°)	87°·6	88°·5	89°·4	89°·3	86°·1	82°·2	81°·1	81°·9	81°·3	83°·4	85°·5	87°·0	85°·3
Mean minimum temperature ...	73°·0	72°·8	75°·3	74°·5	72°·5	71°·2	71°·2	71°·3	72°·0	72°·7	72°·1	72°·3	72°·6
Mean temperature[1] ...	80°·3	80°·6	82°·4	81°·9	79°·3	76°·7	76°·1	76°·6	76°·7	78°·1	78°·8	79°·7	78°·9
Mean daily range[2] ...	14°·6	15°·7	14°·1	14°·8	13°·7	11°·0	10°·1	10°·6	9°·3	10°·8	13°·4	14°·7	12°·7
Mean temperature at 9 a.m. ...	79°·5	79°·8	81°·9	81°·7	80°·5	76°·7	76°·6	75°·5	76°·1	77°·4	79°·3	80°·7	78°·8
„ „ 5 p.m. ...	82°·0	82°·9	83°·5	83°·4	83°·0	78°·1	76°·7	76°·2	76°·6	78°·3	80°·6	81°·5	80°·2
Mean relative humidity[3] (per cent.) ...	84·8	84·2	83·7	83·2	85·5	85·8	89·1	91·0	89·4	88·9	87·6	83·8	86·4

[1] The mean temperature is derived from $\dfrac{\text{max.} + \text{min.}}{2}$.

[2] The daily range is max.—min.

[3] It is not stated in the *Gazette* how the relative humidity mean was arrived at each day.

SÉKONDÍ. Lat. 4° 55' 47" N., Long. 1° 42' 50" W.

Compiled from the weekly numbers of the *Gold Coast Gazette*, 1905, 1906.

	Jan.	Feb.	March	April	May	June	July	Aug.	Sept.	Oct.	Nov.	Dec.	Year
Mean maximum temperature (Fahr.°)	87°·4	87°·1	89°·5	89°·6	87°·0	84°·1	80°·8	80°·0	81°·6	85°·7	88°·5	87°·6	85°·7
Mean minimum temperature ...	72°·9	72°·9	75°·2	74°·0	74°·0	74°·5	73°·6	72°·7	71°·6	72°·9	73°·0	73°·0	73°·3
Mean temperature[1]	80°·2	80°·0	82°·3	81°·8	80°·5	79°·3	77°·2	76°·1	76°·6	79°·3	80°·8	80°·3	79°·5
Mean daily range[2]	14°·5	14°·2	14°·3	15°·6	13°·0	9°·6	7°·2	7°·3	10°·0	12°·8	15°·5	14°·6	12°·4
Mean temperature at 9 a.m. ...	80°·1	80°·8	83°·6	85°·3	82°·3	76°·0	76°·9	76°·2	76°·4	78°·5	81°·5	81°·2	79°·9
„ „ 5 p.m. ...	83°·2	83°·2	84°·7	83°·9	82°·0	75°·4	76°·3	75°·7	76°·1	78°·8	81°·9	82°·3	80°·3
Mean relative humidity (per cent.) ...	66°·1	57·8[3]	64·5	65·0	66·5	66·5	70·0	69·2	70·9	71·8	62·8	61·9	66·1

[1] The mean temperature is derived from $\dfrac{\text{max.} + \text{min.}}{2}$.

[2] The daily range is max.—min.

[3] For one year only.

ACCRÁ. Lat. 5° 32′ 43″ N, Long. 0° 12′ 23″ W.

Means compiled from the Annual Blue Books for 6 years[1].

	Jan.	Feb.	March	April	May	June	July	Aug.	Sept.	Oct.	Nov.	Dec.	Year
Mean maximum temperature (Fahr.°)	85°·6	87°·5	88°·3	88°·9	87°·5	84°·0	81°·0	80°·0	81°·9	84°·0	85°·8	86°·0	85°·1
Mean minimum temperature	70°·6	71°·9	73°·9	72°·5	72°·7	71°·8	69°·6	67°·9	71°·5	67°·5	71°·7	72°·1	71°·1
Mean temperature[2]	77°·5	79°·7	81°·1	80°·7	80°·1	77°·9	75°·3	78°·9	76°·7	75°·8	78°·7	79°·1	78°·5
Mean daily range[3]	15°·0	15°·6	14°·4	16°·4	14°·8	12°·2	11°·4	12°·1	10°·4	16°·5	14°·1	13°·9	13°·9
Mean relative humidity[4] (per cent)	81·0	76·9	79·2	78·2	77·1	82·6	82·8	82·6	80·8	81·4	76·9	76·3	79·5
*Mean annual rainfall for the years 1900—1905 in inches	1·77	1·8	·7	2·8	3·0	4·5	1·3	·79	1·4	2·4	2·0	·75	23·2
*Mean annual rainfall for the years 1886—1900 in inches	·03	·87	2·73	4·25	6·30	7·72	1·35	·12	·87	2·11	1·00	·89	28·24

* During 1903, 1904, 1905 the rainfall was exceptionally low, while during 1887, 1888 it was exceptionally high.

[1] 1900—1905.

[2] The mean temperature is derived from $\dfrac{\text{max.} + \text{min.}}{2}$.

[3] The daily range is max.—min.

[4] It is not stated in the Blue Books for what time or times of day the relative humidity was calculated.

CAPE COAST. Lat. 5° 6′ 5″ N., Long. 1° 13′ 40″ W.

Compiled from the weekly numbers of the *Gold Coast Gazette*, 1901—1906.

	Jan.	Feb.	March	April	May	June	July	Aug.	Sept.	Oct.	Nov.	Dec.	Year
Mean maximum temperature (Fahr.°)	86°·2	85°·3	86°·9	87°·2	85°·7	83°·1	80°·9	79°·8	81°·1	84°·4	85°·7	86°·3	84°·4
Mean minimum temperature ...	73°·7	72°·8	74°·9	75°·2	74°·4	72°·9	72°·1	71°·1	72°·0	73°·6	75°·0	74°·2	73°·5
Mean daily range[1]	12°·5	12°·5	12°·0	12°·0	11°·3	10°·2	8°·8	8°·7	9°·1	10°·8	10°·7	12°·1	10°·9
Mean temperature[2] ...	79°·9	79°·1	80°·9	81°·2	80°·1	78°·0	76°·5	75°·5	76°·6	79°·0	80°·3	80°·2	78°·9
Mean temperature at 9 a.m.	80°·3	79°·0	81°·1	81°·9	80°·9	79°·3	76°·1	75°·5	76°·8	79°·2	82°·5	81°·8	79°·5
„ „ 5 p.m.	81°·7	81°·6	83°·1	83°·0	81°·8	79°·5	77°·1	76°·2	78°·0	80°·0	82°·8	82°·4	80°·6
Mean relative humidity[3] (per cent.) ...	84·1	77·6	80·1	81·5	82·5	84·9	84·8	85·9	86·6	84·4	81·7	82·3	83·0

[1] The daily range is max.—min.

[2] The mean temperature is derived from $\dfrac{\text{max.}+\text{min.}}{2}$.

[3] It is not stated in the *Gazette* for what time or times of day the relative humidity was calculated.

KWITTA. Lat. 5° 55′ N., Long. 1° E. (approx.).

Compiled from the weekly numbers of the *Gold Coast Gazette*, 1901—1906.

	Jan.	Feb.	March	April	May	June	July	Aug.	Sept.	Oct.	Nov.	Dec.	Year
Mean maximum temperature (Fahr.°)	89°·1	88°·9	89°·8	87°·7	89°·1	88°·4	88°·0	86°·7	86°·8	85°·6	89°·4	89°·0	88°·2
Mean minimum temperature ...	75°·8	77°·1	77°·9	76°·3	75°·5	75°·3	74°·4	74°·8	75°·3	75°·2	75°·1	74°·7	75°·6
Mean daily range[1]	13°·3	11°·8	11°·9	11°·4	13°·6	13°·1	13°·6	11°·9	11°·5	10°·4	14°·3	14°·3	12°·6
Mean temperature[2] ...	82°·0	83°·0	83°·8	82°·0	82°·3	81°·9	81°·2	80°·7	81°·0	80°·4	82°·3	81°·8	81°·9
Mean temperature at 9 a.m. ...	81°·5	81°·7	82°·9	82°·4	81°·5	80°·8	79°·0	77°·6	81°·0	81°·1	82°·2	82°·3	81°·2
,, ,, 5 p.m. ...	83°·7	84°·3	84°·6	84°·0	83°·1	83°·6	80°·7	78°·1	80°·1	82°·0	83°·4	83°·4	82°·6
Mean relative humidity (per cent.) ...	77·0	75·7	77·3	76·3	77·3	80·0	80·0	81·1	77·0	75·7	75·7	75·3	77·3

[1] The daily range is max.—min.

[2] The mean temperature is derived from $\dfrac{max.+min.}{2}$.

INLAND.

ABURI (due north of Christiansborg). Lat. 5° 52′ 55″ N., Long. 0° 8′ 21″ W. (approx.).

Compiled from the weekly numbers of the *Gold Coast Gazette*, 1901—1906.

	Jan.	Feb.	March	April	May	June	July	Aug.	Sept.	Oct.	Nov.	Dec.	Year
Mean maximum temperature (Fahr.°)	84°0	83°7	83°5	84°8	83°2	77°8	75°8	77°2	79°7	79°5	90°7[1]	82°4	81°9
Mean minimum temperature ...	73°5	68°3	71°9	71°7	71°2	69°9	69°4	70°7	69°5	70°2	74°4	75°4	71°3
Mean daily range[2]	10°5	15°4	11°6	13°1	12°0	7°9	6°4	6°5	10°2	9°3	16°3[1]	7°0	10°6
Mean temperature[3] ...	78°7	76°0	78°7	78°8	77°2	78°8	72°6	79°0	74°5	74°8	78°5	78°9	76°6
Mean temperature at 9 a.m. ...	77°2	77°5	78°8	78°6	78°0	76°5	74°1	73°4	74°3	75°4	76°8	76°3	76°4
„ „ 5 p.m. ...	77°8	78°5	77°5	77°4	76°0	74°5	73°8	73°7	74°3	75°0	75°2	75°6	75°8

[1] This is probably too high, owing to an abnormally high temperature of 114°8 having been recorded for November 1905. If this figure be disregarded the mean is 82°6; and the daily range for November becomes 8°3. It seems probable, indeed, that 114°8 is an error.

[2] The daily range is max.—min.

[3] The mean temperature is derived from $\dfrac{\text{max.} + \text{min.}}{2}$.

ASHANTI.

Writing in the *Scottish Geographical Magazine* in 1896 (p. 453), Major Barter states: "The climate of Ashanti bears a very bad character, which, I think, it deserves, though probably it is healthier, on the whole, than that of the districts bordering on the coasts." Possibly there have been improvements at Coomassie, and in the immediate vicinity, but, taking Ashanti as a whole, what was true in 1896 is true to-day. There is the great equatorial forest covering the face of the country everywhere, with the increasing decay of vegetable matter, creating an offensive atmosphere, and, in addition, there are the malarious swamps in all low-lying localities. "The temperature throughout, even in the hot season, is not excessive—generally $75°$ to $90°$ Fahr. ($24°$ to $32°$ C.) in the day-time. This fact is principally due to the Harmattan, a steady cool breeze, which blows from a northerly direction during the winter months, and which is considered so healthy that it is often called 'the Doctor.' The nights are fairly cool, as a rule, but the air is never free from the steaming dampness, which weakens the European system, and thereby predisposes it to the effects of fever, a malady, which, in one form or another, is bound, sooner or later, to attack even the strongest European. The evil effects of the climate, in the case of a short stay, do not, as a rule, show themselves until the subject has left the country."

The mean annual temperature compares favourably with the coast stations, being a trifle—though only a very small trifle—lower, but is higher than that at Aburi. The mean annual rainfall is considerably higher than that at any other station, except Axim. There are two rainfall maxima, namely in May/June and in October, with an intervening season of less rain, and a dry season in December/February, the gradual approach of the two wet seasons to one another being noticeable here, on going northwards, until they gradually merge in one in higher latitudes, where the year is divisible into only two seasons, the wet and the dry. The highest rainfall registered for any one month was 17·02 inches (432 mm.) in May, 1906. The highest mean maximum temperature for any month, during the period for

ASHANTI.

COOMASSIE. Lat. 6° 47' N., Long. 1° 37' 30" (approx.).

Compiled from the weekly numbers of the *Gold Coast Gazette*, 1901—1906.

	Jan.	Feb.	March	April	May	June	July	Aug.	Sept.	Oct.	Nov.	Dec.	Year
Mean maximum temperature (Fahr.°)	87°·2	89°·4	90°·9	88°·5	85°·5	84°·9	81°·5	81°·3	83°·5	83°·6	87°·2	86°·4	85°·9
Mean minimum temperature ...	66°·9	69°·9	73°·3	71°·1	71°·1	72°·6	70°·5	69°·4	70°·0	71°·1	71°·0	69°·6	70°·4
Mean temperature[1] ...	77°·1	79°·5	82°·1	79°·8	78°·3	78°·8	76°·0	75°·4	76°·7	77°·4	79°·6	78°·0	78°·2
Mean daily range[2] ...	20°·3	19°·5	17°·6	17°·4	14°·4	12°·3	11°·0	11°·9	13°·5	12°·5	16°·2	16°·8	15°·4
Mean temperature at 9 a.m. ...	74°·6	77°·6	79°·8	79°·8	78°·0	76°·0	74°·7	73°·3	74°·6	75°·4	77°·2	76°·5	76°·5
„ „ 5 p.m. ...	80°·8	85°·4	86°·5	83°·8	82°·6	81°·6	78°·0	77°·7	79°·2	80°·5	82°·1	82°·3	81°·7
Mean relative humidity (per cent.)	82·3	75·5	74·9	84·7	83·7	86·4	90·2	84·1	86·0	87·6	86·2	84·9	83·9

[1] The mean temperature is derived from $\dfrac{\text{max.} + \text{min.}}{2}$.

[2] The daily range is max.—min.

which the table on p. 127 is constructed, was 94°·68 Fahr. (34°·9 C.) in February, 1904, and the lowest mean for any month was 63°·54 Fahr. (17°·6 C.) in January of the same year.

NORTHERN TERRITORIES OF THE GOLD COAST.

As the north of the Northern Territories is approached the four-season equatorial zone is left behind, and there are only two seasons, the wet and the dry. The latter extends from November to March and corresponds with the Harmattan period. The mean annual rainfall varies very considerably; in 1902 it was only 37·80 inches (960 mm.), whereas in 1906 it was 44·67 inches (1134·6 mm.), and in the previous year no less than 55·71 inches (1415 mm.), of which 15·13 and 11·43 inches (384 and 290 mm.) fell in August and September respectively. July, August and September are always the rainiest months, and considerably more than 60 per cent. of the total precipitation falls in that period. There is a discrepancy in the rain records for the year 1905 which needs some explanation. I computed the results from the weekly Government Gazettes; and in the Colonial Reports on the Gold Coast, in which the *totals* for the years alone are given, these *totals* tally exactly with those which I worked out. The total fall for 1905 was 71·71 inches (1820·4 mm.), and this seemed to be so abnormal that I sought further confirmation. During the last few years there have been "Additional Reports" on the Northern Territories, supplementary to the "Colonial Reports" on the Gold Coast, and in these I found that the total for this year was only 55·71 inches (1414 mm.). This figure I have adopted as being the more probable. Taking this total for 1905, the mean annual rainfall is reduced from 48·69 inches (1236·7 mm.) to 45·86 inches (1166 mm.), which, though high, is much more likely to be correct.

The mean relative humidity is much lower than on the coast or in Ashanti, and this renders the heat much less oppressive. The Commandant of the Northern Territories says: "I am decidedly of opinion that the climate of the Northern Territories is infinitely superior to that of the coast, and I consider that there is no reason why officers, who take the ordinary precautions necessary for living in a tropical climate, should not enjoy fairly good health during their tour. Exercise is absolutely essential. The heat of the sun is very great, and helmets are ordered to be worn on all parades from 7 a.m. to 5 p.m.; a thick spine pad is also necessary during the heat of the day. The Harmattan blows for four months of the year, from November to February, giving delightfully cool nights, and generally bracing up one's energies. Except during the rains, the air, generally, of the Northern Territories is a dry one, and free from the load of moisture that is so oppressive on the coast" (*Colonial Report*).

Omitting figures which are evidently inaccurate, the highest mean maximum temperature for any month during the period for which the table below is computed is $100°·09$ Fahr. ($37°·8$ C.) in March, 1906, and the lowest mean minimum temperature for any one month $66°·09$ Fahr. ($18°·9$ C.) in January, 1905. The daily range is sometimes very considerable, in consequence of the distance from the coast, and varies very largely from month to month, being least, however, in the rainiest months.

Lt.-Col. Watherston, the Commissioner of the Northern Territories, writing in the *Journal of the African Society*[1], says: "The rainy season lasts from May to October, and is heralded in April by terrific wind-storms, which often cause a badly made roof to collapse....The rain during the rainy months falls often daily, in violent storms lasting about two to three hours; but seldom continuously throughout the day....The country may be described as a gently undulating plateau rising gradually from Kintampo, and ending abruptly in a sheer scarp running south-west to north-east, and passing six miles north of Gambaga, the drop being about 500 feet, a break existing west of Gambaga to admit the passage of the White Volta river. Politically there is no frontier caused by the scarp, but there seems to be a difference in the climatic conditions that exercises considerable influence on the life, agriculture, and occupations of the inhabitants. In the north-east and north-west corners of the Protectorate the country is more broken, and small hills from 100 to 500 feet high are found, sometimes isolated and sometimes formed into small ranges."

[1] Vol. VII, 1908, p. 347.

K. A.

9

NORTHERN TERRITORIES.

GAMBAGA. Lat. 10° 31′ 38″ N., Long. 0° 25′ W. (approx.).

Compiled from the weekly numbers of the *Gold Coast Gazette*, 1901—1906.

	Jan.	Feb.	March	April	May	June	July	Aug.	Sept.	Oct.	Nov.	Dec.	Year
Mean maximum temperature (Fahr.°)	91°·6	94°·4	94°·4	96°·2	89°·5	85°·1	81°·4	80°·4	82°·9	86°·0	90°·7	91°·8	88°·7
Mean minimum temperature ...	70°·1	70°·7	76°·5	72°·4	74°·2	71°·6	70°·9	70°·9	70°·8	71°·8	73°·3	72°·3	72°·1
Mean temperature[1] ...	80°·8	82°·5	85°·5	84°·3	81°·9	78°·4	76°·1	75°·7	76°·8	78°·9	82°·0	82°·0	80°·4
Mean daily range[2]	21°·5	23°·7	17°·9	23°·8	15°·3	13°·5	10°·5	9°·5	12°·1	14°·2	17°·4	19°·5	16°·6
Mean temperature at 9 a.m. ...	78°·9	82°·6	85°·8	83°·8	80°·0	76°·8	75°·2	74°·7	76°·5	78°·6	81°·8	80°·7	79°·6
,, ,, 5 p.m. ...	87°·4	89°·8	91°·1	91°·3	84°·7	80°·4	78°·3	77°·2	79°·9	85°·6	87°·0	87°·3	85°·0
Mean relative humidity (per cent.)	34·0	36·0	34·5	57·0	68·7	77·2	80·9	84·6	82·8	77·3	57·5	40·0	60·9

[1] The mean temperature is derived from $\dfrac{max.+min.}{2}$.

[2] The daily range is max.—min.

GOLD COAST. MEAN ANNUAL RAINFALL IN INCHES.

Coast Stations from West to East.

	Jan.	Feb.	March	April	May	June	July	Aug.	Sept.	Oct.	Nov.	Dec.	Year
Axim[1]	1·79	2·72	3·67	6·67	15·09	24·90	5·57	1·87	1·09	6·34	7·13	5·39	82·23
Sékondi (only 2 years[2]) ...	1·11	0·23	1·76	4·14	12·81	6·23	3·02	0·57	1·04	1·04	2·44	2·26	36·65
Elmina[3]	0·04	1·93	1·89	3·23	7·40	6·73	1·69	1·06	0·90	2·36	2·13	1·42	30·78
Cape Coast[1]	0·24	1·80	2·79	3·39	9·29	8·15	1·14	0·39	0·85	2·16	2·49	1·26	33·95
Accrá[4]	0·90	1·33	1·72	3·52	4·65	6·11	1·33	0·45	1·14	2·25	1·50	0·83	25·73
Kwitta[1]	0·26	1·57	0·89	2·19	2·74	5·88	0·96	0·34	1·78	1·27	0·82	0·45	19·15
Inland.													
Aburi (N. of Accrá)[1] ...	0·84	1·98	3·14	4·78	5·44	6·82	3·17	1·34	1·95	5·43	4·69	2·31	41·89
Ashanti.													
Coomassie[1]	0·53	1·11	3·29	4·48	8·04	9·05	5·05	1·93	5·32	8·01	2·96	1·61	51·38
Northern Territories.													
Gambaga[1]	0	0·05	0·40	2·34	5·51	5·01	7·77	11·73	8·95	3·71	0·39	0	45·86

[1] For the years 1901—1906.   [2] For 1905, 1906.   [3] 1866—1862.   1886—1905.

9—2

SOUTHERN NIGERIA.

Southern Nigeria is essentially a flat country, a land of forest, intersected by rivers and creeks, and fringed with mangrove swamps, which penetrate the region of the Niger delta, a land of heavy and, in some regions, terrific rainfall, with high humidity at all times, and incessant heat. Notwithstanding this, the riches of the forest are so great that, for ages, the white man has found it profitable to make the "Oil Rivers" his home, and has put up with the discomfort and unhealthy conditions in his pursuit of wealth. In the north-east there are hills, but the highest of these only reach an altitude of 2000 feet.

In a general sense the rainfall in the coastal region increases from west to east, thus Lagos has a mean annual rainfall of 73 inches (1854 mm.), Sapele of 106 inches (2692 mm.), and Bonny of 164 inches (4166 mm.). There is also a decided decrease from south to north. Bonny lies on a very narrow coast belt of exceedingly heavy precipitation, which skirts round the bight to the Cameroons coast, Calabar lies a little way up the river and has a fall of 134 inches (3404 mm.), while further north we have Sapele with 106 inches (2692 mm.), Bende 81 inches (2057 mm.), and Onicha with only 59 inches (1499 mm.). So great is the rainfall in this neighbourhood that some, in all seriousness, actually divide the country into a southern wet zone with a rainfall exceeding 76 inches (1920 mm.) and a *dry zone with less than 76 inches rainfall* in the north. The greatest precipitation recorded, up to the present, in any one month, is 41·98 inches (1066·29 mm.), measured at Forcados in July 1905.

The mean annual temperature in the coast lands is about 78° or 79° Fahr. (26° C.), but increases inland to 81° Fahr. (27°·2 C.) at Onicha. The mean daily range is greatest at Bonny (20° Fahr. or 11° C.) and least at Lagos (9° Fahr. or 5° C.). The mean relative humidity corresponds very nearly with the rainfall, the percentage at Bonny being highest, viz. 85 (Sapele also has 85), Calabar coming next with 83, then Lagos with 80, and Onicha with only 79.

The following table gives the death-rate and invalidings in the central and eastern provinces among Europeans, based on the returns for the years 1903–7.

European population	Number of deaths	Rate per 1000	Number of invalidings	Rate per 1000
546	15	27·47	74	136·19

There are no complete returns for the whole of the western province, but the following table, which includes the deaths of invalids sent down from the interior, gives the death-rate for Lagos town with Ebute Metta, which shows a satisfactory improvement.

Period	European population	Number of deaths	Rate per 1000
1903—5	233	13	55·79
1906—7	400	9	22·50

The Aro country is extremely unhealthy in its southern portion, especially in the neighbourhood of Arochuku. In the north the country is more elevated and better conditions prevail. The proximity of the Cameroons mountains naturally has a powerful effect on the rainfall; and the atmosphere, except when the north wind blows, is very damp.

The meteorological table, given below, for Lagos, is compiled from two different sets of observations, a four-year period, extending from 1887 to 1892, and a six-year period covering the years 1900–5. The former set was registered with the greatest care, and gave the temperature at 8 a.m., at 4 p.m., and the mean of these, as well as the mean maximum and mean minimum temperature, and the absolute maximum and minimum for each month, and also included other particulars. The second set merely gives the absolute maximum and the absolute minimum for each month. In this latter set there are one or two palpable errors, which have been eliminated. Both sets give the relative humidity and the rainfall. The temperatures are very even throughout and the means are good means, that is to say, the figures, from which the mean for any month is derived, differ very little from that mean, so that the results are dependable.

The highest temperature recorded was $93°$ Fahr. ($33°·9$ C.) in February and March 1903 and March 1905, though the thermometer rose to $92°$ Fahr. ($33°·3$ C.) on several occasions, viz. in January 1890, February 1891, April 1900, January, April and May 1903, March and December 1904, and January, April and May, 1905. The absolute minimum was $65°$ Fahr. ($18°·3$ C.), recorded in September, October and November 1889, in August 1891 and in January 1900. So that the range for the whole ten years was only $28°$ Fahr. ($15°·5$ C.), while the mean monthly range was greatest in January ($20°·4$ Fahr. or $11°·3$ C.) and least in August ($12°·2$ Fahr. or $6°·8$ C.), and the mean daily range reached no higher a figure than $11°·5$ Fahr. or $6°·4$ C. at the most, and was as low as $6°$ Fahr. or $3°·3$ C. in July. This is an extraordinary evenness of temperature, but the thermometer is at such a height that the constant heat must necessarily be debilitating, when combined with the perpetual humidity of the atmosphere. There are rains in all months of the year, but those with comparatively little rain, which constitute the so-called dry season, are December, January and February. The daily range is less than at Calabar, and so also are the humidity and the rainfall. During the ten years under consideration, there were only five months which were absolutely rainless, namely, December in 1901 and 1902, January 1902 and 1903, and February 1901, though only 0·01 inch was recorded in December 1905. There are four seasons, the dry mentioned above, the early or heavy wet season, extending from March to July, with the heaviest fall in June, the lesser dry season, with comparatively little rain, in August, and the lesser wet season from September to November. Some people prefer to divide the year into two seasons, viz. a wet season, with two maxima, separated by an interval of less rain, and a dry season. But this is merely a matter of terms. Of the total precipitation for the years for which the table is constructed, the greatest was 112·5 inches (2857·5 mm.) in 1901 and the least 45·9 inches (1165·9 mm.) the next year. In the rainiest month, June, the greatest amount recorded was 25·59 inches (650 mm.) in 1905 and the least 10·62 inches (269 mm.) in 1892, which was altogether a very abnormal year, only 0·55 inch (14 mm.) falling in July, and as much as 14·61 inches (371 mm.) in October, which has a mean of only 7·97 inches (202 mm.).

The daily range of temperature is greatest at Bonny, and is larger than that of any of the stations, especially in the early part of the year; while at Lagos there is very little variation in the temperature. The native portion of the town of Lagos is much congested, and has been described as a "huge cess-pit."

At Calabar the absolute maximum temperature recorded, during the five-year period for which the figures are available, was 98° Fahr. (36°·7 C.) on both the 25th and 26th of February 1905, and the lowest figure that the thermometer reached was 64° Fahr. (17°·8 C.) on the 2nd and 5th of the same month in 1904, and on the 29th of December 1902; so that the extreme range of temperature for the whole period was only 34° Fahr. (19°C.). The relative humidity is always highest in the morning, sinks in the middle of the day to rise again in the evening, thus the mean relative humidity, for the five years, at 7 a.m. was a trifle over 92 per cent., at 1 p.m. it went down to 70·37, and at 9 o'clock in the evening was about 86·25 per cent. The greatest amount of rain in any one year was 140·85 inches (3576 mm.), measured in 1903, and the least 97·04 inches (2464 mm.), recorded in 1898. The months with the lowest rainfall are, as a rule, the least regular from year to year, though in August 1905 the precipitation amounted to over 37 inches (940 mm.), and in the same month in 1900 only 6·39 inches (162·3 mm.) were recorded.

September is a comparatively cool month, and comes in the period of the lesser rains, when the precipitation is not so heavy as, for instance, in June, July and August, and thus forms an average month. The type of weather in this month may be gathered from the entries in the diary of Major Close, C.M.G., R.E.[1], who was the British Commissioner for delimiting the Nigeria-Cameroons boundary, in 1895.

Sept. 2. Raining hard most of day.

　　3. Rain most of the day but not so bad as on Sept. 2.

　　4. Cloudy afternoon.

　　7. Rain most of day.

　　8. Fine most of day.

　　　　.　　.　　.　　.

　21. Fine most of day.

　22. Raining hard.

　23. Raining hard.

　24. Fine.

　25. Raining hard all day.

　26. Raining during most of day.

　27. Fine.

　28. Raining hard.

　30. Fine.

The general direction of the wind at Calabar is from the south-west. This is the invariable rule in July, August, September and October; in November and December it is either S.W. or N.E.; in January the wind is generally in the north, either N.E. or N.W.; in February, March and April it is either N. or S.W. and in the two remaining months either S.W. or W.

At Akassa from December to February 41 per cent. of the wind comes from the S.W. From March to May the prevailing wind is from the south, and the next most frequent wind is S.W., the percentages being 39 and 22 respectively. During July—August 52 per cent. of the wind is from the S.W. and 30 from the south; while from September to November 46 per cent. of the wind is south-westerly, 13 per cent. north-westerly, and 11 per cent. southerly.

The following account of the climate of Southern Nigeria has been written specially for this volume by Captain E. A. Steel, R.A., who was resident in the country for five years:

Properly speaking, there are only two seasons on the West Coast of Africa, namely the wet, with two maxima, and the dry. The former begins in May and continues till

---

[1] Now Lieut.-Colonel Close.

the end of September; the latter occupies the intervening months, and each is ushered in and out by thunderstorms and tornadoes. Those preceding the rainy season are often extremely violent. They first occur from the east, thence veer round by the S. to nearly S.W., and from that point continue during the succeeding four months. Heavy as is the downpour in all parts of the tropics during the rainy season, it is impossible to conceive anything to exceed that which takes place on parts of the coast, accompanied as it is by thunder, lightning, wind and blackness, absolutely terrific while they last. Rain does not continue to fall uninterruptedly during the whole of what is called the rainy season. After the first outburst, clear sunshiny days occur from time to time. When they do, exhalations from the wet earth rise and hang over it like sheets of fog. Wind is very uncertain, and often altogether absent. Malaria accumulates; the European, if walking abroad for the sake of exercise, as he must do daily, experiences a sense of sickness and lassitude, and the chances are that, before many weeks have elapsed, he is "down" with fever. Thus the rainy season advances, heavy downpours, thunderstorms, and tornadoes, varied by short intervals of hot, clear, oppressive, unhealthy weather, till September. The showers then become somewhat lighter; dense fogs now prevail at night, nor are they dispelled till late in the succeeding day. As to the "dry" season, although, by courtesy, it is said to begin in September, it really does not fairly commence till October. In November it is sometimes broken into by what are called "the second rains," which are slight, but with the exception of these the climate remains for the most part agreeable till March, when the hot weather sets in. March, April and May are distinctly hot, the temperature sometimes rising to 98° Fahr. (36°·6 C.) indoors. There are many places in the world which are much hotter, but in the moist air of the coast that degree of heat seems "sweltering"; those who have been unfortunate enough to have been affected by the previous months of rains improve in health —for it is seldom, indeed, if ever, that a white man is altogether unaffected by this ordeal—the spirits rise, life is considered somewhat more secure than it has been : sickness gradually disappears or materially lessens, various means of amusement are thought of; and, for the time, residence on the coast is robbed of half its horrors. Not indeed that the "dry" season is altogether uninterrupted by falls of rain, any more than the "rainy" is by glimpses of dry weather. During a part of January and February the "Harmattan" wind occurs; a dry parching breeze, which sets in from a point a little westward of due north, bringing with it a profusion of impalpable dust, sufficient to cover the furniture in the houses, entering through every crevice, even coating papers on one's desk, while, by its extreme dryness, it curls up corners of books, and, strange enough, sometimes causes glasses to crack and fall to pieces as they stand upon the table. The occurrence of this wind is eagerly looked forward to and welcomed. So long as it lasts, agues disappear, constitutions are for the time being renovated, and among the older residents it becomes known as "the doctor." Among the newly arrived, however, the effects are less favourable, causing heat and dryness of the surface of the body, and, in some instances, inflamed eyes, and attacks of fever, though of a milder nature than in the rainy season. The extreme dryness of this wind is further indicated by the very great extent to which evaporation takes place during its prevalence. It is said to amount to not less than nine and a half inches per day, whereas, in England, the total evaporation per annum is thirty-six inches. No doubt much of this extreme dryness is accounted for by the circumstance that the "Harmattan" blows from the great desert of Sahara; but its moderated temperature, by the time it reaches the coast, so very different from the hot winds of upper India, must arise from the fact of its having, in its course, to cross over a deep tract of rank vegetation. During the dry season, the land and sea breezes alternate with tolerable regularity, the former setting in about sunset and from the N.W., the latter about sunrise and from about S.S.W. In the rainy period of the year, however, such alternation cannot be said to exist. From the end of May till July the sea breeze occasionally occurs, but that from the land is inconsiderable and uncertain, hurricanes and squalls being frequent. In July and August heavy squalls and high winds prevail. In September the weather is inclined to be foggy. In October and November, the breeze blows fresh along the coast from the westward, interrupted only by the tornadoes, which usher in the "little rains." In December the weather is usually clear, the sea and land breezes daily succeeding each other with complete regularity. From January till the end of March the direction of the prevailing wind varies, but is generally from the eastward. In April the breeze shifts round by south to west. The actual range of temperature on the coast is by no means excessive. In fact it is decidedly small. As a rule, indeed, on account of the extreme moisture of the atmosphere the sensations usually are those of cold or at all events chilliness, rather than heat, so that residents usually wear clothing of much heavier description than is sometimes considered necessary so near the equator. Such indeed is the extreme moisture, except during the Harmattan season, that iron rapidly becomes oxidised, bolts and nails are speedily eaten through by rust, salt goes into solution while on the table, glue and paste cease to be adhesive, and dead matter, whether vegetable or animal, rapidly decomposes.

(For Productions *see* under West Coast.)

With reference to the stations for which meteorological tables are given below—

Lagos town stands on an island between Lagos lagoon and the sea, with a creek on the east and Lagos river on the west. It has a river frontage of about a mile and a quarter.

Bonny lies on low ground, scarcely anywhere more than four feet above water, on the east bank of Bonny river, about five miles up from Field Point, on a peninsula formed by that river and a creek. The soil is composed of sand and decayed vegetation, and there are mangrove swamps close by.

Calabar or Duke Town stands on the east side of Old Calabar river, about 27 miles from the mouth, between two high hills. The public buildings lie about 150—200 feet above the river.

Sapele lies on the left bank of the river at the junction of the Ethiope and Jamieson rivers with the Benin river, some 55 miles up from the mouth of the last named.

Onicha lies on the water's edge, about 140 miles up the Niger river, on the left bank, and it is here that the first high ground is met with while ascending the river.

LAGOS. Lat. 6° 26' 39" N., Long. 3° 23' 50" E.

Compiled from 10 years' observations.

	Jan.	Feb.	March	April	May	June	July	**Aug.**	Sept.	Oct.	Nov.	Dec.	Year
Mean day temperature (mean of 8 a.m. and 4 p.m.) Fahr.° 1887—1892	80·7	85°·0	83°·2	83°·5	82°·0	78·8	76·8	76·5	76·7	79·2	81·0	81°·8	80°·4
Mean maximum temperature, 1887—1892	85·5	85·5	86°·0	85°·2	84°·2	81·3	78·2	78·3	80°·0	82·5	85°·5	86°·5	83°·2
Mean minimum temperature, 1887—1892	74·7	76·5	76·2	75·8	75·0	73·0	72·2	71·0	72·3	73·5	74·0	75·0	74°·1
Mean temperature¹, 1887—1892	80·1	81·0	81·1	80·5	79·6	77·1	75·2	74·7	76·1	78·0	79·8	80·7	78·7
Mean daily range², 1887—1892 ...	10·8	9·0	9·8	9·4	9·2	8·3	6·0	7·3	7·7	9·0	11·5	11·5	9·1
Mean of absolute monthly maxima, 1900—1905	89·7	90·9	91·3	90·8	89·5	87·3	84·7	83·6	84·7	87·0	89·0	89·7	
Mean of absolute monthly minima, 1900—1905	69·3	73·0	75·3	74·3	74·0	72·5	71·7	71·4	71·8	71·8	74·0	73·7	
Mean of absolute maxima and absolute minima	79·5	81·9	83·3	82·0	81·8	79·9	77·7	77·0	78·3	79·4	81·5	81·7	
Mean monthly range³, 1900—1905	20°·4	17°·9	16°·0	16°·5	15°·5	14°·8	13°·0	12°·2	12°·9	15°·2	15°·0	16°·0	
Mean relative humidity⁴ (per cent.) 1887—1892 and 1900—1905	78·6	78·2	78·3	79·6	78·9	78·7	84·0	79·2	81·9	81·9	79·1	77·8	79·9

¹ The mean temperature is derived from $\frac{\text{max.} + \text{min.}}{2}$.

² The daily range is max.—min. ³ The monthly range is absolute max.—absolute min.

⁴ The mean relative humidity is derived from $\frac{8a + 4p}{2}$ for the first period and from $\frac{7a + 1p + 9p}{3}$ for the second. Jan.—June 1903 wanting.

For further particulars see above, p. 133.

BONNY. Lat. 4° 3′ N., Long. 7° 10′ E. Sea-level.

Compiled from 3 years' observations[1].

	Jan.	Feb.	March	April	May	June	July	Aug.	Sept.	Oct.	Nov.	Dec.	Year
Mean maximum temperature[2] (Fahr.°)	89°·8	91°·9	93°·5	89°·0	91°·4	91°·1	86°·6	86°·2	86°·1	87°·3	87°·9	88°·7	89°·1
Mean minimum temperature[3] ...	67°·6	67°·6	67°·5	67°·1	69°·2	70°·0	69°·6	68°·7	69°·8	69°·6	70°·3	70°·9	68°·9
Mean temperature[4]	78°·7	79°·8	80°·5	78°·0	80°·3	80°·6	78°·1	77°·5	77°·9	78°·5	79°·1	80°·3	79°·0
Mean daily range[5]	22°·2	24°·3	26°·0	22°·9	22°·2	21°·1	17°·0	17°·5	15°·3	17°·7	17°·6	17°·8	20°·2
Mean relative humidity[6] (per cent.)	84	77	86	85	85	86	84	87	86	86	85	85	85

[1] 1904—1906.

[2] The absolute maximum recorded was 99° on 13th, 14th and 15th March 1904.

[3] The absolute minimum, viz. 60°, was recorded several times early in 1904 and also in January and March 1905.

[4] The mean temperature is derived from $\dfrac{max. + min.}{2}$.

[5] The daily range is max.—min.

[6] For only two years, 1904, 1905.

CALABAR. Lat. 4° 57′ 30″ N., Long. 8° 19′ E., Alt. 158 feet.

Compiled from 5 years' observations[1].

	Jan.	Feb.	March	April	May	June	July	Aug.	Sept.	Oct.	Nov.	Dec.	Year
Mean maximum temperature[2] (Fahr.°)	87°·2	89°·7	90°·6	90°·4	90°·3	86°·6	82°·9	82°·5	84°·1	86°·6	86°·8	88°·1	87°·1
Mean minimum temperature[3]	72°·7	71°·1	73°·6	73°·9	71°·7	71°·5	70°·3	69°·9	70°·0	70°·7	71°·0	71°·3	71°·5
Mean temperature[4]	80°·0	80°·4	82°·1	82°·2	81°·0	79°·1	76°·6	76°·2	77°·1	78°·7	78°·9	79°·7	79°·3
Mean daily range[5]	14°·5	18°·6	17°·0	16°·5	18°·6	15°·1	12°·6	12°·6	14°·1	15°·9	15°·8	16°·8	15°·6
Mean relative humidity[6] (per cent.)	79	77	81	78	83	85	88	87	86	85	83	81	83

[1] 1901—1905.

[2] The absolute maximum, 98°, was recorded on the 25th and 26th of February 1905, while 97° was recorded on the 10th of the same month in 1902.

[3] The absolute minimum, 64°, was recorded on the 2nd and 5th of February 1904, and on the 29th of December 1902.

[4] The mean temperature is derived from $\dfrac{max. + min.}{2}$.

[5] The daily range is max.—min.

[6] The mean is derived from $\dfrac{7a + 1p + 9p}{3}$. The mean annual humidity at 7 a.m. is 92, at 1 p.m. 70 and at 9 p.m. 86 per cent.

SAPELE. Lat. 5° 35' N., Long. 5° 44' E.

Compiled from 3 years' observations[1].

	Jan.	Feb.	March	April	May	June	July	Aug.	Sept.	Oct.	Nov.	Dec.	Year
Mean maximum temperature[2] (Fahr.°)	87°·2	89°·3	90°·6	90°·0	87°·8	86°·4	81°·7	81°·8	83°·2	83°·5	88°·9	88°·1	86°·5
Mean minimum temperature[3] ...	70°·8	72°·1	73°·5	73°·4	72°·4	72°·9	72°·0	71°·7	73°·9	72°·9	73°·5	71°·6	72°·6
Mean temperature[4]	79°·0	80°·7	82°·0	81°·7	80°·1	79°·7	76°·8	76°·8	78°·0	78°·2	81°·2	79°·9	79°·6
Mean daily range[5]	16°·4	17°·2	17°·1	16°·6	15°·4	13°·5	9°·7	10°·1	9°·3	10°·6	15°·4	16°·5	13°·9
Mean relative humidity[6] (per cent.)	83	80	82	83	89	87	90	86	88	86	83	83	85

[1] 1904—1906.

[2] The absolute maximum was 97°, recorded in March 1906.

[3] The absolute minimum, 60°, was recorded in January 1906.

[4] The mean temperature is derived from $\dfrac{\text{max.}+\text{min.}}{2}$.

[5] The daily range is max.—min.

[6] For two years only, 1904 and 1905. The mean is derived from $\dfrac{7a+1p+9p}{3}$. The annual mean at 7 a.m. is 93, at 1 p.m. 72, and at 9 p.m. 86·5 per cent.

ONICHA. Lat. 6° 10′ N., Long. 6° 15′ E. (approx.).

Compiled from 2 years' observations[1].

	Jan.	Feb.	March	April	May	June	July	Aug.	Sept.	Oct.	Nov.	Dec.	Year
Mean maximum temperature[2] (Fahr.°)	89°·3	91°·6	94°·2	92°·7	89°·2	87°·1	84°·3	83°·6	85°·7	87°·7	89°·9	89°·8	88°·8
Mean minimum temperature[3] ...	71°·3	74°·2	77°·5	75°·1	74°·1	73°·7	73°·5	72°·8	73°·4	73°·8	75°·4	74°·2	74°·1
Mean temperature [4]	80°·3	82°·9	85°·8	83°·9	81°·7	80°·4	78°·9	78°·2	79°·1	80°·7	82°·7	82°·0	81°·4
Mean daily range[5]	18°·0	17°·4	16°·7	17°·6	15°·1	13°·4	10°·8	12°·8	12°·3	13°·9	14°·5	15°·6	14°·7
Mean relative humidity[6] (per cent.)	71	62	74	75	80	86	87	84	88	86	83	71	79

[1] 1905, 1906.
[2] The absolute maximum recorded was 98° in March 1905.
[3] The absolute minimum recorded was 62° in January 1906.
[4] The mean temperature is derived from $\frac{max.+min.}{2}$.
[5] The daily range is max.—min.
[6] For 1905 only.

SOUTHERN NIGERIA. MEAN RAINFALL IN INCHES.

	Lat. N.	Long. E.	No. of Years	Jan.	Feb.	March	April	May	June	July	Aug.	Sept.	Oct.	Nov.	Dec.	Year
Lagos	6° 26' 39"	3° 23' 50"	15	1·13	2·15	3·44	5·74	10·92	19·93	11·90	2·89	5·38	8·89	2·12	·98	75·47
Ondo[1]	7° 6'	3° 50'	6	·21	·42	3·85	5·14	6·82	7·01	8·73	3·34	7·54	6·49	1·07	·56	51·18
Ibádan	7° 24'	3° 50'	6	·43	·76	3·77	5·37	5·88	6·47	7·46	1·60	6·15	5·75	·86	·13	44·63
Olokemeji	7° 25'	3° 32'	6	·03	·24	3·37	5·52	6·77	6·42	5·22	1·20	4·74	5·03	1·04	·58	40·16
Badagri	6° 25'	2° 53'	6	·62	2·80	2·83	4·80	8·74	20·64	7·01	·33	3·39	5·35	2·00	·88	57·39
Epe[2]	6° 35'	4° 0'	6	·55	1·03	1·57	5·81	8·73	15·56	13·19	1·59	5·60	4·98	1·52	·41	60·54
Oshogbo[3]	7° 45'	4° 33'	3	·01	·48	2·54	6·52	5·47	6·17	6·47	1·82	8·85	5·67	·46	·18	44·64
Oyo	7° 51'	3° 55'	4	·15	·55	3·41	4·33	6·70	6·74	4·69	·89	4·89	6·46	1·93	1·22	41·76
Saki[4]	8° 41'	3° 24'	6	·22	·76	2·67	3·69	5·80	4·58	4·48	2·66	8·70	6·16	·35	0	40·07
Bonny	4° 27'	7° 10'	5	2·31	2·65	4·77	5·97	13·08	28·94	31·66	13·88	19·96	17·02	13·94	3·93	158·11
Calabar	4° 58'	8° 19'	9	1·94	1·52	8·09	9·99	12·28	20·04	23·32	19·56	14·14	13·16	7·56	2·77	134·37
Bendi[5]	5° 3'	7° 37'	3	·61	1·75	2·05	8·06	10·34	11·62	11·32	9·12	14·08	8·95	0·66	·25	78·81
Afikpo[6]	5° 52'	7° 54'	3	·35	1·43	1·20	6·96	10·64	7·52	12·28	13·47	16·40	10·51	1·68	0	82·44
Forcados	5° 22'	5° 26'	2¼	·80	5·88	4·53	4·35	9·39	12·12	13·95	7·87	13·83	12·40	3·71	1·38	90·19
Sapele	5° 55'	5° 45'	4¾	·49	2·13	4·99	8·70	10·30	16·27	19·08	7·79	15·89	11·35	2·95	·44	100·38
Asaba	6° 21'	6° 42'	4	1·15	1·60	2·15	3·89	4·22	5·67	7·57	4·84	10·16	7·89	·32	·43	49·89
Onicha	6° 20'	6° 43'	2	·85	·84	2·15	5·16	7·52	6·56	7·46	10·99	9·64	6·97	·93	·15	59·22
Benin[7]	6° 17'	5° 36'	4¾	·79	1·45	3·37	4·84	6·35	15·68	21·38	7·93	19·70	15·23	4·66	·77	102·15

[1] 15 months wanting. [2] 4 months wanting. [3] 2 months wanting. [4] 26 months wanting.
[5] 1 month wanting. [6] 2 months wanting. [7] 2 months wanting.

NOTE.—In the above rainfall table the specific years for Onicha were 1905 and 1906, for Calabar 1905 and the years immediately preceding, and for the remaining places, 1907 and the years immediately preceding.

K. A.

10

NORTHERN NIGERIA.

In a report by the Governor, published in 1903, the following appears :—

" The climate of Northern Nigeria is, of course, tropical, but the prevalence of the Harmattan, which blows from N.E. for half the year or more, modifies the temperature in a very marked degree. This wind, coming from the Sahara, is very dry, and the evaporation produced, when it meets the moist air of the Niger valley, and even in the plains to the north, results in a great fall in temperature. When the wind, without having absorbed any moisture, meets the mists and vapours of Lake Chad, the temperature falls sometimes below freezing point. Generally speaking, throughout Northern Nigeria the nights are cold for the greater part of the year. During the rainy season, July—November, the atmosphere is laden with moisture, and a damp heat prevails. For the rest of the year the Harmattan and the total absence of rain render the air extraordinarily dry. The climate of Northern Nigeria is probably far more healthy than that of the coast, to the climate of which it only approximates in the vicinity of the river. The Bauchi highlands enjoy a charming climate, and throughout the greater part of the country the climate is not exceptionally trying."

The highest temperature recorded in Northern Nigeria, during the year 1905, was 118° Fahr. (47°·8 C.) at Maifoni (Bornu) on April 8th, and the lowest 39° Fahr. (3°·9 C.) at Kano on February 2nd, the highest mean temperature for the year being 82° Fahr. (27°·8 C.) at Kontagora, and the lowest 74° Fahr. (23°·3 C.) at Zaria. The greatest rainfall was at Zaria, with 51·27 inches (1302·3 mm.), and the lowest at Sókotó, with 33·32 inches (845 mm.), the maximum fall in one day being 4·04 inches (102 mm.) at Ilórin on June 2nd.

The general direction of the wind throughout the Protectorate was from south-west from June to November, and from north-east during the remaining months of the year. The Harmattan lasted, with slight intermissions, from December to the end of May, the first tornado occurred in March, and the rainy season ended in October.

In the following year, 1906, the highest shade temperature recorded was 120° Fahr. (48°·9 C.) at Dumjeri in North Bornu, on the 26th April, and the lowest 40° Fahr. (4°·4 C.) at Zaria on the 9th January, the highest mean temperature for the year being 82° Fahr. (27°·8 C.) at Yola, while the lowest was, as in the previous year, at Zaria, the figures for the two years working out to exactly the same result, viz. 74° Fahr. (23°·3 C.). The general direction of the wind in 1906 was practically the same as in 1905, and may be taken as the rule for this area; and the same applies to the duration of the Harmattan. The first tornadoes, however, were slightly later in 1906 than in 1905 and did not occur until April. The seasons, in fact, were later throughout the year, for the rainy season did not end till November, whereas in the previous year the season ended in October. The greatest precipitation in 1906 was 61·05 inches (1551 mm.), recorded, as in the previous year, at Zaria; the smallest amount of rain for the year at any one station was 23·49 inches (597 mm.), recorded at Amar. The maximum fall in one day was recorded at Zúngeru on 28th July, and amounted to no less than 7·27 inches (185 mm.), more than would be recorded, as a rule, during three months in London.

The table on p. 148, for the years 1904 and 1905, gives some idea of the prevailing conditions.

Capt. Mockler-Ferryman says of the Ilórin country that it is perhaps one of the most pleasant parts of West Africa south of the Middle Niger, with a climate almost free from the pestilential malaria of the coast; and the same writer declares that "the plateau on which Bauchi is built is remarkable for its excellent climate." At Bukuru, in the province of Bauchi, the temperature in the dry season is stated, in the Report on Northern Nigeria for 1905, to rarely exceed 85° Fahr. (29°·4 C.), while in November and December it falls below freezing point.

At Hadeija and Katagun, in Kano province, the climate is good, but it is necessary here, as in most parts, for Europeans to have properly constructed dwellings.

10—2

SUMMARY FOR THE YEARS 1904 AND 1905.

	1904						1905					
	Temperature (Fahr.)			Rainfall			Temperature (Fahr.)				Rainfall	
	Shade maximum	Shade minimum	Range	Mean	Amount in inches	Degree of humidity per cent.	Shade maximum	Shade minimum	Range	Mean	Amount in inches	Degree of humidity per cent.
Zúngeru	103°	56°	47°	79°	51·1	63·6	106°	56°	50°	80°	41·31	58
Lokója	102°	57°	45°	80°	41·72	—	101°	53°	48°	81°	49·64	72
Yola	107°	60°	47°	80°	33·77	—	108°	58°	50°	81°	42·76	—
Ilórin					Complete statistics not available		106°	58°	53°	78°	47·02	73
Kano							105°	39°	66°	76°	36·69	53
Kontagora							102°	57°	45°	82°	46·28	63
Sókotó							106°	50°	56°	79°	33·32	—
Zaria							102°	43°	59°	74°	51·27	63

In the extreme north of Northern Nigeria the Boundary Commission (1902–4) found that the rains begin in June and last till the end of September. Except in August, however, such a thing as a really wet day is rare, but in that month there appears to be, as a rule, a wet week. Three or four storms a week is the usual average in the rainy season. About two months after the rains have ceased, *i.e.* the end of November, the Harmattan sets in, and blows steadily from the N.E. till April. During this time the air is misty with suspended dust; sometimes the haze is so thick that the largest objects are invisible at a distance of half a mile. Vegetation dries up and everything is covered with the finest powder. The temperatures in December and January are from $100°$ Fahr. $(37°·8$ C.) by day to $60°$ Fahr. $(15°·6$ C.) by night. April is the hottest month; the day temperature reached $114°$ Fahr. $(45°·6$ C.) in the shade, and the lowest it fell to at night was $78°$ Fahr. $(25°·6$ C.).

During April and May the wind blows from other quarters, the air becomes clearer, clouds gather, sandstorms are more frequent, till they occur daily, and a few drops of rain occasionally raise the hopes that are usually disappointed till June is near[1].

Of the region immediately to the north, Gaden gives an account which, confirming, in most respects, that given above by Col. Elliot, differs in some points and gives additional information, and is therefore worth stating. He says that in the Zinder region the seasons are the same as in corresponding latitudes further west, except that rain is rarer, and is never experienced except during the four months May/June—September. In December and January the thermometer falls, as a regular thing, below $50°$ Fahr. $(10°C.)$. This cold is usually accompanied by thick mist. The east winds which prevail during the period from February onwards also bring mists and fogs of a lighter character. These winds are hot and parching, but the radiation is so great that the nights are fairly cool[2]. In this region the rainy season is unhealthy, and in a letter, dated 4th July, it was stated that the rainy season had

---

[1] Col. G. S. M^{c}D. Elliot, *G. J.* xxiv, 1904, p. 514.
[2] Gaden, "Notice sur la Résidence de Zinder" (Extract from *Revue des Troupes coloniales*), p. 17. Paris

begun nearly two months before, and that all the members of the Boundary Commission were or had been ill in consequence.

In the Chad region at the end of November, on one occasion, the thermometer went down to $41°$ Fahr. ($5°$ C.). This is the evidence of Barth[1]. The Harmattan may be expected here about the middle of December. From records taken during $2\frac{1}{2}$ years, the following mean monthly temperature (Fahr.) for Kúkawa may be deduced :

January	$75°·4$	May	$91°·0$	September	$82°·0$
February	$78°·6$	June	$89°·6$	October	$84°·9$
March	$88°·9$	July	$83°·8$	November	$79°·5$
April	$92°·3$	August	$79°·4$	December	$74°·0$

giving a mean annual temperature of $83°·5$ Fahr. ($28°·6$ C.).

The Mandara mountains form the climatic boundary between Adamawa and Bornu, the seasons in the latter being much later than in the former. Thus the Benue river is at its highest in September, and Lake Chad not until December. Mandara is not so hot as the north of Adamawa (Cameroons) owing chiefly to its altitude and forests. To the north of the Mandara mountains, Pavel[2] states that the rains begin in June and end at the end of October. The climate is particularly warm, he having experienced a temperature of $108°$ Fahr. ($42°·2$ C.) throughout the day, during his journey, with a fall to $97°$ Fahr. ($36°$ C.) at night, but the air was exceedingly dry (April, May) and the heat consequently more supportable than in the coast and forest regions.

The death-rate among Europeans in Northern Nigeria, calculated for the 6-year period 1900–5, is 45·19 per thousand, but the figures, on which this calculation is based, are so small, that too much reliance must not be placed on the result. The death-rate among officials is considerably smaller than that among non-officials.

The figures from which the Table (below) of temperature at Lokója was calculated show that the means are good means, and therefore fairly reliable. With regard to the mean maximum temperature the greatest difference between

---

[1] *Travels in Central Africa*, Vol. IV, p. 12.
[2] *Mouvement Géographique*, 1902, p. 629.

the annual results, for any one month, is $5°·5$ Fahr. ($3°$ C.) in November, when the mean for 1901 was $86°·1$ Fahr. ($30°·1$ C.) and that for 1903 was $91°·6$ Fahr. ($33°·1$ C.), but the usual monthly difference throughout the 5 years does not average more than about $2°·5$ Fahr. ($1°·4$. C.). The monthly means of minimum readings are also good throughout, except in February and September, when there was, in the first month, a difference of nearly $9°$ Fahr. ($5°$ C.) between the 1901 and 1903 means, and, in the second, a slightly larger difference between the figures for 1901 and 1902. The highest temperature recorded at Lokója is $102°$ Fahr. ($38°·9$ C.), and this has been measured in three of the years for which the Table is made, namely on 14th February 1900, the 3rd March 1902, and the 11th March 1904. The lowest temperature recorded was $50°$ Fahr. ($10°$ C.) on the 26th November 1902. The minimum registered in 1904 was $57°$ Fahr. ($13°·9$ C.) on the 23rd December, while in the other years the thermometer did not go below $60°$ Fahr. ($15°·6$ C.). The greatest amount of rain measured on one day was 4·3 inches (109 mm.) in 1901 on the 2nd April, 3 inches (76 mm.) in 1902 on the 27th August, and 3 inches in 1904 on the 18th July.

At Zúngeru the highest temperature was $107°$ Fahr. ($41°·6$ C.) recorded on the 7th March, and the lowest, $55°$ Fahr. ($12°·8$ C.), on 30th November. The greatest amount of rainfall on any one day was 7·27 inches (184 mm.) on the 28th July 1906, this being the greatest amount ever measured in one day throughout the whole Protectorate.

Mr F. F. Hopkins, Political Secretary to the Governor and Commander-in-chief of Northern Nigeria, who has been in the country for some ten years, has very kindly written, specially for this work, the following notes on the climate, health conditions, and precautions to be taken in order to avoid disease.

The climate of Northern Nigeria, situated as it is between the 7th and 14th parallels of north latitude, is of course tropical. The hottest months of the year are February, March and April, the heat being incessant, and causing much sleeplessness.

The rainy season extends from April to October, the wettest months being July, August and September. It is ushered in and out by very heavy tornadoes of wind and rain—hail-stones have fallen in the north-east of the size of an African hen's egg—intermittent rains falling from April till July, and from September to October.

The air during the rains is heavily laden with moisture, resulting in a very damp heat, and all clothing becomes saturated (unless kept under closed cover) which necessitates placing it in the sun from time to time, as occasion offers.

The prevalence of the Harmattan wind, which blows hard from about November till February (and intermittently to the end of April), affects the temperature in a very marked and even extraordinary degree. This wind coming from the dry desert of the Sahara is singularly devoid of moisture, and the evaporation produced, when it meets the moist air of the river valleys, and even in the plains to the north, results in a great fall of temperature. It brings with it a very fine white dust, which permeates everything and envelopes the whole country in a white mist, rendering it a matter of impossibility to discern objects clearly at any distance. The extreme dryness of the wind causes everything to contract, to the extent of making wood-work, &c., gape and even split; in extreme cases, namely at Lake Chad and on the Bauchi highlands, the temperature falls below freezing point.

The mornings and evenings are always very cold at this dry season. The trees begin to bud and burst into leaf at this time and the grass, where it has been burnt off in November/December, shows a green shoot.

The general direction of the wind, throughout the Protectorate, is from the south-west from June till November, and from the north-east during the remainder of the year. The tornadoes come from the north-east.

For Europeans this Harmattan season is the healthiest time of year. The natives, on the other hand, suffer more in this dry season, not only from epidemics, but from respiratory and digestive diseases brought on by the cold nights and impure water supply, when it is scarce and polluted. Europeans take proper precautions against this. Their most unhealthy season is during the rains, probably more so at the beginning and end of the rains. The greatest amount of sickness is due to malaria, but blackwater fever and dysentery claim some victims, and many suffer from anaemia, diarrhoea, inflammation of the stomach and indigestion, and there are a few cases of sunstroke; prickly heat is of course general.

Natives suffer principally from malaria, diseases of the respiratory organs and digestive system and of the eyes, and also from parasitic skin and venereal diseases. Epidemics of small-pox and cerebro-spinal meningitis occur at periods. Great efforts are made to induce the natives to improve the sanitation of their towns and mitigate sickness. Native dispensaries have been opened, and much inoculation against small-pox is carried out by the Medical Department.

As a preventive against the many ills arising from the unhealthy climate, the daily use of 5 grains of quinine is strongly advocated, and the boiling and filtering of water, with regular habits and exercise in the shape of polo and tennis—where it is obtainable—is recommended, and shooting, for which there is plenty of opportunity, both big and ground game abounding in the Protectorate.

Helmets are of course always worn and the clothing is light, flannel being mostly adopted as being less liable to convey chills. A cummerbund is usually wrapped round the loins at night outside the pyjamas.

(For Productions *see* under West Coast.)

LOKÓJA. Lat. 7° 47′ 54″ N., Long. 6° 42′ 23″ E. Alt. 320 feet.

Compiled from 5 years' observations[1].

	Jan.	Feb.	March	April	May	June	July	Aug.	Sept.	Oct.	Nov.	Dec.	Year
Mean maximum temperature[2] (Fahr.°)	89°·7	94°·0	96°·9	93°·0	90°·8	91°·9	86°·2	85°·8	87°·2	89°·0	89°·8	91°·2	90°·5
Mean minimum temperature[2] ...	65°·3	72°·4	77°·2	75°·9	70°·6	72°·7	72°·6	73°·3	71°·2	69°·3	66°·3	65°·1	71°·0
Mean temperature[3]	77°·5	83°·2	87°·0	84°·5	80°·7	82°·3	79°·4	79°·6	79°·2	79°·1	78°·0	78°·2	80°·7
Mean daily range[4]	24°·4	21°·6	19°·7	17°·1	20°·2	19°·2	13°·6	12°·5	16°·0	19°·7	23°·5	26°·1	19°·5

ZÚNGERU. Lat. 9° 48′ 59″ N., Long. 6° 17′ 15″ E. Alt. 530 feet.

Compiled from 2 years' observations[5].

	Jan.	Feb.	March	April	May	June	July	Aug.	Sept.	Oct.	Nov.	Dec.	Year
Mean maximum temperature[6] (Fahr.°)	95°·0	96°·3	100°·1	98°·4	91°·7	87°·1	84°·8	83°·4	86°·5	89°·0	94°·0	94°·4	91°·7
Mean minimum temperature[7] ...	68°·0	67°·5	73°·7	75°·2	74°·0	71°·0	71°·8	71°·1	70°·5	70°·0	63°·0	62°·5	69°·9
Mean temperature[3]	81°·5	81°·9	86°·9	86°·8	82°·8	79°·0	78°·3	77°·3	78°·5	79°·5	78°·5	78°·5	80°·9
Mean daily range[4]	29°·0	28°·8	26°·4	23°·2	17°·7	16°·1	13°·0	12°·3	16°·0	19°·0	21°·0	31°·9	21°·8

[1] 1900—1904.

[2] 46 months. The absolute maximum, viz. 102°, was recorded on 14th February 1900, on 3rd June 1902, and 11th March 1904; the absolute minimum, 50°, on 26th November 1902.

[3] The mean temperature is derived from $\dfrac{\text{max.} + \text{min.}}{2}$.

[4] The daily range is max.—min.

[5] 1903, 1904.

[6] The absolute maximum, 107°, was recorded on 7th March 1903.

[7] The absolute minimum, 56°, was recorded on 27th January, 29th November and 15th December, 1904. But subsequently, on 30th January 1906, the thermometer went down to 55°.

NORTHERN NIGERIA. MEAN RAINFALL IN INCHES.

	Lat.	Long.	No. of Years	Jan.	Feb.	March	April	May	June	July	Aug.	Sept.	Oct.	Nov.	Dec.	Year
Lokója	7° 48' 1"	6° 44' 23"	7	·93	·26	1·26	4·34	7·44	5·71	7·89	6·06	11·50	4·67	58	·03	50·67
Ilórin	8° 30' 27"	4° 34' 54"	3	0	·43	1·46	4·45	6·35	7·76	6·09	2·22	12·31	7·44	1·03	·71	50·25
Yola	9° 12' 29"	12° 29' 30"	4	0	0	0	1·35	4·49	5·62	5·11	7·57	7·10	3·35	0	0	34·59
Amar (approx.)	9° 44' 0"	10° 23' 0"	4	0	·03	·34	1·83	5·26	5·60	5·99	5·92	6·11	3·31	0	0	34·39
Zúngeru	9° 48' 32"	6° 9' 42"	5	0	0	·13	1·46	5·26	6·72	8·94	7·84	9·57	4·59	·04	0	44·55
Zaria	11° 6' 10"	7° 42' 37"	3	0	0	0	1·27	6·36	6·73	8·48	11·96	10·00	2·55	·01	0	47·36
Dumjeri¹ (approx.)	13° 1' 0"	11° 54' 0"	1¹⁄₂	0	0	0	0	·48	2·22	3·03	10·28	·41	·70	0	0	17·12
Sókotó	13° 2' 10"	5° 14' 47"	4	0	0	0	·11	1·52	4·50	7·53	9·19	3·96	1·18	0	0	27·99

NOTE.—The specific years are 1907 and those immediately preceding.

¹ The September value for Dumjeri is for the year 1907 only.

FRENCH WEST AFRICA

SENEGAL.

Senegal is a flat sandy area. The fall of the river gives us an idea of the relief. At Kayes the altitude is only 197 feet above sea-level and at Kaedi 98 feet, so that the fall is only about 100 feet in 237 miles, measured direct, without taking into consideration the meanderings of its very tortuous stream. The Cayor country, between St Louis and Dakar, is also flat and sandy, while between the Senegal river on the east and the upper waters of the Marigot Bunu and of the Salum river on the west, and those of the Gambia on the south, lies the desert plateau of Ferlo. Towards the south-east, in the region of the Faleme river, the country rises, but in the greater part of this region the hills do not rise to a much higher altitude than 300—400 feet. A chain of such hills runs from the north, leaving Thiès on the east, and ends at Cape Naze. The Senegal river on the north appears like a barrier beyond which the Sahara cannot pass, though its influence on the climate is very marked. Time was when this was a land of forests, which exist no longer, a mere vestige being left between Thiès and the coast south of Dakar, and a trace here and there. Dr D'Anfreville, writing in the *Revue Coloniale* (April 1908), says that, as a matter of fact, Senegal is nothing but a vast desert for seven months in the year, and its sandy soil appears as though labouring under a perpetual curse, there is no running water, no dense verdure, and its scattered villages look miserable; such trees and shrubs as exist are thorny and sapless. Then come the rains and the whole face of the country is transformed, the rivers are filled, the bare-branched baobabs are covered with thick foliage, the sand is hidden from view by a vigorous growth of grass, and Senegal is a great prairie. The whole change is effected in a few weeks.

The *Service Météorologique* was established throughout French West Africa in 1903–4, and curves, showing the fluctuations in the various climatic elements, have now been published for Dakar and St Louis; and there are also available the figures for 1903. Combining these data we arrive at the tables below. The mean annual temperature of St Louis is 76°·6 Fahr. (24°·8 C.) rising to 80°·6 and 86° Fahr. (27° and 30° C.) in the rainy season, from June to October, when the atmosphere, being charged with moisture, is very hot and trying to Europeans. The barometer is fairly constant, the mean monthly pressure only varying between 30·06 inches (763°·5 mm.) and about 29·92 inches (760 mm.), while at Dakar the corresponding figures are 30·14 inches (765·4 mm.) and 30 inches (762 mm.). At the latter place the mean annual temperature is 76°·1 Fahr. (24°·5 C.) and rises to 80°·6 and 82°·4 Fahr. (27° and 28° C.) in the rainy season.

At Dakar, as at St Louis, there are two seasons, the rainy and the dry. The hot season is the rainy season, and begins towards the end of June, lasting to the end of November. This season is characterised by violent storms or tornadoes, succeeded by stifling heat. The regular rains do not begin till the second fortnight in July, and only last till the middle of September. The temperature varies between 77° and 96°·8 Fahr. (25° and 36° C.) in the shade, and the constant heat is trying. Care should be taken to avoid draughts at night, as chills are sure to produce fever or other ailments. The tornadoes alone bring a certain amount of freshness, but, when they occur at night, sleep is almost impossible owing to the thunder; and if the tornadoes come in the day they are soon followed by oppressive heat. At this season of the year people nearly always grow restive, nervous, and irritable. The wet season, in fact, is disagreeable. As soon as the wind turns to the north, the change is immediately noticeable. The north winds begin in November, but do not become settled till December, and the dry season then lasts till the end of May. The nights in this season are quite cool, and the thermometer sinks to 53°·6 or 55°·4 Fahr. (12° or 13° C.), and sometimes even to 50° Fahr. (10° C.). East winds bring variations of heat and cold, but the north winds are the resuscitators[1].

[1] Ribot-Lafon, *Dakar, ses origines, son avenir*, p. 21. Delmas, Bordeaux, 1908.

In consequence of the introduction of sanitary measures, the death-rate at Dakar, which was 125 per 1000 in 1904, sank to 101 per 1000 in 1905.

Inland the rains become less and less, until the only precipitation is from a few summer storms. In the valley of the Lower Senegal, the heat is even more trying than on the coast, and Podor, which is situated beyond the reach of sea breezes, is described as a very "hell on earth." The thermometer has been known to register no less than 113° Fahr. (45° C.) in the month of May, but the air is relatively dry. On the plateaux, on either side of the river, the north and north-east winds are experienced, and bring some relief. In Senegal, at times when the east winds blow frequently, the temperature is several degrees higher than in French Guinea. The temperature of these east winds is often 107°·6 Fahr. (42° C.), and is, as might be supposed, exceedingly trying.

Cayor, the flat sandy district between Cape Verde and the Senegal river, is almost as hot and dry as the river district, but at Kaolakh on the Salum river, which is said to be the hottest place, next to Podor, the humidity is higher.

The rainfall at St Louis is, to judge by all the evidence, apparently diminishing. The mean annual rainfall for the period 1830–40 was 23·62 inches (600 mm.). Another mean, for 21 years, places it at 17·32 inches (440 mm.). A 33-year mean[1] gives 404 mm., while Courtet in his *Étude sur le Senegal*, published in 1903, and evidently referring to the year 1901 or 1902 or thereabouts says "it is now 400 mm." Further, the 3-year mean now arrived at for the years 1903–4–5 is only 9·36 inches or 237·78 mm. (*see* table below). From this we see that, starting with a 10-year average seventy or eighty years ago, the longer we take the period the less is the mean. The rain cannot be depended on from year to year. As a contrast with all the means given above, the total fall for 1903 was only 4·90 inches or 124·6 mm.

There are two seasons throughout Senegal, and the dry season begins, according to locality, in November or December and lasts till the end of May or even later.

---

[1] *Petermann's Mitteilungen*, 1906, Band LII, p. 80.

SAINT LOUIS.　Lat. 16° 1′ 31″ N., Long. 16° 30′ W.　Alt. 7 feet.

	Jan.	Feb.	March	April	May	June	July	Aug.	Sept.	Oct.	Nov.	Dec.	Year
Mean temperature¹ (Fahr.°) (1903)	69°·4	74°·0	74°·5	69°·6	70°·9	76°·9	83°·2	86°·1	87°·8	85°·1	82°·2	74°·1	77°
Mean maximum temperature　„　…	78°·4	84°·0	85°·6	74°·3	74°·5	81°·3	87°·0	89°·4	90°·7	87°·6	86°·6	79°·9	83°
Mean minimum temperature　„　…	60°·4	64°·0	63°·3	64°·8	67°·3	72°·5	79°·5	82°·8	85°·1	82°·9	76°·6	68°·4	72°
Absolute maximum　„　…	96°·8	97°·5	106°·1	96°·4	78°·8	86°·0	93°·2	93°·2	96°·8	91°·4	96°·8	87°·4	106°·1
Absolute minimum　„　…	56°·8	60°·4	58°·5	50°·0	64°·8	69°·8	77°·0	77°·0	80°·6	75°·2	69°·8	62°·6	50°·0
Mean temperature² (3 years)	71°·8	73°·9	72°·7	70°·7	71°·2	74°·7	82°·6	84°·0	85°·8	81°·9	78°·3	72°·9	76°·7
„　„　³ (5½ years)	70°·1	71°·8	71°·4	70°·5	71°·4	77°·7	81°·0	82°·0	83°·1	82°·2	78°·0	70°·1	75°·8
Mean relative humidity (per cent.) 6 a.m. (1903)	77	64	63	50	83	89	95	93	83	82	64	67	76
Mean relative humidity (per cent.) 2 p.m. (1903)	62	47	48	74	74	83	94	90	69	71	53	66	67·6
Mean relative humidity (per cent.) 9 p.m. (1903)	78	64	59	79	85	85	96	91	77	80	67	69	77·5
Mean amount of cloud 6 a.m. (0—10) (1903)	2·1	3·5	1·8	2·8	1·6	2·3	3·7	3·2	3·2	1·0	1·9	1·8	2·4
Mean amount of cloud 2 p.m. (1903)	1·5	3·8	0·6	2·2	1·8	2·0	1·9	2·0	2·1	1·3	1·0	1·6	1·8
Mean amount of cloud 9 p.m.　„	1·3	4·1	0·8	2·0	1·9	2·5	3·5	2·5	2·5	1·4	1·1	1·6	2·1
Number of days with storms　„	0	0	0	0	0	1	3	9	11	3	0	0	27
Mean number of days with rain⁴ (3 years)	0	0·7	2	0	0	1·3	4	11·7	7·7	0·7	1	1	30·1
Rainfall in inches (1903)　…	0	0	0	0	0	1·30	0·61	0·49	2·50	0	0	0	4·90
Mean annual rainfall⁴ (3 years)	0	0·01	0·14	0	0	0·67	1·58	4·29	2·30	0·15	0·21	0·16	9·51
„　„　⁵ (14 years)	0·06	0·04	0·03	0	0·14	0·98	2·76	4·31	3·26	0·75	0·12	0·08	12·53

¹ The mean is derived from $\dfrac{\text{max.} + \text{min.}}{2}$.

² This mean is derived from that for 1903 and the means for 1904 and 1905, read off the curves published by the *Service Météorologique*. It is not stated how these latter were obtained.

³ These figures appeared in Dr Buchan's *Challenger* Report. They are for the years 1873—78.

⁴ For the years 1903—1905.

⁵ 1892—1905 (8 months wanting).

DAKAR. Lat. 14° 40′ 30″ N, Long. 17° 25′ 22″ W. Alt. 98 feet.

	Jan.	Feb.	March	April	May	June	July	Aug.	Sept.	Oct.	Nov.	Dec.	Year
Mean temperature at 6 a.m. (Fahr.°) (1903)	65°·5	63°·7	66°·2	70°·5	67°·6	74°·3	79°·0	78°·6	79°·0	79°·0	76°·1	67°·5	72°
Mean temperature at noon (1903) ...	69°·6	82°·0	83°·7	80°·0	81°·7	85°·3	87°·8	83°·3	84°·7	87°·6	85°·3	74°·1	82°
Mean temperature at 6 p.m. ,, ...	64°·6	65°·5	64°·8	73°·2	72°·3	76°·8	82°·0	81°·1	81°·1	81°·7	79°·2	71°·2	74°
Mean maximum temperature ,, ...	70°·0	82°·0	84°·2	81°·5	85°·8	87°·3	89°·4	86°·4	85°·1	88°·0	86°·2	76°·1	83°
Mean minimum temperature ,, ...	64°·2	63°·3	65°·1	64°·8	69°·6	71°·0	77°·0	78°·4	77°·9	78°·0	75°·6	66°·2	71°
Mean temperature[1] (3 years) ...	69°·1	70°·0	71°·1	73°·2	73°·2	78°·1	80°·6	80°·8	82°·0	82°·6	78°·6	72°·9	76°
Mean relative humidity (per cent.) 6 a.m. (1903)	71	75	77	81	73	77	77	87	88	83	63	63	76
Mean relative humidity (per cent.) noon (1903)	69	55	56	64	58	61	59	71	76	68	47	47	61
Mean relative humidity (per cent.) 6 p.m. (1903)	72	81	76	60	61	64	73	82	84	82	70	65	72
Rainfall in inches (1903)	0	0	0	0	0	0·34	0·17	11·15	4·47	0·10	0	0	16·23
Mean annual rainfall[2] (3 years) ...	0	0·18	0·08	0	0	0·41	2·24	10·80	3·40	2·26	0·18	0·34	19·89
Mean number of days on which rain was measured[3] (2 years)	0	1	1·5	0	0	2	9	15	7	5	1	1·5	38·5

[1] This mean is derived from $\frac{max. + min.}{2}$ for 1903, combined with the means for 1904 and 1905, which were read off the curves published by the *Service Météorologique*; how these latter means were obtained is not stated.

[2] For the years 1903—1905.

[3] For the years 1904, 1905.

The proximity of the Sahara has a double effect on the climate, it diminishes the humidity and increases the range of temperature.

With reference to the two stations for which meteorological tables are given on pp. 159, 160 :

St Louis stands on a narrow island, not more than a mile long, at the mouth of the Senegal river. The land lies low and to anyone approaching from the sea the town appears to stand on the shore.

Dakar lies on the south of the bay of the same name, in the extreme west. The neighbouring country is open and slightly undulating, with only a few patches of trees, and but little vegetation. Small patches of land are under cultivation. The soil is chiefly sand.

Both these places were created "Stations principales" of the Meteorological Service in 1903, and provided with barometers, thermometers, and psychrometers.

UPPER SENEGAL.

In Upper Senegal, Kayes is the warmest place. The mean annual temperature (mean of maximum and minimum) is 85° Fahr. (29°·44 C.). The mean range is 26° Fahr. (14°·3 C.) being the difference between the mean maximum temperature 98° Fahr. (36°·6 C.) and the mean minimum 72° Fahr. (22°·3 C.). The highest monthly means, 107°·7 Fahr. (42° C.), occur in April and May and the lowest, 61°·8 Fahr. (16°·6 C.), in January. The temperature gradually rises from January to May, begins to fall in June and reaches the mean annual temperature about July, August and September; it increases again in October and then falls again gradually till the minimum is reached. The absolute minimum, 50° Fahr. (10° C.), was recorded on the 2nd of January 1905, and the absolute maximum 119° Fahr. (48°·5 C.) on the 30th April 1902, giving a maximum range, during the five-year period for which observations are available, of 69° Fahr. (38°·5 C.). The temperature in March, April, May and June is always above the mean annual temperature, and in December, January and February it is always below that mean. In the remaining months the temperature is about equal to the annual mean, except in October, when it is generally a little higher.

The relative humidity is highest in July, August and September and lowest in February.

Rain seldom falls throughout a whole day, but, towards the end of the rainy season, a fine rain sometimes falls, when the weather is calm, for several hours. Though the heaviest falls of rain occur in September and October, the rainiest month, as a rule, is August.

The sky is generally the clearest in March, and most overcast at the height of the rainy season, *i.e.* in August.

The most frequent winds come from the north-west, west, west-north-west, east and north-east, the first three blowing most frequently in the period June—October to the exclusion of almost all other winds. The east wind is generally dry and hot. During rain-storms the wind is from the south-east or south-south-east.

The year may be divided into three seasons :

1. The hot dry season, March, April and May, during which the rainfall is negligible, the relative humidity below the mean, and the temperature above the mean.

2. The moist rainy season, July, August and September, during which the temperature is below the mean, the relative humidity much above the mean, and rain is abundant. Storms and tornadoes occur in this season.

3. The dry and comparatively cool season, November, December, January and February, during which the temperature is below the mean and also the relative humidity, the dryest month being February. In this season N.E., E., and E.N.E. winds prevail, and there are no storms.

June is a transition month.

To the south of 14° N. the presence of the mountains gives a sharp rise to the rainfall. In the Kita district the rains are violent and torrential from August to October, and it generally rains, during this period, twice in every three days.

In the Kita or Fuladugu region Europeans cannot work in the burning sun, nor can they even do hard work in the shade, but the climate is not, as was supposed, unhealthy. The months of November and July, which separate the rainy and dry seasons, are the worst for

KAYES. Lat. 14° 29′ 6″ N., Long. 11° 17′ 17″ W. Alt. 197 feet.

Compiled from 5 years' observations[1].

	Jan.	Feb.	March	April	May	June	July	Aug.	Sept.	Oct.	Nov.	Dec.	Year
Mean temperature[2] (Fahr.°)	76°·6	80°·6	88°·0	93°·6	95°·0	90°·7	83°·7	82°·0	83°·5	85°·5	83°·3	77°·7	85°
Mean relative humidity[2] (per cent.)	30·5	23·5	24·4	29·18	43·22	58·76	73·0	78·62	76·4	66·34	44·98	35·86	48·67
Mean rainfall (in inches) ...	0·004	0	0·04	0·09	0·71	4·19	8·12	8·71	4·94	2·24	0·004	0	29·05
Mean number of rainy days ...	1	0	1	1	4	9	11	14·6	10	4	1	0	56·6

NOTE.—Kayes is situated on the left bank of the middle course of the Senegal River, the two nearest points on the coast being Bathurst and St Louis, at a distance of about 350 miles. In 1903 Kayes was fitted up as a "Station principale" of the Meteorological Service with barometer, thermometers, and psychrometer.

[1] 1901—1905.

[2] The annual means, from which the mean temperature for the 5-year period is deduced, were obtained from the *Annales* of the Bureau Central de Météorol. (Paris), and it is not stated how the means were obtained.

Previous temperature means, however, at Kayes, viz. for the years 1895—1899 were obtained from $\frac{max.+min.}{2}$, and it is probable that the above mean temperature was derived from the same formula (see *Meteorol. Zeitschrift,* 1903, p. 231).

Year	Mean annual (absolute)		Monthly means				Absolute min.	Date	Absolute max.	Date	Total fall (in ins.)	Rainy days	Most in a day (in ins.)	Date	Rainiest month	Fall in that month (in ins.)
	min.	max.	highest	month	lowest	month										
1901	73°·0	96°·4	94°·3	April	76°·3	Jan.	55°·4	{11 Jan. 21 Dec.}	113°·0	18 April	29·62	64	2·15	3 Oct	Aug.	10·79
1902	74°·8	98°·8	96°·3	May	76°·5	Dec.	56°·3	31 Dec.	119°·3	30 April	22·98	48·	1·86	13 Oct.	"	7·25
1903	72°·0	98°·2	94°·5	"	76°·8	Jan.}Dec.	56°·3	18 Jan.	112°·1	16 May	28·29	53	3·99	15 Sept.	"	10·23
1904	70°·5	99°·3	96°·4	"	74°·1	Jan.	50°·9	10 Jan.	116°·6	9 May	24·67	62	1·99	1 Aug.	"	7·68
1905	70°·9	97°·3	95°·0	"	76°·8	"	50°·0	2 Jan.	115°·7	19 May	42·22	67	4·72	25 Sept.	July	12·36
means &c.	72°	98°	96°·4	May 1904	74°·1	Jan. 1904	50°	2 Jan. 1905	119°·3	30 April 1902	29·02	58	4·72	25 Sept.	July 1905	12·36

TEMPERATURE AT KAYES | RAIN AT KAYES

Europeans, for it is then that the effect of the decaying vegetable matter in the streams is experienced.

East and south of this the rainy season begins earlier. There are two well-marked seasons, the rainy and the dry, but the natives divide the year into five, as follows :—

(1) That of great heat—March, April, May—mean temperature, maximum $101°·5$ Fahr. ($38°·6$ C.), minimum $83°·5$ Fahr. ($28°·6$ C.). In this season the grass is burned, and there are a few showers.

(2) The sowing season—end of May, June—not quite so hot, mean maximum temperature $91°·9$ Fahr. ($33°·3$ C.), mean minimum $80°$ Fahr. ($26°·7$ C.). In this season the atmosphere is close and charged with electricity.

(3) July, August, September : during this period vegetation is in full growth ; mean maximum temperature $82°·2$ Fahr. ($27°·9$ C.), mean minimum $73°·9$ Fahr. ($23°·3$ C.).

(4) The harvest—October, November.

(5) The cold season—December, January, February ; mean maximum temperature $86°·9$ Fahr. ($30°·5$ C.), mean minimum $69°·3$ Fahr. ($20°·7$ C.).

The above mean temperatures are frequently exceeded, the thermometer sometimes rising to $111°·2$ and $113°$ Fahr. ($44°$ and $45°$ C.), and, on the other hand, falling to $39°·2$— $50°$ Fahr. ($4°$—$10°$ C.)[1].

KASAMANSE.

The climate on the lower Kasamanse resembles that of the Senegal and French Guinea and neighbouring coasts. The dry season lasts from December to May and the rainy season from June to November, these seasons being separated by two tornado periods of some three weeks. The temperature in the rainy season is almost constant; and the mean rainfall in the season is (at Sedhiu) $157·5$ inches (4000 mm.). It is hottest in March and April, when however sea breezes are experienced, and the mean maximum day temperature is about $104°$ Fahr. ($40°$ C.). The lowest temperature is generally recorded in December and at

---

[1] Pérignon, *Haut Sénégal et Moyen Niger*, Paris, 1901.

night, the actual minimum being about 50° Fahr. (10° C.). Consult notes on the neighbouring coasts of Senegal, Gambia, Portuguese Guinea and French Guinea.

FRENCH GUINEA.

In French Guinea the rainy season is at its height in July and August, and the rains, owing to the presence of mountain chains, are greater on the coast than inland, these mountains acting as condensers. At Konakri no less than 7·87 inches (200 mm.) have been measured in 24 hours. The dry season extends from November to March, with perhaps a few showers in February and March, constituting the lesser wet season. There are almost nightly dews, which are very heavy, during the dry season, the humidity being high, and the horizon in the day time never clear: the sun is frequently obscured.

The mean annual temperature at Konakri is about 80°·6 Fahr. (27° C.) or 78°·8 Fahr. (26° C.), and is highest in April and May: it then falls till July or August, rises gradually again, but usually falls in December and January, to rise through February and March. The daily range of temperature on the coast is very small, and amounts to no more than 5°·4 Fahr. (3° C.) at Konakri, and elsewhere is never more than 9° or 10° 8 Fahr. (5° or 6° C.) at most; in the interior, however, it is naturally greater and affects Europeans considerably.

During the hot rainy season tornadoes and thunderstorms are frequent, especially on the coast. At Konakri 83 tornadoes were recorded in 1901, and 48 in 1902: thunder was recorded on 85 days in the former year and on 106 days in the latter. At Kissidugu there were only 17 tornadoes in 1902, and 25 days on which thunder was recorded. The rainy season is always heralded in by tornadoes, and storms also mark its close.

In the dry season the winds are usually between N.N.W. and S.S.W., and in the wet season S. and S.W. At the end of the rainy season or the beginning of the dry season the winds frequently come from the S.W., whilst from February to June they are almost constantly from the

N.W. These winds are not strong, and should a storm arise, it seldom lasts more than two or three days. In January, and sometimes in February, there is a hot east wind (Harmattan), which blows, at intervals, for about 30 days at Konakri, and for some 45 days on the Rio Nunez. Complete calm at Konakri is rare.

The heat of the coast regions would be quite bearable, were it not that the humidity is excessively high. The most trying times for Europeans are the beginning and end of the rainy season. For them the hot season is not unhealthy, for those at least who possess a fairly strong constitution, are moderate in their habits, and live a regular life ; but the wet season is dangerous, for then, especially at the beginning and end, fever is at its highest. But even the dry season is debilitating, and no European should overstay two wet seasons without returning home, preferably in the European summer, so as to avoid too sharp a transition.

The *Service Météorologique* was established throughout French West Africa in 1903–4, and a little volume has been published giving curves which represent the fluctuations of the various climatic elements for the years 1904 and 1905. With regard to the table given below, the decimal figures must be regarded as only approximate, having been read from the curves. The mean barometric pressure rose from 30·088 inches (764·2 mm.) in January 1905 to 30·140 inches or 765·5 mm. in February and remained fairly stationary till May, but it rose abruptly in June and July, reading 30·157 inches (766 mm.), and there was a similar, and higher, rise in 1904, but the corresponding fall was sharper in the former year, and by November the mean reading was 29·936 inches (760·4 mm.), whereas in 1905, after a slight fall in July and August, the average was steady till the end of the year, when the mean reading was 30·140 inches (765·5 mm.). The barometer rises a few millimetres on the approach of a tornado, and is about one millimetre (0·04 inch) above the mean at 10 a.m. and 10 p.m., and one or two below at 2 a.m. and 2 p.m.

The mean monthly temperature was very similar in the two years, the greatest difference occurring in December, when the difference between the means for 1904 and 1905 was as much as $5°$ Fahr. ($2°·8$ C.). The temperature is lowest in August in both years. The relative humidity was lowest in January and February, and highest in July and August. July and August are the rainiest months and from December to March there is usually no rain. In addition to the figures given by the *Service Météorologique* the rainfall for 1901 is also available.

In the neighbourhood of Kade, which is a slightly elevated plateau, and of the Fuladon plateau, the seasons are less marked than on the coast. The rainy season is shorter, and also a little later, but the dry season is hotter. The former only lasts four months, from 15th July to 15th November, and the latter from the end of December to the beginning of June. The intervening transition periods constitute the two tornado seasons, which are very much alike in character, except that at the change from the dry to the wet season these tornadoes are sometimes accompanied by destructive hailstorms. The mean day temperature in the wet season is about $71°\cdot6$ Fahr. ($22°$ C.), and the night mean $59°$ Fahr. ($15°$ C.). In the dry season the corresponding figures are $91°\cdot4$ Fahr. ($33°$ C.) and $62°\cdot6$ Fahr. ($17°$ C.), except in April and May, when the night mean is about $73°\cdot4$ Fahr. ($23°$ C.). The heat in the sun during those months is sometimes almost unbearable, so high a temperature as $147°$ and $156°$ Fahr. ($64°$ and $69°$ C.) having been recorded in April by the Boundary Commission. The tornadoes are perhaps a little less violent and less frequent at Kade than at Fuladon.

The minimum temperature observed at Lembu was $46°\cdot4$ Fahr. ($8°$ C.) at daybreak on 18th January : the maximum at Kebale (near Konakri) $98°\cdot6$ Fahr. ($37°$ C.) on 23rd January, with a hot wind. These are exceptional figures. As a rule the coldest nights average $59°$ Fahr. ($15°$ C.), and it is only on the lofty and denuded plateaux that the difference between maxima and minima is very great.

The factories on the Nunez river are very unhealthy after the wet season, when the river begins to subside.

On the plateau before reaching the actual slopes and Futa Jallon, the mean temperature is about $80°\cdot6$ Fahr. ($27°$ C.), with absolute maxima of $97°\cdot8$ Fahr. ($36°\cdot5$ C.) at Siarea and $92°\cdot3$ Fahr. ($33°\cdot5$ C.) at Kindia.

KONAKRI. Lat. 9° 4' N., Long. 13° 42' W. Alt. 53 feet.

Compiled from 2 years' observations.

	Jan.	Feb.	March	April	May	June	July	Aug.	Sept.	Oct.	Nov.	Dec.	Year
Mean temperature[1] (Fahr.°) ...	78°·8	79°·5	80°·0	81°·3	81°·0	79°·0	77°·4	76°·3	78°·1	79°·0	79°·7	78°·6	79°
Mean relative humidity[1] (per cent.) ...	61	62	68	75	75	82	87	87	81	80	78	73	75·7
Rainfall in 1904 (inches) ...	0	0	—	—	2·84	23·31	70·88	40·33	29·88	16·26	3·11	0	192·61
„ „ 1905 „ ...	0	0	0	0·79	2·48	18·03	57·49	55·68	27·56	27·24	3·90	0	193·97
„ „ 1901 „ ...	0	0	0	0·89	6·87	30·83	49·70	51·94	35·49	15·00	10·12	0·53	201·37
Number of days on which rain was measured (1904)	0	0	—	—	7	22	31	29	18	25	10	0	142
Number of days on which rain was measured (1905)	0	0	0	2	8	23	27	28	27	23	13	0	151

[1] 1904, 1905. The means for the two years were read off the curves published by the *Service Météorologique*: it is not stated how these means were obtained. Konakri was fitted up as a "Station de 1er Ordre" of the Meteorological Service in 1903 with an aneroid barometer, psychrometer, maximum—minimum thermometer, and rain-gauge.

TIMBO. Lat. 10° 40′N, Long. 11° 36′W. Alt. 1900 feet.

Compiled from 3 years' observations[1].

	Jan.	Feb.	March	April	May	June	July	Aug.	Sept.	Oct.	Nov.	Dec.	Year
Mean temperature[2] (Fahr.°)	71°·6	76°·3	80°·8	79°·7	76°·8	73°·4	72°·1	72°·1	72°·0	72°·9	72°·3	71°·1	74°·3
Mean maximum temperature	90°·9	95°·2	98°·1	96°·4	90°·3	85°·3	82°·8	81°·9	83°·5	85°·5	87°·3	89°·2	88°·9
Mean minimum temperature	55°·4	57°·9	65°·3	68°·0	67°·6	65°·3	65°·3	66°·0	65°·5	63°·1	61°·7	56°·1	63°·1
Mean daily range[3]	35°·5	37°·3	32°·8	28°·4	22°·7	20°·0	17°·5	15°·9	18°·0	22°·4	22°·6	33°·1	25°·8
Mean relative humidity[2] (per cent.)	43	40	50	62	72	82	86	88	86	83	77	60	69
Mean number of rainy days[4]	0	0	2·5	5·5	12·5	19·0	19·0	24·3	21·0	16·7	3·7	0	124·2

NOTE.—Timbo lies in the mountainous country, near the source of the Senegal river, about 150 miles from the nearest point on the coast. The annual results appeared in the *Annales* of the Bureau Central de Météorol. (Paris).

[1] 1897—1899 for August—December, the remaining months are for only two years.

[2] The means are derived from $\dfrac{7a + 3p + 10p}{3}$.

[3] The daily range is max,—min.

[4] The rain records are for the full three years.

In the Futa Jallon region, for some five and a half months in the year, between November and April, the wind blows from the north and north-east, and these winds are hot or fresh according to the time of day, but always dry, and frequently laden with red dust. From the month of March there is more moisture in the air than at St Louis or the hinterland of Konakri, and every traveller mentions the violent rains at this period. Plat mentions them at Timbo on the 11th January, and Fras at Dingire between February and April. In the winter the thermometer has been known to register as low as $39°·2$ Fahr. $(+ 4°$ C.). The mean maximum temperature is about $86°$ Fahr. $(30°$ C.). The absolute maximum recorded at Timbo is $100°·4$ Fahr. $(38°$ C.), and at Labe $96°·8$ Fahr. $(36°$ C.). In the middle of April comes the passage from the dry season to the wet, and there are changes of the wind to the south-west and south, accompanied by tornadoes. From the beginning of May, at Timbo, showers and storms are ceaseless, the latter being nearly always accompanied by hail, thunder, and lightning, and deluges of rain. The largest amount of rain falls from July to October, the storms growing less frequent, and the number of rainy days greater. The climate of Futa Jallon is more endurable by Europeans than that of either Senegal or the Guinea Coast[1].

FRENCH NIGER TERRITORY.

The climate of the Bend of the Niger is determined by a dry season with northerly winds, and a wet season with rains and storms coming from the south ; and the region of the Niger is a country of well-pronounced monsoons. During the northern summer, the burning Sahara is a centre of low-pressure, and the winds from the Atlantic blow from S.E. and S.W. bringing with them clouds which part with their moisture on coming in contact with the flanks of the highlands. This is the season of the rains. In the winter this movement is reversed, the ocean being then warmer than the land, and the N. and N.E. winds, blowing from the land, carry no clouds. The rainfall diminishes from S.

---

[1] *Bulletin du Comité de l'Afrique française*, Suppl. No. 7, 1900.

to N. and from W. to E., as the clouds part with much of their rain on the mountains of the coast. Naturally the rains set in later also in these directions. While they begin at Konakri and Tiassale at the end of May or the beginning of June, they do not reach the hinterland of Dahomey and Gurma much before August. Toutée says that two climates prevail in the Niger basin, the equatorial and the tropical. The former reaches almost to the 8th parallel, and has two wet and two dry seasons, while the other embraces a zone of about 10° and has only one wet and one dry season. The axis of the latter is the 13th parallel, along which rain falls almost daily from June to October. The temperature appears to be lower in the rainy season, owing to the constant humidity of the air, than it is during the burning desert winds of the winter. But this damp heat is more difficult for Europeans to endure, whereas the dry season is decidedly more healthy. The nights too are fresh. Decœur recorded 48° Fahr. (8°·9 C.) near Paraku at midnight towards the end of November, and 50° to 53°·6 Fahr. (10° to 12° C.) at about two o'clock in the morning on December 17th[1].

At Timbuktu the months of December, January and February are cold, especially at night. Between July and September thunderstorms are not infrequent. The rainy season, which is the hottest time of the year, extends from May to October, but the greater portion of the rain falls from June to September, and July is the rainiest month. It is during this rainy season, naturally, that there is most cloud. The highest recorded temperature is 117°·3 Fahr. (47°·4 C.) and the lowest 45°·9 Fahr. (7°·7 C.). In December the mean temperature at six in the morning is 57°·2 Fahr. (14°·0 C.), at 2 p.m. it is 84°·9 Fahr. (29°·4 C.), and in the evening at 9 o'clock 69°·1 Fahr. (20°·6 C.). In January and February these figures are slightly higher. In May, June and July the mean morning temperature is 79° or 80° Fahr. (26° or 27°C.), that at midday is from 104° to 109° Fahr. (40° to 43°C.), and in the evening, at 9 o'clock, it sinks to 90° or 91° Fahr. (32° or 33°C.).

In the Zinder region, immediately north of the Northern Nigeria boundary, the seasons are the same as in corresponding latitudes farther west, except that rain is rarer,

[1] *Scottish Geographical Magazine*, 1899, p. 194.

and is never experienced except during four months of the year, from May/June to September. In December and January the thermometer falls, as a regular thing, below 50° Fahr. (10°C.). This cold is usually accompanied by thick mist. The east winds, which prevail during the period from February, also bring mists and fogs of a lighter character. These winds are hot and parching, but the radiation is so great that the nights are fairly cool. On the whole the climate of Demagherim and Damerghu is not febrific, but the bad quality of the water renders dysentery frequent[1].

The rainy season, however, is unhealthy in the Zinder region, and, in a letter dated 4th July, it was stated that the rainy season had begun nearly two months before, and that all the members of the Boundary Commission were, or had been, ill in consequence.

With reference to the table on p. 174, Timbuktu is situated about 9 or 10 miles from the left bank of the Niger river, and only a little above the mean water level. In the rainy season there is connection with the river. The lakes lie to the west.

---

[1] Gaden, "Notice sur la Résidence de Zinder" (Extract from *Revue des Troupes coloniales*), p. 17. Paris.

TIMBUKTU. Lat. 16° 49′ N, Long. 2° 52′ E. Alt. 820 feet.

Compiled from 3 years' observations[1].

	Jan.	Feb.	March	April	May	June	July	Aug.	Sept.	Oct.	Nov.	Dec.	Year
Mean temperature (mean of observations at 6 a.m., 2 p.m. and 9 p.m.) Fahr.°	71°·2	74°·8	82°·6	90°·5	95°·0	93°·7	90°·9	88°·0	89°·2	87°·8	80°·2	70°·5	84°
Mean maximum temperature ...	86°·7	92°·3	99°·1	108°·0	111°·2	109°·0	106°·3	102°·7	105°·1	104°·5	96°·8	85°·6	100°
Mean minimum temperature ...	56°·8	57°·9	66°·7	73°·4	79°·3	80°·2	78°·4	76°·6	77°·0	73°·8	65°·3	56°·5	70°
Mean daily range[2] ...	29°·9	34°·4	32°·4	34°·6	31°·9	28°·8	27°·9	26°·1	28°·1	30°·7	31°·5	29°·1	30°
Mean of absolute monthly maxima ...	95°·2	100°·2	111°·0	115°·7	116°·6	116°·2	114°·1	112°·5	111°·6	110°·5	104°·4	96°·4	
Mean of absolute monthly minima ...	50°·2	49°·8	54°·3	64°·8	72°·9	73°·4	72°·3	72°·7	70°·5	66°·6	58°·3	46°·9	
Mean monthly range[3] ...	45°·0	50°·4	56°·7	50°·9	43°·7	42°·8	41°·8	39°·8	41°·1	43°·9	46°·1	49°·5	
Mean amount of cloud (0—10) ...	1·6	1·6	1·5	1·5	2·3	3·0	3·1	(1·8)	3·1	2·0	1·7	1·8	
Mean number of rainy days ...	0	0·7	3·0	0·5	3·5	8·7	11·0	4·0	5·7	3·5	0	0	40·6
Mean rainfall (inches) (3 years) ...	0	0	0·24	0	0·47	1·18	3·78	1·42	0·98	0·36	0	0	8·43
Mean rainfall* (inches) (from all available observations)	0	0	0·09	0	0·53	1·19	3·37	2·09	0·81	0·49	0	0	8·57

* The figures for January, February and March are for 6 years (1897—1902); April and June 5 years (April—1898—1902; June—1897—1900, and 1902); May, July, September, November and December 4 years (May—1898, 1899, 1901, 1902; July—1897, 1898, 1900, 1902; September—1897—9, 1902; November and December—1897—1900); August and October 3 years (August—1897, 1898, 1902; October—1898, 1899, 1902).

NOTE.—Timbuktu is now (since 1903) a "Station principale" of the Meteorological Service, but previous to 1903 the ordinary thermometers were in use, viz dry and wet bulb, maximum and minimum, and a Richards Thermograph.

[1] January—March, October, November 1896, 1898, 1899.   [2] The daily range is max.—min.

[3] The monthly range is abs. max.—abs. min.

WAGADUGU. Lat. 12° 26′ N., Long. 1° 30′ W. (approx.). Alt. 1060 feet.

Compiled from 4½ years' observations[1].

	Jan.	Feb.	March	April	May	June	July	Aug.	Sept.	Oct.	Nov.	Dec.	Year
Mean temperature (Fahr.°) (mean of 6a, 12 and 9p)	74°·1	74°·8	84°·6	87°·8	86°·2	81°·3	78°·8	77°·4	79°·2	81°·7	80°·4	75°·7	80°
Mean maximum temperature	88°·9	89°·8	99°·0	100°·2	97°·3	91°·2	87°·1	84°·9	88°·5	93°·0	94°·8	91°·0	92°
Mean minimum temperature	61°·3	62°·4	72°·1	77°·2	78°·3	73°·9	72°·1	71°·8	72°·1	73°·0	68°·2	62°·4	70°
Mean daily range[2]	27°·6	27°·4	26°·9	23°·0	19°·0	17°·3	15°·0	13°·1	16°·4	20°·0	26°·6	28°·6	22°
Absolute maxima	98°·6	99°·0	106°·7	108°·5	104°·9	100°·8	95°·9	93°·0	93°·2	98°·6	99°·0	96°·8	108°·5 (for period)
Absolute minima	49°·1	51°·6	59°·7	65°·1	66°·2	66°·0	66°·4	66°·2	64°·4	68°·2	60°·6	53°·8	49°·1 (for period)
Mean relative humidity[3] (per cent.)	32	23	29	49	56	70	79	85	78	68	51	38	55
Mean rainfall (inches)	0	0	0·10	1·81	2·48	4·53	6·20	10·67	5·02	1·26	0·04	0	32·11
Mean number of rainy days	0	0	1·5	4·5	9	11	13·2	18	12·4	7	0·2	0	77·0

NOTE.—Wagadugu lies inland, about 100 miles to the north of the Northern Territory of the Gold Coast Colony. The annual results appeared in the *Annales* of the Bureau Central de Météorol. (Paris).

[1] 1901—1905 (54 months).

[2] The daily range is max.—min.

[3] The mean is derived from $\dfrac{6a + \text{noon} + 9p}{3}$.

MEAN RAINFALL IN INCHES. (SUMMARY.)

	No. of Years	Jan.	Feb.	March	April	May	June	July	Aug.	Sept.	Oct.	Nov.	Dec.	Year
St Louis	14*	0·06	0·04	0·03	0	0·14	0·98	2·76	4·31	3·26	0·75	0·12	0·08	12·5
Bakel[1]	2	0	0	0	0·	0·20	7·83	5·51	6·89	6·42	0·39	0	0·47	27·71
Podor[2]	5	0·04	0·12	0	0	0	0·20	1·61	7·83	1·73	1·18	0	0	12·71
{Goree[3]	12	0	0·04	0	0	0	0·95	3·58	9·88	5·24	0·71	0·12	0	20·52
{Dakar[4]	3	0	0·18	0·08	0	0	0·41	2·24	10·80	3·40	2·26	0·18	0·34	19·89
Kayes[5]	5	0·004	0	0·04	0·09	0·71	4·19	8·12	8·71	4·94	2·24	0·004	0	29·05
Konakri[6]	3	0	0	0	0·83	4·06	24·06	59·35	49·30	30·98	19·49	5·71	0·20	193·98
Timbuktu[7]	(6)	0	0	0·09	0	0·53	1·19	3·37	2·09	0·81	0·49	0	0	8·57
Timbo (Futa Jallon)[8]	3	0	0	0·95	2·44	6·42	8·98	12·40	14·69	10·24	6·70	1·26	0	64·08
Wagadugu[9]	4½	0	0	0·10	1·81	2·48	4·53	6·20	10·67	5·02	1·26	0·04	0	32·11

* The latest 14 years (1892—1905).

[1] Taken from *Petermann's Mitteilungen*, Ergänz. 124, 1898, p. 75, where the years are stated to be 1856—62.
[2] 1892—1896.   [3] 1885—1896.   [4] 1903—1905.   [5] 1897—1899, 1904, 1905.
[6] 1901, 1904, 1905.   [7] For dates, see Meteorol. Table (above).   [8] 1897—1899.
[9] 1901—1905 (54 months).

IVORY COAST.

Speaking generally the seasons are ill-defined, but an approximation may be given to cover the whole coast area, which, however, gives only a rough idea, as the climate varies considerably throughout this region.

1. Dry season, from 15th November to 15th March, with hot day temperature and comparatively low night temperature, the extremes being about 95° and 59° Fahr (35° and 15° C.). In the early morning there is generally a mist, which clears off and is succeeded by a clear sky till about 2 o'clock in the afternoon. In the evening there are light clouds, but no rain. This is the healthiest season.

2. A lesser rainy season from 15th March to 15th May, with high temperature, and dull heavy atmosphere. Violent tornadoes occur, coming from the east, but with little rain. The nights are fresh. This is the worst season for Europeans.

3. A lesser rainy season from 15th May to 15th July, with clear sky and high barometer. Sometimes there are showers, especially in the morning. Every day there is a strongish breeze which lowers the temperature to a very considerable extent.

4. Rainy season from 15th July to 15th November, with high temperature, and atmosphere charged with vapour. Tornadoes occur, and regular and abundant rains fall both day and night.

The prevailing winds between 7° and 9° N. are from the east and north-east.

In the extreme west of the Ivory Coast, in the Lower Cavalla region the sky is always cloudy and the atmosphere saturated with moisture. South-west breezes are almost constant, the direction rarely changing, and, when it does so, it is always to the south-south-east. These breezes begin regularly at 10 a.m. and fall at night. The nights are fresh but moist, and mists—especially in the interior—are very frequent. Theoretically, being in the equatorial zone, there should be two rainy seasons, separated by intervals of comparative dryness; but this is not so, there being here but one rainy season, beginning in mid-April, and ending at the end of October. In this rainy season there is simply tornado upon tornado and the rains are extremely heavy, the rivers rising rapidly, and inundating their banks.

The following table gives the number of rainy days, and the mean maxima and minima of temperature:

Month	No. of rainy days	Mean temperature (Fahr.)	
		minima	maxima
January	2	$71°·1$	$80°·8$
February	3	$72°·3$	$82°·4$
March	7	$73°·9$	$83°·8$
April	13	$73°·9$	$80°·1$
May	16	$73°·6$	$80°·4$
June	23	$71°·2$	$76°·3$
July	11	$71°·2$	$76°·8$
August	15	$72°·1$	$75°·0$
September	21	$71°·6$	$75°·6$
October	16	$74°·3$	$77°·9$
November	9	$79°·2$	$84°·0$
December	3	$80°·2$	$85°·8$

In the dry season the barometer shows a mean height of 30·05 inches (763·25 mm.), the maximum being 30·16 inches (766 mm.), and the minimum 29·88 inches (759 mm.). During the rainy period the mean pressure is 30·16 inches (766 mm.), the maximum 30·28 inches (769 mm.), and the minimum 30·08 inches (764 mm.). The Ivory Coast is perhaps the most unhealthy of all the French possessions, and Europeans can only live there by observing the strictest hygienic rules[1].

Sassandra and Drewin are the two least unhealthy spots on the Ivory Coast, but the air is charged with moisture, and the houses and everything in them become very damp, and rheumatism is prevalent after a year's residence. There are four seasons, but the transition from one to the other is scarcely distinguishable. The dry season extends over January, February and March. The dry Harmattan, called by the natives Kokomlaka, blows for some days in January from the north-east, and lowers the morning temperature. It is not injurious to Europeans, but considerably so among the blacks. April is a very variable month. May, June and July constitute the rainy season. During

---

[1] Repiquet, *A. F.* Suppl. No. 11, 1903, p. 279.

the other months there is the season of less rain and that of tornadoes, which overlap. The rainy season is least oppressive to Europeans, for then the temperature falls, and there is no stagnant water. In the dry season the stagnant waters encourage the dissemination of disease [1].

In the basins of the Bandama and Komoe there are four well marked seasons: from January to March is the first dry season, when it is very warm, the thermometer varying, in the shade, between $82°{\cdot}4$ and $95°$ Fahr. or $28°$ and $35°$ C.: from April to July is the greater rainy season: from August to October the lesser dry season, during which the Harmattan is frequently experienced: from October to December the second rainy season, commonly called winter. The most healthy season throughout the Ivory Coast for Europeans is from the end of November to the end of March, the worst months being April and May [2].

In the upper or northern Ivory Coast there are two seasons, wet and dry. The former begins in May with storms which gradually increase in frequency and violence; in August the rain is, so to speak, continuous, and sometimes does not stop for forty hours. During these storms the rain falls in great sheets of water over all the region, and with such force that it roots up the largest trees, and, as the soil is clayey and thus only slightly permeable, the tiny streams become uncrossable rivers; the roads are inundated continually. After each storm the sun reappears and quickly dries up the soil. In October the fine season begins. Dr Lemasle states that the temperature is not excessively high, the mean being about $100°{\cdot}5$ or $104°$ Fahr., *i.e.* $38°$ or $40°$ C., but, unfortunately, the heat is moist and oppressive, and dysentery, &c., develop rapidly [3].

On the Ivory Coast no European should prolong his first stay over 18 months, and subsequent periods of residence should be limited to 15 months, with breaks of 5 or 6 months for return to Europe. Grand Bassam is very unhealthy, but the following places on the coast are better, viz. Bingerville, Grand Lahu, Assinie, Jacqueville, and, as already mentioned, Sassandra and Drewin; and also the hinterland north of the forest.

[1] Thomann, *A. F.* Suppl. 6, 1901.
[2] *La Côte d'Ivoire*, "Moniteur des Colonies," Paris, 1903, pp. 31, 32.
[3] Van Cassel, *A. F.* Suppl. 5, 1901.

GRAND BASSAM. Lat. 5° 12' N., Long. 3° 44' W. Alt. 10 feet.

		Jan.	Feb.	March	April	May	June	July	Aug.	Sept.	Oct.	Nov.	Dec.	Year
Mean temperature (Fahr.°) (5) years[1]		80°·2	80°·8	82°·4	83°·3	82°·6	80°·1	78°·8	77°·4	78°·8	80°·4	82°·0	82°·2	80°
Mean relative humidity per cent. (2) years[2]		83	84	86	85	88	90·1	90	90·1	90·1	88	84	85	87
Rainfall in 1904* (inches)		1·04	0·16	6·87	5·72	22·80	30·08	1·46	0·91	0·28	0·71	2·05	4·68	76·76
„ „ 1905		2·58	0·36	1·46	3·05	18·00	41·99	3·62	0·51	0·59	5·79	16·89	0·95	95·79
No. of days on which rain was measured (1904)		4	2	10	6	21	18	6	11	7	12	16	11	124
No. of days on which rain was measured (1905)		4	2	4	10	16	28	16	10	10	12	16	5	133
* There is also an old 1859—60 record		4·33	6·58	13·07	19·41	55·64	42·29	24·13	2·05	4·76	25·55	46·74	3·98	248·53

NOTE.—Grand Bassam was equipped as a "Station principale" of the Meteorological Service in 1903 with barometer, thermometers, and psychrometer.

[1] 1858—9, 1863, 1899, 1904, 1905.   [2] 1904, 1905.

For Grand Bassam, since the establishment of the *Service Météorologique* in 1903/4, there are curves of barometric pressure, temperature, relative humidity, and rainfall for the years 1904, 1905. These curves go to show that the pressure was much greater in the former year than in the latter, except in November and December, though the forms of the two curves are very similar, dipping slightly in the early part of the year and rising suddenly from March/April to the highest points in July/August; but the fall from this point to the November/December position was much more gradual in 1905 than in 1904. The two temperature curves are very similar from May to October. To combine with these curves we have the figures for temperature for 1899 published in *La Côte d'Ivoire*, by Villamur and Richaud, and also the figures for the years 1858/9 and 1863. In the year 1899 the temperature was somewhat higher throughout, except in December, than in 1904 and 1905. In 1899 the minimum temperature recorded was 68° Fahr. (20° C.) in January, and the maximum 100°·5 Fahr. (38° C.) in January and April. In the same year nearly 117 inches or 3 metres of rain were measured. In the table given on p. 180 for Grand Bassam, the decimal figures are only approximate, having been read off from the curves. The relative humidity was very even throughout the years 1904 and 1905.

To compare with the temperature at Grand Bassam we have the following figures, derived from three years' observations, for the mean temperature at Assinie :

January	82°·4 Fahr.	July	79°·2 Fahr.
February	82°·0	August	79°·5
March	83°·5	September	78°·6
April	84°·6	October	81°·1
May	84°·6	November	82°·9
June	84°·6	December	82°·8

which give a mean for the year of 82°·7 Fahr. (28°·2 C.).

(For Productions *see* under West Coast.)

DAHOMEY.

Dahomey consists of a narrow strip of country, which, to the north of $10°$ N., broadens out so as to include Gurma. The country rises very gradually from the Gulf of Guinea to the 10th parallel, where it reaches an altitude of some 1200 or 1300 feet, and then slopes down towards the Niger. From the 10th parallel on the west to the Niger runs the Atakora range in a general N.N.E. direction, but the culminating points of these highlands never rise higher than 2500 feet. The eastern and northern parts of Gurma also slope down to the Niger, while the southern, middle, and western portions incline towards the Volta river. Thus there are three slopes, viz. those towards the Gulf of Guinea, the Niger and the Volta respectively.

In Dahomey, as along all this coast, the division of the season is not the same on the coast as in the hinterland. On the coast there are two dry seasons and two rainy seasons. The greater rainy season lasts from 15th March to 15th July: this is the season of violent tornadoes, especially in April and May. The lesser dry season extends from 15th July to 15th September: this season is characterised by thick fogs. The lesser rainy season begins on 15th September and continues to about the beginning of December, and the greater dry season from then to 15th March. The climate is hot and moist, the difference between the day and night temperatures is not very marked, being $5°—7°$ Fahr. or $3°$ or $4°$ C. at most. During the daytime a breeze from the sea usually tempers the heat, while at night the breeze is from the land. The mean temperature is about $79°$ Fahr. or $26°$ C. In Upper Dahomey the rainy season begins in June and lasts till about the middle of October, and the dry season extends from the middle of October to the end of May. During the rainy season the difference between the day and night temperature does not exceed $9°—11°$ Fahr. or $5°$ or $6°$ C. On the contrary, during the dry season the difference amounts to as much as 27° or even 29° Fahr. (15° or 16° C.) in December and January[1].

Speaking generally the climate of Lower Dahomey is moist and hot throughout the year, and Europeans soon feel the results of this. In fact Lower Dahomey is unhealthy. In January the Harmattan blows from the N.N.E. and produces great dryness. It is a fresh wind in the morning, hot in the day, and cold at night, and brings with it clouds of fine sand and dust, partially obscuring the sun. The natives have a saying " When the Harmattan blows it is cold," and they shiver accordingly. This is the most dangerous season for Europeans, and every precaution should be taken.

In Borgu, in the Dahomey hinterland, the year is divided into two seasons, the dry from 15th November to 15th April, and the rainy from 15th April to 15th November. In some years, however, there is a lesser dry season from 15th May to 15th July. During December and January it is not unusual to see the thermometer at 43° or 46°·4 Fahr. (6° or 8° C.) at night, and 89°·7 or 95° Fahr. (32° or 35° C.) the same day in the shade at noon. The mean temperature for the year varies from 78°·8 to 80°·6 Fahr. (26° to 27° C.) and is quite endurable by Europeans. In rooms shut up during the day the mean temperature is only from 68° to 71°·6 Fahr. (20° to 22° C.).

M. Garrerès, *Administrateur des Colonies*, gives the following general account of the climate of Borgu :—" The beginning and end of the rainy season are characterised by frequent violent storms. The sky is then continuously overcast and the dry tornadoes are accompanied by furious thunder claps. But, when the season is fairly set in, the rains are, most frequently, of short duration, though heavy. The sky becomes suddenly darkened, and as suddenly clears again; the sandy soil soon absorbs the water and the air is fresher.

" In the rainy season the temperature is quite supportable with a mean of from 71°·6 to 78°·8 Fahr. (22° to 26° C.). In the dry season, when the thermometer readings are

---

[1] Lorho, *Historique de notre expansion dans L'Hinterland Dahoméen*, Paris, 1904.

$93°·3$ and $96°·9$ Fahr. ($34°$ and $36°$ C.) the heat is more fatiguing. The nights are then stifling, except in January, when readings of $46°·4$ and $50°·0$ Fahr. ($8°$ and $10°$ C.) are not infrequent.

"The Harmattan, as a drying wind from N.N.E., blows in January and February. The air is filled with impalpable dust, through which the sun shines, as if through frosted glass, and the effect is distressing to both animal and vegetable life. The whole country is covered with, as it were, a black mantle. The intense dryness, however, causes rapid evaporation, which, in turn, produces a temperature which is almost cold[1]."

In the extreme north of Dahomey the climate is almost that of the Sudan, *i.e.* tropical, with two seasons, the dry from November to May, and the wet from June to October. The rainy season diminishes sensibly, both in the amount of rain and in duration, on going northwards. The mean temperature varies between $77°$ and $95°$ Fahr. ($25°$ and $35°$ C.) during the day, and between $68°$ and $77°$ Fahr. ($20°$ and $25°$ C.) at night. In December and January there are fresh nights, when the thermometer falls to $53°·6$—$59°$ Fahr. ($12°$—$15°$ C.), but during the Harmattan it rises to $104°$ Fahr. ($40°$ C.). The rain begins to come in storms, at first with long intervals, which gradually become less and less, the storms becoming more frequent and more violent. There is usually a brief drier season towards July, and then the rains become regular. In 1901 and 1902 the rainfall at Karimama was more than double that at Gaya. From May to October the atmosphere is highly charged with electricity, and telegraphic communication suffers in consequence[2].

Since the establishment of the *Service Météorologique* in 1903/4 the following information has been published.

At Porto Novo the relative humidity appears from observations taken in 1904 and 1905 to be fairly constant and to vary between 69 and 79 per cent. It is highest in June and July and lowest in February, and not far from 75 in January, March, April and the first half of May, and also in September; and about 80 in October and November. The following is the rainfall report:

1 *Revue Coloniale*, 1908, p. 515.
2 Drot, *B. S. G.* X. 1904, p. 274.

	1904		1905	
	rainfall (inches)	days	rainfall (inches)	days
January	0	0	0	0
February	0	0	0	0
March	—	—	1·62	6
April	3·16	6	3·94	5
May	8·17	14	4·92	10
June	—	—	11·14	20
July	7·20	16	11·37	10
August	7·10	13	0	0
September	—	—	0·16	11
October	—	—	9·41	19
November	6·39	not stated	2·84	9
December	0·05	,,	0·95	3
Totals	...	...	...	46·35	93

It should be remembered that the western part of the Dahomey coast is included in the peculiarly dry strip of coast land between Cape Three Points and Wida, already alluded to under Gold Coast (see p. 116).

In 1905 the mean temperature ranged between 77°·5 and 86° Fahr. (25°·3 and 30° C.). In January it was 83°·8 Fahr. (28°·8 C.); in April, the hottest month, 86° Fahr. (30° C.); in October, the coldest month, 77°·5 Fahr. (25°·3 C.); and in December, 78°·1 Fahr. (25°·6 C.); and the rise and fall between these points was almost absolutely regular. In the previous year it was a little cooler in April and May, August and December. The records for the other months are wanting.

The following table is compiled for the three years 1897—1899, and it will be noticed that the rainfall is almost identical with that recorded in 1905. The April heat in 1905 was greater than the three-year mean, while the mean temperature for October, the coolest month, in 1905, is higher than that of July, August and September. The highest temperature recorded is 98°·8 Fahr. (37°·1 C.) and the lowest 60°·3 Fahr. (15°·7 C.) giving an extreme range, for the period, of 38°·5 Fahr. or 21°·4 C.

(For Productions *see* under West Coast.)

PORTO NOVO. Lat. 6° 28′ N., Long. 2° 40′ E. Alt. 65 feet.

Compiled from 3 years' observations[1].

	Jan.	Feb.	March	April	May	June	July	Aug.	Sept.	Oct.	Nov.	Dec.	Year
Mean maximum temperature (Fahr.°)	92°·7	93°·6	94°·8	93°·7	92°·3	89°·4	86°·9	86°·0	87°·6	89°·6	93°·0	92°·5	91°
Mean minimum temperature	71°·4	74°·5	75°·9	74°·5	73°·4	72°·0	72°·3	72°·0	72°·7	73°·2	72°·9	72°·5	73°
Mean temperature[2]	82°·0	84°·0	85°·3	84°·1	82°·9	80°·6	79°·6	79°·0	80°·1	81°·4	82°·9	82°·4	82°
Mean daily range[3]	21°·3	19°·1	10°·9	19°·2	18°·9	17°·4	14°·6	14°·0	14°·9	16°·4	20°·1	20°·0	18°
Mean of absolute monthly maxima[4] ...	96°·3	97°·0	97°·3	97°·3	96°·8	94°·5	91°·6	91°·2	91°·2	93°·4	95°·7	96°·6	
Mean of absolute monthly minima[4] ...	63°·0	68°·7	70°·5	68°·0	69°·8	68°·4	68°·5	68°·7	69°·6	68°·2	70°·0	66°·0	
Mean monthly range[5]	33°·3	28°·3	26°·8	29°·3	27°·0	26°·1	23°·1	22°·5	21°·6	25°·2	25°·7	30°·6	
Mean relative humidity (per cent.) at 8 a.m.	86	86	82	81	82	85	85	86	86	85	87	88	85
Mean relative humidity (per cent.) at 4 p.m.	62	67	70	72	74	79	77	78	77	76	73	70	73
Mean number of rainy days	2·3	2·5	6·3	9·2	15·5	13·0	13·3	12·7	14·3	16·0	5·0	2·7	112·8
Mean rainfall in inches	0·59	1·61	2·60	3·86	8·78	8·27	4·92	1·97	3·47	8·86	1·69	0·36	46·98

NOTE.—Porto Novo is situated on the north side of the lagoon of the same name, and about 6 miles from the coast.

[1] 1897, 1898, 1899. Observations were taken at 8 a.m. and 4 p.m.

[2] The mean temperature is derived from $\dfrac{\text{max.} + \text{min.}}{2}$.

[3] The daily range is max.—min.

[4] The absolute maximum recorded was 100°·76 and the absolute minimum 57°·56.

[5] The monthly range is abs. max.—abs. min.

GERMAN WEST AFRICA

TOGOLAND.

In Togoland, during July and August, southerly and south-westerly winds prevail, and, coming, as they do, from the ocean, they render these months the coolest. The winds from the north make the period from January to March the warmest and driest. This rule holds good, not so much for the actual littoral, as for the regions lying behind the coast. During December and in the first half of January, the Harmattan blows, and, in the coast districts, takes the form of a dry, dust-laden wind, which dies away in the evening, but begins afresh early in the morning.

Taking the coastal region as a whole, the mean annual temperature is $78°\cdot8$ Fahr. ($26°$ C.) and that of the interior about $73°\cdot4$ Fahr. ($23°$ C.). Near the coast there are two well-marked rainy seasons, namely from April to July and from September to the end of November; and these are separated by dry seasons. The latter rainy season, in some years, occurs only in a mitigated form, and such years are dry years, and the harvest is correspondingly small. The farther the distance from the coast, that is towards the north, the less marked is the dryness about the middle of the year ; that is to say the two rainy seasons gradually tend to approach one another, merge, and form one. In the north of the four-season zone the autumn wet season brings, as a rule, more rain than that earlier in the year, while in the south the reverse is the case. The rainfall on the coast amounts to some 27·56 inches (700 mm.), annually, while in the interior the figure rises to from 1200 mm. to nearly 1500 mm., that is from about 47 to 59 inches. This yearly rainfall increases fairly quickly from the narrow dry coastal belt, and in the west of the colony it is greater than in the east, evidently because the west is a mountainous region. It should be remembered, in this connexion, that the Togo coast falls within the peculiarly dry—that is, of course, comparatively dry, when the equatorial situation is taken into account—coastal belt, extending eastwards from Cape Three Points through the Gold Coast to the Western parts of the Dahomey coast. In Togo the lines of equal rainfall over 40 inches (1000 mm.) would appear, in the southern portion of the area, to have a S.W.—N.E. direction, so that the poorer rain zone of South Togo, that is the zone of the oil-palms, stretches much farther inland on the east than on the west. Dr Hupfeld traces the abnormally high precipitation at Amejove to local conditions, the comparatively elevated situation (2500 feet), and the character of the vegetation (virgin forest) in the neighbourhood.

A glance at the table of rainfall (p. 192) will show that, as mentioned before, there are maxima of rainfall, corresponding with the two periods when the sun is near the zenith, the first between May and June, and the second in August, September or October. The two minima occur in December, January or February, and between July and August, or in September. But this is no longer the case farther north when the latitude of Sansanne Mangu is reached. Here there is but one rainy season and one dry season, as is also the case at Gambaga, in the northern territories of the Gold Coast, a trifle further north. That is to say these latitudes are beyond the equatorial four-season zone.

The health conditions of Togoland are, as a general rule, satisfactory, and a recent official report states that "the sanitation of the larger towns makes good progress, the appointment of a government doctor at Palime meets a real need,...Four cases of yellow fever remain isolated, though the epidemic has recently broken out in Dahomey and has made quarantine regulations necessary."

At Kpeme the mean of the barometric readings oscillates between 30.06 inches (763.7 mm.) in July and 29.72 inches (755 mm.) in March. The mean maximum temperature is lowest in June, July, August and September, and is over 87°.9 Fahr. (31° C.) for the remainder of the year, reaching 91°.7 Fahr. (33°.1 C.) in March, while so high a figure as 96°.1 Fahr. (35°.6 C.) is sometimes registered. The mean minimum temperature varies between 66°.2 and 73°.4 Fahr. (19° C. and 23° C.), the greatest daily range occurring in November, December, January and February, while the range is least for the period June to September. The highest temperatures registered during the period for which the table below was calculated were in 1902, viz. $95°·7$ Fahr. $(35°·4$ C.), and in 1903, viz. $98°·6$ Fahr. $(37°·0$ C.), and the lowest was $59°·9$ Fahr. $(15°·5$ C.). The hottest time of the day is about 2 p.m. The relative humidity varies, in the morning, between 93 per cent., in March and September, and 97 per cent. in May, and falls from 10 to 30 per cent. in the middle of the day, the greatest fall being towards the end and the beginning of the year, with only 10 to 14 per cent. fall in July. In the evening the relative humidity varies between 83 per cent. in January and 94 to 96 per cent. in July, August and September.

Bismarckburg is situated at an altitude of 2330 feet, and the mean annual temperature is $76°·8$ Fahr. $(24°·9$ C.). The highest and lowest temperatures registered are $99°·6$ Fahr. $(37°·5$ C.) and $52°·7$ Fahr. $(11°·5$ C.). The mean temperature is below $75°$ Fahr. $(24°$ C.) for the months of July, August and September, and during the same months the mean maximum temperature is below $82°$ Fahr. $(28°$ C.). The mean minimum is also lowest during the same period, being about $70°$ Fahr. $(21°$ C.), and during the early months of the year. The mean minimum is highest in May and June.

From December to February the prevailing winds are between N.E. and N.W.; from March to May they are between S. and S.W.; from June to August from S.W. to N.W., and from September to November the prevailing wind is from N.E., though there are also frequent winds from the E., N.W. and W.

Over 5 inches of rain fall in each of the months from April to October (except August with over 4 inches), the heaviest fall, 10·7 inches (272 mm.), occurring in September. The average for March is 3·3 inches (84 mm.) and the dryest month is November with nearly ·8 of an inch (20 mm.).

FREQUENCY OF WIND (PER CENT.).

	N.	N.E.	E.	S.E.	S.	S.W.	W.	N.W.
December—February	25	19	7	9	2	9	8	21
March—May	6	4	7	8	19	27	19	10
June—August	3	2	1	4	7	25	38	20
September—November	9	37	14	5	4	6	11	14

(For Productions *see* under West Coast.)

BISMARCKBURG. Lat. 8° 12′ N., Long. 0° 51′ E. (approx.). Alt. 2330 feet.

Compiled from 6 years' observations[1].

	Jan.	Feb.	March	April	May	June	July	Aug.	Sept.	Oct.	Nov.	Dec.	Year
Mean temperature[2] (Fahr.°) ...	77°.9	79°.7	78°.6	78°.8	78°.1	75°.6	73°.6	73°.6	74°.3	75°.2	78°.3	77°.7	76°.8
Mean maximum temperature[3] ...	86°.4	88°.7	89°.2	88°.7	86°.7	84°.0	80°.2	80°.2	82°.0	83°.5	86°.4	86°.9	85°.3
Mean minimum temperature ...	70°.2	71°.8	70°.9	69°.4	72°.1	70°.7	69°.6	69°.6	69°.8	71°.6	71°.4	71°.2	70°.7
Mean daily range[4]	16°.2	16°.9	18°.3	19°.3	14°.6	13°.3	10°.8	10°.6	12°.2	11°.9	15°.0	15°.7	14°.6
Mean of absolute monthly maxima[5]	90°.0	93°.0	93°.2	94°.8	91°.8	87°.8	85°.6	84°.6	85°.8	87°.8	89°.2	89°.2	
Mean of absolute monthly minima[6]	66°.0	66°.6	66°.7	64°.6	69°.1	68°.0	67°.5	67°.5	66°.0	67°.3	67°.8	60°.8	
Mean monthly range[7]	24°.0	26°.4	26°.5	30°.2	22°.7	19°.8	18°.1	17°.1	19°.8	20°.5	21°.4	28°.4	34°.5 (extreme)

The annual results, from which this and the following Tables were compiled, appeared from time to time in *Mitteilungen aus den Deutschen Schutzgebieten*.

[1] Jan. 1888—June 1893.

[2] The mean temperature is derived from $\dfrac{6a + 2p + 9^p}{3}$.

[3] The mean maximum is wanting for April—July, 1890.

[4] The daily range is max.—min.

[5] The absolute maximum is wanting for April—July 1890, and December 1891. The highest recorded temperature was 99°.68 (March 1892).

[6] The absolute minimum is wanting for December 1891. The lowest recorded temperature was 52°.70.

[7] The monthly range is abs. max.—abs. min.

KPEME. Lat. 6° 13′ N., Long. 1° 32′ E. Alt. 13 feet (on the coast).

Compiled from 3 years' observations.[1]

	Jan.	Feb.	March	April	May	June	July	Aug.	Sept.	Oct.	Nov.	Dec.	Year
Mean maximum temperature[2] (Fahr.°)	90°·3	91°·0	91°·6	91°·2	91°·0	87°·3	84°·6	83°·7	86°·7	89°·4	92°·3	91°·6	89°·2
Mean minimum temperature	70°·2	71°·1	73°·6	73°·2	71°·4	70°·9	68°·9	67°·5	69°·8	70°·3	72°·1	70°·9	70°·8
Mean temperature[3]	80°·2	81°·0	82°·6	82°·2	81°·3	79°·2	76°·6	75°·6	78°·3	79°·9	82°·2	81°·3	80°·0
Mean daily range[4]	20°·1	19°·9	18°·0	18°·0	19°·6	16°·4	15°·7	16°·2	16°·9	19°·1	20°·2	20°·7	18°·4
Mean of absolute monthly maxima[2] ...	94°·3	95°·0	93°·6	95°·0	93°·0	91°·2	89°·1	86°·9	89°·4	92°·1	94°·6	95°·9	
Mean of absolute monthly minima ...	61°·7	62°·1	66°·6	66°·9	66°·0	65°·3	65°·3	64°·6	66°·2	66°·4	68°·5	64°·9	
Mean monthly range[5]	32°·6	32°·9	27°·0	28°·1	27°·0	25°·9	23°·8	22°·3	23°·2	25°·7	26°·1	31°·0	
Mean relative humidity (per cent.) ...	84	81	83	84	86	87	89	90	88	86	87	84	86

NOTE.—Kpeme is fully equipped as a 2nd class meteorological station.

[1] 1902, 1903, 1904.
[2] The records for January and February 1902 are wanting.
[3] The mean temperature is derived from $\dfrac{\text{max.}+\text{min.}}{2}$.
[4] The daily range is max.—min.
[5] The monthly range is abs. max.—abs. min.

MEAN TEMPERATURE AT ZEBE, AMEJOVE AND MISAHÖHE.

		Lat. N.	Alt. (feet)	Jan.	Feb.	March	April	May	June	July	Aug.	Sept.	Oct.	Nov.	Dec.	Year
Zebe	... ...	6° 15′	16	81°·5	82°·2	82°·6	80°·2	80°·2	77°·7	75°·2	75°·4	76°·1	79°·9	81°·9	82°·6	79°·6
Amejove	... ...	6° 50′	2518	73°·0	73°·8	73°·9	72°·7	71°·8	69°·1	67°·8	67°·3	68°·7	70°·3	72°·3	72°·1	71°·1
Misahöhe	... ...	6° 56′	1929	76°·3	77°·9	77°·7	76°·8	76°·5	73°·9	71°·4	71°·1	72°·1	73°·9	76°·1	76°·5	75°·0

The above figures are given in an article on the Climate of Togo, by Dr H. Maurer, which appeared in *Mitteilungen aus den Deutschen Schutzgebieten*, 1907, p. 115, but no mention is made of the period for which the means were calculated, nor by what method they were arrived at.

MEAN ANNUAL RAINFALL IN TOGO (IN INCHES).

According to the results of measurements made at stations, where observations have been taken for at least 5 years.

Stations	No. of Years	Height above the sea feet	Jan.	Feb.	March	April	May	June	July	Aug.	Sept.	Oct.	Nov.	Dec.	Total for Year
Lome	8[1]	15	0·24	1·14	1·69	4·25	5·16	6·18	1·30	0·08	1·42	2·48	0·91	0·32	25·19
Kpeme	5[2]	15	0·55	0·51	1·65	2·72	4·76	5·47	2·64	0·43	2·13	2·56	1·18	0·16	24·76
Anecho	11[3]	15	0·51	0·67	2·72	4·33	6·14	8·58	1·85	0·28	1·02	3·43	1·61	0·39	31·55
Amejove	7[4]	2530	0·47	3·43	3·47	6·46	6·73	9·02	8·74	6·77	8·74	7·60	1·93	1·93	65·29
Misahöhe	12[5]	1935	0·98	2·13	3·66	5·04	5·79	8·74	8·35	5·95	6·18	7·21	2·95	1·85	58·83
Tafie	6[6]	885	1·06	1·02	2·21	3·78	5·79	7·87	6·58	2·91	3·58	6·58	2·91	1·50	45·80
Worawora	6[7]	655	0·28	2·24	3·07	5·16	5·71	6·54	6·26	7·64	7·44	7·36	2·40	1·65	55·75
Atakpame	6[8]	1080	0·43	2·99	2·21	5·24	7·52	7·83	9·41	7·01	6·58	5·47	0·95	0·43	56·07
Kete Krachi	10[9]	360	0·36	1·14	2·01	4·84	7·25	7·05	6·70	5·99	8·31	6·14	1·34	0·51	51·64
Bismarckburg	9[10]	2330	1·38	1·89	3·31	5·39	6·73	6·97	6·10	4·45	10·71	5·75	0·79	1·22	54·70
Sokode	5[11]	1340	0·04	0	2·05	4·21	7·36	6·18	5·12	12·68	8·78	4·65	0·32	0·16	51·55
Basari	5[12]	985	0·08	0·08	0·98	3·78	6·61	5·71	6·50	8·39	9·13	6·85	0·87	0·28	49·26

[1] March—December 1892, January—April 1893, March—December 1900, 1901—1905.
[2] 1901—1905.
[3] June—December 1891, 1892, January—March 1893, 1897, September—December 1899, 1900, January—March and July—December 1901, 1902—1905.
[4] March—December 1894, 1895—1899, January—July 1900.
[5] October—December 1890, 1891, September—December 1892—1894, January—March, July, August 1895, August—December 1900—1902, January—October 1903—1905.
[6] 1900, 1901 (except February), 1902—1905.
[7] May—December 1898, 1899—1903.
[8] March, April, June—December 1899, 1900, 1901, February—December 1902, March—December 1904, 1905.
[9] July—October 1895, April, September 1896, March—December 1898, 1899—1905.
[10] June—December 1888, 1889—1893, January—May 1894, December 1896, January—November 1897.
[11] 1901—1905.
[12] 1901—1905.

CAMEROONS.

In the Cameroons coast region the climate is, of course, tropical, and it is distinguished, moreover, by a continual damp warmth and extraordinary rainfall. Cameroons however is not an offensively hot country, the heat of the coastal regions being tempered by the cold Benguela current coming northward from the Polar regions and skirting the whole of the west coast of Africa as far as the Bight of Biafra. The mean annual temperature is $78°$ Fahr. ($25°·5$ C.); in February, the warmest month, it rises to $81°$ Fahr. ($27°·3$ C.), and falls in July, the coolest month, to $76°$ Fahr. ($24°·4$ C.), the maximum and minimum recorded temperatures being $89°·7$ Fahr. ($32°$ C.) and $66°$ Fahr. ($19°$ C.). At Duala the mean temperature is $77°·5$ Fahr. ($25°·3$ C.), at Barombi $74°·8$ Fahr. ($23°·8$ C.), at Victoria $77°$ Fahr. ($25°$ C.), and at Lolodorf $72°·7$ Fahr. ($22°·6$ C.).

The Cameroons coast is one of the most unhealthy places in Africa, and, indeed, in the world, especially in the forests, but it is considerably better in the high-lying districts to the north. Speaking generally, Europeans cannot live on the coast more than three years without returning to Europe. The rainy season, from a health point of view, is always preferable to the season which intervenes between the rains, during which the sun shines with terrible intensity on the rain-saturated country.

The rainfall amounts to 9 metres (nearly 30 feet) a year at Bibundi, one of the highest records in the world after the 12 metres (over 39 feet) in Assam; at Duala there are 4 metres (13 feet) annually, and even $2·75$ metres (or nearly 9 feet) at Baliburg on the Cross river and $1·53$ metres (5 feet) at Yaunde, which is 150 miles from the sea.

In the savannah country, the mean temperature is only $72°·5$ Fahr. ($22°·5$ C.) at Yaunde, and $64°·4$ Fahr. ($18°$ C.) at Baliburg in north-west Cameroons, and at night the thermometer goes down to $44°·6$ Fahr. ($+ 7°$ C.) and even $42°·8$ Fahr. ($+ 6°$ C.).

The rainy seasons, as might be expected in a country which extends over some $11°$ of latitude and is of such varying relief, are very different in the various provinces, and often vary from year to year. The chief rainy season, generally, falls between May and October. On the coast the rains are abundant in May, continue to increase till the end of August, and by the early part of October they end. Very dense vapours are experienced when the Harmattan blows, especially in November. During December and January there are frequent fogs, lasting from about midnight till ten o'clock in the morning. Frequent and violent tornadoes occur about the end of October and the beginning of November, and also in April and May. These storms generally blow from the land, but the most violent come from seawards. During the dry season there are regular sea-breezes. Towards the end of November there are storms with rain, which sometimes last for several days together. The fine-weather season is from December to February, and it is during these months that the Harmattan is experienced.

Dr Plehn gives the following description, beginning with the turn of the year and the dry season. Behind the dark mist the rising sun remains hidden for a long time. The grass is saturated with dew, which disappears as the sun gets higher. The land-breeze, which has blown all night, with considerable force, grows weaker and weaker and dies away altogether at about eight o'clock. Then begins the most intolerable time in Cameroons. A thick misty greyness hangs over the river and does not disappear when the sun rises, and the Cameroons mountains lie shrouded in thick mist, sometimes for several weeks together. Notwithstanding the decrease in relative humidity, which now varies between 73 and 76 per cent., the air at midday, especially on the banks of the river at ebb tide, has an intolerable, heavy feeling, with its temperature at from 86° to 88° Fahr. (30° to 31° C.), before the sea-breeze springs up. At about one o'clock in the afternoon the south-westerly sea-breeze blows fairly strongly and brings relief, in spite of the heat. This breeze frequently brings with it heavily-laden clouds, which ward off at times and to some extent the intensity of the sun's rays. At this time of the year rain seldom falls, sometimes not for three or four weeks together. In spite of this, vegetation does not suffer much, and the only sign of the lack of rain is found in the sinking of the river-level. Sheet lightning and distant thunder are frequent towards evening, but heavy thunderstorms are rare at this season. The sea-breeze blows till late in the evening. The sunrises and sunsets are of special beauty, the nights are mostly cloudless and a great red-gold halo surrounds the moon. Not long after sunset the sea-breeze dies, giving place, after a few hours, to the easterly land-breeze, which continues till morning. Only for a short time does this description of a typical dry season hold good. For a week before this season begins, and also before its change to the tornado time of spring, days of this kind alternate with such as characterise the so-called transition period, the tornado time, marked by heavy thunder-clouds and sudden downpours.

The thunderstorms, at first irregular, become more frequent and of greater violence in March and April. The summer lightning and distant thunder in the east increase, and full rainy days become more frequent between the hot sunny days. In this transition period the days are most trying, with violent short deluges of rain and intensely hot sunshine alternating.

The further this transition period advances the rarer become the sunny days, and the precipitation assumes a more regular character.

The tornadoes become fewer and weaker, the thunderstorms less violent, the sunsets are clearer, and the nights too, when not obscured by rain-clouds, the easterly land-breezes grow in force, the rains come mainly from the sea, and finally the transition period gives place to the rainy season, which reaches its height between June and August. The sun is now seldom seen and rain falls almost incessantly, but with greater intensity by night than during the daytime. Everything is enveloped in a watery grey cloak, pools are formed, and small dried-up watercourses swell to raging torrents and then to rivers. In spite of the additional humidity and the decrease in the breezes, the atmosphere is not so insupportably heavy, as the temperature is lower at night, though the lack of sunshine is depressing and frequently the forerunner of fever.

With intervals the rains continue till the autumn, and then begins a change, similar to that mentioned above, through a transition tornado period to the dry season again[1].

In the forest the temperature varies little, and the sun never penetrates, the atmosphere is like that of a hot-house day and night, and the daylight, such as it is, is almost always the same, whether the day be bright above, or the sky covered with thick tornado clouds.

In north-west Cameroons at Baliburg the mean annual temperature was $64°·6$ Fahr. ($18°·1$ C.) in 1891 and $64°·4$ Fahr. ($18°$ C.) in 1892. The mean daily variation or range was $13°·1$ Fahr. ($7°·3$ C.) in July and $27°·7$ Fahr. ($15°·4$ C.) in December. The maximum temperature recorded in 1891 was $87°·3$ Fahr. ($30°·7$ C.) and the minimum $45°·7$ Fahr. ($7°·6$ C.); in 1892 the corresponding figures were $90°·6$ and $43°·2$ Fahr. ($32°·5$ and $6°·2$ C.), giving an annual variation of $41°·4$ and $46°·8$ Fahr. ($23°·1$ and $26°·3$ C.). The mean annual rainfall is 2500 mm. or over 98 inches.

Adamawa lies beyond the equatorial four-season belt, and consequently the year is there made up of one wet and one dry season, but the length of these is not invariable throughout the whole district. On the Ngaundere plateau, in South Adamawa, it is sometimes very cold. Mizon complained of the low temperature in January (1892), when it was only $37°·4$ Fahr. ($3°$ C.), and Plehn talks of sleet storms being not at all rare on the western side of this plateau; in fact he experienced no less than twelve in the course of two years, and four of these in one month. On the other hand the temperature at Fort Crampel, *i.e.* the maximum temperature, is higher than at Baliburg, the figure for the former being $113°·2$ Fahr. ($45°·1$ C.). The plateau; even on its western side, has really a climate partaking largely of the continental type, *i.e.* with extremes of temperature; and this is still more marked on the eastern side, for the minimum recorded at Fort Crampel is $49°·3$ Fahr. or $9°·6$ C., giving a range of about $63°$ Fahr. The rainfall at Fort Crampel is 1275 mm. or a little over 50 inches.

This plateau is much more suited to Europeans than any of the Cameroons coast lands. Plehn says that what he suffered from most at Baliburg, $6°$ north of the equator, was cold.

[1] *G. Z.* 1904, p. 74.

The Mandara mountains form the climatic boundary between Adamawa and Bornu, the seasons in the latter being four months later than in the former. Thus the Benue river is at its highest in September and Lake Chad not till December. Mandara is not so hot as the north of Adamawa, owing chiefly to its altitude and forests.

In the Sangoia region, between the upper Benue river and Lake Chad, the rainy season extends from May to September and the dry season from November to March, the months April and October being transition periods. The temperature is high throughout, and during the rainy season is most marked, but the dry season is quite supportable by Europeans[1].

In Lake Chad region during the dry season, *i.e.* for almost nine months out of the year, or *at least* November to May, the N.E. trade winds blow almost continuously and with fair strength.

During the last months of the dry season there is sometimes a breeze from the west and that is the general direction during the rainy season, which is of very short duration, for rain only falls in any quantity from the end of June to the beginning of August. The first storms come from the south, but are frequently not accompanied by rain. In the dry season the wind varies both in direction and intensity according to the time of day : the following example is taken from the month of March :

6 a.m.	N.E., fairly strong.
9 a.m.	N.E., very strong.
12 noon.	E., fairly strong.
6 p.m.	W., slight.
Night.	Calm.

So that the wind follows the sun. On the Chad islands (Kuri archipelago) the winter temperature often sinks to $33°\cdot8$ Fahr. or $1°$ C. at night, and in the summer it sometimes exceeds $104°$ Fahr. or $40°$ C. in the day time. During the dry season the sky is seldom cloudy, though the horizon appears troubled, on account of the sand which is blown about[2].

Bell Town (Duala) and Akwa Town, situated on a

---

[1] Dominik, *D. K.* 1903, p. 131.
[2] Destenave, *R. G.* I. 1903, pp. 483—490.

plain, which stands some 50 feet above the Cameroons river, are reported by business agents, who have long traded there, to be comparatively healthy.

PRODUCTIONS, &c.

In the coastal districts of Cameroons, where the precipitation is exceedingly heavy and the heat excessive, the oil palm and the cocoa-nut palm and rubber flourish. The *Sterculia*, from which kola nuts are derived, thrives on the coast, and one species grows to a height of some 40 feet ; cacao, which requires a hot damp climate, also does well. Coffee is indigenous.

In the middle, more elevated, region, the districts of Kabo and Kongoa seem to be suited for the cultivation of rice, which, however, would not grow near the coast, as it is there actually too rainy for this crop. In these highlands, on the Manenguba plateau, all cereals not only grow, but flourish, in the depressions, and the oil-palm throughout the valleys.

Ground nuts are cultivated, and so also are yams and sweet potatoes, as well as manioc, the plant from which cassava (tapioca) is produced. The *Raphia vinifera* (bamboo palm) is a native, and from this palm-wine is extracted, while the leaves and wood are put to endless uses.

CAMEROONS RIVER. Lat. 4° 5′ N., Long. 9° 45′ E.

From 3 years' observations.

	Jan.	Feb.	March	April	May	June	July	Aug.	Sept.	Oct.	Nov.	Dec.	Year
Mean temperature (Fahr.°)	79°·4	80°·6	79°·8	79°·5	79°·4	77°·7	75°·3	75°·5	76°·3	76°·6	78°·7	79°·3	78°·2
Mean daily range	11°·0	12°·1	13°	12°·2	11°·3	9°·0	7°·0	8°·0	9°·0	10°·0	10°·2	11°·0	10°·3
Absolute maximum	88°·3	89°·4	88°·2	90°·5	91°·2	90°·0	85°·5	85°·3	86°·0	87°·3	87°·8	87°·8	91°·2
Absolute minimum	69°·6	68°·2	67°·8	68°·7	69°·1	70°·5	67°·3	67°·8	68°·2	68°·9	68°·5	68°·4	67°·3
Mean amount of cloud (0—10)	6·9	6·9	7·2	7·5	7·9	9·0	9·3	·9·2	9·1	8·4	7·8	7·1	8·0
Number of rainy days ...	9·7	11·3	15·3	15·7	19·3	22·0	27·0	24·7	25·0	21·3	13·3	8·0	212·6

The above table is taken from the Admiralty publication *Africa Pilot*. No information is given as to which 3 years the table is compiled from, nor as to how the mean temperature was arrived at.

DUALA. Lat. 4° 2′ N., Long. 9° 42′ E. Alt. 40 feet.

Compiled from 12 years' observations[1].

	Jan.	Feb.	March	April	May	June	July	Aug.	Sept.	Oct.	Nov.	Dec.	Year
Mean temperature[2] (Fahr.°) ...	79°·3	79°·9	79°·2	78°·8	77°·3	76°·8	74°·7	74°·5	75°·9	75°·9	77°·9	78°·6	77°·4
Mean maximum temperature[3] (6 years)	88°·0	89°·2	89°·1	89°·5	89°·5	86°·7	82°·9	83°·6	84°·2	85°·8	88°·0	87°·6	87°·0
Mean minimum temperature[4] (6 years)	70°·3	70°·3	69°·1	69°·6	69°·3	69°·1	69°·4	68°·2	69°·3	69°·1	68°·9	68°·2	69°·2
Mean daily range[5]	17°·7	18°·9	20°·0	19°·9	20°·2	17°·6	13°·5	15°·4	14°·9	16°·7	19°·1	19°·4	17°·8
Mean relative humidity (per cent.) ...	85	85	87	87	88	89	92	91	90	90	88	87	88

[1] July 1886—June 1898. See *Meteorologische Zeitschrift*, 1904, p. 544.   [2] The mean temperature is derived from $\dfrac{7a + 2p + 9p + 9p}{4}$.

[3] The absolute maximum recorded was 91°·8.   [4] The absolute minimum recorded was 66°·2.

[5] The daily range is max.—min.

BUEA. Lat. 40° 6′ N., Long. 9° 15′ E. Alt. 3215 feet.

Compiled from 3—4 years' observations[1].

	Jan.	Feb.	March	April	May	June	July	Aug.	Sept.	Oct.	Nov.	Dec.	Year
Mean temperature[2] (Fahr.°) ...	68°·2	68°·0	69°·4	69°·3	68°·2	66°·2	65°·5	64°·4	65°·8	66°·7	68°·9	67°·8	67°·4
Mean daily range[3]	19°·8	16°·9	15°·3	14°·8	14°·4	12°·1	9°·7	9°·0	11°·3	13°·9	17°·1	18°·0	14°·4
Mean relative humidity (per cent.) ...	84	83	84	84	88	93	95	96	95	91	86	88	89

NOTE.—Buea is situated on Cameroons Mt., S.E. of the Peak.

[1] See *Meteorologische Zeitschrift*, 1904, p. 545. Precise years not stated.

[2] The mean temperature is derived from $\dfrac{7a + 2p + 9p + 9p}{4}$.

[3] The daily range is max.—min.

YAUNDE. Lat. $3^\circ 49'$N., Long. $12^\circ 20'$E. (approx.). Alt. 3525 feet.

Compiled from 1 year's observations[1].

	Jan.	Feb.	March	April	May	June	July	Aug.	Sept.	Oct.	Nov.	Dec.	Year
Mean temperature[2] (Fahr.$^\circ$) ...	$73^\circ\cdot8$	$74^\circ\cdot1$	$73^\circ\cdot8$	$73^\circ\cdot3$	$72^\circ\cdot7$	$71^\circ\cdot4$	$69^\circ\cdot9$	$70^\circ\cdot5$	$68^\circ\cdot0$	$71^\circ\cdot4$	$72^\circ\cdot3$	$73^\circ\cdot9$	$72^\circ\cdot1$
Mean maximum temperature ...	$86^\circ\cdot9$	$86^\circ\cdot5$	$85^\circ\cdot5$	$83^\circ\cdot5$	$84^\circ\cdot9$	$82^\circ\cdot2$	$81^\circ\cdot9$	$80^\circ\cdot6$	$79^\circ\cdot9$	$79^\circ\cdot9$	$81^\circ\cdot0$	$85^\circ\cdot3$	$83^\circ\cdot1$
Mean minimum temperature ...	$62^\circ\cdot6$	$65^\circ\cdot3$	$64^\circ\cdot8$	$64^\circ\cdot0$	$64^\circ\cdot9$	$64^\circ\cdot9$	$63^\circ\cdot7$	$63^\circ\cdot1$	$64^\circ\cdot6$	$64^\circ\cdot0$	$64^\circ\cdot0$	$64^\circ\cdot9$	$64^\circ\cdot2$
Mean daily range[3]	$24^\circ\cdot3$	$21^\circ\cdot2$	$20^\circ\cdot7$	$19^\circ\cdot5$	$20^\circ\cdot0$	$17^\circ\cdot3$	$18^\circ\cdot2$	$17^\circ\cdot5$	$15^\circ\cdot3$	$15^\circ\cdot9$	$17^\circ\cdot0$	$20^\circ\cdot4$	$18^\circ\cdot9$
Absolute maximum temperature ...	$88^\circ\cdot7$	$90^\circ\cdot5$	$87^\circ\cdot8$	$87^\circ\cdot8$	$87^\circ\cdot8$	$86^\circ\cdot0$	$86^\circ\cdot9$	$84^\circ\cdot2$	$85^\circ\cdot1$	$82^\circ\cdot4$	$86^\circ\cdot0$	$87^\circ\cdot8$	$90^\circ\cdot5$
Absolute minimum temperature ...	$54^\circ\cdot5$	$60^\circ\cdot8$	$58^\circ\cdot1$	$60^\circ\cdot8$	$59^\circ\cdot9$	$62^\circ\cdot6$	$59^\circ\cdot0$	$59^\circ\cdot9$	$60^\circ\cdot8$	$59^\circ\cdot9$	$60^\circ\cdot8$	$62^\circ\cdot6$	$54^\circ\cdot5$
Mean amount of cloud (0—10) ...	$2\cdot9$	$5\cdot5$	$5\cdot9$	$5\cdot4$	$4\cdot7$	$5\cdot6$	$6\cdot1$	$5\cdot6$	$6\cdot7$	$6\cdot8$	$5\cdot1$	$3\cdot3$	$5\cdot3$
Number of days with rain[4] ...	3	10	17	18	16	14	7	7	20	18	15	6	151

NOTE.—Yaunde is situated inland, the nearest point on the coast being about 125 miles distant in a S.W by W. direction.

[1] 9th—31st December 1889, January—November 1890.
[2] The mean temperature is derived from $\dfrac{7a+2p+9p}{3}$.
[3] The daily range is max.—min.
[4] The number of days with over one inch of rain was 16, viz. one in February, two each in March, April, May, June and September, four in October, and one in November.

CAMEROONS. MEAN RAINFALL IN INCHES.

	Lat.	Long. E.	No. of Years	Jan.	Feb.	March	April	May	June	July	Aug.	Sept.	Oct.	Nov.	Dec.	Year
Cameroons river[1] ...	4° 5′	9° 45′	3	2·40	3·31	8·39	7·40	10·59	22·28	36·00	22·00	19·29	20·20	4·61	3·58	160·05
Debunja[2]	4° 8′	9° 0′	12	7·99	10·91	17·13	17·28	24·80	59·74	64·46	57·73	65·21	45·25	26·61	15·12	412·23
Duala[3], alt. 40 feet	4° 3′	9° 45′	16	1·18	3·15	7·72	8·90	12·99	21·46	28·07	27·64	19·17	16·61	5·87	2·87	155·63
Yaunde[4], alt. 3525 feet	3° 49′	12° 20′	6	1·02	2·80	5·67	8·58	8·07	4·61	2·24	2·76	8·27	8·94	5·83	1·97	60·76
Victoria[5], alt. 265 feet	4° 0′	9° 13′	(6)	0·71	2·27	5·71	7·83	10·04	27·09	48·44	33·63	16·93	12·40	4·41	1·81	171·27
Bwea[6], alt. 3215 feet	4° 9′	9° 16′	(6)	0·26	1·18	3·39	8·15	8·15	11·38	19·53	21·93	18·07	9·53	3·23	0·91	105·71

[1] Taken from the *Africa Pilot*: the specific years are not stated.

[2] December 1894, 1895—1905, except July and August 1900, and January 1901, which are wanting.

[3] November, December 1885, January—September 1886, April—December 1888, 1889—1891, January—October 1892, March—December 1893 (only 15th—31st March), 1894 (only 1st—20th Dec.), 1895—1899, January—March and June—December 1900, January—September, and 1st—16th October 1901, 13th—30th September, October—December 1903, January—March, 1st—15th April, October—December, 1904, 1905.

[4] April—November 1889, January—November 1890, 1891—1894, January, February 1895.

[5] April—December 1893, January—March 1894, May—December 1896, May—September 1897, January, February, July—December 1898, 1905.

[6] May—October 1891, April—December 1896—1898 (only 26 days in January, 25 in February, 27 in May, 30 in July, 28 in August, 27 in September, 29 in October, 27 in November, 30 in December), May—December 1905.

PORTUGUESE GUINEA

There are no series of meteorological returns available for Portuguese Guinea, but a general summary of the climatic conditions, so far as the coastal regions are concerned, is given by Vasconcellos in his work *As Colonias Portuguezas*, from which most of the following facts are gleaned. The Colony lying north of the parallel of $10°$ N. is well beyond the limits of the four-season equatorial area, and has two well-defined seasons. On the coast, the dry season, which is least unsuitable for Europeans, extends over December, January, February and the early part of March, when the days are bright, the sky clear, and the relative humidity extremely low. In this season the Harmattan, in the form of a hot desert wind, is frequently experienced. The high temperature is, however, in part alleviated by sea breezes, and the nights are comparatively cool. In April come the first indications of rain, and the sky becomes more or less cloudy; there are heavy dews, and the mean temperature is about $89°·7$ Fahr. $(32°$ C.). In May the humidity increases, great banks of clouds obscure the horizon, and the tornado season begins. The period immediately preceding the tornadoes has been found to be very trying to the respiratory organs, and most people suffer from cold in the head. The rain, which accompanies the tornadoes, has the effect of lowering the temperature. In June and July the humidity is higher and the rains are more frequent and torrential, and the tornadoes reach their maximum of frequency, but cease altogether in August, which, however, is also a rain month. During September, October and November the weather gradually moderates and eventually there occurs a transition period preparatory to the commencement of the dry season in December.

At Bolama, which lies on the south of the estuary of the Geba river, the mean temperature varies, during the year, between $86°$ Fahr. $(30°$ C.) and $77°$ Fahr. $(25°$ C.), the thermometer falling with the approach of the tornadoes. At Bissan, which lies at the north of the estuary, the mean for the year is about $78°·8$ Fahr. $(26°$ C.). At the mouth of the Cacheo river, still farther to the north, there is usually a sickly heaviness about the atmosphere, but at Farim, some 85 miles up the river, the air is comparatively fresh and invigorating, and the mangroves, which occupy the whole of the low-lying, swampy, coastal areas, begin to disappear.

In the interior, in the neighbourhood of the eastern boundary, it is stated in the report of the Franco-Portuguese Boundary Commission, as given in the *Revue Coloniale*, that the seasons are less marked than on the coast, the rainy season being less severe, shorter, and a little later; on the other hand the dry season is hotter than on the coast.

The rainy season only lasts four months, from about the 15th July to about the 15th November. The rain, though abundant, is not continuous, and generally falls at night, and it is not uncommon to have two or three fine days with a blue sky and clear atmosphere. The mean day temperature is about $71°·6$ Fahr. ($22°$ C.) and that at night about $59°$ Fahr. ($15°$ C.). The total rainfall during this season of four months is about 3500 mm. or nearly 138 inches.

The passage from season to season is marked by a period of tornadoes, which generally lasts till near the end of December.

The mean day temperature in the dry season is about $91°·5$ Fahr. ($33°$ C.), and the corresponding night temperature about $62°·6$ Fahr. ($17°$ C.), except in April and May, when it is $73°·4$ Fahr. ($23°$ C.). From noon to 2 p.m. the heat in the sun is sometimes terrific. On May 2nd, 1904, the Boundary Commission recorded, at 1.45 p.m., the thermometer being one metre from the ground, a temperature of $143°·6$ Fahr. ($62°$ C.) when the sky was clear, and on the 18th of the same month, at the same time of day, $152°·6$ Fahr. ($67°$ C.) when it was slightly cloudy.

When the season changes again, *i.e.* from the beginning of June to 15th July, the tornado period occurs again, and all nature appears to become revivified. The two tornado periods have very similar characteristics, except that during the last mentioned, hail storms sometimes occur, which are extremely destructive.

LIBERIA

All that is really known of Liberia, so far as the matter in hand is concerned, is that it is situated in one of the rainiest regions of Africa, and is, in general, a land of forest, especially in the east. In the west there are several gaps in the forest, especially near the coast and in the north. Groups of hills, the highest of which may reach 1500 feet (500 metres), separate the coastal basins from one another. Other hills, which attain a higher elevation, separate these basins from that of the Niger. Travellers and others have ascended the coastal rivers to the points where they are cut by rapids, and in some cases higher, but record nothing relating to the climate. Delafosse, in his monograph on Liberia, which appeared as Suppl. No. 9 to the *Bulletin du Comité de l'Afrique française*, 1900, says not a word about climate. The Boundary Commissioners have not taken the public into their confidence, so far as climate is concerned. Chesneau, in his account of the Woelffel-Mangin expedition, gives no information as to climate, and I am officially informed by the representatives of Liberia in this country that no meteorological records exist.

Hostains and d'Ollone tell us that, in the region of the sources of the Cavalla river, on the 25th November the rains had stopped eight days, while from the Mission Blondiaux we learn that, in the latter part of February and the beginning of March, the rivers were dry. Now, at Sierra Leone, the rainy season begins in March with 2·23 inches; the precipitation increases up to July and August; and the season ends in December, with about an inch of rain. We may conclude, therefore, that if the information vouchsafed by the two French missions is not for abnormal areas, the rainy season in the hinterland of Liberia is not so long as it is at Sierra Leone. The rainfall is greater in the western coastal regions of the Ivory Coast, and also, we may presume, in the southern parts of Liberia, than it is farther east. For information as to the seasons, temperature, &c., we must refer to the section on the "West Coast," and compare also with the notes on Sierra Leone and the Ivory Coast.

Hübner, writing in *Petermann's Mitteilungen*, 1903, p. 174, tells us that at Boporu, where the savannahs and highlands begin, the climate is healthy and similar to that experienced in Adamawa in Cameroons, and those who have an interest in the country say that the coast of Liberia is probably less unhealthy for Europeans than the Ivory Coast, the Gold Coast or Lagos (Western Province of Southern Nigeria), and that the remarkable absence of mosquitoes ought to stand in its favour, as it generally coincides with a less pronounced prevalence of malarial fever; but that from a European point of view the northern part of the territory, the Mandingo plateau, is probably more suitable for settlement.

Mr A. Whyte, who was in Liberia for part of the year 1904, considers that there are two climates. In the southern regions below 6° Lat., the rainiest time of year appears to be the months of March to June, and August to December. North of this—round Monrovia, for example—the specially rainy months are April to the end of July, September and October[1].

Sir H. H. Johnston considers that the heavy rainy season begins in April and lasts till the end of July; then, he says, there is a pause of a month or six weeks with less rain, the heavy rains beginning again in September and lasting till the middle of November; from mid-November till the end of March is the dry season, at any rate in the northern half of Liberia; between November and April is the worst season for storms[1].

(For Productions *see* under West Coast.)

[1] *G. J.* Vol. XXVI, 1905, p. 132.

WESTERN TROPICAL AFRICA

SOUTH SECTION.

SPANISH GUINEA.

For this small enclave, bounded on the north by the German colony of Cameroons, and on the south and east by French Congo, the sections on those two possessions should be consulted. Being in the neighbourhood of the Equator it may be assumed at once that the climate is both hot and damp. Lying on the seaboard, the daily range of temperature is naturally small. The coastal settlements are malarious, and it is unsafe for Europeans, even with the greatest precautions, to remain there for any considerable length of time, without returning home to recuperate.

No series of meteorological observations are available for this area, but such particulars as may be gleaned from authorities on the subject are appended.

Ricardo, writing in the *Bulletin* of the French Geographical Society, now called *La Géographie*, Vol. VIII, 1900, p. 232, states that, in the basin of the Lower Muni river, the rains are abundant and torrential, even in what is called the dry season, *i.e.* in the neighbourhood of the equator, the season of less rain. The humidity is at all times excessive, and the temperature is almost constant, ranging merely between $82°\cdot4$ and $86°$ Fahr. ($28°$ and $30°$ C.).

In the borderland traversed by the Franco-Spanish Boundary Commission, from the beginning of August to mid-October, the mean temperature varied between $73°\cdot4$ and $80°\cdot6$ Fahr. ($23°$ and $27°$ C.), and the difference between the maximum and minimum of daily temperature was slight, never exceeding $25°\cdot2$ Fahr. ($14°$ C.). With the exception of a slight breeze on a few days, a complete calm prevailed. The rains began on August 19th, and fell every day, generally in the form of a storm, up to and including October 5th, when they suddenly ceased entirely. During that rainy season a mean of $1\cdot02$ inches or 26 mm. fell daily, that is to say, a total of a little over 48 inches (1222 mm.) fell in a period of 47 days.

FRENCH EQUATORIAL AFRICA

FRENCH CONGO.

The very extensive area embraced in the new administrative title French Equatorial Africa includes French Congo, the Ubangi-Shari region, and the Military District of Lake Chad, and reaches from about $5°$ S. to $13°$ or $14°$ N., *i.e.* it extends over something like $18°$ or $20°$ degrees of latitude, and therefore, as might be expected, the climatic conditions vary considerably at places in the south and north respectively, though a very large portion of this area lies in what is usually called the equatorial belt, in the great equatorial forest zone, in the four-season region. In connection with this huge area very little precise information is available, except for the coastal regions. De Bellay, writing some years ago in the *Tour du Monde*, described the climate of the coast as *terrible*, "not because the heat is excessive, but because it is constant, the thermometer only varying between $77°$ and $89°·7$ Fahr. ($25°$ and $32°$ C.), so that the heat is uniform and debilitating," and these figures are confirmed by what appears below. The official report on the colony, published in Suppl, No. 3 of the *Bulletin du Comité de l'Afrique française* for the year 1900, states that the year may be divided into four periods :—

1. From May to October, the greater dry season.
2. From October to December, the lesser rainy season, the rains being heavier in the latter half of the period.
3. From late December to early February, the lesser dry season.
4. From February to May, the greater rainy season.

It would be more correct, perhaps, to call the lesser dry season a season of less rain, for rain is more or less abundant throughout the whole period from October to May, and some people prefer to describe the conditions throughout the year as comprising two maxima of rainfall with a dry season from May to October, the dryest period in equatorial regions being almost invariably during July / August. Thus Bestion, writing in the *Archives de médicine navale*, XXXVI, pp. 241–278, 401–421, says that the year is divided into two seasons, the dry and the wet, the latter extending from September to May, though it is interrupted about January by a short dry season or season of less rain.

The dates of the seasons given above are only strictly applicable to the Gabun region, including Libreville. The seasons are more or less in advance of these further inland.

The following results are derived from observations made at Libreville in 1898, which was a fair average year[1] :—

Temperature mean ... ... $78°\cdot1$ Fahr. *i.e.* $25°\cdot6$ C.
" maximum (1st March) ... $90°\cdot6$ Fahr. *i.e.* $32°\cdot5$ C.
" minimum (15th July) ... $63°\cdot9$ Fahr. *i.e.* $17°\cdot7$ C.
" mean of the hottest month (January) } $81°\cdot3$ Fahr. *i.e.* $27°\cdot4$ C.
" mean of the coolest months (July and August) } $75°$ Fahr. *i.e.* $23°\cdot9$ C.
Rainfall for the year $103\cdot5$ inches (2629 mm.).
Number of rainy days 153.
Mean fall on a rainy day $0\cdot67$ inches ($16\cdot9$ mm.).

In 1903–4 the absolute maximum was $89°\cdot5$ Fahr. ($31°\cdot9$ C.), which was recorded on 30th March and 7th April, while $89°\cdot1$ Fahr. ($31°\cdot7$ C.) was recorded on 14th January and 27th February, and $87°\cdot9$ Fahr. ($31°$ C.) on 1st May. In this year $67°\cdot3$ Fahr. ($19°\cdot6$ C.) was the absolute minimum (11th June), so that the extreme range of temperature during the year was only $22°\cdot2$ Fahr. ($12°\cdot3$ C.), as compared with $26°\cdot7$ Fahr. ($14°\cdot8$ C.) in 1898. The number of rainy days in 1903–4 was 168, as compared with 153 in 1898, though the total rainfall was less by $6\cdot77$ inches (172 mm.).

In the Gabun region the thunderstorms are very heavy, but do not last long, and are usually accompanied by rain.

Cape Lopez, at the mouth of the Ogowe river, is reported to be most healthy and comparatively cool during the months of July, August and September.

---

[1] Official Report, *A.F.* Suppl. 3, 1900.

Farther south, on the coast north of Loango, the rains begin in September and continue through October, November, and part of December. This rainy period is followed by hot weather, which usually extends to the end of February. Throughout March, April, May, and part of June tornadoes and storms are experienced, and also in September and October. From the latter part of June to the end of August is the dry season, as it is also further north.

At Brazzaville, on Stanley Pool, the absolute maximum temperature is about $100°\cdot5$ Fahr. ($38°$ C.) and the absolute minimum $59°$ Fahr. ($15°$ C.), the former being usually recorded in February, and the latter in August, giving an extreme yearly range of $41°\cdot5$ Fahr. ($23°$ C.). The mean maximum temperature varies between $91°\cdot2$ Fahr. ($32°\cdot9$ C.) in October, and $81°\cdot1$ Fahr. ($27°\cdot3$ C.) in July, while the mean minimum temperature oscillates between $70°\cdot9$ Fahr. ($21°\cdot6$ C.) in March and $63°\cdot5$ Fahr. ($17°\cdot5$ C.) in July, and the mean annual maximum and minimum are $87°\cdot5$ and $68°\cdot5$ Fahr. ($30°\cdot8$ and $20°\cdot3$ C.). On the whole, March is the hottest month, followed by April, with a difference of only $1°\cdot4$ Fahr. ($0°\cdot8$ C.) between the mean temperatures. July is felt as a real winter month by Europeans. The excessive heat of the warmest months is tempered by the frequent rains, and a refreshing breeze usually springs up about the hottest part of the day. In the very early hours of the morning, and also in the evening, the sky as a rule is overcast; in the forenoon it is generally clear, and also in the early afternoon, and for two or three hours after nightfall.

M. Marc Bel, who, at the instance of the Minister of Public Instruction, went on a scientific mission in the southern portion of French Congo, from Loango to Brazzaville, divides the country into four zones.

1. A littoral zone, consisting of a sandy plain, standing at an average altitude of 160–330 feet, with tall grasses and some trees. In this zone most tropical crops can be grown.

2. A forest-clad mountain region, extending for some 60 miles. This is the outer fringe of the rubber area, and though not plentiful, the plants are found. All the hard timbers of commerce abound—mahogany, ebony, &c.

3. East of the above is the African *platforme*, with the Niori basin.

4. Between that basin and the right affluents of the Congo is a mountainous zone, and here begins the region of grass-rubber.

The death-rate of Europeans at Libreville, the only place for which there are any statistics, is about 137 per 1000. It is highest in the two months which are coolest and when there is least rain, namely June and July and it is lowest in March / April and September. It is probable that the rate at Brazzaville is lower. No one should go to French Congo under 25 years of age, and a sound constitution is absolutely necessary, which alone, combined with temperance, activity and careful attention to hygienic laws, can render any prolonged stay possible. Of course the northern regions are more healthy, but no one should attempt to stay near the coast without at least annual visits to Europe to recuperate.

LIBREVILLE.　Lat. 0° 23′ N., Long. 9° 26′ 23″ E.　Alt. 101 feet.

	Jan.	Feb.	March	April	May	June	July	Aug.	Sept.	Oct.	Nov.	Dec.	Year
Mean morning temperature at 8 a.m. Fahr°. (1 year[1])	79°·3	79°·7	79°·2	78°·4	77°·7	73°·8	73°·4	73°·9	76°·3	75°·9	76°·6	77°·2	76°·8
Mean maximum temperature (1 year[1])	86°·5	86°·5	86°·9	86°·4	84°·9	81°·7	80°·6	82°·0	83°·8	83°·3	83°·5	84°·2	84°·2
Mean minimum temperature (1 year[1])	75°·6	75°·4	74°·7	73°·6	74°·5	70°·9	70°·5	71°·2	73°·2	72°·9	73°·0	73°·6	73°·2
Mean daily range[2]	10°·9	11°·1	12°·2	12°·8	10°·4	10°·8	10°·1	10°·8	10°·6	10°·4	10°·5	10°·6	11°·0
Mean temperature (5 years[3]) ...	79°·9	80°·1	79°·9	80°·1	79°·3	77°·4	74°·5	75°·9	78°·1	78°·3	78°·3	78°·9	78°·7
Mean relative humidity per cent. (1 year[1])	86·5	84·8	85·2	85·1	82·8	77·5	73·4	78·0	80·1	85·4	84·3	84·6	82·7
Mean amount of cloud (0—10) (1 year[1])	6·0	5·9	6·9	7·0	7·2	6·5	7·5	7·9	7·9	8·2	7·1	6·4	7·0

The relative humidity is always higher in the morning than in the afternoon; the mean morning humidity for the year is 86·1, and that at 4 p.m. 79·3. The barometer is also always higher in the morning and appears to oscillate between 760·48 mm. in the morning in July and 753·96 at 4 p.m. in March. It is always cloudier in the morning than at 4 p.m., but between 4 and 6 p.m. is the usual time for rain in the rainy season.

[1] 1903.

[2] The daily range is max.—min.

[3] 1903—1907, the mean being derived from $\dfrac{\text{max.} + \text{min.}}{2}$.

UBANGI-SHARI-CHAD REGION.

Of the climate of the region lying west of the Ubangi river and east of the Cameroons boundary little is known with accuracy, but Dr Herr, who accompanied Clozel on his journey from Bania (Lat. $4°$ N., Long. $16°$ E. approx.) through Carnot and northwards to above $6°$ N., took observations from 26th November 1894 to 3rd January 1895, though unfortunately he says nothing about the useful facts which may be deduced from non-instrumental observations. From 26 readings taken at 6 a.m. it appears that the early morning temperature was at that time of year $58°·6$ Fahr. ($14°·8$ C.); the midday temperature derived from 24 readings at noon was $88°·9$ Fahr. ($31°·6$ C.); and the evening temperature deduced from 22 observations taken at 9 p.m. was $63°·3$ Fahr. ($19°·6$ C.). These results give a mean temperature of about $73°·4$ Fahr. ($23°$ C.), corresponding to about $80°·6$ Fahr. ($27°$ C.) at sea-level. The very large daily range indicated by the above figure, which would be larger if the observations had been taken at 2 p.m., the hottest time of the day, instead of at noon, points clearly to a continental type of climate, which also exists in the eastern parts of Cameroons.

The relative humidity, from such observations as were taken, appears to vary, at midday, between 57 per cent. and 20 per cent., with a mean of 35 per cent. Apparently there was no rain during the period under consideration. There are a few particulars available for southern Adamawa, on the west of this region (*see* under Cameroons) and for the Ubangi district on the east (*see* under Belgian Congo) from which a general idea of the climate may be derived.

The northern portion of the above area borders on the Military District of Chad, which Gentil, in *La Chute de l'Empire de Rabah*, divides into three climatic zones. The first extends from $5°$ N. to about $6°\,45'$ N., the second northwards to $9°$ N., and the third from $9°$ to $13°$ N. Further north the climate is of the Saharan type. In the first or southernmost zone, he says, there is a dry season of four months, during which it scarcely rains at all, extending from 15th November to 15th March. This was the season during which Dr Herr took his observations and, therefore, we need not be surprised to find no mention of rain in his report. This dry period is followed by an intermediate season, extending up to the 15th June, during which moderate rains are experienced. The principal rainy season occurs during the period 15th June to 15th October. Finally, from 15th October to 15th November is a transition period separating the rainy season from the dry. It will be found that the few returns there are for the rainfall at Gribingi (Fort Crampel), which is situated immediately to the north of this zone in 7° 0′ 50″ N. Lat. bear out this division of the seasons to a nicety.

The second zone differs from the first only in the quantity of rain, and has a dry season of five months, instead of four. In the third zone there are eight months in the dry season and four in the wet, as these latitudes are north of the four-season equatorial belt. The rainy season extends from 15th June to 15th October. This last zone is by far the most healthy of the three; the heat certainly is greater, but it is a dry heat. At Bir-Alali, N.E. of Lake Chad, there are only 14 rainy days in the year.

From a series of observations taken at Fort Crampel (Gribingi) in the extreme south of the Military District in 1903–4 it may be gathered that though the relative humidity is low at the end and the beginning of the year (the northern winter) it rises to 84·9 per cent. and 83·7 per cent. in May and June. The mean morning temperature (8 a.m.) was here 78°·8 Fahr. (25°·98 C.) in April but lower in May and June. The mean maxima and minima of temperature and the absolute extremes are given below, for the short period during which the observations were taken.

	Mean maxima	Mean minima	Absolute maxima	Absolute minima
February (Fahr.°)	100°·7	56°·7	105°·9	48°·6
April	88°·7	71°·2	107°·6	57°·4
May	92°·5	72°·3	97°·2	68°·4
June	89°·1	70°·5	94°·6	65°·8
August	88°·3	67°·3	95°·0	64°·4
September	90°·9	67°·3	96°·1	62°·1

For Fort Archambault and Lai, the former on the Shari

river, the latter on the Logone river, there are some particulars, which are given in the tables below (pp. 219, 220).

The mountainous region lying north of Kunde, on the Franco-German boundary, enjoys an extremely healthy climate, the thermometer frequently goes down to $42°·8$ Fahr. ($6°$ C.) and seldom registers more than $95°$ Fahr. ($35°$ C.). The rainy season extends from March to October[1].

The climate of Chekna in Bagirmi is typical of the transition between Sudan and Sahara—very hot in the dry season, and with very little rain during the remainder of the year. The Harmattan blows with violence almost every day from December to March. The rains do not make their appearance till June, and then the showers are few and far between, about once a week. Many violent wind storms bring no rain at all. The chief rains begin about the second fortnight in July and end about the second fortnight in September. From the 1st to the 13th August, 1903, says Chevalier[2], of the Shari-Chad mission, the sky was almost constantly overcast, and showers frequently occurred, without being preceded by wind. Sometimes the rain lasts for several hours. The relative humidity, at this season, is very high, and everything becomes damp. On the 14th August it was dry with a hot and sometimes strong wind from the south-east, and lightning; in the evening great clouds became massed in the south, the wind turned S.W. and a few drops of rain fell. On two or three occasions between the 14th and 20th August there were tornadoes at daybreak with dust clouds, but no rain fell, or only a few drops. These dry tornadoes are common in the extreme north of the Sudan.

Speaking of Kanem, further north, the same writer says that the rains begin about the 20th July, though there may be a few drops previous to that date, and end on the 15th September, so that they occur during some two months. From the 15th to the 20th September there is a transition period of dry tornadoes. On the 25th September the Harmattan, or N.E. wind, begins, usually lasting from 7 a.m. to 6 p.m., and continues till April, frequently accompanied

[1] Moll, *A.F.* 1907, p. 392.

[2] *Mission Chari-Lac Tchad*, 1902—4: *L'Afrique Centrale française* par Aug. Chevalier. Challamel, Paris, 1908.

by dust clouds. On the whole, the climate of Kanem is healthy and well suited to Europeans. Mosquitoes, which are so troublesome at Lake Chad, are not numerous in Kanem.

In the neighbourhood of Ndelle (Lat. $8° 23' 30'' N.$, Long. $20° 45'$ approx.) during December, January and February, the wind is between N. and E., frequently N.E. The east wind is cold and damp, the north, on the other hand, is hot, and is strongest in January. This north wind blows strongly all night, increases in intensity between 7 and 10 a.m., and falls off towards midday to rise again in the evening. The N.E. wind brings slight sand clouds, and has a bad effect on vegetation. The tornadoes, which precede the lesser rains, come with a sudden change of direction from the west; in 1902–3, however, there was no rain worth speaking of from November to February inclusive. In Dar Kuti the very dry season corresponds with spring, and from February the rivers only flow near their sources. The rainy season begins in the early days of March, with west winds and a covered sky. The first rains at Ndelle fell on 16th March with a few drops and on the 3rd April there was half an hour's downpour in the evening, and again on the 4th, so that the season really begins early in April[1].

The sultanate of Zemio (Upper Ubangi) has a very constant climate, varying very little from year to year. There is a dry season extending from November to the end of March, which is the most healthy for Europeans. The thermometer reaches $86°$ to $89°·7$ Fahr. ($30°$ to $32°$ C.) in the day time, and falls to $59°$ Fahr. ($15°$ C.), or sometimes even to $50°$ Fahr. ($10°$ C.) in the late night or small hours of the morning. The first rains are experienced towards the latter end of March, and these rains are always accompanied by storms and tornadoes from the east or south-east, which begin about 3 p.m., and usually follow after high temperatures; while the storm is at its height there is, not unfrequently, hail. The rains leave off partially at the end of April, and then begins the lesser dry season, which lasts till July. The air is moist, the thermometer rises to $102°·3$ Fahr. ($39°$ C.), and it is very unhealthy for Europeans.

[1] Chevalier, *Mission Chari-Lac Tchad.* Challamel, Paris, 1908, p. 214.

In July it rains almost every evening, and the fall increases, and reaches its maximum in October, when even the smallest streams become torrents, and the temperature varies between 77° and 86° Fahr. (25° and 30° C.) day and night. During this rainy season the most frequent winds come from the south-east and sometimes from the south-west. On the contrary, during the dry season, it is the east wind which scorches up everything, but there are refreshing dews, which, however, are evaporated by the fierce sun by about 7 a.m.[1]

At Mobaye on the Upper Ubangi the atmospheric pressure varies only ·59 inch (15 mm.) in the year, *i.e.* between 24·41 and 25 inches (720 and 735 mm.): in December 1899 the variation was only ·12 inch (3 mm.), and the same in the following January and July. The minimum temperature was recorded in January, viz. 49°·3 Fahr. (9°·6 C.), and the maximum in March, viz. 113° Fahr. (45°·1 C.). The highest temperatures were recorded during the first five months in the year, and the lowest from November to March. November, December, January and February were dry, the rainy months being June, July and August. The greatest rainfall in one day was 73·4 mm. or 2·89 inches on the 14th June. The annual rainfall is twice that of Paris.

It is stated by Lamothe in the *Bulletin du Comité de l'Afrique française*, 1902, p. 392, that the climate of Wadai is such as to render the country susceptible of colonisation by Europeans, at least from 10° N. Advancing northwards the climate is relatively healthy and more supportable. The French mission, he adds, in three years only lost two men from sickness.

PRODUCTIONS, &c.

Cacao, or cocoa, and coffee are both indigenous in French Congo. Rubber of the grass or root kind is found near Brazzaville, the principal variety being *Landolphia tholloni*, and rubber vines occur in the Ubangi forests further north, both of the *Landolphia* and *Clitandra* kinds. The forests also produce excellent hard and ornamental woods, such as

---

[1] *M. G.* 1907, p. 434.

ebony, mahogany, &c. Maize and other cereals do well in the middle region. Cotton should do well, but not on the coast, as there the rainfall is probably too heavy; rice should pay in the low-lying districts, while oil-palms grow in the valleys of the middle region. Manioc plantations are found in almost every native village, and bananas and plantains thrive. Cocoa-nut palms are a forest production, especially near the coast, tobacco and sugar have both been tried and found to succeed, and so, too, has vanilla. Sweet potatoes and yams both produce good crops, and pineapples grow wild in the forest. European vegetables, such as potatoes, cabbages, carrots, &c. are always grown around the French stations and do well.

FORT ARCHAMBAULT (ON SHARI RIVER).

Compiled from 16 months' observations[1].

	Jan.	Feb.	March	April	May	June	July	Aug.	Sept.	Oct.	Nov.	Dec.	Year
Mean maximum temperature (Fahr.°) ...	95°·2	96°·3	100°·4	100°·0	95°·7	90°·2	89°·1	85°·5	86°·9	91°·4	94°·6	95°·0	93°·4
Mean minimum temperature	61°·5	61°·0	70°·7	75°·4	74°·8	71°·6	70°·2	70°·3	70°·3	69°·1	66°·2	63°·4	68°·7
Mean temperature[2]	78°·4	78°·6	85°·6	87°·7	85°·3	80°·9	79°·6	77°·9	78°·6	80°·2	80°·4	79°·2	81°·0
Mean daily range[3]	33°·7	35°·3	39°·7	24°·6	20°·9	18°·6	18°·9	15°·2	16°·6	22°·3	28°·4	31°·6	24°·7
Absolute maximum temperature ...	99°·9	105°·1	106°·3	104°·9	99°·5	98°·1	92°·8	89°·8	92°·7	95°·5	98°·1	98°·6	
Absolute minimum temperature ...	55°·8	56°·3	63°·5	69°·6	68°·7	66°·2	66°·4	66°·6	65°·3	63°·9	61°·7	68°·6	
Mean morning temperature at 8 a.m. ...	69°·3	72°·1	83°·1	83°·7	80°·8	77°·5	75°·4	74°·7	74°·5	76°·1	75°·9	73°·0	76°·2

[1] March 1903—July 1904. In the former year the March records are only for 10th—29th, those for May from 1st—13th, and for July from 3rd—31st; while the June records are entirely wanting. In the latter year the July records are for 1st—8th.

[2] The mean temperature is derived from $\frac{\text{max.} + \text{min.}}{2}$.

[3] The daily range is max.—min.

LAI (ON LOGONE RIVER).

Compiled from one year's observations[1].

		Jan.	Feb.	March	April	May	June	July	Aug.	Sept.	Oct.	Nov.	Dec.	Year
Mean temperature at 8 a.m. (Fahr.°)	...	$67°.8$	$50°.5^2$	$77°.0$	$79°.3$	$75°.6$	—	$75°.6$	$72°.5$	$70°.7$	$69°.4$	$70°.7$	$72°.1$	
Mean minimum temperature	...	$56°.5$	$53°.6$	$55°.3$	$68°.9$	$69°.8$	—	—	$63°.1$	$55°.6$	$52°.2$	$56°.8$	$61°.2$	
Absolute minimum temperature	...	$52°.0$	$49°.1$	$51°.8$	$61°.2$	$69°.3$	—	—	$53°.6$	$53°.6$	$49°.1$	$53°.6$	$55°.8$	
Number of rainy days	... ...	0	0	3	3	14	11	15	18	18	6	0	0	88

[1] July 1903—May 1904.
[2] This and the previous table are taken from *L'Expansion coloniale au Congo française*, by Rouget, p. 274, where the 8 a.m. temperature for February is given as $10°.3$ C., *i.e.* $50°.5$ Fahr., but it seems probable that this is an error, and is meant for $20°.3$ C., which would give $68°.5$ Fahr.

MEAN RAINFALL IN FRENCH EQUATORIAL AFRICA IN INCHES.

	Jan.	Feb.	March	April	May	June	July	Aug.	Sept.	Oct.	Nov.	Dec.	Year
Libreville (6 years)[1]	6·50	9·49	13·82	14·72	6·10	0·24	0·08	0·51	1·65	12·40	15·28	9·61	90·4
Gribingi (Fort Crampel) (1 year)[2] ...	0·02	0·15	2·28	3·60	5·63	8·50	10·79	10·08	4·01	5·08	0	0	50·2
Fort Archambault, on Shari river[3] (1 year)}	0	0	0·14	0·39	2·87	5·00	12·64	7·99	11·14	0·29	1·22	0	41·7
Lai, on Logone river (1 year)[4]	0	0	1·38	0·83	5·43	0	5·90	12·04	17·67	7·48	0	0	50·7
Brazzaville combined with Leopoldville (Stanley Pool) (25 months)[5] }	6·30	4·88	7·36	7·05	4·29	0·55	0·01	0·04	2·16	5·35	11·50	8·42	57·9
Mayumba (4 years)[6]	2·71	3·03	8·31	4·96	2·13	0·16	0·04	0·39	1·18	6·61	14·92	5·98	54·4

[1] 1901—1906.
[2] 1903, 1904.
[3] July 1903—June 1904.
[4] June 1903—May 1904.
[5] Leopoldville, March 1886—February 1887, and November 1893—May 1894; Brazzaville, April—June 1893, October—December 1894, September, October 1895.
[6] 1899—1902.

BELGIAN CONGO

GENERAL.

The mean annual temperature at the mouth of the Congo in about $6°$ S. (Banana) is no higher than that in the middle of the continent in $30°$ S., in consequence of the action of the cold Benguela current which washes all the western coast, in its northerly course, as far up as the "waist" of Africa and beyond. And it follows that in going eastwards towards the middle of the continent the temperature rises from $78°·8$ Fahr. ($26°$ C.) at Banana to $80°·6$, $82°·4$ Fahr. ($27°$, $28°$ C.) and higher in equatorial regions. Further south, in the highlands of Katanga, the mean annual temperature is only $73°·4$ Fahr. ($23°$ C.). In the southern regions of the great Congo area the mean temperature varies but little, but, going northwards, it increases gradually, till, in the north of the area, in $4°$ N., it reaches $84°·2$ Fahr. ($29°$ C.). Forming a general mean from all the available data, the result arrived at is $80°·6$ Fahr. ($27°$ C.) for the whole area.

July is the coolest month and February the hottest, though near the Congo mouth the warmest month is March, the months of February, March and April forming the hottest season, and having approximately the same mean temperature throughout. The mean monthly temperatures for October, November, December, January, February, March, April, and May are higher than the mean annual temperature, and those for the remaining months lower. The maximum temperature is reached earlier in the year in the north than in the south, and there is a second maximum in the south which occurs about September to October. Over the greater part of the area there is a distinct fall of $5°·4$ Fahr. ($3°$ C.) from the May temperature to that of June, corresponding with the change from the wet season to the dry.

The daily range is about $15°·3$ Fahr. ($8°·5$ C.) in the west, but increases eastwards to $19°·3$ Fahr. ($10°·7$ C.) at

Kimuenza and to 23°·4 Fahr. (13° C.) at Luluabourg. The thermometer is generally lowest between 5 and 6 a.m., and highest about 1 p.m., and begins to fall before 2 o'clock is reached. The mean maximum temperature works out to between 84°·2 and 86° Fahr. (29° and 30° C.), the mean for March being about 89°·7 Fahr. (32° C.), and for July between 77° and 78°·8 Fahr. (25° and 26° C.). The lowest average temperature, in situations of comparatively low elevation, is between 68° and 69°·8 Fahr. (20° and 21° C.), but 73°·4 and 75°·2 Fahr. (23° and 24° C.) in Feb./April, and 64°·4 Fahr. (18° C.) in July, and thus the difference between the warmest and coolest nights is some 10°·8 Fahr. (6° C.).

In the equatorial portions of the Congo area it rains almost daily throughout the year, and the seasons resolve themselves into periods of more rain and less rain, but in the Lower Congo there are well-marked seasons, so well marked, indeed, that the change from wet to dry, and *vice versa* may be depended upon almost to the day. The rainy season begins in October and ends in May, and from June to September is the dry season. The former begins with light rains at intervals for a couple of days, becoming gradually more frequent and copious, and accompanied almost always by thunder and lightning. November and December are the rainiest months, but the precipitation decreases in the latter month, and a short dry season intervenes, to be followed by more rain, which lasts from February to mid-May, and is most copious in the beginning of April.

The rains are frequently accompanied by violent wind squalls from no fixed direction, and generally fall between 2 and 9 p.m., and in the night up to 7 in the morning, the rainstorms being of short duration.

The total annual rainfall, near the coast, varies considerably from year to year, but becomes more constant as stations further inland are reached. In the dry season, near the coast, there is no rain, except possibly an occasional shower of 10 minutes' duration between 5 and 9 a.m. The further the distance from the coast the shorter the dry season becomes, and in equatorial regions it vanishes altogether, and what elsewhere are dry seasons become, as mentioned above, seasons of less rain. The annual rainfall thus increases from south to north and from west to east.

During the cooler months in the Congo area the relative humidity is lowest and during the hot season it is at its highest, reaching its maximum at about the same time as the heat maximum, and thus the humidity is greatest in April/May and lowest in August. Comparing the interior with the coast it is found that in the morning the humidity is greater in the interior than on the coast, while, in the middle of the day, the reverse is the case, this being especially marked in the dry season.

During the rainy season, at sunrise, the sky is always completely clouded, but gradually clears between eight and ten o'clock ; between one and two o'clock the storm-cloud begins to appear, and the sky remains overcast till evening. In the dry season the sky gradually becomes clearer till early in the afternoon, when the clouds disperse altogether, and the sky remains clear, although it may be misty, till late in the evening (say nine or ten o'clock), when it again becomes cloudy[1].

TEMPERATURE, WINDS, &c.

On the Lower Congo the prevailing winds are from the west, west-south-west, and south-west ; in the rainy season the vane turns slightly eastwards, but in the dry season the winds are almost uniformly from the west or south-west. In the interior, in the central and eastern portions of the area, the prevalent winds are from the south-east[1].

The general *régime* of the temperature on the Lower Congo has already been alluded to. It is worth while mentioning, however, that the forenoon is, generally, the coolest time of day ; after midday it becomes hot, particularly when there is any sun. Cloudy days, however, are characteristic of the climate, and this makes the heat quite endurable, particularly as a breeze often blows from the west, that is, from the ocean. It is said there are not more than a hundred days of sunshine in the year. A peculiar fact has been pointed out by Dr von Danckelmann with regard to this Lower Congo region, namely, that, along the river, reaching up through the Cataracts district as far as Stanley

---

[1] *Revue Scientifique* 12 (1899), Paris.

Pool, immediately at sunset, or a quarter of an hour after, there is always a violent blast of wind from the west or north-west, which lasts for from ten minutes to half an hour, and then dies down to be replaced by a steady wind gradually turning to south-west. The period between mid June and the beginning of September is the pleasantest, as it is also the healthiest, on the Lower Congo, and the afternoons are frequently, if not mostly, cloudless.

Great improvements have been, and are being, carried out at Boma, and latterly the health of the town has been much better. The neighbouring lands are being cultivated, and some of those parts which were annually flooded by the river have been filled in, and drainage works carried out—all of which tend to better health conditions.

Matadi lies in a hollow, surrounded by mountains, and there are no sanitary arrangements; there is, moreover, a large native population; it is extremely hot, and, no wonder, very unhealthy.

Banana is situated on a sandy peninsula between a creek and the sea, and, thanks to the sandy soil, is a fairly healthy station. The country round is low and flat within a radius of five miles, and, could the marshes be got rid of, it would doubtless be more healthy.

In the Cataracts district, in the neighbourhood of Tumba and Lukungu the absolute maximum temperature never exceeds $96°\cdot8$ Fahr. ($36°$ C.) and the corresponding minimum is never below $62°\cdot6$ Fahr. ($17°$ C.). The sea or west wind is regularly felt at the latter place between 4 and 5 o'clock in the afternoon, and is always strong. Dews are frequent on the hills, but not in the valleys. The seasons are very marked and regular in this region.

In the Stanley Pool district, at Leopoldville, the mean maximum temperature varies between $91°\cdot3$ Fahr. ($32°\cdot9$ C.) in October and $81°\cdot1$ Fahr. ($27°\cdot3$ C.) in July, and the corresponding minimum between $70°\cdot9$ Fahr. ($21°\cdot6$ C.) in March and $63°\cdot5$ Fahr. ($17°\cdot5$ C.) in July, with annual means of $87°\cdot5$ Fahr. ($30°\cdot8$ C.) and $68°\cdot5$ Fahr. ($20°\cdot3$ C.). March is the hottest month, followed by April (with a difference of $1°\cdot44$ Fahr. *i.e.* $0°\cdot8$ C.). July is felt as a real winter month by Europeans. The excessive heat in the warmest months is tempered by the frequent rains and a refreshing breeze at about the hottest time of the day. In the very early morning and also in the evening the sky is overcast; in the later morning it is generally clear, and also in the early afternoon, and two or three hours after nightfall.

The climate of Kwamouth is a kind of transition between that of Leopoldville and that of Bolobo.

At Bolobo the means of five years' observations show that the mean monthly maximum temperature only varies between $89°$ and $83°·8$ Fahr. ($31°·7$ and $28°·8$ C.) and the minimum between $73°·4$ and $70°$ Fahr. ($23°·0$ and $21°·1$ C.). The relative humidity is very high, except in the period from June to August, which coincides with the cooler months. The nebulosity is greater than on the Lower Congo. The mean annual rainfall is $65·96$ inches ($1675·5$ mm.) and the days on which rain is measured vary in number from 90 to 119; the greatest amount measured in 24 hours is $6·44$ inches, or $163·6$ mm., in February.

In the Kwango district at Popokabaka, during the intense heat of the rainy season, the maximum temperature reaches $93°·3$ Fahr. ($34°$ C.) and the minimum is as low as $71°·6$ Fahr. ($22°$ C.) while, during the dry season, the extremes are $82°·4$ and $57°·2$ Fahr. ($28°$ and $14°$ C.). The prevalent winds are from the west, and blow especially in the afternoon; they are not strong winds. The rain storms generally come from the E. and N.E., but when they come from the south they are accompanied by tornadoes. Fogs and mists are very frequent during the dry season, but they generally disperse by 9 o'clock in the morning.

The climate further south, at the Francis Joseph Falls, is very similar to that of Luluabourg in the Lualaba district.

In the Lac Leopold II district, at Bolongo or Bolondo, the temperature is very regular; the nights are relatively cool and damp. It rains regularly every four or five days, but not violently, and the rains do not last more than 24 hours at most. Storms are fairly frequent, and there are mists at all seasons. The sky is often overcast, or partially so, in the morning.

In Equateur district, in the neighbourhood of Equateur-ville and Coquilhatville, the absolute maximum temperature is about $94°·2$ Fahr. ($34°·5$ C.) early in April and $93°·3$ Fahr. ($34°·0$ C.) in March, and the absolute minimum about

$63°\cdot5$ Fahr. $(17°\cdot5$ C.) in early June, giving an annual extreme range of $30°\cdot7$ Fahr. $(17°$ C.); but it is not usual for the thermometer to rise above $89°\cdot7$ Fahr. $(32°\cdot0$ C.) on more than a dozen days in the year. The daily range is sometimes very large, and, on the occasion of a tornado coming from the west, the thermometer has been noticed to fall from $88°\cdot2$ Fahr. $(31°\cdot2$ C.) in the early afternoon to $77°$ Fahr. $(25°$ C.) in the evening, and to reach $70°\cdot2$ Fahr. $(21°\cdot2$ C.) by six o'clock next morning. The mean monthly minimum temperature is not subject to such great variations, and the greatest range is not more than $3°$ Fahr. $(1°\cdot7$ C.), that is the difference between $71°\cdot4$ and $68°\cdot4$ Fahr. $(21°\cdot9$ and $20°\cdot2$ C.). There is in the mean monthly maximum temperature a range of about $5°\cdot7$ Fahr. $(3°\cdot2$ C.) while the range at Banana is as much as $12°\cdot6$ Fahr. $(7°$ C.). The great heat is tempered by breezes, which blow up the river, and the rain, too, has a cooling effect, and there are about 130 rainy days in the year.

Lemaire divides the year at Equateurville as follows ·

Latter part of January, February, March, April	Hottest season.
May...	Season of fewer storms.
June, July, August...	Season of fresh breezes.
September, October	Mean season.
November, December, beginning of January	Season of rains and storms.

During the "Fresh Breeze" season the wind generally gets up about three in the morning, and falls in the afternoon. Storms are frequent, and about half of them are accompanied by rain. Dew is almost unknown. For the northern part of the district *see* Nouvelle-Anvers in Bangala district.

In the Aruwimi district, at Basoko in the north, the hottest time of day is between one and three o'clock in the afternoon, the absolute maximum being $98°\cdot6$ Fahr. $(37°$ C.) at three o'clock; and the lowest record at the same hour $66°\cdot2$ Fahr. $(19°$ C.) in the middle of August, after torrential rain; while the lowest temperature measured is $62°\cdot6$ Fahr. $(17°$ C.) at six in the morning in July. The greatest annual range is therefore $36°$ Fahr. $(20°$ C.). The coolest period is the dry season and extends from mid December to mid March when the mean temperature seldom exceeds $82°\cdot4$ Fahr. $(28°$ C.). The general direction of the

15—2

winds is from the west; they are light in the dry season, more intermittent and stronger, sometimes almost amounting to a tempest, in the rainy season. Flamme speaks of its being so cold near the Aruwimi river, on the 17th July, during the rains, between seven and eight o'clock in the morning, that he had to put on a thick overcoat[1]; and again, farther east, on the Ituri river, on 15th August, he speaks of the night being cold and damp.

At Isangi, at the confluence of the Lomami, and therefore S.E. from Basoko, no very trying temperature is experienced, and it rains all the year round every four or five days; storms and tornadoes are frequent, the latter generally coming from the east; dews are of daily occurrence.

In the Bangala district, at Nouvelle-Anvers, in the west of the district, an absolute maximum temperature of $100°·5$ Fahr. ($38°$ C.) was measured on 19th January, 17th and 18th May and 9th June, and an absolute minimum of $64°·4$ Fahr. ($18°$ C.) on 11th October, giving a yearly range of $36°·1$ Fahr. ($20°$ C.). The mean maximum temperature oscillates between $89°·3$ Fahr. ($31°·8$ C.) in December and $93°$ Fahr. ($33°·9$ C.) in May. The mean monthly temperature is greatest from November to June, and appreciably less from July to October, February being the hottest month, with a mean of $80°·2$ Fahr. ($26°·8$ C.), and July and August the coolest months, with a mean of $76°·8$ Fahr. ($24°·9$ C.). The greatest range of temperature on one day was registered in May, when there was a range of $25°·2$ Fahr. ($14°$ C.) between the maximum and minimum.

At Bokula, in the northern part of the district, March is the hottest month, with a maximum shade temperature of $96°·9$ Fahr. ($36°$ C.). During the evenings of the rainy season there is invariably a moderate wind from the south. It rains about every three days; when the rain is fine and gentle it generally lasts about 36 hours, but these rains are rare; the torrential showers are more frequent, but do not last so long. Most storms are recorded in March, August and September.

The eastern part of the district has a climate similar to that at Nouvelle-Anvers.

[1] Flamme, *Dans la Belgique Africaine*, pp. 100, 121. Bruxelles, 1908.

In the Ubangi district near the Zongo Gorge the mean daily temperature is about 80°·6 Fahr. (27° C.) in the hot season and 78°·8 Fahr. (26° C.) in the dry season. The night temperature varies between 60°·8 and 68° Fahr. (16° and 20° C.). At all times of the year occur strong winds from N.E. and E., accompanied by storms and tornadoes, but generally there is a light W.S.W. breeze in the middle of the day from March to May, and a north-westerly breeze from June to September.

At Yakoma, in the east of the district, at the Mbomu river junction, dews are frequent at all seasons of the year.

At Banzyville, also on the Ubangi, but lower down, in the evenings of the dry season there is usually a strong breeze from N.N.E.

At Mobaye, on the Upper Ubangi, on the French side of the river, the mean atmospheric pressure varies only 15 mm. in the year, viz. between 24·41 and 25 inches (720 and 735 mm.); in December 1899 the variation was only 0·12 inch (3 mm.) and the same in January and July 1900. The minimum temperature in 1900 was recorded in January, viz. 49°·3 Fahr. (9°·6 C.) and the maximum in March, 113°·2 Fahr., *i.e.* 45°·1 C. The highest temperatures are recorded during the first five months of the year, and the lowest from November to March[1].

At Bomokandi, in the Welle district, the rains are both frequent and abundant, lasting about two or three hours, and, at the most, six. It rains every four or five days. Storms are very frequent and there are daily dews. There are mists on four or five days of the week.

At Jabbir, farther west, the absolute maximum temperature is 95° Fahr. (35° C.) and 62°·6 Fahr. (17° C.), the lowest, giving a range of 32°·4 Fahr. (18° C.). The former record is for September, the latter for April. When the rainy season is at its highest the heat is greatest. The greatest heat is in November. It rains throughout the year, with possibly a slight cessation at the beginning of the year, but by April the rains are frequent and copious, and the greatest precipitation is registered from July to November, when it rains every two or three days at least. Storms are frequent and violent, May and October being

[1] *M. G.* 1902, p. 437

especially marked in this respect, and hail is not rare. Mists are frequent, especially at sunrise, and clear off about 9 or 10 o'clock in the morning, otherwise the early morning is generally clear.

In the north of the Province Orientale, at Stanley Falls or Stanleyville, rains are frequent, and often abundant, throughout the year, and generally last from one to four hours. The only difference in the seasons is the greater or less amount of rain.

At Ponthierville, farther south, the temperature is pretty much the same all the year round, but there is a marked difference between the day and night temperatures. Rains are both frequent and copious, occurring every three or four days and lasting four or five hours. Storms are frequent in February and March. Mists occur in October, and there are dews all the year round. The sky is rarely quite covered. During February and March there is regularly a wind from N.E., which drops towards noon.

South of $4°$ S. in the neighbourhood of Nyangwe the prevalent wind in the dry season is from the west, which falls generally about 2 o'clock in the afternoon. The mean maximum temperature varies between $90°$ Fahr. ($32°·2$ C.) in October and $85°·8$ Fahr. ($29°·9$ C.) in January, and the mean minimum between $69°·1$ and $71°·6$ Fahr. ($20°·6$ and $22°·0$ C.); the mean monthly temperature varies between $77°·9$ Fahr. ($25°·5$ C.) in August and $80°·4$ Fahr. ($26°·9$ C.) in March, giving a mean annual temperature of $79°·2$ Fahr. ($26°·2$ C.). The mean range is about $18°$ Fahr. ($10°$ C.). The hottest season is later than on the lower Congo and extends from March to May.

There are rains throughout the year, lasting from half an hour to six hours, but they are very light in August.

At Kasongo the wind in the rainy season rises at about 11 a.m. and falls at 3 p.m. The N.E. winds are called "Lake breezes," coming from Tanganyika. There are sometimes strong westerly winds and the rainy season is from September to May. In the dry season there are mists.

In the Lualaba district at Luluabourg there is a marked uniformity in the monthly means of temperature, but, standing at a considerable altitude—2035 feet—this uniformity is not extended to daily readings of the thermometer, and there is a large daily range for the entire year varying between $42°\cdot8$ Fahr. ($23°\cdot8$ C.) and $48°\cdot8$ Fahr. ($27°\cdot1$ C.). The mean maximum temperature for the year is $90°\cdot2$ Fahr. ($32°\cdot3$ C.) and the minimum $66°$ Fahr. ($18°\cdot9$ C.). The mean monthly maximum only varies between $89°\cdot5$ Fahr. ($31°\cdot9$ C.) in October, November, February, and March, and $92°\cdot9$ Fahr. ($33°\cdot8$ C.) in July. The mean monthly minimum varies between $62°\cdot2$ Fahr. ($16°\cdot8$ C.) in July and $67°\cdot6$ Fahr. ($19°\cdot8$ C.) in April. On the other hand the mean monthly temperature is almost constant, as alluded to above, varying between $77°\cdot5$ Fahr. ($25°\cdot3$ C.) in June, July, August and September, $78°\cdot4$ Fahr. ($25°\cdot8$ C.) in March and December, and $78°\cdot6$ Fahr. ($25°\cdot9$ C.) in January and April.

The most prevalent wind is from the east ($41\cdot2$ per cent.), and the next from the south-east ($22\cdot4$ per cent.). It is seldom calm; winds from the N. and S. are rare; and winds from the remaining points are distributed in about equal proportions of from 7 to 8 per cent.

At Lusambo the prevalent wind in the dry season is from the west, and this generally dies away at about 2 o'clock in the afternoon. The mean maximum temperature varies between $90°$ Fahr. ($32°\cdot2$ C.) in October and $85°\cdot8$ Fahr. ($29°\cdot9$ C.) in January, and the mean minimum between $69°\cdot1$ Fahr. ($20°\cdot6$ C.) and $71°\cdot6$ Fahr. ($22°\cdot0$ C.); the mean monthly temperature varies between $77°\cdot9$ Fahr. ($25°\cdot5$ C.) in August and $80°\cdot4$ Fahr. ($26°\cdot9$ C.) in March. The mean annual temperature is $79°\cdot2$ Fahr. ($26°\cdot2$ C.). The mean range is about $18°$ Fahr. ($10°$ C.). The hottest season is later than at Banana and covers the period from March to May.

Buttgenbach, writing of his experience of the Katanga district, in the region of the Kambove mines, says that the thermometer, during June and July, did not rise higher than $80°\cdot6$ Fahr. ($27°$ C.), and that, generally, at the hottest time of the day, it was no higher than $73°\cdot4$ or $75°\cdot2$ Fahr. ($23°$ or $24°$ C.); in the evening, says he, it falls rather quickly, and, at 8 o'clock, the readings were $50°$ or $46°\cdot4$ Fahr. ($10°$ or $8°$ C.). Every morning there is a south-east wind, sometimes very strong. Europeans suffer rather from the cold than the heat. In the rainy season, of course, things are slightly different[1].

At Kansonso mines, the situation seems favourable as regards health. The mean maximum temperature during April was 89° Fahr. or 31°·6 C. and the minimum 69° Fahr. or 20°·6 C. The whole territory of the Tanganyika Concessions, Ltd., is reported on by the medical officer thus: " ... a country which, I believe, after two and a half years' observation, to be in every way well adapted to offer excellent health to any Europeans, who are willing to take simple and reasonable precautions[2]."

"In the Ituri forest," says Major Powell-Cotton, "the seasons are very ill-defined. Generally rain falls on four or five days of every week, while seven days without a thunderstorm was the longest dry period I experienced. In any big clearing, it was curious to hear a storm coming up, for the sound of the drops pattering on the leaves of the trees reached us long before the rain. The roar of a hurricane through the forest was an experience not to be forgotten. Our camp was nearly wrecked on one occasion, and a passage, several hundred yards wide, was cleared through the trees for a distance of some miles[3]." The Major was in the forest from the end of June to mid August in 1905 and from the last week in January to the first days of August in 1906, in practically the same districts, making in all some ten months, the wettest of which was July 1905, while July 1906 was one of the driest. Mosquitoes are almost unknown in the forest. The native from the plains or the white man usually suffers severely after a few months' residence in the damp atmosphere of the forest, rheumatism, dysentery, and bilious fevers being the most common complaints.

In the region between the Luru hills and the Ituri forest, and on the upper Ituri river, "the climate is splendid, and well suited for the white man's occupation[4]."

The coast region of Tanganyika is described by Mr Beringer as "ferociously hot[5]."

(For the Bangweulu district *see* under Rhodesia.)

---

[1] *M. G.* 1902, p. 577.   [2] Tanganyika Concessions *Report*, July, 1906.
[3] *Geographical Journal*, Vol. XXX, 1907, p. 371.
[4] Browne, *Scottish Geographical Magazine*, 1907, p. 86.
[5] *Geographical Journal*, Vol. XXIX, 1907, p. 370.

THE SEASONS.

In the Lower Congo district, from observations at Luozi, Kimoko and Lemba, it appears that the rainy season begins towards the end of the first fortnight in October and sometimes a little later (as in 1900 at Luozi, where it began on the 29th), and lasts till the 15th May, being interrupted by a lesser dry season, or a season of less rain, of about a fortnight, which occurs between the last days of December and early February. At Kimoko, however, this short dry season is said not to exist. The rains from February to May are intermittent.

In the Cataracts district, from observations at Lumba, Kusu, in Bembo region, at Tumba post, Kitobola, and Luvituku, the rainy season appears to begin a few days earlier (10th October), and at Lumba and Kusu so early as 24th—29th September. At these two places the lesser dry season only lasts 11 days but at the other places it is longer and lasts for 3 weeks or a month[1].

At Leopoldville the rainy season begins about the 20th September and lasts till the 20th May, with a short intermediate dry season in January, never exceeding 3 weeks, but variable. At Kutu (Lac Leopold II district) the rainy season lasts from mid September to mid May, with a short dry season from 15th January to 15th February. At Popokabaka (Kwango district) the rainy season extends from 15th October to 15th March, with an intermediate short dry season. At Lusambo (Lualaba-Kasai district) the rains begin between the middle of September and the middle of October and last till somewhere between the end of May and the end of June, with a short dry season between the end of January and the end of February. At Nouvelle-Anvers (Bangala district) the rains continue from August to the end of December; the dry season extends from December to some time in March; the period between March and August is alternately rainy and dry. At

[1] *La Belgique Coloniale*, 1902, p. 304.

Umangi, in the same district, the rainy season is said to extend from 15th March to 15th December, without an intermediate dry season; during the dry season, from December to March, it rains from time to time. In Bangala district, at Bokula, however, the rainy season is described as lasting from March to October, the remainder of the year being a season of less rain.

At Basoko (Aruwimi district) the rains begin at the end of August, and continue till the beginning of December; there is no dry season, as it rains, on an average, one day in every four for the remainder of the year. In Ubangi district, as a whole, the rainy season extends from the 20th April to early November, there being no intermediate dry season, but one of less rain, for from ten to fifteen days[1].

But in Ubangi district, at Yakoma, in the east, the rainy season begins during the latter part of March and lasts till the early days of November. December is a transition month, and there are fewer and lighter rains in January, February and early March. The rains are most frequent from May to August.

And at Banzyville, which, like Yakoma, also lies on the Ubangi, but lower down, the rainy season lasts from the end of May to the end of November.

In Equateur district, in the neighbourhood of Equateurville and Coquilhatville the seasons are only differentiated by the relatively greater and smaller amounts of rain. It rains most in November and December, and least in July.

At Bomokandi, in the Welle district, the rainy season lasts from May to October, the remainder of the year being the dry season, or season of less rain.

Though the latitude of Banana is approximate to that of Luluabourg, in the Lualaba district, the *régime* of the rains is perfectly different. At the latter station rain occurs on about 120 days in the year, and the total fall is some 60·82 inches (1544·5 mm.). May, June, July and August are the dry months, June and July being rainless; and there are two maxima, in March and November respectively, the total for March being 7·92 inches (201·3 mm.), and for November 8·68 inches (220·6 mm.).

[1] *La Belgique Coloniale*, 1902, p. 316.

At Mobaye, on the Upper Ubangi, on the French side of the river, November, December, and January are dry, the rainy months being June, July and August. The greatest fall on one day was 2·8 inches (73·4 mm.) on the 14th June[1].

In the north of the Province Orientale, on the Nepoko river, the dry season begins at the end of August, and lasts till March, while the rainy season extends from the end of April to July.

At Mawambi the dry season extends from the end of November to the end of February, while the rainy season lasts from the end of February to the end of November. At Avakubi there is no dry season, though in January (7 days), and in June (9 days) and July (8 days) there are seasons of less rain. During the four months from February to May there were in 1900, 18, 15, 16 and 14 days' rain respectively; during August, September, and October, 12, 11 and 12 days, and in November and December, 23 and 20 days, making altogether 165 rainy days in the year[2].

In the Lado Enclave, at Kiro (Kero), the dry season begins at the end of November and lasts till the end of March; the rainy season begins in April and ends in November, with an intermediate dry season extending from the end of April to the beginning of June. Major Powell-Cotton, writing of the Lado Enclave, says that all the Belgian stations along the Nile are malarious and swarm with mosquitoes; Kiro, the most picturesque of them all, being literally infested. " In fact the Enclave, generally, must rank among the most unhealthy districts of Central Africa; in one year the death-rate among the Europeans rose to over 20 per cent.[3]"

In the Ru Nzori region the heaviest rains occur in October, November, and December.

Briart, writing of the Katanga area, says that the climate is not only supportable, but in a way almost temperate. The region of the Kambove mines lying in the extreme west of the Lufila river basin enjoys a climate at least as healthy as any in Katanga. In the region of the

---

[1] *Mouvement Geographique*, 1902, p. 437.
[2] *La Belgique Coloniale*, 1902, p. 341.
[3] *Geographical Journal*, Vol. XXX, 1907, p. 371.

Kambove and Ruwe mines, the rainfall in 1904–5 was (in inches):

	Kambove	Ruwe
September	0	·53
October	3·21	2·16
November	5·28	4·73
December	11·69	11·63
January	9·37	4·41
February	7·31	4·45
March	5·80	4·11
April	·38	0
May	·77	0
Totals	43·81	32·02

With reference to the stations for which meteorological tables are given below :

Banana is on a peninsula, flat and sandy, on the right bank of the Congo mouth, with the sea on one side and a creek on the other. The nearest high ground is four or five miles off, in rear, and rising to a height of about 250 feet[1].

Luluabourg is situated twelve or thirteen hundred yards from the left bank of the Lulua river, and about 200 feet above its level. The plateau on which the town stands is undulating savannah, with forest in the hollows. The soil is sandy clay[2].

Bolobo stands on the left bank of the Congo river, some 60 feet above water-level, on the slope of a gently rising hill. The soil is sandy clay. The neighbouring country is savannah with forest. There is a marsh only a mile off[3].

---

[1] This station was started in 1890, with a complete set of instruments, as a station of the second order.

[2] Dr Lancaster makes the following statement : " Les thermomètres étaient installés favorablement, dans une maisonnette speciale et dégagée ; le baromètre était un baromètre de voyage Fortin. En ce qui concerne les corrections des thermomètres, le résultat de trois comparaisons avec un thermomètre normal a été noté, et il a été tenu compte dans la réduction des observations."

[3] The figures in this table were taken from the work of Dr Lancaster (*Rapports présentés au Congrès national d'Hygiène et de Climatologie médicale de la Belgique et du Congo*), who obtained them from the Reports of the British Association, and who appends the following note :—"Dans les considérations qui suivent, pour avoir un terme de comparaison avec la température moyenne des autres stations du Congo, nous nous sommes basés sur les moyennes déduites des maxima et minima moyens, tandis que dans nos tableaux résumés, nous avons conservé les moyens données dans les Reports de la *British Association* et obtenues par une combinaison que nous ignorons."

I have adopted $\dfrac{\text{max.} + \text{min.}}{2}$.

Kimuenza is on a plateau some 15 miles S.S.E. of Leopoldville. It stands a little more than 400 feet above a little stream, the Lukaya, some 20 minutes' walk from the station. The neighbouring country is accidented and the station is shut in by hills from S.S.E. to N. by W[1].

PRODUCTIONS, &c.

The great Congo area, with the exception of the extreme east and south, is essentially a land of rubber, rubber trees growing in some regions, rubber vines, the large woody climbers (*Landolphia*), in others, and what is known as "grass-rubber," in which the rubber is extracted from the roots, in others, more especially in the Kwango basin and to the west as far as the Monts de Cristal.

The oil-palm grows everywhere east of these mountains. The hot damp climate is suitable for the production of cocoa, the *Raphia vinifera* or bamboo-palm, from which palm-wine is extracted, flourishes in the riverain marsh lands, coffee is indigenous not only in Mayumba, but also in the Kasai valley. Bananas are most prolific.

Among other crops which can be grown with certain success in this climate are manioc or manihot, a shrubby plant growing to six or eight feet in height, from which cassava (tapioca) is produced, several varieties of maize or Indian corn, the bow-string hemp (*Sanseviera longiflora*), millet, tobacco and sugar-cane, provided of course that suitable localities be chosen. Ground nuts are produced and cotton will grow. In the clearings of the Ituri forest, in the north-east, there have been seen growing bananas, millet, rice, maize, sweet potatoes and manioc. In the peripheral regions, where the altitude is greater, there appears to be no reason why European vegetables should not flourish.

---

[1] Dr Lancaster writes as follows :—"La station météorologique de Kimuenza, qui fonctionne depuis le 1er octobre 1894, est installée dans d'excellentes conditions et suivant toutes les exigences scientifique. Son outillage comprend les instruments d'une station complète de deuxième ordre, tous de bonne qualité et dûment vérifiés."

BANANA. Lat. 6° 0′ 23″ S., Long. 13° 30′ 40″. Alt. 6 feet.

Compiled from 6 years' observations[1].

	Jan.	Feb.	March	April	May	June	July	Aug.	Sept.	Oct.	Nov.	Dec.	Year
Mean maximum temperature[2] (Fahr.°)	88°·5	89°·4	89°·6	88°·8	85°·8	82°·2	79°·3	78°·8	82°·0	83°·1	87°·1	87°·6	85°·2
Mean minimum temperature[3]	74°·7	74°·9	75°·2	73°·9	72°·5	68°·0	65°·8	66°·4	69°·6	72°·9	73°·8	74°·1	71°·8
Mean temperature[4]	81°·7	82°·0	82°·4	81°·1	79°·2	75°·2	72°·7	72°·5	75°·4	78°·8	80°·4	80°·8	78°·5
Mean daily range[5]	13°·8	14°·5	14°·4	14°·6	13°·3	14°·2	13°·5	12°·4	12°·4	10°·2	13°·3	13°·5	13°·4
Mean relative humidity (per cent.) [2] years (1890, 1891) }	77·7	77·2	76·4	79·4	79·7	76·9	78·9	78·4	79·0	78·2	78·7	80·1	78·5

Wind. Over 55 per cent. of the wind was in the half-quadrant between S.W. and W., and 67 per cent. in the quadrant between S. and W., the next most frequent wind was S.E. with 64 per cent.

Calms. These are most frequent in July and least frequent in March.

[1] 1890, 1891, (January, February, March, and July, August, September, 1891), (June—December, 1893), 1894, (January—March, 1895).
[2] The absolute maximum temperature recorded was 96°·8 on 28th February 1891.
[3] The absolute minimum temperature recorded was 59°·9 on 29th July 1894.
[4] The mean temperature is here the mean of the monthly results where both maximum and minimum were available, and not the mean of the mean maximum and mean minimum; hence the slight discrepancy in the figures.
[5] The daily range is max.—min.

LULUABOURG. Lat 5° 56′ S., Long. 22° 50′ E. Alt. 2035 feet.

Compiled from 2½ years' observations[1].

	Jan.	Feb.	March	April	May	June	July	Aug.	Sept.	Oct.	Nov.	Dec.	Year
Mean maximum temperature[2] (Fahr.°)	90°·3	89°·4	89°·4	89°·8	89°·8	91°·6	92°·8	90°·5	89°·6	89°·4	89°·4	89°·8	90°·1
Mean minimum temperature[3]	67°·3	66°·6	66°·6	66°·6	66°·2	63°·7	62°·2	64°·8	65°·7	66°·7	67°·3	67°·3	65°·9
Mean temperature[4]	78°·6	77°·9	78°·4	78°·6	77°·9	77°·5	77°·5	77°·5	77°·5	78°·1	78°·3	78°·4	78°·0
Mean daily range[5]	23°·0	22°·8	22°·8	23°·2	23°·6	27°·9	30°·6	25°·7	23°·9	22°·7	22°·1	22°·5	24°·2
Mean number of rainy days	12	11	16	14	9	1	1	5	6	15	18	13	121

[1] 1885, 1886, (January—June, 1887).

[2] The absolute maximum recorded was 103°·8 in February 1887.

[3] The absolute minimum recorded was 55°·0 in August 1885.

[4] The mean is derived from $\dfrac{7a + 2p + 7p}{3}$.

[5] The daily range is max.—min.

BOLOBO. Lat 2° 10′ 5″ S, Long. 16° 13′ 3″. Alt. 1080 feet.

Compiled from $3\frac{1}{2}$ years' observations extending over a period of 5 years[1].

	Jan.	Feb.	March	April	May	June	July	Aug.	Sept.	Oct.	Nov.	Dec.	Year
Mean maximum temperature[2] (Fahr.°)	86°·4	87°·6	88°·7	88°·5	87°·3	85°·8	86°·5	87°·3	86°·9	86°·4	85°·5	85°·3	86°·9
Mean minimum temperature	71°·8	71°·8	71°·9	72°·1	71°·8	71°·4	70°·5	71°·1	72°·3	71°·4	70°·9	70°·9	71°·5
Mean temperature[3]	79°·1	79°·7	80°·3	80°·3	79°·5	78°·6	78°·5	79°·2	79°·6	78°·9	78°·2	78°·1	79°·2
Mean daily range[4]	14°·6	15°·8	16°·8	16°·4	15°·5	14°·4	16°·0	16°·2	14°·6	15°·0	14°·6	14°·4	15°·4
Mean relative humidity[5] (per cent.) ...	82·1	81·0	79·8	80·5	81·8	77·3	72·2	72·0	76·4	81·5	83·5	83·8	79·3

[1] 1891, (January—June, 1892), (October—December, 1893), 1894, 1895.

[2] The absolute maximum temperature recorded was 97°·2 in April 1895, and the minimum 64°·9 in October 1895.

[3] The mean temperature is derived from $\dfrac{\text{max.} + \text{min.}}{2}$.

[4] The daily range is max.—min.

[5] The mean relative humidity is derived from $\dfrac{7a + 2p + 9p}{3}$.

KIMUENZA. Lat. 4° 29′ S., Long. 15° 22′ 30″ E. Alt. 1570 feet.

Compiled from $3\frac{1}{2}$ years' observations[1].

	Jan.	Feb.	March	April	May	June	July	Aug.	Sept.	Oct.	Nov.	Dec.	Year
Mean maximum temperature[2] (Fahr.°)	86°·9	87°·4	88°·5	88°·7	86°·4	82°·9	89°·3	83°·8	86°·4	86°·7	86°·7	86°·5	86°·7
Mean minimum temperature[3] ...	68°·7	68°·5	68°·5	68°·4	68°·0	62°·6	61°·5	62°·9	66°·0	67°·8	67°·5	68°·2	66°·6
Mean temperature	77°·7	77°·9	78°·6	78°·4	77°·2	72°·7	70°·5	73°·4	76°·1	77°·2	77°·0	77°·4	76°·1
Mean daily range (max.—min.) ...	18°·2	18°·9	20°·0	20°·3	18°·4	20°·3	27°·8	20°·9	20°·4	18°·9	19°·2	18°·3	20°·1
Mean relative humidity[4] (per cent.) ...	80	79	78	79	81	78	76	62	67	75	79	79	76
Mean number of rainy days	9	9	11	11	8	0	0	0	1	7	12	11	79
Maximum of rain in one day during the $3\frac{1}{2}$ year period (inches)	2·36	2·21	4·45	3·23	1·89	0	0	0	0·79	2·99	3·35	3·78	4·45

See note under Banana as to mean temperature.

Wind. Over 55 per cent. of the wind was between W.S.W. and W. The next most frequent wind was W.N.W. with 7·5 per cent.

[1] (October—December, 1894), 1895, 1896, 1897 (January—March, 1898).
[2] The absolute maximum recorded was 96°·8 in February and March 1896, while in the same year 96°·4 was noted in November.
[3] The absolute minimum was 54°·3 in August 1895, but this is extraordinarily low.
[4] The mean relative humidity is derived from $\dfrac{8a + 1p + 6p}{3}$.

K. A.

16

MEAN RAINFALL IN BELGIAN CONGO IN INCHES.

	Years	Lat.	Long.	Alt. feet	Jan.	Feb.	March	April	May	June	July	Aug.	Sept.	Oct.	Nov.	Dec.	Year
Banana[1]	$4\frac{6}{12}$	6° 0′ 23″ S.	12° 23′ 40″ E.	6	2·08	2·28	3·74	5·31	1·49	0·02	0·02	0·07	0·12	1·57	5·43	5·07	27·2
Luluabourg[2]	$2\frac{6}{12}$	5° 56′	22° 50′	2035	7·17	5·43	7·95	6·06	3·07	0·15	0·12	2·48	6·45	6·57	8·70	6·61	60·7
Congo da Lemba[3]	2	5° 53′	14° 43′	1666	0·94	1·26	3·58	4·80	1·73	0	0	0	0·04	1·26	2·91	1·34	17·8
Vivi[4]	$1\frac{3}{12}$	5° 40′	13° 49′	370	3·62	1·42	5·67	9·09	1·94	0	0·03	0	0·02	0·51	11·33	8·93	42·6
Lusambo[5]	$1\frac{9}{12}$	4° 57′	23° 28′	1380	8·22	7·51	8·50	5·51	3·92	—	—	2·04	6·14	5·86	8·38	9·84	65·9
Kabambara[6]	2	4° 37′	27° 40′	2770	8·54	4·48	7·16	8·26	6·85	0·98	0	0	0·71	5·74	6·73	8·91	58·3
Kimuenza[7]	$3\frac{6}{12}$	4° 29′	15° 22′ 30″	1570	4·02	6·02	6·93	8·34	5·34	0	0	0	0·39	2·95	7·63	7·28	48·9
Leopoldville[8] } Brazzaville	$2\frac{5}{12}$ $1\frac{5}{12}$	{4° 19′ 36″ {4° 17′ 2″	15° 19′ 11″ 15° 21′ 20″	1115} 1080}	6·30	4·88	7·36	7·04	4·29	0·55	0·02	0·04	2·16	5·35	11·49	8·42	57·9
Bolobo[9]	$3\frac{8}{12}$	2° 10′ 5″	16° 13′ 30″	1080	5·00	6·97	4·61	7·16	5·63	0·55	0·02	2·60	3·93	6·53	9·57	10·23	62·8
Irebu[10]	$2\frac{8}{12}$	0° 32′	17° 48′	1210	6·49	3·82	11·10	5·82	3·46	3·50	4·80	5·28	6·69	9·48	7·67	4·41	72·5
Basoko[11]	$1\frac{5}{12}$	1° 13′ 49″ N.	23° 39′ 20″	1380	2·16	3·66	6·02	6·10	3·46	7·40	6·29	4·44	6·73	3·35	6·02	4·37	60·0
Nouvelle-Anvers[12]	$1\frac{10}{12}$	1° 35′ 36″	19° 9′ 12″	1230	4·09	3·46	4·09	5·55	6·18	4·17	6·29	6·29	6·26	6·61	2·75	9·38	65·1

[1] 1890, 1891, January—March, July—September 1892, June—December 1893, 1894, January—March 1895.
[2] 1885, 1886, January—June 1887.
[3] October 1892—August 1894.
[4] May 1882—July 1883.
[5] August 1896—May 1897.
September 1869—August 1871.
[7] October—December 1894, 1895—1897, January—March 1898.
[8] Derived from combining the observations taken at both places. Leopoldville, March 1886—February 1887 and November 1893—May 1894. Brazzaville, April—June 1893, October—December 1894, September, October 1895.
[9] 1891, January—June 1892, October—December 1893, 1894 (except June), 1895.
[10] May—December 1897, 1898, 1899.
[11] November 1893—March 1895.
[12] February 1890—December 1891.

ANGOLA, WITH KABINDA.

"In Angola the rainfall, and with it the luxuriance of the vegetation, diminishes in intensity from north to south. Down to Ambriz the forests extend to the shore, then they gradually recede, never passing the line of greatest elevation, and cease almost completely in the district of Mossamedes. The Congo district, very warm, damp, and covered with dense vegetation, and the coastal zone to a few degrees south of Benguela, are considered unhealthy. At San Salvador, where regular records have been kept, the mean temperature from January to April is 75° Fahr. (23°·9 C.), with a maximum of 81° (27°·2 C.), and a minimum of 61° (16°·2 C.). It then falls gradually to 66° (18°·9 C.) in July or August, rising again to 79° and 77° (26°·2 and 25° C.) in November and December. The mean annual temperature is 76° (24°·5 C.). The humidity is not very high, if the mean, 77 per cent. be taken as a test. But the variations are very great, and in summer 99 per cent. is reached. The sky is rarely clear," and the rainfall is about 40 inches (1000 mm.)[1].

The above is contained in an article contributed to the *Scottish Geographical Magazine* of 1896, and is an abstract of a paper which appeared in the *Bulletin Soc. Géog. Belge*, and fairly represents the circumstances of the case, in a general kind of manner, as is shown by an examination of the figures from which the table (below) is compiled, the results of such examination being as follows.

At San Salvador the barometer is highest in the dry season, which is also the coolest season (715·3 mm. or 28·16 inches in July), and lowest in the rainy hot season (712·1 mm. or 28·03 inches in March). March is the hottest month and July the coolest, though there is but little difference in the temperature for the three months February —April; and though the maximum temperature is generally in March, it sometimes occurs a little earlier, sometimes

---

[1] *Scottish Geographical Magazine*, p. 560, 1896. The figures here given do not quite agree with the meteorological table below.

16—2

a little later. There is no variation, however, about the coolest month. It is always July. The change from the hot to the cooler season is abrupt, there being usually a sharp fall of $4°·5$ Fahr. ($2°·5$ C.) from the May temperature to that of June. The mean daily variation in temperature varies from $20°$ to $19°·2$ Fahr. ($11°·1$ to $10°·7$ C.), but the greatest range is about $21°·6$ Fahr. ($12°·1$ C.) in the dry season (July to September).

The relative humidity is about 78 per cent. from November to June and 72 from July to October.

The greatest amount of rain falls in April, when the precipitation is about a quarter of the total yearly amount; and there is a second heavy fall in November. July and August are absolutely rainless and there are slight showers in June and September. Thunderstorms occur in the period from January to May, but are most frequent in April, when they are sometimes of almost daily occurrence.

The sky is most cloudy at the beginning of the rainy season and in April.

Westerly and south-westerly winds are the most prevalent, 68 per cent. of the winds coming from these directions; about 12 per cent. come from the south.

The plateau or highland belt of Angola is undoubtedly healthy and its climate invigorating, a suitable locality, in short, for European colonization.

In the Malanje and Cassanje districts, at altitudes of 3786 and 3100 feet respectively, the temperature is more variable. The mean temperature at the latter is $77°$ Fahr. ($25°$ C.) for the months October to December, and $75°$ Fahr. ($23°·9$ C.) in January and February. Between October and January the hygrometer has indicated from 45 to 91 degrees of humidity. Travellers have extolled the salubrity of this region, but observations have not yet been made throughout the year. Much the same has been said of the Bihe country, but Capello and Ivens noticed great ranges of temperature; not infrequently it freezes at night, when during the day the temperature rises to $82°$, or even to $86°$ Fahr. ($27°·8$ or $30°$ C.).

Cunninghame, writing in the *Geographical Journal*[1], says that, in Angola, the Loanda-Benguela-Mossamedes

---

[1] *G. J.* Vol. XXIV, 1904, p. 154.

coast belt, varying in width from 150 to 30 or 40 miles, is enervating, relaxing, malarious, and intensely hot in the wet season ; but the average maximum temperature at midday in the dry season is 70° Fahr. (21°·1 C.), whilst at night it falls sometimes as low as 58° Fahr. (14°·4 C.). The great heat of the day is tempered on the coast by an almost constant sea-breeze, which generally dies down at sunset. The coast has a very bad reputation for unhealthiness, not wholly undeserved, but this could be considerably ameliorated by attention to sanitary rules.

The coast of southern Angola has the reputation of enjoying proverbially fine weather, for Benguela is the northern point of the Kalahari Climatic Province, and on the coast rain never falls, though there are very heavy dews, these peculiarities being shared by the coast lands of German South-West Africa further south.

The Belgian writer, in the article quoted above, says that Loanda consists of two quarters—the native town, situated on the shore, and the European, built on the nearest elevation, 200 feet above the sea. The surrounding country is arid, and there are no marshes near the town, but the banks of the Peixo and Bungo, which have been receptacles for refuse from time immemorial, exhale pestilential vapours when the water is low. Loanda has a bad reputation for intermittent fevers, says another authority, especially in the period covered by February, March and April. A virulent form of smallpox also attacks the native population at periodic intervals, notably between October and April.

Being situated on the coast, Loanda lies in a zone of low pressure. The mean barometric measurement for twelve years is 29·764 inches (756 mm.). The nights are often fresh, and, during six months of the year, it is pleasant to sleep under a blanket. The mean humidity varies between 82 and 92 per cent. with a maximum of 95·2 and a minimum of 75·7. The sky is generally cloudy, being absolutely clear on only ten days or so in the year, but yet the rainfall is small. The rainy days, that is, days on which rain can be, or has been, measured, are only about 42 in the year, and the quantity of water that falls does not exceed 10·6 inches (270 mm.). The rainy season is from

November to the beginning of May with the heaviest fall in March and April, but August is no drier, though it owes its humidity not to the rain, but to the *kassimbo*, a mist which rises in the night and is not dispersed till long after sunrise. In general the season of the *kassimbo* is most favourable to immigrants from Europe, while, on the contrary, acclimatised colonists then feel the cold at night. The town has been considerably improved, looking back to what it was twelve or fifteen years ago, the sand having been removed from the streets, and the shore cleared of its impurities.

At Loanda the south-west, west, and south winds are the most prevalent, in the order named; those most seldom experienced being from the north and north-east. From a study of the figures from which the table (below) has been calculated, it may be gathered that the rainfall is very variable, but that the period from May to October may be regarded as almost sure to be dry, though on one occasion $2 \cdot 22$ inches ($56 \cdot 3$ mm.) of rain fell in May. November is, as a rule, fairly dry, though as much as $3 \cdot 77$ inches ($95 \cdot 8$ mm.) once fell in that month, and the same with December. January is also a dry month, but $2 \cdot 43$ inches ($61 \cdot 9$ mm.) have been recorded in that month. February, March and April are the rainiest months, but in 1891 no rain fell in February, and only $0 \cdot 13$ inches ($3 \cdot 2$ mm.) ten years before; and somewhat similar vagaries have been recorded concerning March and April.

February is the hottest month, while July and August are the coolest. The mean temperature is $74^{\circ} \cdot 3$ Fahr. ($23^{\circ} \cdot 5$ C.). June, July, August and September are below the mean, October very near the mean, and the remaining months above it. The absolute maximum and minimum temperatures registered are 92° and 56° Fahr. ($33^{\circ} \cdot 3$ and $13^{\circ} \cdot 3$ C.).

Ambriz bay is very unhealthy, owing to the existence of extensive tracts of marshes, reaching several miles inland, over which, at night, dense fogs and mists hover, and there are also very heavy dews. The one redeeming point is the cool east wind.

The climate of Massangano, about 100 miles up the Kwanza river, is unhealthy, but Dondo and Kambamba, about 16 miles higher up the river, at the limit of navigation, are reported to be healthy settlements. Novo Redondo, at the most easterly point of the Angola coast, is, like most of the towns along this littoral, unhealthy, especially in the rainy season.

The winds from the swamps at Lobito Bay are very unhealthy, but the evil effects produced by them may be partially avoided by imitating the custom of many of the fishermen, and living on Lobito Spit, where the full benefit of the sea-breezes may be enjoyed. The town of Benguela is also unhealthy, though the inhabitants appear to find no fault with the climate. It is built on a marshy plain, and the exhalations from this, when inundated in March and April, the rainy months, have a very bad effect[1].

At Mossamedes the temperature in the hot season is about 86° Fahr. (30° C.), while from June to September the thermometer frequently sinks to 42° Fahr. (5°·5 C.). Near the marshes of the Bero river it is naturally unhealthy, but otherwise Mossamedes enjoys a salubrious climate, and the period from June to September is said to be both cool and bracing. The district of Mossamedes is, in fact, the only district of which it can, with certainty, be said that it possesses a climate suitable for Europeans, and here alone have white families permanently established themselves.

PRODUCTION.

Two climates prevail in Angola, the one in the north the other in the south. The northern parts form the southern coastal limit of cacao; and coffee, which is indigenous all along the west coast, from Sierra Leone, grows also in Angola, but only in the north; again, the bamboo palm (*Raphia vinifera*), from which palm-wine is extracted, whence its scientific name, and which loves hot coastal or riverain marshes, though flourishing in the north, ceases suddenly on reaching the eighth parallel of South Latitude. On the other hand, sorghum—the dhurra of Egypt, the Kafir corn or Guinea corn of other places— does well in the south and middle, but not in the north.

[1] *Africa Pilot.*

Rubber grows in the north, this being also the southern coastal limit, on the west, of the large woody climber, *Landolphia owariensis*, which, like coffee, is indigenous within the limits mentioned above.

The Lower Kwanza valley is well suited for the cultivation of the sugar cane.

Among other crops which may be produced are bananas, figs, manioc, from which cassava (tapioca) is produced, millet, maize, sweet potatoes, various kinds of pulse and sesame. Tobacco also will grow, and cotton, there being plenty of what is known as black cotton soil, while many gum-bearing trees thrive in the south, especially copal. In the forests are many valuable woods, such as *Diospyros mespiliformis*, an ebony, called *omuati* by the natives, and the mopani tree (*see* under North-Eastern Rhodesia, Productions) and oil palm. The south of Angola is the southern limit of the baobab.

LOANDA. Lat. 8° 49′ S., Long. 13° 7′ E. Alt. 192 feet.

Compiled from 12 years' observations, 1880—1891.

	Jan.	Feb.	March	April	May	June	July	Aug.	Sept.	Oct.	Nov.	Dec.	Year
Mean barometric pressure in inches[1]	29·722	29·706	29·710	29·718	29·756	29·836	29·868	29·852	29·824	29·760	29·724	29·722	29·764
Mean maximum pressure ...	29·756	29·764	29·772	29·784	29·816	29·944	30·008	30·000	29·932	29·816	29·772	29·788	29·828
Mean minimum pressure ...	29·678	29·670	29·670	29·658	29·722	29·792	29·804	29·808	29·792	29·732	29·698	29·662	29·732
Mean temperature[1] (Fahr°.)	77°·0	78°·8	78°·6	78°·6	75°·4	70°·2	67°·6	67°·5	70°·5	74°·1	76°·8	76°·8	74°·3
Mean maximum temperature[2]...	78°·1	81°·3	81°·1	80°·2	79°·7	74°·5	71°·4	70°·2	72°·3	76°·6	79°·2	77°·9	76°·5
Mean minimum temperature[2]...	74°·7	75°·6	75°·2	75°·0	72°·1	66°·6	65°·5	65°·3	68°·5	72°·3	75°·2	75°·0	71°·8
Mean daily range	3°·4	5°·7	5°·9	5°·2	7°·6	7°·9	5°·9	4°·9	3°·8	4°·3	4°·0	2°·9	4°·7
Mean relative humidity[3] (per cent.)	86	83	85	87	88	87	86	87	85	84	85	87	86
Mean amount of cloud[3] (0—10)	6·2	6·4	6·6	7·2	5·4	4·6	5·1	5·6	6·0	6·3	6·9	6·8	6·1

NOTE.—Loanda is a port on the Atlantic coast, on a bay, sheltered on W.N.W. by a sandy island. The town stands on a hill, and here was the meteorological observatory, equipped with all the usual instruments. The observations were published annually in the *Annaes* of the Lisbon Observatory.

Wind. The most frequent winds are S.W., W., and S., in the order named, and these winds account for 73 per cent. of the whole.

[1] The mean pressure and temperature are both derived from $\dfrac{9a + 3p + 9p}{3}$.

[2] The absolute maximum temperature recorded was 99°·1 on 22 March 1888, and the absolute minimum 56°·3 on 7 and 22 August 1880, 7 September 1882, 8 July 1891.

[3] The cloud and humidity observations were only taken for 5 years, 1881—1885.

SAN SALVADOR. Lat 6° 17′ S., Long. 14° 53′ E. Alt. 1828 feet.

Compiled from 4 years' observations, 1884—1887.

	Jan.	Feb.	March	April	May	June	July	Aug.	Sept.	Oct.	Nov.	Dec.	Year
Mean temperature[1] (Fahr°.)	74°·5	75°·6	75°·7	75°·2	73°·6	69°·4	67°·3	68°·4	70°·9	73°·6	73°·9	73°·8	72°·7
Mean maximum temperature	85°·3	87°·1	88°·0	91°·2	84°·7	82°·0	80°·1	81°·3	83°·7	85°·5	84°·6	83°·3	84°·7
Mean minimum temperature	66°·2	67°·1	67°·5	67°·5	66°·7	62°·2	59°·2	59°·9	62°·1	65°·3	66°·4	66°·6	64°·7
Mean daily range[2]	19°·1	20°·0	20°·5	23°·7	18°·0	19°·8	20°·9	21°·4	21°·6	20°·2	18°·2	16°·7	20°·0
Mean relative humidity[3] (per cent.)	76	76	75	80	80	77	74	71	70	74	79	80	76
Mean maximum humidity	82	81	77	83	85	80	77	74	72	76	81	84	79
Mean minimum humidity	69	71	72	78	78	75	70	66	69	69	76	77	72·5
Mean nebulosity (0—10)	6·5	6·4	5·8	7·5	6·2	3·8	4·0	4·5	5·5	6·8	7·3	7·2	6·0

NOTE.—San Salvador lies in an elevated region, at a distance of about 150 miles from the Atlantic coast, near the water-parting between the Mpozo and Kwelu rivers, on a plateau isolated on all sides by ravines, whose waters flow to one or other of these two streams. The observations were made regularly three times a day, at 9 a.m., 3 p.m., and 9 p.m., and appeared, at the time, in the *Annaes* of the Lisbon Observatory (*Postos meteorologicos*).

[1] The mean temperature is derived from $\dfrac{9a + 9p + \text{max.} + \text{min.}}{4}$, and is probably too low.

[2] The daily range is max.—min.

[3] The mean relative humidity is derived from $\dfrac{\text{max.} + \text{min.}}{2}$.

ANGOLA. MEAN RAINFALL IN INCHES.

	Jan.	Feb.	March	April	May	June	July	Aug.	Sept.	Oct.	Nov.	Dec.	Year
San Salvador (4 years[1])	3·76	3·87	4·23	9·59	2·34	0·17	0	0	0·08	3·38	7·62	4·44	39·48
Loanda (12 years[2])	0·39	1·58	2·26	4·05	0·42	0	0	0·02	0·04	0·18	0·98	0·70	10·62
Comber Station[3] (Kibokolo), Lat. 6° 16′ S., Long. 15° 17′ E. Alt. 3100 feet.	7·16 6·98	5·06	12·22	5·66	0	0	0	0·81	8·19	7·13	7·25	60·46

[1] 1884—1887. The season 1886—1887 was the rainiest, when, from October to May, 48·24 inches fell: the greatest quantity on one day was 4·33 inches in October. The mean number of rainy days in a year is 80.

[2] 1880—1891. The rain is very variable, the annual mean of 10·62 inches being formed from such figures as 21·46 inches in 1883 and 3·98 inches in 1891.

[3] November, December 1899, January—November 1900.

KABINDA.

This small sea-board territory lies north of the Congo mouth and it is recommended that enquirers should consult the paragraphs on the Lower Congo under the "Belgian Congo." At Chinchoxo, on the coast of Kabinda, the mean annual temperature is $75°\cdot9$ Fahr. $(24°\cdot4$ C.), the mean maximum is $83°\cdot3$ Fahr. $(28°\cdot5$ C.), and the mean minimum $71°\cdot8$ Fahr. $(22°\cdot1$ C.), giving a yearly range of only $11°\cdot5$ Fahr. $(6°\cdot4$ C.), which of course is due to its position on the coast. The absolute maximum recorded is $96°\cdot9$ Fahr. $(36°$ C.), and the absolute minimum $58°\cdot1$ Fahr. $(14°\cdot5$ C.), giving an absolute range of $38°\cdot8$ Fahr. $(21°\cdot5$ C.). The mean relative humidity is $84\cdot8$ per cent., that at 6.30 a.m. is $90\cdot8$ per cent., and that at 2 p.m. 76 per cent. The coolest months are June, July, August and September. The warmest time of day is between eleven o'clock and noon, after which there is a sea-breeze which keeps the temperature down. The nebulosity diminishes from morning towards evening, except in the months of March, April, and November, when there are evening storms. Heavy rains are always accompanied by storms of thunder and lightning. There are very heavy dews during the dry season. November and December are remarkable for the purity of the air. The sea-breezes, mentioned above, are generally from the S.S.W., and become gradually stronger till four or five o'clock and then the direction changes to west. In the evening the wind dies off and turns south. Between sunset and 9 p.m. it is usually beautifully calm, and then the land-breeze springs up and lasts till midnight. This rule for the winds is fairly constant during the dry season, but cannot be depended on in the rainy season. Very violent winds are both rare and of short duration.

At Malemba there is no forest, and the atmosphere and soil are both dry; hence the climate, as compared with other West African coast towns, is salubrious, and it is reported to be well suited for Europeans. At any rate, the settlers there enjoy excellent health.

CHINCHOXO. Lat. 5° 9′ S, Long. 12° 5′ 30″ E. Alt. 39 feet.

From 2 years' observations[1].

	Jan.	Feb.	March	April	May	June	July	Aug.	Sept.	Oct.	Nov.	Dec.	Year
Mean temperature² (Fahr.°)	78°·3	77°·4	79°·2	79°·2	77°·7	75°·7	72°·3	71°·1	71°·2	73°·8	76°·5	78°·3	75°·9
Mean maximum temperature³	86°·4	86°·5	87°·6	86°·4	84°·9	82°·8	78°·8	78°·4	77°·5	80°·1	82°·9	86°·2	83°·2
Mean minimum temperature	73°·9	75°·4	75°·0	73°·4	72°·7	70°·9	67°·3	65°·3	66°·7	70°·2	73°·0	75°·7	71°·6
Mean daily range⁴	12°·5	11°·1	12°·6	13°·0	12°·2	11°·9	11°·5	13°·1	10°·8	9°·9	9°·9	10°·5	11°·6
Absolute maximum temperature	93°·9	93°·4	96°·6	93°·7	90°·1	89°·1	83°·5	82°·4	81°·9	85°·1	90°·9	93°·9	96°·6 (highest)
Absolute minimum temperature	69°·1	67°·5	70°·3	70°·2	69°·1	62°·4	58°·3	59°·4	60°·4	61°·9	69°·8	67°·3	58°·3 (lowest)
Mean relative humidity per cent (6.30 a.m.)	90	93	93	91	92	91	92	91	91	88	89	90	90·8
Mean relative humidity per cent (2 p.m.)	73	76	74	72	77	77	76	75	80	76	76	77	76
Mean number of days cloudy (8—10)	4	6	3	8	6	5	6	6	15	15	9	10	93
" " clear (0—2)	4	0	5	1	1	1	6	4	1	0	0	0	23
Mean cloudiness (0—10)	5·1	5·1	5·0	6·5	6·5	5·8	4·9	5·2	7·1	7·6	6·8	6·9	6·0
Mean rainfall in inches⁵	12·24	4·72	7·32	3·98	2·13	0	0	0·20	0·32	0·87	8·54	2·09	42·41
Mean number of rainy days	13	10	12	10	3	3	0	3	5	14	12	6	88

[1] 1874, 1875, *Die meteorologischen Beobachtungen der Güssfeldt'schen Loango-Expedition, nebst einem Anhang: Resultate der Beobachtungen von Dr O. Lenz am Ogowe.* Dr von Danckelmann, Leipzig, 1878.

[2] The mean temperature is derived from $\dfrac{6.30\,a + 2p + 10p}{3}$.

[3] The hottest time of day is between 11 a.m. and noon.

[4] The daily range is max.—min.

[5] February—December 1874, 1875, January—March 1876.

EASTERN TROPICAL AFRICA

THE WINDS AND RAINS OF EAST AFRICA[1].

In the south, during the summer months, *i.e.* from October to April, south-easterly winds prevail. These are the "South-East Trades," and they follow the trend of the coast, blowing in a direction almost normal to the general line of the curve. At False Bay they are S.S.E., and gradually change to S.E. at Cape Agulhas, while between Natal and Delagoa Bay they have an almost easterly direction. This gradual easting of the wind is due to the gradual development of a V-shaped baric depression inland, corresponding roughly with the form of the southern part of the continent, within which a cyclonic area is eventually formed, to be dispersed as the summer wears on. In the winter months, *i.e.* from April to October, the prevailing winds are westerly. All the eastern portion of South Africa is in the region of summer rains.

In Southern Rhodesia the prevailing winds for the greater part of the year are in the quadrant between N.E. and S.E. In the early summer months, however, the wind backs through N.N.E. to N. and N.W.; and these backing winds bring the early rains to the whole of this region. The change in direction is due to the diminution in pressure, from the steady pressure area south-eastwards, during this period, the changes increasing rapidly in that direction from 0 inch through -0.05 inch to -0.10 inch off the coast from October to November and November to December. During the later summer months cyclonic disturbances, coming from over the Indian ocean, are experienced, and these bring with them the later rains, which are heaviest in the eastern portion of the area and gradually diminish in intensity westwards.

[1] The principal authorities for this Section are the *Challenger* Reports, *Africa Pilot*, Part 3, the *Meteorological Reports* of the Egyptian Survey Department, and *Science in South Africa*, 1905.

Monsoon effects are not experienced much south of $18°$ S., but in the Mozambique Channel north of this the winds depend on the northerly and southerly monsoons of the Arabian Sea, but are not so regular and are also somewhat modified in direction. Here the northerly monsoon begins between mid September and mid October, and the southerly, which varies between S.S.E. and S.S.W., between mid March and mid April, and has its greatest westing in May and June. It is also noticeable that the southerly monsoon is the period of greater, while that of the northerly is the period of less, wind. Abreast of the Channel, in Nyasaland and neighbouring regions, the South-West Monsoon begins in April, and, blowing over the land, brings no rain with it, in fact the beginning of the dry season coincides, as nearly as possible, with the arrival of the monsoon. In North-Western Rhodesia the prevailing wind is from the north-east. In North-Eastern Rhodesia, from April to October the North-East Trades are the prevailing winds. In summer the direction is variable, but the most general direction is from the north.

North of the Mozambique Channel, the monsoons of the Arabian Sea prevail, in their ordinary form. The North-East Monsoon begins in the Arabian Sea about mid October, but sometimes it is not experienced on the African coast till the middle or end of November, and it usually lasts till the middle of March, acquiring its greatest strength just after the turn of the year. The North-East Monsoon is succeeded, in the region between Cape Delgado and the Equator, by a short period of usually light winds from E.N.E. and E.S.E. This variable period extends to the end of April. From the end of April to the beginning of October is an uncertain period, but one of two southerly winds always blows in the more southerly coast districts, either the South-East Trades, or the South-West Monsoon. The wind then veers from N.E., through E., to South. In the more northerly coastal districts the South-West Monsoon is the more regular wind. In the southern portions of this district the southerly wind has more east than west in it at the beginning of the season, and hence brings copious rains in April and May as, near the Anglo-German boundary, it impinges on the coast, which trends slightly eastwards.

Abreast of this coastal district, in the interior uplands of the German sphere, strong South-East Trades, deprived of their moisture by Madagascar, and therefore dry, prevail from April to September, while, for the remainder of the year, variable light local winds, with no fixed direction, are the rule.

Further north, in the neighbourhood of Kilimanjaro, the North-East Monsoon prevails from December to February. Nothing can be said with certainty with regard to March, this being a very irregular month. During April the wind is nearly always in the east, sometimes directly from the east, at others from the south-east. In the following month the wind steadies down and the South-East Trades may be counted on, as a rule, till December.

North of the zone of variable airs and calms, the whole area of North-East Africa south of 18° N. is subject to the influences of the North-East Trades and the South-West Monsoon, which regulate the seasons. In the Sudan and Abyssinian areas, the N.E. winds are experienced for more than half the year, and are essentially dry winds coming over the land. In February the African baric depression begins to move northwards, gradually becomes steeper, and settles over the Eastern Sudan in May, to become linked up with the Siberian depression in June, passing N.E. in company with that cyclonic area throughout July and August, the minimum pressure in the last month lying over the Persian Gulf. On the eastern Somali coast the influence of this monsoon depression is felt earlier than in the inland areas (the Sudan, &c.), and also earlier on the southern part of that coast than at Guardafui, where the South-West Monsoon is not experienced till about the end of April. It is during this South-West Monsoon period that the rains are experienced, but the underlying causes which produce the rain are quite different inland and near the east coast respectively. In the Sudan and Abyssinia and W. Somaliland the precipitation is caused by the uplifting of the wind to scale the lofty plateaux, while in the extreme east, in the Italian sphere of Somaliland and further inland, the rain is caused by the backing of the S.W. winds. These at first blow strongly from S.S.W. parallel with the general trend of the coast, but owing to the development of the deep depression, alluded to above, back round through south to east, and bring the rain from over the Arabian Sea.

In the extreme north, in the Lower Egypt area, Captain Lyons[1] states, "rain is comparatively frequent on the coast, but it rapidly diminishes as the sea is left behind. It is almost invariably cyclonic in character, and falls when the trough of the storm has passed, so that it is accompanied by westerly or north-westerly wind, and it is almost entirely confined to the winter months." These winds are due to the almost invariable rule that the direction of the Mediterranean cyclones is from west to east.

[1] Survey Department, Egypt. *Meteorological Report* for the year 1904. Part II, p. viii.

K. A.

17

ANGLO-EGYPTIAN SUDAN[1]

GENERAL.

Taking a general view of the Egyptian Sudan, it may be said that during the greater part of the year the climate is dry and healthy. From about the 17th parallel northwards the Nile valley is rainless throughout the year. Southwards of this parallel the summer rains gradually increase (1) with decrease in latitude, and (2) with approach to either the sea or the Abyssinian uplands, when two or three months in the year are wet.

In the winter months there are usually a few rainy days between Halfa and Dóngola, but on the coast, at Suakin, there are heavy intermittent rains between August and January, and now and then rain is measured in spring.

In the regions of the Upper Blue Nile and Átbara the rains begin towards the middle or end of May and last till mid September; and there are also light rains in January and February, and somewhat heavier rains in October and November; so that the really dry months are usually March, April and December.

Such rain as occurs at Khartoum is measured, on only about fifteen days, between the latter part of May and the beginning of October.

The following list gives the rainy season in various regions:

Region	Rainy Season
Bahr-el-Ghazal	April—October.
South Kordofán	June—October.
Darfur	June—October.
Sobat region	May—end of October.
Upper White Nile	September—November.
Equatorial Nile region	February or March—October or mid November.

[1] This section is compiled mainly from

1. *Anglo-Egyptian Sudan.* By Colonel Count Gleichen, K.C.V.O., D.S.O.
2. *Physiography of the Nile and its Basin.* By Captain Lyons, F.R.S., Surveyor-General of Egypt.
3. "Die Hydrographie des obern Nil-Beckens." *Zeitschrift der Gesell-schaft für Erdkunde,* 1897, p. 303.
4. The annual Meteorological Reports of the Egyptian Survey Department.
5. The annual Reports on the Rains of the Nile Basin issued by the same Department.

In the Upper White Nile region, in the hill country about Rejaf and the district north of it, where the river is characterized by the presence of sudd, the rains are usually accompanied by thunderstorms.

The most unhealthy period is that immediately after the rains, when the Bahr-el-Ghazal and Upper White Nile regions have a bad reputation, partly shared by the Upper Blue Nile country. The least healthy period in Kordofán, at Kássala, and at Suakin is also that following the rains.

As to the temperature in the Egyptian Sudan, Count Gleichen, in his book, *The Anglo-Egyptian Sudan*[1], says that this "is, in summer, undoubtedly hot, the thermometer having on rare occasions risen to as much as $126°$ Fahr. ($52°\cdot2$ C.), but, on the other hand, except in the rainy season, the nights are always cool and refreshing. During the rest of the year the temperature naturally varies considerably, but it may be said that the winter is bright and invigorating throughout and not too hot. There are even unpleasantly cold winds at times. Except during the rainy season, the prevailing winds are, in the Nile Valley, always from the north. In other parts the wind varies according to the season, but all get their share of the north wind." In the northern part of the area (Anglo-Egyptian Sudan) the N.E. Trades blow uniformly throughout the year, and southerly winds are practically unknown, while rain only comes in occasional falls when there are thunderstorms. In the southern portion, in summer, the winds of the S.W. Monsoon blow from May to October, and the N.E. Trades for the remainder of the year. The extreme south is in the region of equatorial calms and the direction of the wind is uncertain.

WIND.

At Wadi Halfa the prevailing winds throughout the year are from the north-west, the observations for ten years showing 51 per cent. from that direction, against 22 per cent. from the north, the next in order of frequency.

At Mérowe in 1905 the prevailing wind was from the north-east, but during the period May—September it blows from the south.

[1] Published by H. M. Stationery Office.

17—2

The observations for three years at Berber show that the most frequent direction is from the north, though in July, August and September there are winds from other directions.

At Suakin 62 per cent. of the wind is from the north and north-east, but during June, July and August the winds are variable ; the direction is frequently from the north-west in the early months of the year.

At Tokar, during the first five or six months of the year and the last three, the winds are from N.N.E. or N.E.; for the remaining months the direction is S.W.

During the first and last quarters of the year, *i.e.* the dry season, the wind at Dueim is nearly always from the north or north-west ; the prevailing winds for the remaining six months are southerly.

In the Sobat plains the winds are north and north-east in the dry season and southerly in the rains.

At Lado the winds are northerly in the dry season and southerly in the rainy season, and similar conditions prevail at Mongalla, Gondókoro and Wádelai.

In the Upper Bahr-el-Jebel region the prevalent winds are from the north during the period November to March, and from the south from May to September, being variable in April and October.

At Wau, in the Ghazal region, during December, January, February and March the wind is almost entirely, from the north ; in April chiefly from the north and between S.E. and S.W.; in May entirely from the south, in June mostly from the south, while it is variable in July and August ; in September it is generally from the south-west.

RAINFALL.

In the Nile region of the Nubian Desert, roughly from Wadi Halfa to Berber, including the Dóngola province, there is very little rain, but showers may generally be expected in September.

In the Atbai region, east of Wadi Halfa, though the air is dry, there are heavy dews and mists on the west of the watershed, especially in December, in the northern parts, and violent local showers in July and August, and, in some years, in May and June, though some districts are rainless for years together. On the coast the dews are very heavy. In the southern part of this district there are no dews to the west of the watershed, but on the east there are heavy dews in autumn and winter.

In and around Suakin from October to February heavy intermittent rains are experienced, and there are occasional storms even up to the end of March. In the adjacent desert regions to the west, torrential rains occur occasionally in the period from July to September.

At Kássala such rain as falls is always brought from the east and usually falls in the evening between 6 and 8 o'clock, being ushered in by violent dust storms; but the total fall is very small.

In the Gedáref district the rainy season lasts from May or June till October.

In the Blue Nile region the rainy season begins in May and lasts till the end of October, in which month there are frequent local thunder-storms with torrential rain.

In the district extending south of this to the Sobat the rainfall is very heavy, increasing from north to south, and the season lasts from the end of April to the beginning of November.

In the White Nile plain, in the Baro river, Ajuba river, and Sobat river regions, the dry season lasts from October to the middle of June. There is scarcely ever a rainstorm during this period. Towards the end of June the wet season is heralded by almost daily storms, which continue through July; and in August and September the rains are abundant.

In Kordofán the rainy season lasts from mid June to the end of September. At the beginning, every two or three days, there are storms of rain from the south, ushered in by storms of dust and sand, thunder and lightning. Occasionally there is a steady rain for twenty-four hours, but usually it comes in the form of storm showers.

In the Bayuda Desert, west of Khartoum, the rainy season is uncertain, the usual season is in June and July or August, but there may be showers in May. In fact the climate is Saharan in type; there may be several years without rain, and then a very deluge, flooding the country, may be experienced.

In the highlands east of the Bahr-el-Jebel, Capt. Lyons[1] estimates that the total annual rainfall must amount to from 55 to 60 inches (1400 to 1500 mm.).

At Wau in the Ghazal region in 1905 there was no rain in January, February, March, or December, and the last-named month was also rainless in 1904. The means for the two years were in April 0·87 inches (22 mm.), May 4·29 inches (109 mm.), June 6·77 inches (172 mm.), July 3·94 inches (100 mm.), August 7·60 inches (193 mm.), September 4·41 inches (112 mm.), October 2·09 inches (53 mm.) and November 2·13 inches (54 mm.).

In the area lying between Lake No and Kodok, usually known as the Shilluk country, from its inhabitants, the greatest amount of rain falls in August. There were in 1901, in that month, eighteen rainy days, not successive, the precipitation occurring rather in terrific storms, preceded, accompanied, and followed by violent thunder and lightning. In September there were fourteen days on which rain fell[2].

TEMPERATURE, &c.

In the Nile region of the Nubian Desert, roughly from Wadi Halfa to Berber, including Dóngola province, the climate is healthy all the year round, though from June to the end of September the weather is very hot and enervating, but the air is dry and the wind almost always from the north.

The Atbai climate has been pronounced to be the best in the Egyptian Sudan, the air being absolutely pure, the shade heat mitigated by the altitude (1500—3000 feet), but the heat of the sun is intense in warm weather, owing largely to the glare from the sand and rocks. In winter the cold is quite severe. In the southern parts of this region, east of the watershed, an almost perfect winter climate is enjoyed, the range of temperature being less than in the north.

[1] *The Physiography of the Nile and its Basin.* Capt. Lyons, F.R.S.
[2] Tappi, *Bulletin de la Société Khédiviale de Geographie*, Vᵉ série, 1904 No. 4, pp. 178 *sqq.*

At Suakin, from June to September, the heat is intense and the relative humidity high. Sandstorms are frequent in the summer. In January the means of the day and night temperatures are respectively $77°$ and $73°$ Fahr. ($25°$ and $22°·8$ C.). The means of the absolute maxima for the two years 1904, 1905, for the months of June, July, and August, are $117°·5$, $117°·7$ and $117°·5$ Fahr. or $47°·5$, $47°·6$, and $47°·5$ C.; and the corresponding means of absolute minima for the same months are $72°·1$, $77°·2$ and $79°·3$ Fahr. or $22°·3$, $25°·1$ and $26°·3$ C. Sir Charles Watson, who registered the temperature at Suakin during these months in 1886, found that the highest temperatures registered at 1 p.m. were respectively $100°$, $109°$ and $111°$ Fahr. ($37°·7$, $42°·8$ and $43°·9$ C.), but the thermometer was fixed up in a cool room in Government House. The *mean* temperature at 2 p.m., from the 1904–5 returns, is, in the same three months, $94°·8$, $98°·2$ and $98°·8$ Fahr., or $34°·9$, $36°·8$ and $37°·1$ C. When our troops were at Suakin in 1896 it was found that, on the 22nd June, when the wind veered from W. to N., at 4 p.m., the thermometer suddenly dropped from $106°$ to $94°$ Fahr. ($41°·1$ to $34°·4$ C.) at 5 p.m.

When the sandstorms are rife in the summer time it is difficult to conceive a more disagreeable climate.

At Kássala the climate is healthy for eight months in the year, though very hot in the period from March to May. Malaria is frequent from July to October, especially in very rainy years.

In the Gedáref and Gállabat districts the climate is healthy from December to May, but in September and early October there is a good deal of malaria, especially, as at Kássala, when there has been a good rainy season.

In the Blue Nile region March, April, and May are hot, August, September and October, the last months of the rainy season, hotter, and damp as well; the heavy rains may cool the air temporarily, but the subsequent heat is moist and enervating. December, January, and February are both cool and healthy. At Sennar, Karkoj and farther south malaria is prevalent, and increases with the density of the vegetation. On the whole September and October are the worst months. After the October storms are over the climate gradually improves.

In Kordofán the year is divided as follows:

Name	Equivalent	Duration
Kharif	Rainy season	Mid June—end of September
Shita	Cold weather	October—February
Seif	Hot weather	March—Mid June

In the latter part of the rainy season a general lassitude is felt by day, the nights are characterized by a steamy moisture, and fever is prevalent. By the end of September the north winds set in, the air becomes cool and refreshing, and the breezes bracing. The heat during "hot weather" is not so great as in some other parts of the Sudan.

The climate of Fasher in Darfur is said to be healthy and that of Jebel Marra both healthy and cool.

Mr Bulpett, in *A Picnic Party in Wildest Africa*, writes: "Khartoum has a glorious climate during the winter months. The desert air has a crispness which must be experienced to be appreciated; and though the rays of the midday sun are powerful, the freshness of the early mornings in winter is delightful."

Mrs McMillan, who formed one of the same party as Mr Bulpett, says: "It is very hot in the daytime, and the heat is modified only by a strong wind which blows dust and sand into one's eyes...but the nights are beautiful beyond words."

In the Shilluk country, that is between Lake No and Kodok, the minimum and maximum temperatures are in June $69°\cdot8$ and $82°\cdot4$ Fahr. ($21°$ and $28°$ C.), in July $66°\cdot2$ and $86°$ Fahr. ($19°$ and $30°$ C.) and in August $64°\cdot4$ and $86°$ Fahr. ($18°$ and $30°$ C.). In September the corresponding temperatures are $66°\cdot2$ and $87°\cdot9$ Fahr. ($19°$ and $31°$ C.); in October the figures are practically the same, and in November the thermometer oscillates between $66°\cdot2$ and $91°\cdot5$ Fahr. ($19°$ and $33°$ C.). The rainy season is unhealthy, and fever is prevalent, October and November being the worst months; but in the latter month the north winds begin, the Nile falls, and, by the latter half of December, the country reassumes its normal appearance[1].

In the Lower Bahr-el-Jebel region the highest tempera-

---

[1] Tappi, *Bulletin de la Société Khédiviale de Géographie*, v^e série, 1904, No. 4, pp. 178 *sqq.*

ture is reached in March and April, and the lowest about September, in the rainy season.

The northern portion of the Egyptian Sudan, Latuka, Makraka, and the Niam-Niam region, Felkin considers to be regions where Europeans can avoid malarial fever, in a severe form, provided that they are not predisposed to it before arrival[1].

While Sir Charles Watson was at Gondókoro and Rejaf with Gordon, in 1874, during November and December, he took daily temperature observations at 9 a.m., 3 p.m. and 9 p.m., and from his diary, which he kindly placed at my disposal, it appears that thermometer readings of $93°$, $94°$ and $95°$ Fahr. ($33°·9$, $34°·4$, $35°$ C.) were not at all uncommon at 3 p.m., while $96°$ Fahr. ($35°·5$ C.) was registered on the 16th December, and $97°$ Fahr. ($36°·1$ C.) on both the 19th and 26th of that month. From recent observations at Nimule (1904) the mean maximum is given as $91°·8$ Fahr. ($33°·2$ C.). The lowest temperature recorded in the diary is $76°$ Fahr. ($24°·4$ C.) at nine o'clock in the evening, the mean minimum temperature at Nimule being a trifle over $59°$ Fahr. ($15°$ C.). This would appear, therefore, to indicate that the nights, at this time of the year, are very much cooler than the evenings.

PRODUCTIONS.

In riverain regions in the Anglo-Egyptian Sudan wheat and barley should do well, while dhurra, beans, lentils, sesame, onions, and melons are at present largely produced. Cotton and sugar are both crops which could be grown successfully in these regions. Date-palms flourish especially in the region of the Nile bend, near Ámbukól and Dóngola. In the more elevated regions of Kordofán wheat and tobacco do well, while the forests of this district and those of the Blue Nile produce gum, ebony, mahogany and other hard and ornamental woods. Rubber is indigenous in the Bahr-el-Ghazal region as well as in Kordofán. Cattle are said to thrive in the Upper White Nile districts. The regions where cultivation is at present most carried on

---

[1] Felkin, *Journal of Balneology and Climatology*, January, 1899.

are along the Nile, in the Gezira, between the White and Blue Niles, lying south of Khartoum, and on the upper Átbara river in the Gedáref district, but there is ample scope elsewhere, especially for cotton, as the black cotton soil abounds in many districts.

With reference to the tables below, *see* quotation under Egypt, at p. 66.

The stations for which meteorological tables are given are climatological stations of the Anglo-Egyptian Sudan service.

WADI HALFA. Lat. 21° 54′ 49″ N., Long. 31° 19′ 3″ E. Alt. 420 feet.

Compiled from observations recorded from February 1890 to December 1905.

	Jan.	Feb.	March	April	May	June	July	Aug.	Sept.	Oct.	Nov.	Dec.	Year
Mean temperature[1] (Fahr.°) ...	59°·0	62°·9	71°·4	79°·5	86°·9	89°·9	89°·6	89°·1	86°·5	81°·7	70°·3	62°·8	77°·5
Mean maximum temperature ...	73°·9	79°·9	90°·3	96°·8	103°·5	106°·5	105°·4	104°·0	101°·3	97°·2	85°·5	78°·4	93°·5
Mean minimum temperature ...	48°·2	49°·6	55°·8	63°·5	71°·4	74°·5	74°·1	75°·0	73°·4	68°·7	58°·3	51°·6	63°·7
Mean daily range [2]	25°·7	30°·2	34°·6	33°·3	32°·0	31°·0	31°·3	28°·9	27°·9	28°·4	27°·2	26°·8	29°·8
Mean relative humidity[1] (per cent.) ...	45	38	32	24	19	21	27	32	32	36	43	46	33

NOTE.—During the 10-year period extending over 1891—1900, the absolute maximum temperature registered was 119°·9 Fahr. in the month of June 1899 and 1900, and the absolute minimum 33°·98 Fahr. in January 1899 and February 1891.

[1] Both the mean temperature and mean relative humidity are derived from $\frac{8a+8p}{2}$ from 1901 to 1905, and from $\frac{9a+9p}{2}$ from 1890 to 1900.

[2] The daily range is max.—min.

MÉROWE. Lat. 18° 28′ 26″ N., Long. 31° 48′ 57″ E. Alt. 805 feet (approx.).

At Mérowe the mean maximum temperature is about two degrees higher than at Wadi Halfa in April, May, July, August and September, and about four degrees higher in June, October, November and December.

The mean minimum temperature is also higher than that at Wadi Halfa by about four degrees in April and May, a little more than five degrees in June, July, August and December, by nine degrees in October, and nearly eleven degrees in November. The mean relative humidity is lower than that at Wadi Halfa by between seven and thirteen points in the hundred.

BERBER. Lat. 18° 1′ N., Long. 34° 4′ E. Alt 1145 feet.

Compiled from observations recorded from March 1902 to December 1905. Observations were taken twice daily, at 8 a.m. and 8 p.m.

	Jan.	Feb.	March	April	May	June	July	Aug.	Sept.	Oct.	Nov.	Dec.	Year
Mean temperature[1] (Fahr.°)	67°·3	70°·0	76°·8	84°·0	92°·3	93°·9	91°·9	92°·5	90°·7	86°·2	78°·6	70°·0	82°·9
Mean maximum temperature ...	86°·4	88°·5	95°·5	102°·6	109°·4	111°·6	108°·3	109°·2	107°·8	103°·6	95°·9	90°·0	100°·7
Mean minimum temperature[2] ...	50°·7	55°·9	63°·3	68°·2	77°·5	79°·9	79°·7	83°·8	79°·5	74°·3	68°·8	58°·8	70°·0
Mean daily range[3]	35°·7	32°·6	32°·2	34°·4	31°·9	31°·7	28°·6	25°·4	28°·3	29°·3	27°·1	31°·2	30°·7
Mean relative humidity[1] (per cent.) ...	45	40	29	16	18	19	30	29	31	32	35	35	30

[1] Both the mean temperature and the mean relative humidity are derived from $\dfrac{8a+8p}{2}$.

[2] The minimum for 1903 was rejected.

[3] The mean daily range is max.—min.

DUEIM. Lat. 13° 59′ 31″ N., Long. 32° 20′ 0″ E. Alt. 1253 feet.

Compiled from observations recorded from March 1902 to December 1905. Observations were taken twice daily, at 8 a.m. and 8 p.m.

	Jan.	Feb.	March	April	May	June	July	Aug.	Sept.	Oct.	Nov.	Dec.	Year
Mean temperature[1] (Fahr.°) ...	69°·3	70°·3	78°·4	85°·3	88°·5	88°·0	83°·3	82°·6	83°·1	84°·7	80°·6	72°·9	80°·6
Mean maximum temperature ...	91°·2	92°·7	99°·0	104°·9	105°·3	103°·8	99°·0	97°·0	100°·8	103°·8	99°·1	92°·1	99°·0
Mean minimum temperature ...	54°·3	55°·4	63°·1	69°·3	75°·6	76°·8	74°·5	73°·2	73°·6	73°·0	68°·0	58°·5	67°·9
Mean daily range (max.—min.) ...	36°·9	37°·3	35°·9	35°·6	29°·7	27°·0	24°·5	23°·8	27°·2	30°·8	31°·1	33°·6	31°·1
Mean relative humidity[2] (per cent.) ...	29	28	21	19	30	41	61	67	58	42	32	34	38

[1] The mean temperature is derived from $\dfrac{8a+8p}{2}$.

[2] The mean relative humidity is derived from $\dfrac{8a+8p}{2}$.

KHARTOUM. Lat. 15° 36′ 33″ N., Long. 32° 33′ E. Alt. 1247 feet.

Compiled from observations recorded from July 1900 to December 1905. Observations were taken thrice daily, at 8 a.m., 2 p.m., and 8 p.m.

	Jan.	Feb.	March	April	May	June	July	Aug.	Sept.	Oct.	Nov.	Dec.	Year
Mean temperature[1] (Fahr.°)	69°.1	73°.2	78°.6	86°.0	91°.6	91°.6	89°.4	89°.1	88°.5	88°.2	81°.3	72°.9	83°.3
Mean maximum temperature	80°.8	89°.2	95°.7	102°.4	107°.2	107°.8	102°.6	101°.5	103°.3	102°.4	96°.4	88°.7	98°.2
Mean minimum temperature	58°.3	61°.5	64°.2	70°.3	77°.4	79°.0	77°.5	77°.5	76°.8	74°.7	69°.6	61°.2	70°.7
Mean daily range[2]	22°.5	27°.7	31°.5	32°.1	29°.8	28°.8	25°.1	24°.0	26°.5	27°.7	26°.8	27°.5	27°.5
Mean relative humidity[3] (per cent.)	30	29	22	17	21	30	43	49	43	32	31	34	32

[1] The mean temperature is derived from $\dfrac{8a + 2p + 8p + \min.}{4}$.

[2] The daily range is max.—min.

[3] The mean relative humidity is derived from $\dfrac{8a + 8p}{2}$.

EL OBEID. Lat. 13° 11′ 0″ N., Long. 30° 14′ 0″ E. Alt. 1913 feet.

Compiled from observations recorded from July 1901 to December 1905. Observations were taken thrice daily, at 8 a.m., 2 p.m., and 8 p.m.

	Jan.	Feb.	March	April	May	June	July	Aug.	Sept.	Oct.	Nov.	Dec.	Year
Mean temperature[1] (Fahr.°) ...	69°·1	72°·5	79°·7	85°·6	87°·8	87°·3	81°·9	81°·0	83°·1	84°·0	78°·8	70°·9	80°·2
Mean maximum temperature[2] ...	83°·7	87°·8	96°·1	99°·7	101°·7	98°·2	94°·3	90°·1	94°·1	96°·3	93°·2	88°·3	93°·6
Mean minimum temperature ...	52°·2	54°·3	57°·6	66°·2	70°·5	72°·9	72°·7	72°·3	71°·8	69°·1	62°·1	54°·1	64°·7
Mean daily range (max.—min.) ...	31°·5	33°·5	38°·5	33°·5	31°·2	25°·3	21°·6	17°·8	22°·3	27°·2	31°·1	34°·2	29°·0
Mean relative humidity[3] (per cent.) ...	26	17	14	16	29	42	62	71	62	42	29	26	36

[1] The mean temperature is derived from $\dfrac{8a + 2p + 8p + \text{min.}}{4}$.

[2] There were no maximum records for January, February, March, April, May, and December 1905.

[3] The mean is derived from $\dfrac{8a + 8p}{2}$.

MONGALLA.　Lat. 5° 11′ N., Long. 31° 46′ 42″ E.　Alt. 1455 feet.

Compiled from observations recorded from April 1903 to December 1905.　Observations were taken thrice daily, at 8 a.m., 2 p.m., and 8 p.m.

	Jan.	Feb.	March	April	May	June	July	Aug.	Sept.	Oct.	Nov.	Dec.	Year
Mean temperature[1] (Fahr.°) …	81°·3	81°·1	81°·7	80°·4	79°·0	79°·0	76°·6	76°·8	79°·3	78°·8	80°·2	79°·9	79°·5
Mean maximum temperature[2] …	100°·2	95°·9	90°·3	90°·7	89°·8	90°·7	87°·4	88°·0	92°·5	93°·4	94°·8	97°·7	92°·6
Mean minimum temperature …	66°·9	68°·5	72°·5	72°·9	72°·0	70°·5	68°·9	68°·5	70°·0	69°·1	69°·3	66°·9	69°·7
Mean daily range[3] … … …	33°·3	27°·4	17°·8	17°·8	17°·8	20°·2	18°·5	19°·5	22°·5	24°·3	25°·5	30°·8	22°·9
Mean relative humidity[4] (per cent.) …	56	59	66	73	80	86	88	72	82	82	75	69	74

[1] The mean temperature is derived from $\dfrac{8a + 2p + 8p + \text{min.}}{4}$.

[2] The maximum temperature was not recorded in February and March 1905.

[3] The mean daily range is max.—min.

[4] The mean relative humidity is derived from $\dfrac{8a + 8p}{2}$.

MEAN RAINFALL IN INCHES.

EASTERN SUDAN PLAINS.

	Alt. in ft.	No. of Years	Jan.	Feb.	March	April	May	June	July	Aug.	Sept.	Oct.	Nov.	Dec.	Year
Khartoum[1]	1247	6½	0	0	0	0	0·16	0·98	0·95	1·34	0·39	0·39	0	0	4·21
Kássala[2]	1664	6	0	0	0	0·08	0·28	0·95	2·72	4·25	2·60	0·67	0·04	0	11·59
Gedáref[3]	—	2⅜	0	0	0·24	0	1·46	3·39	6·97	6·34	4·21	0·39	0·59	0	23·59
Gállabat[3]	—	2½	0	0	0	0·04	1·81	5·91	7·52	9·73	5·63	2·05	0·24	0	32·93
Wad Médani[4]	1341	4	0	0	0	0	1·34	2·05	5·16	5·32	3·19	0·08	0·16	0	17·30
Roseires[5]	1472	2¼	0	0	0	0	1·34	3·03	9·17	7·44	5·87	1·06	0·43	0	28·34

COAST.

	Alt. in ft.	No. of Years	Jan.	Feb.	March	April	May	June	July	Aug.	Sept.	Oct.	Nov.	Dec.	Year
Suakin[6]	16	16	1·02	0·39	0	0·04	0·04	0	0·12	0·20	0	1·54	3·82	1·89	9·06

WHITE NILE, KORDOFÁN AND SOBAT RIVER.

	Alt. in ft.	No. of Years	Jan.	Feb.	March	April	May	June	July	Aug.	Sept.	Oct.	Nov.	Dec.	Year
Dueim[7]	1253	3¹⁰⁄₁₂	0	0	0	0	0·55	0·32	2·91	2·17	1·77	0·16	0·04	0	7·92
El Obeid[8]	1913	3½	0	0	0·24	0	0·04	0·95	4·25	4·33	1·89	0·71	0	0	12·41
Doleib Hilla[9]	1292	2½	0	0	1·02	1·26	3·39	4·76	6·10	5·87	5·35	3·27	1·54	0	32·56
Nasser[10]	1537	2¼	0·08	0	0·55	0·55	3·19	6·18	5·75	7·64	3·82	1·97	0·51	0	30·24

BAHR-EL-JEBEL.

	Alt. in ft.	No. of Years	Jan.	Feb.	March	April	May	June	July	Aug.	Sept.	Oct.	Nov.	Dec.	Year
Ghaba Shamba[11]	—	2¼	0	0·04	0·04	0·71	3·07	1·65	3·39	3·94	5·35	1·73	0·32	0·04	20·28
Mongalla[12]	1455	2⅝	0·08	0·47	2·21	4·41	3·74	5·55	2·84	7·21	4·41	5·59	1·73	0·20	38·54
Lado	—	1½	0	0	1·06	5·35	3·43	5·95	8·58	5·08	4·84	2·24	0·79	0·08	37·40
Gondókoro	1464	5	0·16	0·36	2·13	3·19	6·70	4·17	5·20	5·87	5·59	5·67	2·13	0·24	41·41
Nimule	—	2	0·08	0·16	3·62	2·05	5·67	2·05	4·80	4·88	4·84	6·26	6·06	1·58	42·05
Wádelai	2001	7	1·10	0·75	4·37	3·82	4·88	3·43	3·90	4·61	4·02	6·61	4·49	0·87	42·85

BAHR-EL-GHAZAL.

	Alt. in ft.	No. of Years	Jan.	Feb.	March	April	May	June	July	Aug.	Sept.	Oct.	Nov.	Dec.	Year
Wau[13]	—	3	0·16	0·98	0·59	0·79	4·02	6·77	3·78	7·28	4·57	2·84	1·46	0	33·24

1 July 1900—December 1906. 2 1900—1905, except January—July 1900. 3 June 1903—December 1905.
4 1902—1905, except January, February 1902. 5 1900, April—December 1904, 1905. 6 1890—1905.
7 March—December 1902, 1903—1905. 8 July 1902—December 1905. 9 April 1903—December 1905.
10 October 1903—December 1905. 11 September 1903—December 1905. 12 April 1903—December 1905.
13 1904—1906, except January and March 1904.

ERITREA.

Eritrea presents a great variety of climate and of temperature, from the oppressive heat of the Red Sea littoral to the exhilarating breezes of the lofty plateau, which reaches an elevation of over 8500 feet (2592 m.). As regards temperature Eritrea may be divided into two regions, viz. (1) the temperate, including the elevated plateau and the mountain districts of the eastern mountain zone over 3000 feet (915 m.) in altitude and the undulating country of the north-west, and (2) the torrid, comprising the lower lying districts of the south-east, and the maritime zone. The temperate region includes Hamasen, Serae (with Kohain, Deki Terfa, &c.), Akele Guzai, the territories of Mensa and Bogos, the less elevated regions of Habab, the territory of Maria, of Halhal and Bet Takue. The natives of the high plateau divide the country into three zones, as in Abyssinia, according to climate and the productions of the soil, as follows :

Dega, over 7500 feet.

Woina Dega, between 7500 and 5000 feet.

Kwolla, under 5000 feet.

But this is not a rigid division. For instance the inhabitants of Serae make Kohain a portion of the Kwolla, though the more elevated portions of the region lie at an altitude of 5900 feet and over, and therefore ought to be included in the Woina Dega. In the temperate zone the seasons are :—the cold moist rainy season from 20th June to 20th September, which the natives call *Keremti*; the cold dry harvest season from 20th September to 20th January, called *Keuei*; the hot dry season from 20th January to 20th March, called *Hagai*; and the hot moist season from 20th March to 20th June, called *Tseddia*.

The mean annual shade temperature of the Dega varies from 50° to 75° Fahr. (10° to 24° C.), that of the Woina Dega from 53°·6 to 79° Fahr. (12° to 26° C.), and that of the Kwolla from 59° to 95° Fahr. (15° to 35° C.). The temperature of the temperate region is thus mild, but there are great differences between the day and night temperatures. During the *Keuei*, in more elevated regions, the thermometer sinks below freezing point, and frosts are frequent. During the *Keremti* cloudy weather is experienced here, especially on the edge of the eastern plateau. Serae has the mildest climate of all the elevated districts. Deki Terfa is healthy, but the Mai Ambessa valley and those of the tributaries of the Barka, Mareb and Obel are warmer, and malarial fever is frequent, especially in the rainy season and during September, May and June. In the Kwolla the climate is fairly warm, the soil dry and siliceous, and water scarce; abundant storm-rains fall, especially in July and August, in the afternoon and at night. In the temperate region the rain is not equally distributed, Serae being more favoured than Bogos, Mensa, Maria and Habab, in which dry periods are more frequent; and in Mensa, Bogos and Habab the rains begin somewhat later than in the rest of the high country.

The torrid region comprises the Red Sea littoral, Sahel, Samhar, the islands, and Danakil on the east, and the Barka region, Baria, Bazen, Algeden, Elit, and Sabderat on the west. In the eastern areas the hot season lasts from March to October, interrupted only by violent storms in August; the rainy season extends from November to February. In the western areas, the rainy season coincides with that of the temperate regions. In the torrid region the mean temperature during the cooler season is about $73°·4$ Fahr. ($23°$ C.), and during the hot season $86°$ Fahr. ($30°$ C.), with a very slight diurnal variation. At Massaua, Assab and other places on the littoral, during the hot season, the temperature reaches $113°$ Fahr. ($45°$ C.) in the shade, but the heat is somewhat mitigated by the sea-breezes. In the maritime zone there is little rain. Samhar enjoys a dry healthy climate. In the Barka, Baria, Bazen, Elit, Sabderat and Algeden districts the climate is hot, but the rain, which falls more abundantly than on the coast, and the proximity of the highlands, render the heat more endurable. The Barka region is the hottest and most subject to fever. The region situated between the eastern edge of the plateau and the region of 3000 ft. (915 m.) altitude enjoys the rains of the temperate zone (June— September), and also those of the torrid region (November —February), and is covered with dense vegetation[1].

[1] Mulazzani, *Geografia della Colonia Eritrea*, Firenze (1904).

The mean annual temperature of the narrow coast belt is 88°·6 Fahr. (31°·4 C.), the mean for January, which is the coolest month, is 77°·9 Fahr. (25°·5 C.), and that of June, the hottest month, 98°·5 Fahr. (36°·9 C.). The Woina Dega, or plateau to the west of the coastal zone, is a temperate region and peculiarly fitted for European colonization, the thermometer seldom rising above 89°·7 or 91°·5 Fahr. (32° or 33° C.), and frequently falling as low as 44°·6 or 46°·4 Fahr. (7° or 8° C.). This zone includes the provinces of Hamasien and Akele Guzai[1].

Another authority (Professor Isaja Baldrati) still further differentiates between the climatic zones. His remarks may be summed up as follows. Eritrea has two different rainy seasons, one in winter in the eastern portion of the country, the other in summer on the central plateau and the western plains. The real rainy season, called the "great rains" (*grandi pioggie*), is preceded by a short period of irregular rain, called the "little rains" (*piccolo pioggie*), though this rule is modified by local circumstances.

In the eastern parts of the colony the winter rains are considerably less frequent and more irregular than is the case further inland. Mounting towards the plateau the country becomes accidented with hills and high mountains, alternating with small plains and valleys, and not only are the winter rains more regular, but, from its proximity to the central plateau, the region also enjoys the summer rains, the *piccolo pioggie* and *grandi pioggie* mentioned above.

In all, six climatic zones can be recognized in Eritrea.

(1)   The coastal zone with hot climate, where rain is scarce and torrential (Sahara or desert type).

(2)   The zone of the winter rains.

(3)   The zone with two rainy seasons.

(4)   A temperate zone with summer rains, including Asmara, Saganeiti, Addi Ugri, a part of Keren and a considerable portion of Habab.

(5)   A hot zone with summer rains.

(6)   A torrid zone with less rain than the plateau, where the greatest heat occurs immediately after the rains[2].

---

[1] Saint-Yves, *B. S. G.* v. 1902, p. 475.
[2] *Atti del Congresso Coloniale Italiano in Asmara*, Vol. I, pp. 145 sqq.

At Massaua the mean monthly maximum temperature is below the mean maximum for the year from November to April; in May and October it is nearly equal to it; and it is above the yearly figure in June, July, August and September. The mean annual minimum is $79°·3$ Fahr. $(26°·3$ C.); from September to April the mean monthly minimum is below this; in October, November and April it is almost equal to it; and in June, July, August and September the mean monthly minimum is above the figure for the year.

In 1898 it was hotter than in any other year in the period 1885—1902 in the months of April, June, July, August and September, the mean maxima for those months in that year being $97°·6$, $110°·1$, $112°·7$, $110°$ and $105·2°$ Fahr. $(36°·4$, $43°·4$, $44°·3$, $42°·8$ and $40°·6$ C.) respectively; and in the same year it was colder than in any other of the years in January, February and June, the mean minimum temperature being $67°·1$, $68°$ and $81°$ Fahr. $(19°·5$, $20°$ and $27°·2$ C.) respectively. With the exception of that year, the monthly temperatures are fairly regular, and the means given in the table (below) are consequently reliable. The total annual rainfall appears to vary between $2·8$ inches (72 mm.), the amount measured in 1901, and $19·7$ inches ($500·7$ mm.), the total for 1891. In most years there is no rain in May, June or July. In the wettest year of the period (1891) there was practically no rain from June to October inclusive, but heavy falls in January and December. In only one of the years, viz. 1894, was there no rain in January, while in 1893 the total fall, viz. $11·7$ inches ($297·4$ mm.), with the exception of $0·04$ inch (1 mm.), fell in the period January—March, the remainder of the year being rainless.

At Massaua and Assab, though there can scarcely be said to be summer rains, in some years a small amount of rain falls in July and August at the time when the S.W. winds reach this locality. In January, on the coast, the winter rains fall, and on the crest of the plateau, at Addi Ugri, a small amount of rain is experienced. On the coast the rains end in March (at Massaua), in April in the first valleys, and in May before the plateau is reached. As a rule, when abundant rains are coming, they are preceded by wind and lightning, but rarely at Massaua, where the rains are accompanied by a light S.E. wind, and during that season the neighbourhood is healthy. But after the rain the hot sun decomposes the vegetation and fevers follow. During the dry season the wind is generally N.E. or N.W., and is fairly strong. Without this wind the heat would be unbearable, for the thermometer frequently goes up to 113° Fahr. (45° C.) in the shade, in July and August.

In the Addi Ugri district the extremes of temperature are not more felt than in Italy. The winds which prevail in the warmer months, without being too strong, tend to moderate the temperature. The whole district is most healthy, and Europeans can easily become acclimatised. The climate naturally is far better than nearer the Red Sea on the one hand, and the Sudan on the other[1].

An Italian Commission has lately visited Eritrea for the purpose of finding out whether the climate and other conditions were favourable to the employment of white labour, and has come to the conclusion that below 5900 feet (1800 m.) the climate is not suitable, and that over 6500 feet (1980 m.) it is too cold for industrial plants, but that between these two altitudes the climate is as salubrious and as suitable for agriculture as any in Italy.

PRODUCTIONS.

The coast lands of Eritrea are unproductive, but the plateaux inland are exceedingly fertile. Barley and wheat do uncommonly well, and so too do many leguminous plants such as peas, beans and lentils. Vetches are grown for fodder. Onions, pepper and mustard thrive. Nearly all European vegetables can be grown, including the more delicate kinds, such as asparagus. Flax gives good results, and from an indigenous plant called *niugh* (*Guizotia oleifera*) the natives extract large quantities of oil. In the lower country sorghum or dhurra produces good crops, both cotton and tobacco do well, and also an oil-producing seed called by the natives *selit*. A cereal called *taf* (*Poea abyssinica*) also flourishes in the Woina Dega. The forests produce good ebony and other woods, and date-palms grow in the Assab district; gum-producing acacias, indigo, tamarinds and bananas also thrive in the lower regions.

[1] Tancredi, *B. S. G. I.* 1903, p. 76.

MASSAUA. Lat. 15° 36′ 40″ N., Long. 39° 27′ E. Alt. 63 feet, on the Red Sea.

Compiled from 18 years' observations[1].

		Jan.	Feb.	March	April	May	June	July	Aug.	Sept.	Oct.	Nov.	Dec.	Year
Mean temperature[2] (Fahr.°)	...	78°·1	78°·6	81°·1	84°·6	87°·8	91°·8	93°·9	93°·4	91°·6	88°·3	83°·1	80°·2	86°·0
Mean maximum temperature[3]	...	84°·9	85°·3	88°·9	91°·8	95°·4	101°·5	103°·1	101°·7	98°·2	93°·6	90°·1	87°·1	93°·4
Mean minimum temperature[3]	...	72°·5	72°·1	73°·6	77°·4	80°·4	84°·2	87°·8	87°·1	84°·6	81°·1	77°·7	71°·6	79°·2
Mean daily range[4]	... ...	12°·4	13°·2	15°·3	14°·4	15°·0	17°·3	15°·3	14°·6	13°·6	12°·5	12°·4	15°·5	14°·2
Highest mean maximum temperature during the 18 years		88°·9	89°·1	90°·3	97°·5	100°·0	110°·1	111°·7	109°·0	105°·1	99°·5	94°·6	91°·2	—
Lowest mean minimum temperature during the 18 years		67°·1	68°·0	69°·8	69°·9	74°·8	80°·9	83°·5	78°·3	77°·2	74°·5	73°·4	69°·3	—
Mean relative humidity[5] (per cent.)	...	73	74	72	70	66	52	56	57	59	60	66	69	65
Mean rainfall[6] (in inches)	... ...	1·69	0·67	0·55	0·24	0·32	0	0·09	0·43	0·16	0·43	1·02	1·65	7·25

[1] 1885—1902.

[2] The mean temperature is derived from $\dfrac{max. + min.}{2}$.

[3] The maximum and minimum records are wanting for January—April 1885, May—December 1893, August and December 1897, January—June 1899.

[4] The daily range is max.—min.

[5] The humidity observations are wanting for January—April 1885, February—April 1886, May—December 1893, August and December 1897, January—June 1899, and also for the years 1901, 1902.

[6] The rain records are wanting for January—April 1885, April 1895, March 1896, August and December 1897, February 1898, January—May 1899.

GINDA. Lat. $15° 26' 13''$ N., Long. $39° 7'$ E. (approx.). Alt. 3153 feet.

Compiled from observations recorded from August 1891 to April 1895 and communicated to Capt. Lyons by the Observatory, Rome.

		Jan.	Feb.	March	April	May	June	July	Aug.	Sept.	Oct.	Nov.	Dec.	Year
Mean temperature[1] (Fahr.°)	...	$64°.9$	$67°.3$	$68°.9$	$75°.2$	$80°.4$	$84°.9$	$85°.3$	$82°.6$	$82°.9$	$76°.8$	$73°.0$	$68°.4$	$75°.9$
Mean maximum temperature	...	$71°.9$	$72°.9$	$74°.1$	$82°.4$	$88°.3$	$94°.3$	$93°.4$	$88°.9$	$92°.3$	$85°.1$	$80°.2$	$74°.3$	$83°.2$
Mean minimum temperature	...	$57°.9$	$61°.3$	$63°.3$	$67°.8$	$72°.5$	$75°.6$	$77°.4$	$76°.1$	$74°.8$	$68°.2$	$65°.8$	$62°.1$	$68°.5$
Mean daily range[2]	...	$14°.0$	$11°.6$	$10°.8$	$14°.6$	$15°.8$	$18°.7$	$16°.0$	$12°.8$	$17°.5$	$16°.9$	$14°.4$	$12°.2$	$14°.7$

[1] There is no statement as to how the mean temperature was obtained, but it is apparently, from internal evidence, derived from $\dfrac{\text{max.} + \text{min}}{2}$.

[2] The daily range is max.—min.

ADDI UGRI. Lat. 14° 53' N., Long. 38° 49' E. Alt. 6634 feet.

Compiled from 7 years' observations, extending from 1899 to 1905.

	Jan.	Feb.	March	April	May	June	July	Aug.	Sept.	Oct.	Nov.	Dec.	Year
Mean temperature[1] (Fahr.°) ...	65°.1	67°.5	69°.4	71°.1	71°.4	71°.4	65°.8	65°.7	67°.1	66°.4	64°.2	64°.8	67°.4
Mean maximum temperature[2] ...	77°.9	80°.8	84°.6	85°.3	83°.7	80°.7	74°.7	73°.9	77°.5	79°.9	78°.3	78°.6	79°.6
Mean minimum temperature[3] ...	53°.6	57°.5	57°.6	58°.8	59°.7	59°.4	56°.5	56°.3	57°.0	55°.8	53°.2	52°.2	56°.4
Mean daily range[4]	24°.3	23°.3	27°.0	26°.5	24°.0	21°.3	18°.2	17°.6	20°.5	24°.1	25°.1	26°.4	23°.2
Mean relative humidity[5] (per cent.) ...	28.8	30.1	24.4	38.6	30.6	36.4	65.9	75.2	51.7	44.5	45.1	38.4	42.5
Mean amount of cloud (0—10) ...	1.2	1.9	2.0	3.3	3.8	4.2	6.4	6.1	3.5	2.2	1.7	1.4	3.1
Mean number of rainy days	1	3	6	12	11	14	25	24	9	5	3	1	114

NOTE.—The Meteorological Observatory is fitted with all the instruments usual in a first class Italian Observatory.

[1] Temperature observations are wanting for April 1903 and April and May 1904. The mean temperature is derived from $\frac{9a + 9p + \text{max.} + \text{min.}}{4}$. Observations are also available for the period April 1894 to December 1895, but here the mean was derived from $\frac{\text{max.} + \text{min.}}{2}$, and they have not been included in this table.

[2] The absolute maximum recorded was 92°.30 Fahr. on 23rd March 1899, but 91°.76 was recorded on 12 December 1900 and also on 4 April 1900 and in May 1902.

[3] The absolute minimum recorded was 41°.54 on 25 December 1901; 41°.90 was recorded on 13 February 1899.

[4] The daily range is max.—min.

[5] The relative humidity is for the years 1901—1905 and is derived from $\frac{9a + 3p + 9p}{3}$; the relative humidity in 1899 was derived from $\frac{9a + 9p}{2}$ and that in 1900 from $\frac{\text{noon} + 9p}{2}$ and these have not been included in the table.

KEREN. Lat. 15° 46' 44" N., Long. 38° 28' E. (approx.). Alt. 4790 feet.

Compiled from observations recorded in 1892 and from May 1894 to April 1895.

	Jan.	Feb.	March	April	May	June	July	Aug.	Sept.	Oct.	Nov.	Dec.	Year
Mean temperature¹ (Fahr.°)	64°·9	66°·5	71°·4	74°·1	78°·1	73°·0	69°·1	66°·0	67°·3	67°·8	67°·3	65°·5	69°·3
Mean maximum temperature	77°·2	—²	84°·6	81°·3	88°·2	79°·2	73°·8	68°·5	76°·1	77°·5	79°·3	79°·5	78°·6
Mean minimum temperature	55°·5	—²	57°·4	61°·3	65°·5	59°·7	57°·2	59°·5	57°·0	59°·2	59°·4	57°·2	58°·8
Mean daily range³	21°·7	—	27°·2	20°·0	22°·7	19°·5	16°·6	9°·0	19°·1	18°·3	19°·9	22°·3	19°·8

¹ Extracted from *The Physiography of the Nile and its Basin.* There is no indication as to how the mean was arrived at.
² The maxima and minima for February are wanting.
³ The daily range is max.—min.

ASSAB, on the Red Sea Coast, Lat. 12° 59' N., Long. 42° 44' 30" E. (approx.). Alt. 20 feet.

	Jan.	Feb.	March	April	May	June	July	Aug.	Sept.	Oct.	Nov.	Dec.	Year
Mean temperature (Fahr.°)	78°·1¹	78°·3¹	81°·9¹	82°·5¹	86°·7¹	91°·6¹	95°·9²	95°·0²	92°·8³	86°·4³	81°·3³	77°·9³	85°·7
Mean maximum temperature	82°·8	82°·9	86°·0	89°·4	93°·9	99°·0	102°·7	100°·2	97°·3	91°·8	86°·4	82°·8	91°·3
Mean minimum temperature	72°·9	72°·7	75°·4	77°·0	77°·4	81°·1	86°·2	85°·3	84°·2	79°·2	75°·6	71°·2	78°·2
Mean daily range⁴	9°·9	10°·2	10°·6	12°·4	16°·5	17°·9	16°·5	14°·9	13°·1	12°·6	10°·8	11°·6	13°·1

¹ The figures for these months are throughout for the years 1886 and 1887.
² ,, ,, ,, ,, 1885—1887.
³ ,, ,, ,, ,, 1885 and 1886.
⁴ The daily range is max.—min.

ERITREA. MEAN RAINFALL IN INCHES.

	Alt. in feet	No. of Years	Jan.	Feb.	March	April	May	June	July	Aug.	Sept.	Oct.	Nov.	Dec.	Year
Keren[1]	4790	$5\frac{1}{4}$	0	0·12	0·08	0·36	0·98	4·17	4·21	11·81*	3·11	0·08	0·24	0	25·16
Asmara[2]	7782	$\frac{25}{12}$	0·08	0·12	0·63	0·71	1·42	2·44	4·61	4·49	0·71	2·21	0	0	17·42
Addi Ugri[3]	6634	$8\frac{1}{2}$	0	0·24	0·51	0·87	1·65	2·60	6·18	7·05	1·46	0·39	0·36	0·20	21·51
Massaua[4]	63	18	1·69	0·67	0·55	0·24	0·32	0	0·09	0·43	0·16	0·43	1·02	1·65	7·25

* This mean is probably too high, owing to excessive rain in 1905 in that month.

[1] September—December 1890, 1891, 1892, 1893 except October, 1894 except August, 1895 except January and February, 1896 up to April.

[2] During the period 1894—1905, viz. May—November 1894, April 1895, December 1901, March—October 1904, January—April and June—September 1905.

[3] May 1894—December 1895, 1899—1905 except April and May 1904.

[4] 1885—1902,

ABYSSINIA.

The Abyssinian area consists, in the main, of a huge elevated plateau, cleft by deep-cut valleys, and enjoys a typically tropical climate, modified, however, by the effect produced by the presence of great mountain peaks, which exert a very considerable influence, and differentiate the climate from that of the Sudan plains on the west, and that of the low-lying belt of country below the plateau on the east.

The period from October to March/April is practically rainless over the greater part of the plateau, except for a brief interval, generally in February, and the winds are from the north and east. In the northern and western parts the prevailing winds are from the north and north-east, while in the east and south they are from the north-east and east.

The dry season, alluded to above, is followed by the short season of the "lesser rains," usually of about three weeks' duration, which occurs, as a rule, in March/April or in April/May. In all south-western Abyssinia, as well as the greater part of the plateau, the weather in January is dry, and the sky beautifully clear, but on the crest of the plateau, at Addis Abbaba, Ankober and Magdala a small amount of rain usually falls.

On the plateau the main rainy season begins in July and ends in October; in Kaffa and Shoa it begins in June and lasts until mid September, the heaviest precipitation occurring in July and August. In the southern provinces, in Sidamo, Walamo, &c., the rains usually begin in May and leave off about mid September, but they are neither so regular nor so copious.

In the more elevated districts, the heat of the summer months is not excessive, in fact, at very high altitudes, the weather is distinctly cold, especially at night; and the damp oppressive heat of the tropics is not experienced over the greater part of the country, though in the valleys and plains, as might be expected, tropical conditions prevail through-out the year. Hence there follows a very peculiar result, almost confined to this area, namely, that side by side there may be, at no great distance from one another, two essentially different climates.

Michel, in *Vers Fachoda*, p. 529, writes somewhat to the following effect:—On the Abyssinian plateau the dry season begins between the 20th and 30th September and ends on the 1st June; but there is generally a short rainy break of 20 days towards the end of February. It never rains in October, November, and December, except at altitudes exceeding 9514 feet, and in the Kaffa Motsha and Ilu-Babor mountains above 8200 feet. During March, April, and the first half of May, there are generally three or four showers a month, and when these do not occur it is a bad year for crops, though the February rains are more important. In October, November, and December the thermometer is lowest, and then, above 9156 feet (2800 m.), it sinks to $24°\cdot8$ or $21°\cdot2$ Fahr. ($-4°$ or $-6°$ C.) at night. The second half of May is marked by evening storms at about sunset, generally accompanied by thunder. In June these storms are of daily occurrence, but less severe, and about the middle of the month they are of longer duration. In July it rains, especially at night and in the morning, while during August and September the rains are continuous, and leave off abruptly between the 20th and 30th September.

Lefebvre, who was in Abyssinia for five years, and was most energetic in recording meteorological observations and also non-instrumental matter with regard to weather and climate, summarises his information in Vol. III of his great work *Voyage en Abyssinie pendant 1839—1843*, in some such terms as the following:—The main rainy season generally begins, on the plateau, in July and ends in October, but from April to May there are lesser intermittent rains called *Azmera* or seed-time rains. Thus there are three seasons in Abyssinia,

> *Baga*, the dry season,
> *Keremti*, the rainy season,
> *Azmera*, the lesser rainy season.

In some localities the rains begin at regular hours, for instance, at Adua and in all that part of Tigre between the Mareb and the Weri, except the crest or rim of the plateau (*e.g.* Mágdala, Addis Ábbaba, Ankober), they begin between noon and 1 p.m., and the sky is almost always overcast except in the mornings and evenings. This is true for the whole Woina Dega or region of mid-elevation. On the high plateau, or Dega, the rains are almost continuous, while the rainy season begins earlier and ends later. In the lesser rainy season hail and thunder are frequent, but in the Woina Dega thunder is seldom heard.

In the Samien region the prevailing wind is west, on the eastern border of the plateau from $16°$ to $14°$ N. it is north-west, and beyond that the prevalent wind is north-east.

In Doba, Wojerate and Bugena there are rains at all seasons, but with frequent intervals, and the prevailing wind is from the south-east. The rain is fine, while, on the other hand, in those regions where the rains are regular, it comes down in torrents.

Immediately before the rains in the low-lying valleys it is very dry, but there is a cold moist feeling on the high ground.

The yearly rainfall is not regular; during the five years that Lefebvre was in the country, three of the years were wet years and two dry years.

To the west of Addis Ábbaba, on the broken uplands between Tulu Dimtu and the Baro river water parting, in April the climate is bracing, but violent rain-storms are experienced, which at the beginning come on soon after midday. As the month advances, however, these storms become more violent and last longer.

The Kaffa region of Abyssinia is described as a forest-land of romantic beauty, and its mild and moist climate renders it exceptionally fertile.

At Harar for the whole year the extremes of temperature are $63°$ and $82°$ Fahr. ($17°\cdot2$ and $27°\cdot8$ C.); during eight months the variation rarely exceeds the limits $71°$ and $75°$ Fahr. ($21°\cdot7$ and $23°\cdot9$ C.). The town stands at an altitude of about 6000 feet, and the air is better than that of many Alpine health resorts. The rainy season lasts for 3 or 4 months, and is heralded by several days of mist, then come the heavy storms, generally at night, lasting about two hours out of the twenty-four. The climate of Harar is very suitable to Europeans.

Major Gwynn, C.M.G., D.S.O., R.E., who knows more of Abyssinia as a whole than any other authority, has kindly contributed the following notes on the "Seasonal Changes in Southern Abyssinia," *i.e.* approximately south of the 10th parallel of north latitude.

The seasons in Southern Abyssinia may be roughly divided into wet and dry, but the division is not so distinctly marked as in most tropical countries, and important variations are found in different localities.

Generally speaking, throughout the country the height of the dry season is in the months December, January, February, and the height of the wet season June, July, August. The heaviest and most prolonged rainfall probably occurs in the south-west, *i.e.* in Kaffa and the neighbouring districts, in which the Omo, the Sobat (and its tributaries), and the Didessa rise.

The eastern and south-eastern districts are drier although the Harar and Arussi plateaux, being affected by the coastal spring rains, have a well-distributed rainfall.

In the country round Addis Ábbaba, which may be taken as typical of the main plateau, it is customary to talk of the light rains and "*the* rains." The former commence in March or April, and continue intermittently until they merge into the regular rainy season, which sets in in June and continues to the middle of September, or, in some years, a month later.

During the light rains several wet days may occur together, but there are dry intervals which prevent the ground becoming saturated. No difficulty is experienced in travelling a few hours after rain has ceased.

During the regular rains, on the other hand, although rain is by no means continuous, and the total fall not great, the air is always saturated with moisture and the ground becomes absolutely waterlogged. There is a great deal of heavy clay soil, which forms a heavy clinging mud even on the steepest slopes. Travelling, where practicable at all, becomes a penance, and only the slowest footpace is possible. At this season there seems to be no prevailing wind, and the majority of the rain falls in local thunderstorms.

The dry season is announced by a fairly strong and

steady east wind which soon picks up the moisture in the soil. Even during the height of the dry season rainy days occur, but they produce little effect, and bright, windy weather is the prevailing condition.

In the south-western portion of the plateau the light rains commence earlier and are much heavier than at Addis Ábbaba. It is probable that the height of the rainy season is reached earlier than further east, *i.e.* in May or June.

Anything like continuous records are lacking, but the following has been my own experience, or the experience of expeditions I have known of :—

Early in March, 1900, at Gidami, on the western escarp between the Dabus and Baro rivers clouds had formed and some showers fell. The natives stated that the rains were late.

At the beginning of April, 1901, a week of heavy rain was experienced in Walega, between Didessa and Dabus rivers, and this extended at the same time into Beni Shangul. Although the weather cleared somewhat subsequently, the sky was generally overcast and rain frequent as far north as the line of the Blue Nile at Fazokli. North of that line the sky remained clear till the middle of May, the line of division being sharply marked.

The same year Major Austin experienced rain at Boma at the head waters of the Akobo in March, and found the country around the Lower Omo flooded in April. He was prevented by rain from attempting to reach the high ground.

In 1904 Capt. Maud found the rain had set in at the north end of Lake Rudolf in May, and was prevented from going further north.

In the latter part of February and beginning of March, 1909, the highest points of the plateau visible from the north end of Lake Rudolf were frequently visited by thunderstorms, though there was practically no rain in the Kibish valley or on the extreme south-western Tirma plateau (elevation 6000 feet), till the third week in March. The sky was frequently overcast and rain threatened some weeks earlier. The natives stated the rains were nearly a month late. On gaining the main plateau (elevation 8000 feet) in the last week of March, it was found that a good deal of rain had already fallen, and the native merchant caravans had given up travelling on account of the rains.

Travelling from Maji to Addis Ábbaba as far as the Gojeb river rain fell every day, and there was much cloud. Spring growth was also advanced.

North of the Gojeb river, and between the Upper Omo and Hawash there was little rain, and growth in the country was more backward. Some heavy rain was experienced in the Upper Hawash valley just before reaching Addis Ábbaba, April 20th, but the weather cleared again in a few days.

In the Harar province the rainy season is markedly different from that of the main plateau. The effect of the November and February Somali coast rains is felt as far as the foot of the main plateau on the west side of the Hawash. Thus, in May, 1909, at Addis Ábbaba, and on the main plateau as far as the Balchi escarp, the grass was only beginning to grow, whereas at the same date on the Hawash plain below it was at its full height, and was in fact beginning to dry. Similarly at Dire Daua and along the foot of the Harar escarp in July, 1908, there was little grass, and everything was dry and parched by the dry weather of May and June, while at the same time the grass crop at Addis Ábbaba was reaching its maturity. The high ground of the Harar plateau appears to catch some, at least, of the rainy season of the main plateau, but the intervening Hawash valley follows the Somali coast seasons as far as the growth of vegetation is concerned.

In some years the valley will get the benefit of both rainy seasons, in others of neither, and at all times the western side of the valley as far as the Balchi escarp is very dry, not getting the full benefit of either.

I have now traversed the Assobat road between Dire Daua and the Hawash at three different seasons, namely, early in December, 1900, early in August, 1908, and early in May, 1909. On each occasion there was a fair amount of rain, though more in August than at the other times.

Even in August the highest points on the Harar plateau were not continuously in cloud to at all the same extent as the main plateau on the opposite side of the Hawash.

The neck connecting the Harar plateau to the Arussi plateau narrows till it partakes of the character of a mountain range, which, in August, 1908, received less rain than either the Harar or main highlands. The Arussi plateau rises practically to the same height as the main plateau at Addis Ábbaba, and the mountains which separate it from the Zwai and Lake valley attain a height of 12,000 feet. When the Addis Ábbaba districts were experiencing heavy rain in September, 1908, the Arussi districts had comparatively dry weather. Later, in October and November, when the rains had ceased at Addis Ábbaba the Arussi plateau became again wrapt in cloud. At Ginir, on the southern slopes of the plateau, early in November heavy rain was experienced, and local reports stated that November was one of the wettest months of the year.

Thus this district appears to receive the Somali coast November rain. Later in November, in the lower southern districts, near the junction of the Daua and Ganale, little rain fell, though the sky was frequently overcast, and the air moist.

In the frontier districts where the southern slopes of the Abyssinian plateau merge into the great Golbo plain, the rainy season begins in March and culminates in May, corresponding in character with the light rains of the main plateau. Thus when the light rains are poor in the north, the rains in the frontier region will be deficient.

The main rains of the northern districts do not, apparently, extend as far south as the frontier. The total rainfall in the frontier districts must be much less than on the plateau. In December there is, in some years, a short period of local thunderstorms. The rain that then falls is of great value in filling the semi-perennial wells found at the bases of the detached hills in the Golbo. The natives name these "The Rains of God."

PRODUCTIONS.

The vegetation of Abyssinia is governed by elevation, and the three zones, the Kwolla, the Woina Dega and the Dega, alluded to above, may be regarded as the three vegetation zones. In the Kwolla cotton does well, and is the principal cultivation of Gardulla, in the south, and of most of the countries between this and the left bank of the Omo river, and also of Kosha and Konta on the right bank. Further north in Kaffa, coffee is indigenous, deriving its name from the province, and excellent crops are produced here, as well as in Jimma, though the quality is not so fine as that grown in Harar (*see* under Somaliland). Sugar can also be grown in the Kwolla zone, while indigo is indigenous and so too is the banana. The forests produce ebony and mahogany and other hard and ornamental woods, as well as the acacia, whence gum is derived, while in Sidamo and Walamo there are bamboo forests, which occur also along most of the streams. The Arussi grow barley in the clearings of their juniper woods, the north-west of their country is a famous place for cattle, and the forests produce palms. The Woina Dega is the most productive belt, and here wheat, barley, millet and all European cereals flourish, and all British vegetables can be grown, and many fruits, such as peaches, apricots, oranges, &c. The Dega is too elevated for agriculture, except in the case of very hardy cereals, but affords excellent pasture for goats and the hardy, long-haired breeds of sheep.

ADDIS ABBABA[1]. Lat. 9° 2' N., Long. 38° 44' E. Alt. 8005 feet.

Compiled from observations recorded from January 1901 to December 1904.

	Jan.	Feb.	March	April	May	June	July	Aug.	Sept.	Oct.	Nov.	Dec.	Year
Mean temperature[1] (Fahr.°)	61°9	59°4	64°0	61°2	63°1	59°2	56°7	58°8	57°9	60°1	61°3	60°1	60°3
Mean minimum temperature ...	44°9	47°5	50°0	50°4	49°3	49°6	50°0	50°0	49°6	47°1	44°2	44°4	48°1
Mean of absolute monthly maxima	73°2	73°6	75°9	73°2	74°7	71°6	68°4	67°6	68°5	71°6	71°6	71°1	71°7
Mean of absolute monthly minima	39°6	41°7	42°8	46°4	44°6	46°4	46°6	46°0	46°2	42°1	39°2	38°7	43°3
Mean monthly range of temperature[2] ...	33°6	31°9	33°1	26°8	30°1	25°2	21°8	21°6	22°3	29°5	32°4	32°4	28°4
Mean relative humidity[3] (per cent.) ...	48	58	52	65	51	74	86	84	74	48	45	53	61

[1] Converted into Fahr. degrees from the figures given in *The Physiography of the Nile and its Basin*. No indication is given as to how the mean temperature was arrived at.

[2] The mean monthly range of temperature is the difference between the mean of absolute monthly maxima and the mean of absolute monthly minima.

[3] Derived from $\frac{7a+1p+9p}{3}$ from July 1900 to December 1904, except during July—September 1902, February, April—July, September and October 1903.

ADDIS ALEM. Lat 9° 0' N., Long. 38° 22' E. Alt. 7873 feet.

Compiled from observations recorded from June 1902 to May 1903[1].

	Jan.	Feb.	March	April	May	June	July	Aug.	Sept.	Oct.	Nov.	Dec.	Year
Mean maximum temperature (Fahr.°) ...	81°·1	81°·1	80°·1	81°·3	78°·3	77°·2	77°·2	76°·5	76°·5	78°·6	78°·8	78°·8	78°·8
Mean minimum temperature	42°·8	46°·8	50°·9	50°·9	51°·3	49°·3	50°·4	50°·0	49°·6	44°·4	41°·9	40°·1	47°·3
Mean temperature[2]	63°·3	64°·0	65°·5	65°·7	64°·8	63°·1	63°·9	63°·3	62°·9	61°·5	60°·4	59°·7	63°·2
Absolute maximum temperature ...	82°·4	86°·0	84°·2	87°·8	84°·2	82°·4	78°·8	81°·5	80°·6	83°·3	81°·5	82°·4	82°·9
Absolute minimum temperature ...	39°·2	41°·0	45°·5	42°·8	39°·2	44°·6	46°·4	46°·4	44°·6	40°·1	36°·5	—[3]	42°·4
Number of rainy days	8	4	13	9	15	24	30	28	23	5	3	3	165
,, of days with storms	1	—	—	—	—	28	30	21	3	—	—	—	83
,, of fine clear days	8	21	9	7	6	3	—	—	2	14	22	24	116

[1] Recorded by L. De Castro, see *Bol. Soc. Geog. Ital.* 1905, p. 19.

[2] The mean temperature is derived from $\dfrac{\text{max.}+\text{min.}}{2}$.

[3] The absolute minimum for December is wanting.

ANKOBER[1]. Lat. 9° 35' N., Long. 39° 53' E. approx. Alt. 6202 feet.

	Jan.	Feb.	March	April	May	June	July	Aug.	Sept.	Oct.	Nov.	Dec.	Year
Mean maximum temperature ...	58°·3	59°·9	62°·8	59°·0	64°·4	66°·2	62°·6	60°·6	60°·1	57°·6	57°·7	58°·5	60°·6
Mean minimum temperature ...	45°·5	48°·6	51°·4	51°·3	55°·0	57°·9	53°·4	50°·0	50°·5	46°·2	46°·0	44°·6	50°·0
Mean temperature[2] (Fahr.°)	51°·8	54°·7	57°·2	55°·2	59°·7	62°·1	58°·1	55°·8	55°·2	52°·2	51°·9	51°·8	55°·5
Mean daily range of temperature[3]	12°·8	11°·3	11°·4	7°·7	9°·4	8°·3	9°·2	10°·6	9°·6	11°·4	11°·7	13°·9	10°·6

[1] Recorded by Major Harris from August 1841 to July 1842.

[2] The mean temperature is derived from $\dfrac{max.+min.}{2}$.

[3] The daily range is max.—min.

MÁGDALA. Lat. 11° 23' N., Long. 39° 24' 22" E. Alt. 9050 feet approx.

Compiled from observations recorded by Dr Blanc from August 1866 to March 1868[1].

	Jan.	Feb.	March	April	May	June	July	Aug.	Sept.	Oct.	Nov.	Dec.	Year
Mean temperature at sunrise (Fahr.°) ...	46°·5	48°·7	50°·4	50°·6	55°·6	53°·1	50°·7	51°·1	51°·5	46°·7	43°·8	45°·2	49°·5
Mean temperature at 9 a.m. ...	58°·2	55°·9	61°·8	64°·5	66°·3	63°·4	58°·5	58°·5	60°·5	59°·7	57°·5	56°·6	60°·1
,, ,, 2 p.m. ...	67°·0	66°·9	72°·2	77°·8	78°·2	74°·6	68°·5	67°·5	66°·1	65°·9	66°·4	67°·3	69°·9
Mean temperature at sunset ...	55°·2	56°·2	61°·7	60°·7	62°·7	62°·8	59°·6	58°·8	59°·7	56°·3	55°·1	55°·1	58°·6

From the 18th to 21st September 1866 and from 4th to 6th December 1867 the thermometer was in the sun when the 9 a.m. and 2 p.m. observations were taken and these have therefore been discarded.

[1] See *The Abyssinian Expedition*, by C. R. Markham, 1869.

ABYSSINIA. MEAN RAINFALL IN INCHES.

In the following table the values for Addis Ábbaba were compiled by Captain Lyons from all available sources.

	Alt. in feet	No. of Years	Jan.	Feb.	March	April	May	June	July	Aug.	Sept.	Oct.	Nov.	Dec.	Year
Addis Ábbaba	8005	8[1]	0·32	1·30	3·50	3·15	1·73	6·14	11·58	10·67	6·73	0·55	0·43	0·55	46·65
Gondar ...	6247	3[2]	—	—	—	—	2·64	4·80	11·42	14·65	4·02	1·81	0·55	0	39·89
Gambela ...	—	$1\frac{6}{12}$[3]	0·67	1·46	4·84	2·36	5·83	7·40	4·57	9·69	8·82	2·91	2·76	0·08	51·39
Addis Alem ...	7678	1[4]	1·28	0·98	3·70	3·47	10·55	5·71	8·07	6·58	5·12	0·67	0·36	1·58	48·07

[1] The observations cover the period 1898—1905, except March—December 1899, January—May 1900, April, May 1903, April—September 1905.

[2] May—December 1770, May—September 1771, July—October 1838. Quoted by Capt. Lyons from *Travels to discover the Source of the Nile*, J. Bruce, F.R.S., Edinburgh, 1790.

[3] In 1905/6.

[4] June 1902—May 1903.

THE HORN OF AFRICA, OR SOMALILAND

(ABYSSINIAN, BRITISH, FRENCH AND ITALIAN.)

Somaliland falls naturally into three distinct zones :

1. The arid maritime plain, extending from Obok, in French Somaliland, through British Somaliland, where it reaches a breadth, in some parts, of 60 miles, to Guardafui, where it is only a couple of hundred yards broad. It then turns southwards, and widens out until it merges in the plains in the neighbourhood of Mogdishu. This maritime zone receives only about three or four inches of rain in the year, and the northern parts are skirted by the thermic equator.

2. The maritime ranges, which, in a general sense, run parallel with the coast, limit the maritime plain, and are themselves frequently interrupted by inland plains, and connected by plateaux and low-lying valleys, the sites of numerous tugs or wadis, *i.e.* watercourses, which are for the most part dry except in the rains. The annual precipitation in this zone varies between 10 and 20 inches per annum, and the climate is comparatively equable.

3. The raised plateaux, which culminate in the Harar highlands, with lofty peaks of from 8000 to 10,000 feet above sea-level, and are continued southward in the mountainous country which forms the upper basins of the Fafan and Webi Shebeli, while to the east, south of the northern maritime ranges, lies the Haud, an elevated undulating plateau falling gradually from north-west to south-east, from some 6000 feet to 2000 feet, and covered in the southern portion by thorn bush. In the dry season the Haud is most inhospitable and entirely waterless, but in the rains this area affords excellent pasturage.

From a climatic point of view yet a fourth area may be added, viz. the valley of the Webi Shebeli, which is 'damp, trying and sultry from March to September," with a heavy rainfall.

The main determining factors in the climate of the whole of north-eastern Africa south of about 18° N. are the North-East Trades (or in the east the North-East Monsoon), and the South-West Monsoon. The latter is generally experienced from May to September, and the former from November to March, April and October being transition months characterized by variable airs and calms. The actual dates, however, vary with the locality.

In the south-east of this area, in Italian Somaliland, the Somalis divide their year into four seasons, of about three months each, starting from the middle of August (their new year's day), namely (1) *Der*, in which the S.W. Monsoon still prevails, and which is rainy, until the N.E. winds set in; (2) *Jilal*, a dry season with a constant N.E. wind; (3) *Gu*, a wet fertile season, in which the N.E. winds prevail, until the S.W. Monsoon sets in. The latter wind backs round by south and then up to east. It is this wind that brings the rain. It is what is called a "large" wind, and is known on the coast as *tanganbili*, an expression implying that it serves coasting vessels both from Zanzibar to Benadir and *vice versa*; (4) *Agai* (*Hagai*), the hot season with lesser rains at intervals, in which the S.W. Monsoon is constant. On this coast a heavy swell presages the arrival of the S.W. Monsoon. At Illig the wind is not so high in November and December as it is in January and February, after which it again improves till the burst of the S.W. Monsoon. The mornings too are usually calmer than the afternoons, and should be chosen for landing purposes.

In the north, in British Somaliland, and the northern parts of Abyssinian Somaliland, the seasons are a trifle later, as might be expected, since they follow the sun. The *Der*, which is the cool season, frequently extends into January, which is here the first month of *Jilal*, whereas in the south that season begins a little after the middle of November. The end of *Jilal* is frequently sultry and calm. The S.W. Monsoon is usually ushered in by rain somewhere about the second week in April, *i.e.* in *Gu* (in the north), and this extends over a couple of months. At the departure of the monsoon there are heavier rains in the interior, but on the northern coast the second or more regular rains do not occur till November or December. The *Der* rains begin near the end of September and last for a month and a half. The fall is heavier in the west than in the east, and in the interior the *Der* rains are often heavier than the *Gu* rains. The second interior rains, at the end of the monsoon, alluded to above, occur at the end of *Agai* (*Hagai*), and are confined to the western highlands, not being experienced farther west than the Wagger mountains.

The rains however cannot be depended upon. In April and May, 1901, there was a bad drought. From 28th April to 15th May, 1903, there were no rains in the south, near the Webi Shebeli abreast of Obbia. Again, in the northern Haud, the autumn or second interior rains failed altogether in the same year. The heaviness of the rain (*Gu*) at times may be gathered from the following, contained in a despatch from Sir C. Egerton to the Secretary of State, during the late operations against the Mullah, and dated 12th April: "If the rain sets in in the Nogal, the withdrawal of the troops there will be an impossibility owing to the deep nature of the soil."

With regard to temperature, it may be stated generally that throughout the maritime zone it is hot and varies between $60°$ and $108°$ Fahr. ($15°·6$ and $42°·2$ C.), though of course these figures are merely an approximation and apply mainly to the northern coast, and the temperature varies according to locality. In this zone the wind is constant, when in the N.E., and is cool. This refers to the period from November to March, while from May to September, during the period of the S.W. Monsoon, the weather is squally and hot.

In the maritime ranges of western Somaliland the temperature varies between $40°$ and $91°$ Fahr. ($4°·4$ and $32°·8$ C.), while towards the east the thermometer may fall as low as $25°$ Fahr. ($-3°·9$ C.). Throughout there is a large daily range, *i.e.* a large difference between the maximum in the day and the minimum at night. The winds are very similar to those in the maritime zone, but are not dependable from April to September.

The extremes of temperature in the uplands are $56°$

and 108° Fahr. ($13°·3$ and $42°·2$ C.), and the temperature varies greatly according as the country is well wooded, or bare arid plain, or grass country. Here, as in most inland tracts, the daily range is greater. This zone is not infrequently visited by drought, especially in the region lying south of the eastern portion of the Haud, and these droughts have been found to correspond, in point of time, with those experienced in India. From June to September the northern highlands are characterized by very strong winds. In the basin of the Webi Shebeli, on the contrary, the winds are very light almost all the year round, except near the coast.

Italian and Abyssinian Somaliland.—On the coast of Italian Somaliland the climate is healthy almost throughout the whole stretch from the Jub river to Guardafui and beyond. Except in a few places there is no malaria. The monsoons mitigate the heat of spring and summer. From November to April the N.E. Monsoon blows strongly; from May to October a fresher wind blows from the south. These are the prevailing winds. At intervals there are winds of variable direction, while some days are perfectly calm and these are hotter. The climate is rather that of the southern than that of the northern hemisphere, as the country lies to the south of the thermic equator, and the summer is cooler than the spring. From observations taken at Mogdishu it appears that in spring the morning temperature varies between $77°$ and $82°·4$ Fahr. ($25°$ and $28°$ C.); at three o'clock in the afternoon the thermometer usually registers about $82°·4$ Fahr. ($28°$ C.), and at nine in the evening about the same; whereas in summer the morning temperature is only about $69°·8$ or $71°·6$ Fahr. ($21°$ or $22°$ C.), the temperature at 3 p.m. is two degrees, on the average, lower than in spring time, and the same in the evening. The barometer varies but little.

In the north the climate is hotter than on the Benadir coast, as it is much closer to the thermic equator, but the climate is never actually torrid, as it is in the Red Sea and Danakil, and therefore on the Benadir coast Europeans are less likely to be affected by the heat[1].

At Gobwein, at the mouth of the Jub river, as at Kismayu on the British side of the river, the heat is dry, and there

---

[1] Mucciarelli, *Bolletino della Società Geografica Italiana*, Serie IV, Vol. IX No. 4, 1908, p. 348.

LUGH. Lat. 3° 48·3′ N., Long. 42° 39′ E.

Compiled from observations recorded from January 1896 to January 1897.

		Jan.	Feb.	March	April	May	June	July	Aug.	Sept.	Oct.	Nov.	Dec.	Year
Mean temperature[1] (Fahr.°)	...	89°·4	88°·3	90°·3	90°·7	88°·3	86°·4	84°·0	85°·3	85°·5	85°·8	85°·1	88°·5	87°·3
Mean maximum temperature[2]	...	101°·3	105°·4	102°·7	101°·8	98°·1	94°·8	93°·6	94°·5	94°·3	94°·1	93°·0	97°·5	97°·6
Mean minimum temperature[3]	...	77°·5	71°·2	73°·6	78°·1	77°·2	76°·3	74°·3	75°·0	74°·8	77°·0	75°·7	78°·1	75°·7
Mean daily range of temperature	...	23°·8	34°·2	29°·1	23°·7	20°·9	18°·5	19°·3	19°·5	19°·5	17°·1	17°·3	19°·4	21°·9

[1] The mean temperature is derived from $\dfrac{9a + 9p + \text{max.} + \text{min.}}{4}$. In the few instances when the 9 p.m. observations were not recorded, the mean of the three preceding and three succeeding days at 9 p.m. has been used in constructing the table.

[2] The absolute maximum recorded was 113° Fahr. in February 1896.

[3] The absolute minimum recorded was 64°·4 Fahr. also in February 1896.

[4] The mean daily range is max.—min.

are no swamps in the neighbourhood, as the river does not overflow its banks. When Commander Dundas surveyed the river in 1892, there was not a single case of fever among his party.

Brava is reported to enjoy a healthy climate.

At Lugh, on the Jub river, Ferrandi took temperature observations six times a day, at 6 a.m., 9 a.m., noon, 3 p.m., 6 p.m. and 9 p.m., during the whole of 1896, and January 1897, and from these observations the table on p. 300 is constructed.

At Lugh the following was recorded by Ferrandi[1] for the same period as the above table.

		Number of days			Number of days with		
		Fine	Overcast or cloudy	Varied	Little rain	Much rain	Storm
1896	January	23	—	8	2	—	2
	February	20	—	9	1	—	3
	March	14	—	17	6	—	7
	April	8	4	18	2	4	8
	May	3	5	23	4	2	5
	June	1	2	27	3	—	—
	July	1	3	27	—	—	—
	August	2	5	24	1	—	1
	September	8	3	19	3	—	1
	October	2	11	18	6	3	3
	November	5	10	11	5	8	8
	December	20	4	7	3	1	—
1897	January	21	4	6	4	—	3
	February (1—15)	9	—	6	3	—	2

From the above tables it appears that March and April are the hottest months on the whole, though similar high temperatures may be recorded in January and February, and July is the coolest month. The daily range is considerable, but variable, at least so far as January is concerned, and this is the only month for which we have observations for two years. There are two rainy seasons, and heavy rains, especially in the later season, usually come with storms. December, January, February, and March are the finest months.

On the way from the coast to Lugh, the second Bottego expedition recorded, in October, three days overcast, eight varied and six with a little rain; the remaining days were fine. At and near Lugh, in November, the weather was

[1] Ferrandi, *Lugh*, Rome, 1903, p. 384.

not good, only five days being really fine; of the remainder five were overcast and fifteen varied, while on the other five no observations were recorded, but there were copious rains with storms about the 1st December. In the non-mountainous regions at and beyond Lugh, December, January, and February are dry months. In following the notes taken on this journey, it should be remembered that from the 27th December to the 24th January the party was going to the junction of the Web river with the Ganale river; from that date to the 18th February they were proceeding thence to Salole, and from the 29th February to the 28th March to Bwigi, the last two sections of the route being really in British East Africa borderlands. From the 29th March to the 30th September they were making for, and travelling in, the country to the north of Lake Rudolf. In February there were only twelve fine days, three being overcast and eleven varied, with two rainy days at the end of the month. In March the expedition was on higher ground, and eleven fine days are recorded, seven varied and twelve overcast, nine of these being rainy days, and the first ten days of the month being the worst. In April there were four fine days, while sixteen were varied and nine overcast, rain being recorded on eleven of these, the heaviest falls occurring in the early part of the month. Leaving the more elevated districts and descending to the lower lake region, twenty-six rainy days are recorded in May, and twenty-two in June, there being no really fine days between the 1st and 26th of the latter month. In July, proceeding to the Omo river, the weather was better, there being fourteen rainy days, but only four on which it was fine throughout the day, thirteen being varied. August in the Omo valley was very similar to July. In the Lake Rudolf region, in September, there were fourteen fine days, the remainder being varied or overcast[1].

(*See* also under Abyssinia at pp. 287–290.)

French Somaliland.—From the official report of 1903 it appears that, from a health point of view, the intense summer heat is the only thing against Jibuti. Some 28 Europeans are annually treated for various diseases, and

---

[1] Vannutelli e Citerni, *L'Omo.* Ulrico Hoepli, Milano, 1899, pp. 566 *sqq.*

90 per cent. of these are cases coming from the East and landed at the port. The average number of deaths among Europeans is five, and the total European population 400, which gives an annual death-rate of only $12\cdot5$ per thousand. These figures are from the means of three years. As mentioned above, the summer is very hot, but the air is absolutely dry, and therefore the heat is the easier to put up with. In the winter months from October to April the Khamsin, the burning desert wind (S.W. Monsoon) gives place to north-easterly breezes, which render the air comparatively fresh. The mean temperature in summer varies between 88° and 97° Fahr. (31° and 36° C.) with an absolute maximum of 107°·7 Fahr. (42° C.). Between October and April the mean temperature oscillates between 70° and 84° Fahr. (21° and 29° C.), with an absolute minimum of 68° and 70° Fahr. (20° and 21° C.).

It seldom rains at Jibuti, but there are sometimes short showers, and these only in the winter months.

In the Danakil desert there are short severe storms rather than regular rain. These occur chiefly in August and September, but, throughout the whole year, there are short local storms here and there, and the wadis become full, rushing down violently, without notice. It is therefore advisable to avoid camping in the beds of wadis.

(See also p. 275.)

British Somaliland.—Bérbera is reported to be a comparatively cool place during the prevalence of the north-east winds, but intensely hot during the South-West Monsoon. The climate is fairly healthy except when the Kharif wind blows. In the whole coastal area, however, fever is prevalent, and Colonel Swayne, during the 1901 operations in Somaliland, wrote as follows: "the fever from which both British men and officers suffered, whilst in the low-lying country, disappeared when once the force had left Burao."

Mr Consul Keyser writes of the climate of Zeila: "In the summer months the coast is extremely hot. Most natives leave for the interior, and very little business is carried on. The people abandon themselves to the effects of the weather, and consider this as an excuse for contented idleness. In the winter season the climate is not unpleasant, but a damp wind frequently blows, and fever is prevalent among all classes."

In the Haud and Nogal plain districts throughout the year it is hot in the day time, but from November to February it is cold at night. Rain, in this part of the country, may be expected twice in the year, namely, about April and May, and also about September and October. The climate is very healthy, enteric fever being unknown. Dust storms are experienced, especially in dry years, and where the country is bare of wood, during April, May and June, and these storms and haze prevented heliographic signalling during those months in 1901.

Parkinson, writing in the *Geographical Journal* (Vol. XI, 1898, p. 18), says that at Dongorre the air was splendidly cool and bracing in early November, the thermometer sinking to 65° Fahr. (18°·3 C.) in the early morning. And Aylmer, in the same publication, at page 40, says that Somaliland (*i.e.* British Somaliland) in the main is most healthy, and the hill-country, for Europeans, particularly so, but care must be taken to guard against the great variation of temperature between the midday heat and the cold after sunset.

The absolute maximum temperature recorded at Bérbera is 115° Fahr. (46°·1 C.), and the absolute minimum 41° Fahr. (5° C.); the corresponding figures for Zeila are 106° and 63° Fahr. (41°·1 and 17°·2 C.); the Bérbera maximum was recorded in June, and that at Zeila in July; the Bérbera minimum was recorded in March, and the Zeila minimum in November.

PRODUCTIONS.

The coast lands of Somaliland are unproductive, being arid and desert-like. But, in the uplands, cattle, sheep and goats all thrive; and ghee, a clarified butter, as well as hides are articles of export. The districts of Isa and Gadabursi are useless for agricultural purposes, but numbers of camels are bred there, and mules do well in Harar. This last district produces the finest coffee, and three crops a year are gathered. The aloe *Sanseviera ehrenbergii*, from which fibre is obtained, grows in profusion in many parts of British

Somaliland, while from the Ógaden woods gum and resin are brought down to the coast. In the region of the lower Jub river and the lower Webi Shebeli valley dhurra grows at an unprecedented rate, being harvested from 115 to 120 days after the sowing, and, when cut down, the same roots produce a second crop, better even than the first, while a second sowing is sometimes made subsequently. Sheep and cattle do well, and some of the cows are especially fine, being excellent milk-givers. Ghee forms one of the principal productions. Maize and sesame, as well as dhurra, are grown, and cotton will thrive. On the Somali steppe all European cereals flourish, as well as such leguminous plants as beans; and cotton will grow here too. In addition there are numerous native aromatic plants.

BÉRBERA. Lat. 10° 26′ 7″ N, Long. 45° 0′ 52″ E.

Compiled from 3—5 years' observations.

	Jan.	Feb.	March	April	May	June	July	Aug.	Sept.	Oct.	Nov.	Dec.	Year
Mean temperature (Fahr.°) (mean) of $7a+1p+9p$) (3 years[1])	76°·0	77°·3	82°·7	85°·3	89°·3	95°·3	97°·3	97°·0	92°·7	86°·0	80°·7	76°·7	86°·3
Mean maximum temperature (5 years[2])	82°·2	84°·2	85°·6	88°·6	94°·8	106°·6	108°·3	107°·3	100°·8	91°·2	85°·4	81°·8	93°·1
Mean minimum temperature (4 years[3])	62°·0	65°·5	74°·1	74°·6	76°·2	87°·7	89°·3	89°·6	78°·0	69°·2	65°·5	60°·0	74°·3
Mean daily range	20°·2	18°·7	11°·5	14°·0	18°·6	18°·9	19°·0	17°·7	22°·8	22°·0	19°·9	21°·8	18°·8
Mean relative humidity (per cent.) (3) years[1])	66	67	74	73	68	45	44	39	48	67	60	62	59

[1] 1904—1906.   [2] 1902—1906.   [3] 1903—1906.

ZEILA. Lat. 11° 21′ N, Long. 43° 28′ 20″ E.

Compiled from 2—3 years' observations.

	Jan.	Feb.	March	April	May	June	July	Aug.	Sept.	Oct.	Nov.	Dec.	Year
Mean temperature (Fahr.°) (mean) of $7a+1p+9p$) (2 years[1])	73°·5	78°·0	79°·0	80°·0	83°·5	85°·0	91°·5	91°·5	88°·5	82°·5	78°·5	74°·5	82°·2
Mean maximum temperature (3 years[2])	82°·5	84°·5	85°·5	88°·0	91°·0	96°·6	99°·3	98°·0	95°·5	88°·7	86°·0	83°·0	89°·9
Mean minimum temperature (2 years[1])	69°·0	71°·0	72°·0	74°·0	76°·5	77°·5	84°·5	85°·0	83°·5	77°·5	71°·0	65°·5	75°·6
Mean daily range	13°·5	23°·5	23°·5	14°·0	14°·5	19°·1	14°·8	13°·0	13°·0	11°·2	15°·0	17°·5	14°·3
Mean relative humidity (per cent.) (2) years[1])	79	73	77	81	80	75	65	72	77	74	75	75	75

[1] 1905, 1906.   [2] 1904—1906.

BRITISH SOMALILAND. MEAN RAINFALL IN INCHES.

	Jan.	Feb.	March	April	May	June	July	Aug.	Sept.	Oct.	Nov.	Dec.	Year
Bérbera (43 months[1])	·24	·16	1·60[2]	·42	·01	0	0	·04	·16	0	·29	1·06[3]	3·98
Zeila (31 months[4])	·07	·40	·33	·07	0	0	·05	·01	·07	0	1·96	·10	3·06

[1] December 1901, January—July 1902, November, December 1904, April—December 1905, 1906, 1907.
[2] Entirely due to a fall of 4·80 inches in March 1906.
[3] Largely due to a fall of 3·87 inches in December 1901.
[4] January—March, and September—December 1904, 1905, 1906.

UGANDA PROTECTORATE.

In the extreme south-west of the Uganda Protectorate is Ankole, on the slopes of Mfumbiro (14,683 ft.), which lies between Lake Kivu (4829 ft.) and Albert Edward Nyanza[1] (3004 ft.). To the north of the latter, and between it and Albert Nyanza (2037 ft.), is the huge mountain mass or range of Ru Nzori (16,619 ft.), "the water producer," with Toro on its slopes. On the east of the Protectorate rise the wooded volcano Elgon (14,152 ft.) and the heights of Karamojo, reaching right away north to the eastern side of Lake Rudolf. In the south is Victoria Nyanza, which stands at an altitude of 3720 feet above sea level. The streams from Ankole flow to this. The remaining waters all drain north-westwards. The Victoria Nile has a general north-west direction; Namatara river from the slopes of Elgon flows into Lake Choga (Kioga); the Siroko and Kiboko, from its northern slopes, enter Lake Salisbury; the Aswa and its numerous feeders from the Karamojo heights, after reaching the lower country, all turn north-west to join the Nile. Similarly the Gomoro, the Khoz and the Tu all flow north-west.

The country south of $1°$ N. all lies at an elevation of between 4000 and 6000 feet above the sea, and then gradually falls to the north-west till at Gondókoro, at the northern end of the Wádelai-Gondókoro gorge, the altitude is 2000 feet below the area lying immediately north of Victoria Nyanza, the general level of which is a trifle under 4000 feet.

The great lake covers so large an area that it acts, in a measure, like the ocean; it has a tempering, a moderating effect on the climate of the country in its vicinity. It causes land and sea-breezes, and keeps the range of temperature low.

Thus at Entebbe the morning temperature (7 a.m.) only varies between $72°\cdot4$ and $64°\cdot5$ Fahr. ($22°\cdot4$ and $18°$ C.), giving a total range of $7°\cdot9$ Fahr. ($4°\cdot4$ C.); again at midday (2 p.m.) the highest figure registered by the thermometer is $83°\cdot3$ Fahr. ($28°\cdot5$ C.), and the lowest $74°\cdot8$ Fahr. ($23°\cdot8$ C.), a difference of only $8°\cdot5$ Fahr. ($4°\cdot7$ C.), and in the evening the corresponding figures are $75°\cdot8$ and $68°\cdot6$ Fahr. ($24°\cdot3$

[1] Now called Lake Edward.

and $20°·3$ C.), giving a range of $7°·2$ Fahr. ($4°$ C.). The absolute maximum temperature recorded at Entebbe is $90°$ Fahr. ($32°·2$ C.), and the absolute minimum $55°$ Fahr. ($12°·8$ C.). The relative humidity is greatest in the morning, decreases in the middle of the day, and rises again towards evening. The mean number of rainy days is 118.

The mean annual temperature at Mumia's is a degree higher than at Entebbe. In December, January, February, March and April it is considerably higher, while in July and August it is lower. In the remaining months the temperatures at the two places are about the same.

In a general sense, the annual rainfall decreases from south to north. At Mbarara in Ankole it is 69 inches ($1752·6$ mm.), at Entebbe on the Victoria Nyanza, and at Fort Portal in Toro it is 59 inches ($1498·6$ mm.), at Ripon Falls 45 inches (1143 mm.), at Wádelai nearly 44 inches ($1117·6$ mm.), and at Gondókoro 42 inches ($1066·8$ mm.).

At Mbarara, in Ankole, at an altitude of 4840 feet, the mean temperature is considerably less than at either Entebbe or Mumia's throughout the whole year, but the range is much greater. The average maximum temperature at Mengo, for seven years, was $83°·8$ Fahr. ($28°·8$ C.), and the average minimum $59°$ Fahr. ($15°$ C.).

On the whole the climate of Uganda is mild, neither excessively hot nor cold. But the greater part of the Protectorate is malarious; certain districts, notably the Nile valley, appear to be more so than others.

Albert Nyanza and its shores are certainly unhealthy, says Capt. Vandeleur, to people coming from the high ground, and the cold wind off the lake produces chills and induces fever. Mruli[1] is low and unhealthy, says the same authority.

The climate of the uplands of Unyoro[2] is a fairly good one, and is drier than that of Uganda. "The maximum temperature registered at Fort Hoima, while I was there" (quoting again from Vandeleur), "was $89°$ Fahr., and the minimum $49°$ Fahr. ($31°·7$ and $9°·4$ C.); as a rule the thermometer did not go below $60°$ Fahr. ($15°·6$ C.) at night. The rains are very variable. In September, October and November there is usually most rain; and July was the coolest month."

---

[1] Now called Buruli.   [2] Now called Bunyoro.

In the region of Kioga (Choga), Kwania and the other great backwaters of the Victoria Nile, the country is for the most part low, and exceedingly swampy, and therefore very unhealthy for Europeans.

The rains collected by Ru Nzori render the climate of Toro excessively moist. In the dry season the range is nearly always shrouded in a haze, a fact which led Stanley, who passed within a few miles, to doubt at first of its existence. Katwe, on the north of Albert Edward Nyanza, is regarded as one of the hottest and most unhealthy neighbourhoods; and the Semliki plain is anything but a sanatorium.

The north part of Ankole possesses a very healthy climate, and is suited for European settlement. The southern part is somewhat drier and less equatorial in climate; it falls in altitude, and has a more parched appearance.

The Nile Province is generally looked upon as the most unhealthy portion of the Protectorate. North of the Aswa river, in the dry season the swamps disappear and the rivers nearly run dry. The rainy season usually begins in April, though sometimes not till May. In normal seasons it rains for about two months, and should recommence about September, but the second rains are not so heavy as the first.

The Nile districts are considerably warmer than Uganda (proper), or Buganda as it is now called, in consequence of the general elevation being 2000 feet lower. It is sometimes very hot in the valley of the river and at Gondókoro the temperature occasionally exceeds $100°$ Fahr. ($37°\cdot7$ C.); but, speaking generally, the country cannot be called hot in the sense of the hot weather of India. When the stations can be placed further east on higher ground, delightfully cool and healthy sites will be available; in the Latuka country, for instance, the mountains reach a height of 9000 feet above the sea level[1].

In the neighbourhood of the Anglo-German boundary west of Victoria Nyanza, the gales from E. and S.E., coming from the open lake, are extremely violent, and the climate is similar to that of Entebbe and other places on the lake.

[1] *Geographical Journal*, Vol. XXVI, 1905, p. 481.

The Commissioners found that the night temperature in the open air at 10 p.m. during December and January averaged 68° Fahr. (20° C.), while the day temperature at Mizinda in the sun during the same months averaged 78° Fahr. (25°·6 C.). At the western end of the boundary the night temperature at 10 p.m. in June and July averaged 60° Fahr. (15°·6 C.). Near the shores of the lake mosquitoes are troublesome, but at Mizinda, at a little distance from the lake, they were rarely seen. At Mulema, towards the end of the dry season the days were warm, but the nights pleasantly cool. After the rains begin the heat is never excessive.

At Burumba, 5200 feet above the sea, the temperatures ruled lower. On dull days the thermometer did not rise above 58° Fahr. (14°·4 C.) in the huts, and at night and in the early morning the air felt very cold ; and moreover cold S.E. winds were experienced blowing up the Kagera valley. At Rukirra the night temperatures were low and the days pleasantly warm. On the Ruchigga mountains there was a decided fall of temperature, the altitude being 7000 feet and over. The air was invigorating, and the situation might prove suitable for a sanatorium. When the sun shone it was never unpleasantly cold, but when the hill-tops were covered with a cloud the effect was chilling, and, on these occasions, fires at night were needed[1].

The following is from an official report:—"A comparison of the rainfall records, taken at Entebbe for eight years, leads one to the conclusion that there are no very well defined seasons, dry or wet. The whole of Uganda, especially the lake districts, is particularly liable to thunderstorms, and it is very seldom that a month passes without rain. At Entebbe there is not a month on record which has not had one or two wet days, or days on which rain has fallen. Usually January and February are fairly dry months, with an average of 3 inches (76 mm.) each. March, April and May are the wettest months of the whole year, and occasionally the wet season extends into June, as was the case in 1902, when 10·26 inches (269 mm.) of rain fell during that month. July, August, September and October are the driest months of the year. It is rarely

[1] *Geographical Journal*, Vol. XXVI, 1905, p. 616.

the rainfall during either (*sic*) of these months exceeds 4 or 5 inches (100 or 127 mm.). November and December are both unreliable months; occasionally the rainfall has amounted to 12 inches (305 mm.), in other years it has been as low as 1·63 inches (40·6 mm.); but, as a rule, if a heavy rainfall occurs in November, much less falls during December, and if deficient in November, it is usually made up in December. These remarks are only applicable as far as Entebbe district is concerned. As the rain falls throughout Uganda more or less in the form of thunderstorms, it does so usually within very limited areas, and the records of any one particular station cannot be taken as strictly representative of the whole district or county, much less that of the whole country. It is not possible to compare the mean daily sunshine of the various stations, as Entebbe is the only one provided with a sunshine recorder. The highest daily average for any one month during the year at Entebbe was that of February 1905, when the daily sunshine amounted to 7 hours 32 minutes. The lowest daily average was 4 hours 5 minutes, and was registered during the month of November. The daily average for the whole year amounted to 5 hours 52 minutes. These registrations are of bright sunshine only.'

PRODUCTIONS, &c.

In the Uganda Protectorate rubber is indigenous, but it is of poor quality. The fact, however, showed that the conditions were suitable for its growth, and imported varieties are now producing better results than the native species. Coffee does well, being a native of the southern Abyssinian uplands. The climate is excellently suited for the production of cotton, while the low swampy lands in the vicinity of the Victoria Nyanza, around Choga, and on the lower Kagera, are well adapted for the growth of rice. Chillies and indigo are indigenous. Dhurra (or North African millet) is regularly planted in the dry season, which serves for the collection of other millet crops. A species of *Ficus* yields an excellent fibre which is known to trade as "Uganda bark cloth." Of fibre plants the *Agave sisalana*, or Sisal hemp, and the *Sanseviera ehrenbergii*, a stemless perennial, one of the "Bow-string" hemps, should do well; and, of oil-producing plants, oil-palms flourish on the shores of the lakes, ground-nuts and sesame (semsem) form part of the ordinary agricultural produce of the country, which also includes bananas, sweet potatoes, maize, beans and tobacco. Linseed does well in the central province. The eastern part of Ankole consists of rolling downs with good pasture land, in fact Ankole is essentially the ranch-land of the Protectorate, and the cattle are exceptionally fine; goats, too, and sheep also thrive. "In Toro," says the Rev. A. B. Fisher[1], a missionary resident in Toro, "strawberries can be had for nine months out of the twelve, and green peas, cauliflowers, potatoes, beans, marrows, &c., all the year round. On the slopes of Ru Nzori tea could be most successfully grown; the coffee is of an exceptionally fine flavour, and the marshy character of the country makes it valuable for rice fields." The forests yield many most valuable hard and ornamental woods.

[1] *Geographical Journal*, Vol. XXIV, 1904, p. 249.

ENTEBBE, ON VICTORIA NYANZA, NORTH SHORE.　Lat. 0° 3′ S., Long. 32° 30′ E.
Alt. 3906 feet.

Compiled from 2 years' observations[1].

	Jan.	Feb.	March	April	May	June	July	Aug.	Sept.	Oct.	Nov.	Dec.	Year
Mean temperature[2] (Fahr.°) ...	72°·7	72°·3	71°·8	70°·2	71°·2	70°·9	71°·8	70°·5	71°·8	72°·5	72°·1	71°·4	71°·6
Mean maximum temperature[3] ...	81°·8	80°·6	80°·2	78°·8	77°·7	78°·1	77°·5	79°·5	81°·7	81°·8	81°·3	80°·2	79°·9
Mean minimum temperature[4] ...	64°·0	63°·8	63°·3	63°·5	64°·4	62°·8	61°·5	61°·3	62°·2	62°·4	62°·9	62°·2	62°·9
Mean daily range[5]　...　...	17°·8	16°·8	16°·9	15°·3	13°·3	15°·3	16°·0	18°·2	19°·5	19°·4	18°·4	18°·0	17°·0
Mean relative humidity (morning[6]) ...	86	88	91	95	89	88	89	88	86	90	92	92	89
„　„　(2 p.m.[7]) ...	66	72	74	76	80	79	75	79	75	68	68	70	73
Number of hours' sunshine (1902) ...	205·5	188	243	167	200	198	150·5	207	201	161	221	185	2324

NOTE.—The instruments in use at this station were :—Barometer (N. and Z.) 2025; Dry Bulb (Casella) B.T. 2244; Wet Bulb B.T. 2245; Maximum thermometer (N. and Z.) M.O. 1665; minimum thermometer (N. and Z.) M.O. 1670. Instruments were first kept under a verandah, but after 15th July in a lath screen, under a grass-shed in open ground.　The figures, from which this table is compiled, appeared in *Meteorological Office: Climatological Observations at Colonial and Foreign Stations, I. Tropical Africa*, E. G. Ravenstein, 1904

[1] 1901, 1902.
[2] The mean temperature is derived from $\frac{max. + min.}{2}$ up to June 1901, and subsequently from $\frac{7a + 2p + 9p + 9p}{4}$.　In July 1901 the 2 p.m. and 9 p.m. observations are for 12th—31st only.　Up to July 1901 the recording times were 7 a.m., 1 p.m., 4 p.m. subsequently 7 a.m., 2 p.m., and 9 p.m.
[3] The absolute maximum recorded was 90° in January.
[4] The absolute minimum recorded was 55° in January.
[5] The daily range is max.—min.
[6] At 7 a.m.
[7] Up to July 1901 the recording time was 1 p.m, subsequently 2 p.m.

MBARARA, IN ANKOLE. Lat. 0° 39' S., Long. 30° 49' E. Alt. 4500 feet.

Compiled from observations recorded in 1902.

		Jan.	Feb.	March	April	May	June	July	Aug.	Sept.	Oct.	Nov.	Dec.	Year
Mean temperature [1] (Fahr.°)	...	66°.0	66°.7	66°.8	66°.8	66°.4	63°.7	66°.9	67°.9	68°.6	66°.5	67°.0	66°.3	66°.6
Mean maximum temperature	...	79°.0	80°.2	79°.0	77°.5	76°.6	79°.5	79°.5	79°.1	78°.6	76°.6	77°.5	77°.5	78°.4
Mean minimum temperature	...	55°.6	56°.8	55°.9	55°.9	56°.3	55°.1	56°.1	57°.6	55°.1	58°.1	55°.6	52°.2	55°.9
Mean daily range [2]	...	23°.4	23°.4	23°.1	21°.6	20°.3	24°.4	23°.4	21°.5	23°.5	18°.5	21°.9	25°.3	22°.5
Mean relative humidity (morning [3])	...	82	87	85	87	86	85	88	82	76	82	87	87	84
,, ,, (evening [4])	...	77	74	86	85	84	77	81	82	82	85	88	84	82

NOTE.—The instruments in use at this station were:—Dry Bulb, B.T. 5826; Wet Bulb, B.T. 5829; maximum thermometer, M.O. 1901; minimum thermometer, M.O. 1875. Rain-gauge, M.O. 437; grass thermometer, B.T. 5825. Also Barograph. *See* note under Entebbe (above) for authority.

[1] The mean temperature is derived from $\dfrac{8a+8p}{2}$ for January and February; $\dfrac{9a+9p}{2}$ for March, April, May, June; and from $\dfrac{7a+2p+9p+9p}{4}$ for the remaining months.

[2] The daily range is max.—min.

[3] At 8 a.m. in January and February; at 9 a.m. in March, April, May and June; and at 7 a.m. in the remaining months.

[4] At 8 p.m. in January and February; and at 9 p.m. in the remaining months.

UGANDA PROTECTORATE. MEAN RAINFALL IN INCHES.

	Alt. in feet	No. of Years	Jan.	Feb.	March	April	May	June	July	Aug.	Sept.	Oct.	Nov.	Dec.	Year
Butiaba (L. Albert)[1]	2320	2	0.71	0.35	5.16	5.04	5.47	2.84	2.21	4.41	3.03	5.91	5.47	3.27	43.87
Fort Portal (Toru)[2]	4740	36 months	1.93	2.28	6.38	5.43	6.30	4.84	4.17	4.53	6.42	8.98	5.67	2.56	59.49
Mbarara (Ankole)[3]	4480	(4)	4.53	3.50	11.02	4.25	3.15	1.34	0.91	4.84	8.35	9.13	10.35	7.99	69.36
Masaka (Buddu)[4]	4080	(4)	3.07	2.68	4.69	8.43	6.38	0.75	1.10	2.40	4.33	5.51	5.04	5.20	49.58
Jinja (Ripon Falls)[5]	3710	(4)	3.27	2.13	3.78	6.69	4.96	2.44	2.28	4.17	2.60	4.33	5.32	3.62	45.59
Natete (Mengo)[6]	4090	7	2.36	4.33	4.49	7.21	4.80	2.83	3.35	2.87	3.90	5.24	3.62	1.65	46.65
Entebbe (Victoria Nyanza)[7]	3740	(9)	3.03	3.39	7.21	8.70	6.77	5.08	2.84	2.60	3.30	3.54	6.77	5.87	59.10

[1] 1904, 1905.
[2] May—November 1902, February—December 1903, 1904, January—April and October—December 1905.
[3] 1902, February and May—December 1903, 1904, 1905.
[4] July—December 1902, March—December 1903, 1904, 1905.
[5] 1902, January—April, June, August—December 1903, 1904, 1905.
[6] April—June 1878, 1879, January—March 1886, 1881—1885, January—June 1886.
[7] April—December 1896, January—July 1897, December 1899, 1900—1905.

ON THE BAHR-EL-JEBEL.

				No. of Years	Jan.	Feb.	March	April	May	June	July	Aug.	Sept.	Oct.	Nov.	Dec.	Year
Wâdelai[1]	...	...	...	(8)	0.95	1.42	4.49	3.70	4.69	3.70	3.90	4.61	4.02	6.61	4.45	1.26	43.80
Nimule[2]	...	...	...	3	0.08	0.47	4.06	2.01	5.47	2.68	4.76	4.61	5.16	6.02	4.69	1.06	41.07
Gondókoro[3]	...	...	...	6	0.20	0.75	2.72	3.23	6.46	3.94	5.79	5.55	5.99	5.55	1.69	0.47	42.34

[1] August—December 1885, 1886, 1887, 1888, March—December 1902, 1903, 1904, 1905. [2] 1904—1906. [3] 1901—1906.

EAST AFRICA PROTECTORATE.

The territory known as the East Africa Protectorate rises, like the rest of Africa, from the ocean to the interior by terraces. The coastal plain or Temborari varies in width, being about two miles wide at Mombasa. At Takaungu the hills come right down to the sea. Between Malindi and Lamu the plain is some ten miles wide, while near the Jub river it extends inland for 100 miles. Above this coastal plain rise the foot-hills and the barren Nyika plateau, which in turn are succeeded by the fruitful Rangatan plateaux which combined form the Lake Plateau. Only an eastern portion of the highlands is included in the East Africa Protectorate. If the towering peaks such as Elgon and Kenia, &c., be left out of account these highlands have an average altitude of from 5000 to 6000 feet, and are cleft by the Rift valley, which extends north and south.

The climate naturally varies with the elevation, and, speaking in a general way, there are in this area three climatic zones.

1. The Coast, including under this title a band of about 100 miles in width extending along the whole coast from the Anglo-German boundary to the Jub river. Here the thermometer rarely falls below 70° Fahr. (21° C.), and seldom rises above 90° Fahr. (32° C.), the monthly mean of the maximum temperature at Mombasa, in the hottest months, namely March and April, being about 87° Fahr. (30°·6 C.), and that of the minimum from 78° to 80° Fahr. (25°·5 to 26°·7 C.), but at Malindi both the mean maximum and the mean minimum temperature are considerably lower. The temperature is therefore high, but not excessive, the relative humidity is high, and the climate, owing to this fact, is damp and enervating and difficult for Europeans to endure.

2. The Highlands, including the greater part of the Protectorate, the Nyika and Rangatan. Here the thermometer rarely falls below 50° Fahr. (10° C.) and seldom rises above 80° Fahr. (26°·7 C.), and here the white man can live and flourish; this is a land eminently suited for colonization.

At Machakos the mean maximum temperature for the years 1901 and 1902 was $74°·8$ Fahr. ($23°·8$ C.), and the mean minimum $53°·4$ Fahr. ($12°·1$ C.). The hottest months had a mean maximum temperature of $76°$ Fahr. ($24°·4$ C.), and a mean minimum of from $56°$ to $57°$ Fahr. ($13°·3$ to $13°·9$ C.), while the mean minimum of the coolest months was $49°$ Fahr. ($9°·4$ C.).

3. The Lake district, *i.e.* the country bordering on the Victoria Nyanza. Here the temperature is equal to, if not greater than, that on the coast, and the climate is less agreeable than in any other part of the Protectorate.

The statistics show, says Mr Buckley, writing in the *Geographical Journal*, Vol. XXI, 1903, p. 350, that near the coast the S.W. Monsoon causes heavy rain in April and May, some 20—30 inches (500—760 mm.), and more falling during these months; light falls occur from June to October, and in November the fall is usually heavier; December to March are dry months. In the higher land above 5000 feet the heaviest falls are usually in March, April and May; a dry season follows from June to October, when there is usually a second period of rain in November and December. Falls of more than 5 inches (130 mm.) in 24 hours occur very rarely; falls of more than 4 inches (100 mm.) are not frequent.

Speaking generally of the highlands through which the railway passes, Mr Buckley says they have a pleasant temperate climate well suited to Europeans, and Commander Whitehouse, who carried out the hydrographic survey of Victoria Nyanza, adds that the climate of the district the line passes through is pleasant from Sultan Mamoud, mile 238, to the lower part of the Nyando valley, mile 550. From above mile 33, bad nights, such as have to be endured on the tropical coasts of Africa and India, are never experienced.

In these highlands the wet seasons vary in length, and sometimes in date, but, generally speaking, there are two— one from November till the end of December, and the other from March to May. The remaining months are usually dry, but there have been rains right up to the end of June, and at times in February. July, August and September are, as a rule, the cold months. The mean temperature is about $67°$ Fahr. ($20°$ C.) at 9 a.m., and $78°$ Fahr. ($25°·5$ C.) at noon, while at night it is much colder, and the thermometer often goes down to $45°$ Fahr. ($7°·2$ C.) in the early morning. The mean rainfall at Machakos for six years was $34·76$ inches (883 mm.), including two dry years with an average of only 25 inches (635 mm.) each. At Kikuyu the mean for six years was $36·11$ inches (917 mm.), including two dry years. The heat increases sometimes to about $80°$ Fahr. ($26°·7$ C.) at noon, and remains stationary till about 2.30 p.m., when it cools down, but the average at that hour is somewhat lower.

Sir H. H. Johnston, writing of the western districts, that is of the area which used to form the Eastern Province of Uganda, says that "the climate is extremely varied. On the extreme south, towards the German frontier, down the depression of the Rift valley, it tends decidedly to sterility, and on the north, too, the tendency is to a diminution in the rainfall, and consequent deficiency in vegetation. South of Lake Naivasha there is a marked decrease in the annual rainfall, and a tendency in the Rift valley towards a desert appearance. This is the case north of Lake Baringo, and many parts of the country near Lake Rudolf offer a striking resemblance to the Sahara."

In the uplands behind Naivasha and east of the Rift valley the climate is temperate. In the district near Njoro, where the Mau escarpment begins to rise from the plains, the climate is even more bracing than the eastern Kikuyu slopes.

The Kikuyu hills enjoy a cool climate, excelled probably by few in the world; and here the white man can not only live in comfort, but work in comfort. In fact those accustomed to a tropical climate often find it too cool. In June and July, *i.e.* in the dry season, there is often a fog, like a Scotch mist, until midday, or drizzling rain. Here there are two rainy seasons called the *maize* rains and *millet* rains respectively. The first begins early in March and lasts till the end of May. The second begins about the middle of October, and lasts till about the middle or end of December. Only very slight variations of temperature are experienced, and both extreme heat and severe cold are unknown, except of course on the summit of Kilinyatha, and there is an abundant and usually constant rainfall. There is no malaria at altitudes above 7000 feet. In the lower country, in the region of the Upper Athi and Tiva rivers, the rainy seasons are apparently earlier than in the highlands, for Capt. Aylmer found that in this country his expedition was hampered by want of water about a month after the end of the highland rainy season.

The Nandi country lies at an average altitude of over 6000 feet above sea-level, and is excellently adapted for colonists. The climate is described as "very good." It is cold at night on the high ground, and the thermometer rarely rises to over 80° Fahr. (26°·7 C.). There is usually a good deal of rain at night in the rainy season. On the whole the Nandi escarpment is said, by its admirers, to possess, even in a higher degree, the excellence of the climate of Nairobi and Njoro.

"In Ukamba," says Mr Ainsworth, the sub-commissioner of the district, writing in the *Journal of the Manchester Geographical Society*, Vol. XVIII, 1902, p. 56, "the rainfall for the years 1890–1900, excepting 1898 and 1899, was over 50 inches (1270 mm.) a year ; for 1898 it was 24 inches (610 mm.), and for 1899 it was 25 inches (635 mm.). These two years seemed to complete a cycle of semi-drought, and from native information these droughts appear to occur with more or less severity every ten years, and generally the result has been a famine before the rains have reached their normal again. The mean average temperature is about 64° Fahr. (17°·8 C.) at 9 a.m. ; at midday it ranges to about 75° Fahr. (23°·9 C.). Just before the rains it is hotter, but, generally speaking, the districts to the north of the south end of Ulu are blessed with a climate that white men can live in."

The whole of the Kenia forest region is healthful and invigorating, but there is a great contrast, during at least half the year, between the wet and misty southern slopes and the dry bracing plateau country of northern Kenia. Messrs Hutchins and Ross experienced hail on numerous occasions ; in fact, on the wetter southern side there was a severe hail-storm daily. A real snow-storm was experienced on one occasion only. The above notes apply to the wettest time of the year, viz. April, May and part of June. December, January and February form the dry season; March, April and May the rainy season; June, July and August the misty season. The weather on the S. side, when visited by Messrs Hutchins and Ross, was a striking contrast to what Messrs Mackinder and Hausberg experienced. Mr Mackinder remarks that the air was usually dry, the relative humidity falling on one occasion to as low as 52 per cent. "There was no considerable rain or snow-fall during our stay at high altitudes (August 20th to September 20th), but a little fine wind-driven hail on most afternoons, preceded by a sharp drop in temperature about 2 p.m."

Mr C. W. Hobley, now the Assistant Deputy Commissioner of the Protectorate, says the air of Kavirondo is far drier and more bracing than that of Uganda (proper) and Usoga; the heat of the sun is greater during the day, but the nights are deliciously cool. The greatest range of temperature is during the dry season, when the early morning temperature sometimes falls as low as $55°$ Fahr. ($12°{\cdot}8$ C.), and the shade temperature at 2 p.m. will occasionally reach $90°$ Fahr. ($32°{\cdot}2$ C.). The dry season practically commences about the close of December, and continues until the middle of March. From March to December it rains with great regularity; in the afternoon of most days there occurs a rain-storm of from one to three hours' duration, accompanied by strong winds and a heavy thunderstorm; rain rarely occurs in the morning. Seasons of drought are of extreme rarity, but the rainfall (total yearly) varies greatly. In two consecutive years 80 and 59 inches (2032 and 1499 mm.) were registered. The rains reach their maximum about August. The strip of country bounding Victoria Nyanza in this region usually has a much lower rainfall than the rest of Kavirondo, due probably to the fact that, the bulk of the storms coming from E. and N.E., their force is spent on the Samia and Maragolia hills, and the clouds discharge their moisture before reaching this belt. The area enjoying the highest rainfall is probably Ketosh, which lies immediately at the foot of Mount Elgon[1].

In Jubaland January and February are the hottest

---

[1] *G. J.* Vol. XII, 1898, p. 363.

months. In April and May and again in September are the rains. Lying under the equator there is little variation in the seasons. The highest temperature, registered in February 1901, was 104° Fahr. (40° C.) at Afmadu. The nights are always cool. During the heat of the day the sun is powerful, but the temperature compares favourably with that of Indian hot weather. The climate generally is healthy, and there is little fever or dysentery. There is not much rainfall in Jubaland. During the break of the monsoon there is no general downpour at Kismayu, but merely rain-storms and squalls. Further inland, at Kurkumes, the rain falls frequently with great regularity, from 1 p.m. to nightfall, daily, in April and May. During the rainy season the country becomes impassable to laden men and animals, every slight hollow becoming first a bog and then a lake, which holds water for two or three months according to its depth.

The climate of the Tana river valley up to Golbanti is unhealthy, and, when the river is in flood, the extensive swamps breed legions of mosquitoes, but above that point the climate is pleasant.

Judging from a residence of three months during the rainy season, Mr Wakefield said, in 1879, that the country between the Tana and the Sabaki was exceedingly healthy, as well as being very fertile.

Great things are expected of Mombasa, and it is to be one of the healthiest places in the tropics, when the country round has been duly put under cultivation, and improvements in sanitation effected. These improvements are well under weigh, and the town has a bright outlook. There is a sufficient range of temperature to render the locality healthier than many places further south.

The mean annual rainfall at Malindi is 41·43 inches (1052 mm.); in 1901 the total was only 36·35 inches (923 mm.), of which 12·69 inches (322·6 mm.) fell in 17 days in May; in 1902 the total was 47·25 inches (1201 mm.), of which 27 inches (686 mm.) fell in the period May to July; the greatest precipitation in 24 hours was 3·96 inches (100 mm.) in May 1901 and June 1902. The number of rainy days does not necessarily bear any relation to the total, for in 1902, when the total was over 47 inches

(1193 mm.), it only rained on 39 days, while in 1901, with a total fall of 36·35 inches (925 mm.), it rained on 97 days. The mean temperature at 9 a.m. varies but little, ranging from 79° to 85° Fahr. (26°·1 to 29°·4 C.). There is a good supply of water from deep wells.

At Kismayu, on the coast, near the mouth of the Jub river, under the equator, fevers are said to be unknown. The air is at all times dry, and the nights are almost cool. When Commander Dundas surveyed the river, there was no case of fever among his party, including the crew of his ship. During 1900 the mean monthly temperature only varied between 79° Fahr. (26°·1 C.) in July and 85°·6 Fahr. (29°·8 C.) in April.

At Machakos the mean temperature for the year, at 9 a.m., is 65°·4 Fahr. (18°·5 C.), and in no month in the year does it differ more than 3° or 4° Fahr. from this. The mean maximum temperature only reaches 80° Fahr. (26°·7 C.) in one month in the year, viz. October, and the mean for the year is 74°·8 Fahr. (23°·8 C.), while the nights are decidedly cool, the mean minimum being below 50° Fahr. (10° C.) in June, July and August, and just 50° Fahr. in September. The relative humidity is decidedly low, being only a little over 50 per cent. in any month. The rainfall cannot be depended on from month to month, as will be seen from the table below. The most frequent winds are from the south-east, the strongest and warmest from N.E. and N.N.E. and the coolest from S.W.

Speaking generally of the rainfall in the Protectorate, there are three very rainy areas, namely, on the coast towards the south, a large area north of Nairobi, and the coast regions of Victoria Nyanza, where the heaviest fall is experienced. In the first mentioned area the total annual precipitation reaches 50 inches (1270 mm.); between this area and that north of Nairobi it sinks to 30 inches (762 mm.), and along the Anglo-German boundary and in the Rift valley it sinks to considerably less. The region north of Nairobi, including Kenia, has a total annual rainfall of between 40 and 50 inches (1000 and 1250 mm.), while in the region near the lake it amounts to 60 and 70 inches (1500 and 1750 mm.). Indeed, in 1906 the yearly total at Nandi was no less than 99 inches (2514 mm.).

PRODUCTIONS, &c.

In the lowlands the climate is suitable for the production of maize and various native grains, while in parts, where the rainfall is sufficient, rice may be grown. The cocoa-nut palm thrives, and along the coastal belt rubber does well. Cotton and tobacco may also be grown.

In the highland regions coffee may be produced, and such European products as potatoes, wheat and barley thrive, and there is also extensive grazing for cattle. Castor-oil beans are produced. European fruits do well in Ukamba, and, in the neighbourhood of the lake, cotton, ground-nuts and rice yield good results ; in fact, of the first named there are several native species; and if rice does well, so also should jute, if tried.

One of the great drawbacks, however, in the East Africa Protectorate is the uncertainty of the rainfall.

In the forests are mines of wealth—ebony, mahogany, and various hard and ornamental woods.

MOMBASA. Lat. 4° 4' S., Long. 39° 42' E. Alt. 60 feet.

Compiled from 7 years' observations[1].

	Jan.	Feb.	March	April	May	June	July	Aug.	Sept.	Oct.	Nov.	Dec.	Year
Mean temperature at 9 a.m. (Fahr.°) ...	82°9	82°7	84°7	83°8	80°5	78°6	77°5	77°9	79°5	80°5	83°9	82°8	81°3
Mean maximum temperature ...	85°0	85°8	87°1	86°7	83°9	84°1	82°1	81°8	82°3	83°5	84°4	85°4	84°3
Mean minimum temperature ...	77°8	78°8	77°6	79°9	77°4	75°9	74°9	75°3	76°5	77°2	78°4	78°9	77°4
Extremes of the mean maxima and { max.	88°5	90°0	98°0	91°3	89°0	86°3	92°3	86°3	86°0	88°0	88°3	89°0	98°0
minima during 7 years { min.	71°9	70°0	60°0	68°0	71°0	71°1	69°4	69°7	72°0	71°9	70°0	70°0	60°0
Mean daily range[2] ...	7°2	7°0	9°5	6°8	6°5	8°2	7°2	6°5	5°8	6°3	6°0	6°5	6°9
Mean temperature[3] ...	81°4	82°3	82°3	83°3	80°7	80°0	78°5	78°6	79°4	80°4	81°4	82°1	80°9
Mean relative humidity (per cent.) ...	88	85	87	87	90	89	89	89	88	88	88	87	88
Extremes of rainfall in each month { max. in inches	4·09	2·23	6·62	15·42	25·40	4·37	6·13	4·91	5·18	6·22	27·77	4·86	
{ min.	0·03	0	0·52	1·03	4·30	0·38	2·77	0·75	1·48	0·33	0·51	0	
Mean number of rainy days ...	2·3	2·1	7·0	9·6	14·3	6·4	10·6	8·6	8·6	7·7	7·7	6·1	91·0

NOTE.—Up to August 1901 no observations, except rainfall, were recorded on Sundays. The readings of the minimum thermometer must be accepted with caution, being suspiciously high. The relative humidity, too, must be accepted with caution. The instruments used were by Negretti and Zambra and were Dry Bulb 4590, September 1889: up to 72°, 0·0; at 82°, +0·1; at 92°, +0·1. Wet Bulb 4596, September 1889: at 32°, −0·1; at 42°, 52°, 62° and 72°, 0·0; at 82° and 92°, +0·1. Maximum thermometer 1358, June 1890: at 62°, −0·7. Minimum thermometer 1458, January 1891: up to 32°, 0·0; at 42°, 52°, 62° and 72°, −0·1. The instruments were placed in a hall under the Sub-Commissioner's office, five feet from the floor. See note under Machakos for authority.

[1] 1896–1902. [2] max.—min. [3] $\dfrac{\text{max.}+\text{min.}}{2}$.

MACHAKOS. Lat. 1° 31′ S., Long. 37° 18′ E. Alt. 5400 feet.

Compiled from 7 years' observations[1].

	Jan.	Feb.	March	April	May	June	July	Aug.	Sept.	Oct.	Nov.	Dec.	Year
Mean temperature at 9 a.m. (Fahr.°) ...	67°·3	68°·2	68°·0	67°·2	66°·1	62°·3	61°·3	61°·8	64°·8	66°·4	66°·3	65°·7	65°·4
Mean maximum temperature (2 years)[2]	77°·5	76°·1	76°·0	75°·8	75°·1	72°·3	72°·2	73°·8	77°·8	80°·2	76°·5	73°·6	74°·8
Mean minimum temperature (2 years)[2]	51°·9	56°·4	56°·4	57°·2	54°·8	49°·8	49°·6	49°·0	50°·1	53°·0	57°·2	55°·7	53°·4
Mean relative humidity (per cent.) ...	50·5	51·2	55·0	56·3	54·3	53·2	52·5	52·9	53·5	56·6	57·0	54·4	53·9
Extremes of rainfall in each month { max.	8·11	8·10	10·15	15·48	5·89	0·88	1·05	1·25	1·13	4·78	12·00	12·42	
in inches { min.	0	0·30	1·03	3·66	1·32	0	0	0·02	0	0·41	3·98	0·32	
Mean number of rainy days	3·4	8·3	10·6	16·0	9·7	3·11	2·9	2·6	1·6	6·9	1·8	15·9	82·8

NOTE.—The above table is compiled from tables given in *I. Tropical Africa* issued by the Meteorological Office. It is there stated that a new set of instruments came into use in September 1899. The old instruments had been set upon a board hanging on the wall of a stone house on the verandah, while the new ones were placed in a grass shed, with the side walls open, and at some distance from houses or trees. Certain corrections were, in consequence, required to be applied to the older observations. This however has not been done, as from 1896—1900 the figures for the separate years are not given. The new instruments were Dry Bulb (N. and Z.) 5779 (correction +0·1 at 32°, 42° and 52°, 0·0 at 62°, 72°, 82°, and +0·1 at 92°); Wet Bulb (N. and Z.) 5778 (correction 0·0 at 32°, +0·1 at 42°, +0·2 at 52°, +0·1 at 62°, 0·0 at 72°, +0·1 at 82° and 92°). These corrections were carried out.

[1] 1896—1902. [2] 1901 and 1902.

EAST AFRICA PROTECTORATE. MEAN RAINFALL (IN INCHES).

	No. of Years[1]	Jan.	Feb.	March	April	May	June	July	Aug.	Sept.	Oct.	Nov.	Dec.	Year
Along the Railway from East to West														
Mackinnon Road	3	2·64	2·17	5·16	6·26	4·29	2·40	1·18	0·87	1·50	0·71	5·39	4·96	37·53
Athi R.	3	0·75	1·89	10·63	5·63	1·93	0·28	0·08	1·22	0	3·07	3·90	1·38	30·76
Nairobi	7	1·81	4·72	5·16	7·91	5·20	2·13	0·75	1·02	0·59	2·32	4·88	3·62	40·11
Kikuyu	13	1·50	3·50	6·61	9·57	7·21	2·68	1·02	0·75	1·46	1·81	5·24	3·70	45·05
Nakuru	3	0·67	0·67	5·55	3·39	3·90	2·13	2·52	3·47	1·85	3·11	2·36	1·65	31·27
Molo	3	2·68	2·72	7·83	8·98	7·91	4·96	3·78	14·37	4·33	4·88	3·62	2·87	68·93
North of the Railway														
Machakos	12	1·30	2·56	5·63	8·50	2·52	0·71	0·20	0·39	0·20	2·17	8·17	4·65	37·02
Fort Hall	6	0·91	3·39	5·99	12·17	4·41	2·60	0·95	0·87	0·83	5·35	7·36	2·76	47·59
Eldama Ravine	6	1·30	3·86	4·25	5·08	4·61	3·94	3·15	3·47	2·36	1·93	1·97	1·61	37·53

[1] The years are 1905 and those immediately preceding.

EAST AFRICA PROTECTORATE. MEAN RAINFALL (IN INCHES).

Coast Stations, South to North		No. of Years[3]	Jan.	Feb.	March	April	May	June	July	Aug.	Sept.	Oct.	Nov.	Dec.	Year
Shimoni	... ...	13	1·38	0·87	3·74	10·32	15·75	5·08	4·45	2·44	2·24	1·93	6·02	2·28	56·40
Mombasa	... ...	16	0·87	0·95	2·68	7·32	13·82	3·98	3·19	2·21	2·09	2·87	5·43	2·28	47·69
Takaungu	... ...	14	0·39	0·67	2·28	5·24	14·88	4·17	4·02	1·93	1·93	2·56	4·33	0·95	43·35
Malindi	... ...	15	0·32	0·24	0·98	6·46	14·21	5·12	3·82	1·42	1·93	2·52	3·90	0·91	41·43
Kismayu	... ...	12	0·04	0·04	0·24	1·73	5·35	3·62	1·93	0·79	0·75	0·16	0·75	0·28	15·68
Lake District															
Mumia's[1]	... ...	9	3·11	3·70	5·08	9·06	8·23	7·64	6·58	6·12	6·90	5·99	6·10	4·25	72·76
Kisumu[2]	... ...	7	3·15	4·37	4·88	5·98	4·49	2·95	1·77	1·97	2·68	3·50	5·71	7·09	48·54
Port Florence	... ...	4	3·07	6·46	8·50	5·95	5·16	4·41	1·81	3·31	2·60	3·54	3·82	9·02	57·65

[1] January and February 1899 are wanting.
[2] January, February and August 1899 are wanting.
[3] The years are 1905 and those immediately preceding.

GERMAN EAST AFRICA

German East Africa is a great block of country roughly about twice as broad in the north as it is in the south. A general idea of the "lie" may be gathered from the altitudes of the water surfaces. The Indian Ocean on the east washes its 600 miles of coast. In the extreme north-west lies Lake Kivu at an altitude of 4829 feet, and from the mountains in the neighbourhood of this the Kagera runs down to Victoria Nyanza below, which, lying in the north, stands 3720 feet above sea level; while on the west, after a sudden drop from Kivu, we have Tanganyika with an altitude of 2624 feet and Nyasa with 1645 feet. In the centre lies a great tableland with a general altitude of between 3000 and 4000 feet. This plateau is bounded on the east by mountains, many of which rise to a great height above the country to the east. The mass of Kilimanjaro is connected by the Pare mountains with the Usambara Highlands. In a general south-west direction from the last-named there are the mountains of Nguru, Usagara and Uhehe, continuing the plateau border towards Nyasa, and this mountain fringe sinks gradually eastwards, till the coastal plain is reached. From the plateau there flow towards the north the minor feeders of Victoria Nyanza; to the west, into Tanganyika flow the many constituent streams, which go to form the Malagarasi; while to the east flow the northern members of the Rufiji system and the northern coastal streams. The southern members of the Rufiji and the Ruvuma come from the highlands near Nyasa and flow through the coastal plain, which is broad to the south of the Rufiji, but narrows down towards the north to some 30 miles.

In the coastal regions the North-East Monsoon is experienced, as a rule, from mid November to mid March, being strongest in January. Thence to the end of April there are usually light winds from the east. From the end of April to the beginning of October is an uncertain period, but one of two winds nearly always blows, the South-East Trades or the South-West Monsoon. The wind thus works round with the clock, or veers gradually from north-east through east to south. From the beginning of October to mid November, on the contrary, though somewhat variable, it usually backs towards the east; but these winds are very light.

In the highlands of the interior, in Ugogo and westward, the South-East Trades, strong and dry, are the rule from April to September, and consequently this region is one of small rainfall. From October to March there are usually light local winds with no fixed direction.

In the Kilimanjaro district the North-East Monsoon blows from December to February. March is a very irregular month, but during April the wind is nearly always in the east, blowing sometimes directly from the east, and at other times from the south-east. In May the wind steadies down, and the South-East Trades are experienced generally till December[1].

For the southern inland districts there are no dependable statistics available with regard to the wind. It is recorded, however, that in Kondeland, on the extreme north-east of Nyasa, the prevailing wind is over the lake from the south.

The humidity is always high in the coastal regions, especially in the north, and also on the shores of Victoria Nyanza, except in the neighbourhood of Shirati and to the south, but, in the interior, on the plateau, in the centre of German East Africa, it is much lower, except in December and January, when it is at its maximum.

Speaking generally of the rainfall, it is less than in the German colonies in the west of Africa, in Togo and in Cameroons. The precipitation, on the whole, decreases from north to south in the coastal regions. In the central seaboard districts the greater rainy season extends from March to May, and the lesser occurs in November and December. In the northern coast-lands the rainy seasons are in April and May, from August to September, and again in November, while in the southern coastal districts the chief rains fall in December, January and March, with a dry season extending from June to October.

Along the area of the coastal range, or plateau borderland, the rains are heavier on the east than on the west, and the seasons set in earlier than on the coast. At Kwamkoro, in Usambara, in one very rainy year no less

---

[1] *Die Deutschen Kolonien in Wort und Bild.* Seidel, Stuttgart, 1903.

than 133·5 inches (3390 mm.) of rain were recorded. On the inland plateau, in Masailand, Ugogo, and the eastern parts of Unyamwezi the precipitation is small. Northern Unyamwezi is more favoured, but the eastern shores of Victoria Nyanza are comparatively dry. The western shores of the lake have a heavy rainfall; at Bukoba, for instance, the mean annual fall is 75·4 inches (1914 mm.), while at Shirati on the east side it is only 25·43 inches (646 mm.) annually. In these interior plateau regions the dry season extends from July to October, while from the beginning of November to the end of April is the rainy season, but with very little in January. In the neighbourhood of the Kagera river valley, too, the fall is much heavier than in the Tanganyika district.

In the Kilimanjaro district the eastern and northern parts are much drier than the south-west.

With regard to temperature, and again, of course, speaking generally, the mean annual thermometric record along the coastal area varies between 77° and 82° Fahr. (25° and 28° C.). Along the coastal range the mean annual temperature is about 64·5° to 70° Fahr. (18° to 21° C.), the maximum and minimum being 97° and 41° Fahr. (36° and 5° C.). For the interior no general statement can be made, the variation between different districts being so great, that a general mean would have no practical value. In the Kilimanjaro district, however, at Moshi, at an altitude of 3750 feet, the mean annual temperature is 69°·3 Fahr. (20°·7 C.), and at Marangu (1344 feet) it is 62°·4 Fahr. (16°·9 C.), the range at the former being 32°·4 to 43°·2 Fahr. (18° to 24° C.), and at Marangu 25°·2 to 36° Fahr. (14° to 20° C.).

The following tables (p. 332), computed by Maurer[1] from five years' observations, show very clearly the general conditions as to temperature and rainfall in the Usambara district.

Vanga, near the Anglo-German boundary, in the same district, and on the coast, is unhealthy owing to the presence of a mangrove swamp immediately behind it.

At Tanga the hottest month is February, when the mean temperature is 82° Fahr. (27°·8 C.), the range of temperature during the month being 13°·7 Fahr. (7°·6 C.).

---

[1] *Geographische Zeitschrift*, 1903, p. 1.

The coldest month is July with a mean of $73°\cdot6$ Fahr. $(23°\cdot1$ C.) and a range of only $11°\cdot7$ Fahr. $(6°\cdot5$ C.). The mean annual temperature is $78°\cdot2$ Fahr. $(25°\cdot7$ C.), and the mean monthly range for the year is $13°$ Fahr. $(7°\cdot2$ C.). The absolute maximum temperature recorded during the six years for which there are observations was $95°\cdot2$ Fahr. $(35°\cdot1$ C.), and the absolute minimum during the same period $63°\cdot7$ Fahr. $(17°\cdot6$ C.).

TEMPERATURE IN USAMBARA.

Station ...	Mazindi	Buloa	Ambangulu	Kwai
Position in district ...	W.	E.	S.W.	W.
Altitude in feet ...	1870	3018	4100	5280
Warmest {mean ...	Feb. $78°\cdot3$ F.	Feb. $73°\cdot6$ F.	Feb. $69°\cdot6$ F.	Feb. $65°\cdot8$ F.
month {fluctuation	$27°\cdot2$	$19°\cdot4$	$15°\cdot5$	$22°\cdot1$
Coldest {mean ...	July $68°\cdot4$	Aug. $64°\cdot8$	July $57°\cdot7$	July $56°\cdot1$
month {fluctuation ...	$19°\cdot6$	$13°\cdot9$	$4°\cdot9$	$10°\cdot3$
Year {mean ...	$73°\cdot4$	$69°\cdot6$	$64°\cdot0$	$61°\cdot3$
{fluctuation ...	$21°\cdot6$	$16°\cdot0$	$10°\cdot6$	$17°\cdot3$
Absolute maximum ...	$99°\cdot9$	$88°\cdot9$	$81°\cdot5$	$87°\cdot1$
Absolute minimum ...	$56°\cdot3$	$49°\cdot5$	$51°\cdot6$	$41°\cdot9$

RAINFALL IN USAMBARA.

Station ...	Kwamkoro and Buloa	Sakarra and Ambangulu	Kwai	Mtai
Position ...	S.E. and E.	S.W.	W.	N.W.
Altitude in feet ...	3220 and 3020	4100	5280	5350
Wet year ...	146 inches	80 inches	43 inches	—
Dry year ...	43 ,,	31 ,,	18 ,,	—
Mean year ...	67 ,,	61 ,,	27·5 ,,	18 inches
Dec.—Feb. ...	8 ,,	8 ,,	5 ,,	4 ,,
March—May ...	35 ,,	35 ,,	16 ,,	10 ,,
June—October ...	20 ,,	20 ,,	4 ,,	2 ,,
November ...	4 ,,	2 ,,	3 ,,	2 ,,
Wettest months in} whole period {	April '97: 36·2 in. May '99: 23·9 ,,	May '97: 21·8 in. —	Nov. '96: 12·6 in. May '99: 7·5 ,,	May '99: 3·9 in. —
Wettest monsoon} months in whole} period {	July '99: 13·4 ,, ,, '97: 10·6 ,,	July '99: 12·8 ,, ,, '97: 9·7 ,,	July '97: 4·7 ,, --	July '99: 1·5 in. —

At Kwai, in West Usambara, the mean evening temperature at 8 p.m. is very slightly lower than, but almost identical with, the morning temperature at 7 a.m. The year 1902 was, however, warmer in the mornings than the two preceding years, throughout all the months, and at the same time the morning relative humidity was somewhat lower than the average. The mean relative humidity is 71·3 per cent. for the year, and it is higher in the evening than in the morning; the absolute minimum for the period during which observations were taken is 23 per cent. July is the coldest month and January the warmest. The average percentage of sunshine is fairly regular from year to year, the greatest difference occurring in February, when it was 65·4 per cent. in 1900, and only 40·2 per cent. in 1902, and in November, when it was 22·4 per cent. in 1901, and 41·9 per cent. in 1902. The rainfall is exceedingly variable. The figures for the six years, for which rainfall observations are available, show that June, August and September are the only months not liable to a precipitation of 4·75 inches (120 mm.). In July the fall was 4·75 inches in 1897, and only ·2 inches (5·6 mm.) in 1900. October is a comparatively dry month, but 6·2 inches (158 mm.) fell during that month in 1902, and only ·18 inches (4·6 mm.) in 1898. Again, the usual fall in November varies from about 1·6 to 4 inches (40 to 100 mm.), but in 1902 no less than 8 inches (213 mm.) were measured; while in December 12·7 inches (322·4 mm.) fell, the highest record previously for that month being 3·36 inches (85·6 mm.). The average number of days, on which rain is measured, is 140—180, and on about 100 of these the fall is over ·2 inches (5 mm.), and on a dozen or so over 1 inch (25 mm.).

Msalabani-Magila is situated at the south foot of the Magrotto mountains, in the east of Usambara province, in the north-east of the colony, where there are three rainy seasons:

1. The little rainy season generally in November but sometimes beginning late in October and sometimes not closing till early in December.

2. The great rainy season which lasts from March to May.

3. The season of the South-East Monsoon, generally in July, but sometimes in August.

Owing to the proximity of the mountains there is no really very dry season.

Pangani, at the mouth of the Pangani river, in the province of the same name, lies low on a narrow strip of sand, in close proximity to extensive mangrove swamps, and bears an evil reputation; but the neighbouring hill country to the north, in Mundo district, is more favoured, and is considered to be not unhealthy.

At Kilwa Kisiwani and Kilwa Kivinje, the former on an island and the latter on the coast, in Kilwa province, the presence of mangroves renders the neighbourhood unhealthy, and malaria is prevalent.

At Livale, in the south of Kilwa province, the rain falls most frequently at night, and, next to this, in the afternoon, there being very little rain in the morning. There is only one rainy season, which is ushered in by gentle showers in the middle of November or the beginning of December, and lasts till the end of April. There are frequent thunderstorms in all months, except during the period from May to August. The dry season is very marked.

At Dar-es-Salam it is warmest at the beginning and end of the year, while June, July, August and September are the coolest months. The early morning (7 a.m.) temperature varies between $79°\cdot5$ Fahr. ($26°\cdot4$ C.) in the warmest month and $68°$ Fahr. ($20°$ C.) in the coolest; the midday temperature similarly varies between $84°\cdot5$ and $80°$ Fahr. ($29°\cdot2$ and $26°\cdot7$ C.), and the evening temperature between $81°$ and $73°$ Fahr. ($27°\cdot3$ and $22°\cdot8$ C.). The greatest range of temperature in the year, that is the difference between the highest recorded maximum and the lowest recorded minimum is, taking the average for five years, about $30°$ Fahr. ($16°\cdot4$ C.), the maximum being $91°\cdot8$ Fahr. and the minimum $62°\cdot2$ Fahr. ($33°\cdot2$ and $16°\cdot8$ C.). It rains heavily at Dar-es-Salam in November, December and January, not so heavily in February, and then, increasing, reaches its maximum in April and May. This is the general rule, but the rainfall varies from year to year, and from month to month. For instance, in December the fall has been as low as ·14 inches (3·8 mm.), and as high as 9·69 inches (246 mm.); in November the record has been as low as ·16 inches (4 mm.) in one year and as high as 10·55 inches

(268 mm.) in another. Similarly, in the early months of the year, it is found that the precipitation varies, January being credited with ·95 inches (24 mm.) in 1902, and as much as 10·24 inches (260 mm.) in 1901; and so with February and March. April and May, however, seem sure of a heavy fall, amounting in some years to as much as 17·7 inches (450 mm.); the lowest record for either of these two months is 6·5 inches (165 mm.) in May 1897. At Dar-es-Salam it rains on an average 170 days in the year, of which 101 have a precipitation of ·2 inches (5 mm.) and over, 10 of 1 inch (25 mm.) and over. The heaviest rainfall on any one day was 4·5 inches (116 mm.).

During the period from December 1900 to April 1905, there were 22 occasions on which more than 2 inches (50 mm.) of rain fell, the average duration of each fall being about 9 hours. Eight of these occasions were in April, five in December, four in March, two in May, and one each in January, February and November; so that these heavy falls of rain may be said to occur chiefly in March—April and December. The two heaviest falls were both in April. On the 13/14th of that month, in 1901, 4·3 inches (110 mm.) were measured and the rain lasted for nearly 25½ hours; and again on the 27/28th of April 1905 the precipitation measured 4·6 inches (117 mm.), and it rained from 7 o'clock in the evening to half-past five the following morning, or about 10½ hours only.

A recent account of the rise and progress of Dar-es-Salam concludes as follows. The town is surrounded with beautiful pleasure-grounds, and a network of shady roads. On the whole Dar-es-Salam has become a beautiful, healthy, high-class town, which increases and becomes more beautiful every day; and, in any case, as soon as the railways and steamship undertakings, and, with them, commerce and agriculture are still further developed, there will be a great future before it.

At Lindi, on the coast, at the mouth of the river Lukuledi, in the south of the colony, the mean annual temperature, derived from nine years' observations, is 79°·5 Fahr. (26°·4 C.), and the fluctuation from year to year amounts to 19°·6 Fahr. (10°·9 C.), which is the difference between the greatest and least annual means for the nine years. The hottest month is December, and the coolest August, the mean temperature of the former being $82°·2$ Fahr. ($27°·9$ C.), and the fluctuation $20°·0$ Fahr. ($11°·1$ C.), while in August the mean temperature is $76°·6$ Fahr. ($24°·8$ C.), with a fluctuation of $21°·78$ Fahr. ($12°·1$ C.) between the warmest and coolest August in the nine years. The absolute maximum temperature, which has, up to the present, been recorded, is $96°·9$ Fahr. ($36°$ C.), and the absolute minimum $56°·7$ Fahr. ($13°·7$ C.).

At Ulanga station, in Uhehe, in the southern part of the colony to the north-east of Nyasa, standing at an altitude of 754 feet, the summer, as in other parts of this tropical region, is the rainy season, but there is a short break in the month of February. The mean annual rainfall is $55·12$ inches (1400 mm.). Storms in summer are frequent. The coolest month, July, has a mean temperature of $71°·6$ Fahr. ($22°$ C.). The absolute extremes for the year are $57°·2$ and $93°·2$ Fahr. ($14°$ and $34°$ C.), and the daily variation in winter $27°$ Fahr. ($15°$ C.)[1].

At Tosamaganga, near Iringa, which stands 5578 feet above sea-level, the mean annual temperature is $63°·5$ Fahr. ($17°·5$ C.), November being the warmest month, with a mean of $68°·4$ Fahr. ($20°·2$ C.), and June the coolest with $57°·7$ Fahr. ($14°·3$ C.) as a mean. The extremes for the year are $87°·2$ and $43°·2$ Fahr. ($30°·6$ and $6°·2$ C.). The mean daily variation is $19°·3$ Fahr. ($10°·7$ C.), and the greatest range recorded on any one day $32°·4$ Fahr. ($18°$ C.). The cool period, from May to October, is rainless. During the fifteen months, June—October, 1897, 1898, 1899, only $·55$ inches ($14·4$ mm.) of rain fell altogether, and of this small amount $·39$ inches ($10·2$ mm.) fell in October, 1899. The summer months are rainy, especially January and February.

At the mission stations of Paramiho and Ngombe (alt. 4265 feet) in Songea, the mean annual rainfall is 40 inches (1000 mm.), but little falling in the months December to April, while the winter months are rainless. The mean annual temperature is $68°$ Fahr. ($20°$ C.), the warmest month being November, with a mean of $75°·4$ Fahr. ($24°·1$ C.), and the coolest month June, with a mean temperature of $61°·3$ Fahr. ($16°·3$ C.)[2].

---

[1] *Geographische Zeitschrift*, 1903, p. 213.
[2] *Ibid.*

The province of Mahenge lies adjacent to, and to the west of, Kilwa province; the capital, Mahenge, occupies a lofty situation on the Muhulu mountains, standing above the alluvial plains of the Ulanga river, and here there were, between August and December 1903, 60 days with mist, and no less than 98 days in 1904, and 45 between January and May. There is apparently no such frequency of thunderstorms as at Livale[1].

At Kisaki in the south of Morogoro province at the south foot of the Uluguru mountains, the one rainy season begins in December and ends with May. The dry season is more markedly dry than at Mahenge but not so pronounced as at Livale. Between October 1902 and July 1903 there were 125 downpours of rain, of which 66 occurred at night, 30 in the afternoon, and 29 in the morning. The thunderstorms are pretty evenly distributed throughout the year[1].

At Kondoa-Irangi, in the northern part of the Mpwapwa province, in the central highlands, there is only a relatively small rainfall. The one rainy season begins near the end of November or the beginning of December and lasts till the end of April, though it is sometimes prolonged into May. The dry season, which extends from mid May to mid October is very marked, there being a perfect absence of rain[1].

At Tabóra, which is situated in Lat. 5°3′S. and Long. 32°53′ E. at an altitude of 4035 feet, the mean barometric record varies between 667 mm. (26·26 inches) calculated for both June and December, and 655 mm. (25·79 inches) recorded in December; and the thermometer oscillates between 96°·9 Fahr. (36° C.), which was registered in December and 50°·7 Fahr. (10°·4 C.) recorded in June. The mean temperature for June, July, and August, given in the table below, is slightly higher than would probably have been obtained from observations for a longer series of years, in consequence of the means for those months having been abnormally high in 1902, and similarly the mean for December is a trifle lower than it would have been had not the figure for the same year been abnormally low in that month.

[1] M. D. S. XIX, 1906, pp. 164—169.

K. A.   22

Situated on the dry highlands, the mean annual rainfall amounts to no more than about 25·6 inches (650 mm.), and falls almost entirely in the period from November to April. During the remaining months, the South-East Trades are experienced, with the result that the sky is clear, and the atmosphere dry; and this period is practically rainless.

In the extreme west of the colony, at Ujiji, on Tanganyika, in the province of the same name, the one rainy season lasts for six months from November to April. The remaining months form the dry season, July and August being perfectly rainless. From a little south of Ujiji to the north of the lake is a comparatively low precipitation area, in the midst of a country, on both sides of the lake, which receives over 40 inches (1000 mm.) per year, lying close, however, to the region of small rainfall in the centre of German East Africa, which coincides with the plateau area. Mr Beringer, who surveyed the telegraph route up the east coast of Tanganyika, describes the coast region as "ferociously hot."

In the extreme south-west, to the north-west of Nyasa, lies Lake Rukwa, and Johnston described the region of the lake as "possessed of seven devils—hunger, thirst, a scorching and skinning wind, a blazing sun, venomous flies, and wicked and sullen men." It had not rained, when he was there, for two years. Things are very different, at least so far as the moisture is concerned, on the Nyasa-Tanganyika plateau, and at Neu Langenberg and Manow, where the precipitation amounts to no less than 87·95 inches and 96·70 inches (2234 mm. and 2456 mm.) respectively.

The whole of the Konde country is rich in rainfall, owing to the prevailing southerly winds, which bring up the moisture-laden clouds over the lake, to be arrested and cooled by the heights situated to the immediate north of Nyasa. The larger proportion of the rain falls between January and June, April and February being the rainiest months. Ikombe, on the extreme north-east of Nyasa, holds the record for rainfall in German East Africa. There, from the 16th to the 18th June, 1898, 18·62 inches (473 mm.) of rain fell in the 48 hours, of which no less than 12·40 inches (315 mm.) fell in 24 hours. At Ikombe November, with a mean temperature of 79°·1 Fahr. (26°·2 C.) is the hottest month, and June, with a mean of $68°\cdot9$ Fahr. ($20°\cdot5$ C.), the coolest. In the highlands of the Konde country, the meteorological observations recorded at the three stations, Wangermannshöhe (2900 feet), Rutenganio (4250 feet), and Manow (5200 feet), go to show that here also November is the warmest month, though at the last-mentioned the mean temperature barely reaches $68°$ Fahr. ($20°$ C.), while in July the thermometer sinks below $57°\cdot2$ Fahr. ($14°$ C.). In these highlands the difference between the temperature at 2 p.m. and that recorded in the mornings and evenings averages between $10°\cdot5$ and $18°$ Fahr. ($6°$ and $10°$ C.). The comparatively dry winter extends from June to October. Severe thunderstorms are frequent at most times of the year, and small earthquake shocks are also quite frequent, though no great damage has, as yet, been recorded.

On the Livingstone mountains there is abundance of rain and mists, and, when the days are quite hot, the nights are frequently very cold, and night-frosts are quite common at the colder season of the year. The forests here are confined to gorges, and slopes protected from the wind. The lands bordering on the lake enjoy the advantage of land and sea breezes.

To turn now to the north-west of the colony, to the neighbourhood of Victoria Nyanza, it is found that at Muanza, in the district of the same name, on the southern shores of the lake, the rainy seasons are April—May and October—November, the absolute minimum occurring in July. Daily land and sea breezes are experienced here, as they are also at Bukoba, on the western shore—off the lake in the afternoon, and off the shore in the morning; and the effect of these is so marked as to deprive the South-East Trades of their influence, except in the months of June and July. In Usiba it rains almost every day, and the climate is exceptionally wet; thunderstorms, too, are of frequent occurrence. In Karagwe, in Bukoba province, there are heavy rains in March and April; and, in the western part, heavy dews are experienced, followed by fine mornings, the rain and the usual thunderstorm coming in the afternoon, to be succeeded by clear evenings. In Ruanda the wet seasons also merge and form one, and there is no actual dry season, rain occurring almost daily.

On the east of the lake, however, even the shores are included in the low precipitation area. At Shirati April and May are the rainiest months, and there are sometimes heavy rains in December and January, and also in March, but the amount is very irregular, and varies from year to year. For instance, in December 4·73 inches (120·3 mm.) were measured in one year, and only 0·43 inches (10·9 mm.) in the next; and again in April the recorded monthly precipitation varies between 10·43 and 2·87 inches (265 and 73 mm.), and in May between 8·07 and 2·8 inches (205 and 71 mm.). July, August, and September are here the dryest months, sometimes, though in such close proximity to the equator, being actually rainless. At Shirati it rains on an average 87 days in the year, and of these thirty-five have a fall of 0·2 inch (5 mm.) and over, and eight of over 1 inch (25 mm.). The greatest rainfall in one day, recorded up to the present, is 5·67 inches (144 mm.). In the neighbourhood of Stuhlmann Sound, in Muanza province, nearly all the rain comes from the north and north-east. The greatest amount of rain appears, from such observations as exist, to fall in January and March, 5·88 inches (149·3 mm.) having been measured in the latter month in 1904, and 5·09 inches (129·1 mm.) in January, 1905. In June and July, in those years, there was no rain; and in August, 1904, only 0·30 inch (7·7 mm.). The greatest amount measured in one day was, in 1904, 1·52 inch (38·7 mm.) in January, and 1 inch (25·5 mm.) in May and December. In the first half of 1905 there was 0·95 inch (24 mm.) in one day in March, and 0·98 inch (25 mm.) in April. The total precipitation for the year 1904/5 was 21·71 inches (551·8 mm.).

PRODUCTIONS, &c.

In so large an area as German East Africa, with such varying soil and relief, with alluvial plains and grassy plateaux, towering heights and wooded uplands, it might be expected that a *habitat* could be found for almost any crop, and within certain areas this is strictly true. But, as has been estimated by high German authorities, only about one-fifth of the country is really cultivable, with any prospect of reasonable returns. This is not due to the climate, but to the soil; for the remaining four-fifths of the area are occupied by barren steppe land, savannah and bush. Rubber does well near the coast and on the shores of Victoria Nyanza, where ground-nuts and rice should thrive, especially on the eastern side. Of fibre plants, Sisal-hemp (*Agave sisalana*) and bow-string hemp (*Sanseviera ehrenbergii*) are both natives. Cotton gives good results, especially in the districts of Morogoro, Mpwapwa, Tabóra, Ruanda, Muanza, and in the country lying east of Tanganyika, south of Ujiji. In addition to the lake shores, rice has been found to produce paying crops in Morogoro, Tabóra and Uha. Coffee is grown in Irangi, Usambara, Ruanda, and also near Bukoba. Both oil palms and cocoa-nut palms flourish, the latter forming, with copra, copal, rubber, and sesame seed, the principal exports. Maize and several native grains thrive, and there is no reason why wheat and other European cereals should not do well in the uplands. Tobacco has been tried with satisfactory results, especially in Tabóra, Urundi and Ngorongoro. European vegetables, such as potatoes, peas, beans, onions, and fruits such as apples, bananas, &c., grow well, and so too do yams, mangoes and chillies, while sugar is now being tested in Tabóra and Uha.

With reference to the stations for which meteorological tables are given below—

Dar-es-Salam lies on a flat sandy stretch of coast behind extensive coral reefs. The Pugu hills lie about 12 or 13 miles off in a W.S.W. direction and rise to a height of about 1200 ft.[1]

[1] Dr P. Heidke makes the following statement as to the Instruments in use :—An Instrumentem wurden benutzt das Stationsbarometer Hechelmann Nr. 623 (Korrektion + 0·3, Korrektion des Thermometers am Barometer -0·6), trockenes Thermometer R. Fuess Nr. 671a (Korrektion 0·0) bis Dezember 1900; von Januar 1901 bis Dezember 1902 sind die Angaben der beiden Aspirations-Thermometer R. Fuess Nr. 2341 (Korrektion 0·0) und R. Fuess Nr. 2342 (Korrektion 0·0) verwandt worden. Als Maximum-Thermometer war in Gebrauch bis Dezember 1900 Meissner Nr. 1 (Korrektion 0·0), von Januar 1901 bis September 1901 R. Fuess Nr. 3629 (Korrektion +0·1), Oktober 1901 R. Fuess Nr. 3644 (Korrektion 0·0), von November 1901 bis Dezember 1902 R. Fuess Nr. 3887 (Korrektion -0·3), als Minimum-Thermometer Meissner Nr. 1 (Korrektion +0·4) bis Dezember 1900, R. Fuess Nr. 3508 von Januar 1901 bis Dezember 1902 (Korrektion -0·1). An Registrierinstrumenten wurden bedient bis Dezember 1900 ein Bohnescher, später ein Richardscher Barograph, ein Richardscher Thermograph, ein Anemograph ein Sonnenscheinautograph und ein Hygrograph. (*M. D. S.* XIX. 1906, p. 49.)

Lindi is situated on the S.E. bank of the Lindi creek. The land falls abruptly to the water on the N.W., while to the S. it rises gently. At the back of Lindi, to the north, rises the Makonde plateau. The slopes are thickly covered with bush. The lagoon is 120 feet broad, with mangrove swamps 150—200 feet broad[1].

Bagamoyo is situated on the sandy coralline coast. There is no high ground in the neighbourhood, and it stands some 150 yards from the beach[2].

Kwai lies in an open valley, running E.S.E., on the north side of West Usambara. To the west the country rises in terraces, and the slopes are covered with grass and low bush[3].

Tabóra is situated on an open plain, from which, at a distance of from a mile and a half to three miles, rise granite hills to a height of 150 to 180 feet. The observing station is situated to the S.W. of the town. The hills are covered with scrub[4].

Muanza lies on the eastern side of Bukumbi gulf, a deep southern inlet in Victoria Nyanza, and is situated on an extensive plain sloping down to the lake. On the east are high mountains. Soil, red laterite[5].

[1] The instruments in use were :—A Hechelman barometer, No. 756 (correction −0·1), dry bulb thermometer R. Fuess No. 601 (correction 0·0), wet bulb thermometer R. Fuess No. 603 (correction 0·0), maximum thermometer R. Fuess No. 2963 (correction + 0·2), minimum thermometer R. Fuess No. 2935 (correction +0·1).

[2] The instruments in use were :—A dry bulb thermometer R. Fuess No. 2791 (correction −0·1), wet bulb thermometer R. Fuess No. 2806 (correction + 0·1), maximum thermometer R. Fuess No. 504 (correction −0·2) and a minimum thermometer R. Fuess No. 477 (correction 0·0).

[3] The instruments in use were:—A barometer R. Fuess No. 1312 (correction −0·1), a dry bulb thermometer R. Fuess No. 234 (correction +0·2), a wet bulb thermometer R. Fuess No. 519 (correction 0·0), a maximum thermometer R. Fuess No. 3093 (correction 0·0), a minimum thermometer R. Fuess No. 2633 (correction +0·5), and a Richard's Barograph and Thermograph.

[4] The instruments in use were:—A Hechelman barometer No. 2386 (correction +3·6), dry bulb thermometer R. Fuess No. 608 (correction +0·1), wet bulb thermometer R. Fuess No. 609 (correction +0·1), maximum thermometer R. Fuess No. 3081 (correction + 0·2) up to November 1899, from March 1901 R. Fuess No. 478 (correction −0·2), minimum thermometer R. Fuess No. 3050 (correction + 0·5) up to November 1899, from March 1901 R. Fuess No. 469 (correction +0·3), and a Richard's Barograph and Thermograph.

[5] The instruments in use were :—A dry bulb thermometer R. Fuess No. 2947 (correction +0·1), wet bulb thermometer R. Fuess No. 2948 (correction +0·1), maximum thermometer R. Fuess No. 449 (correction −0·1 at 19°·6 and 26°·3, and +0·1 at 31°·1 and 39°·9), minimum thermometer R. Fuess No. 381 (correction −0·1).

Bukoba is situated on the west side of Victoria Nyanza, four or five hundred yards from the shore, and some feet above the water-level, at the S.W. end of a sand-hill about 30 feet high. The shore is partly sandy and partly rocky. On the S. side of the station the Kanoni stream flows eastwards. The plateau on the west rises to an altitude of some 4000 feet and more above sea-level, and is covered with banana plantations[1].

[1] The instruments in use were:—A maximum thermometer R. Fuess No. 3099 (correction $-0\cdot1$), and a minimum thermometer R. Fuess No. 1733 (correction $+0\cdot4$).

DAR-ES-SALAM. Lat. 6° 45′ S., Long. 39° 20′ E. Alt. 46 feet.

From 5 years' observations[1]. Compiled from annual tables which appeared in *Mitteilungen aus den Deutschen Schutzgebieten* XIX. 1906, pp. 51—53; XXI. 1908, pp. 52—57.

	Jan.	Feb.	March	April	May	June	July	Aug.	Sept.	Oct.	Nov.	Dec.	Year
Mean temperature at 7 a.m. (Fahr.°)	79°·5	77°·9	76°·8	75°·2	73°·2	69°·1	68°·4	68°·4	70°·2	72°·7	76°·3	78°·1	73°·8
,, ,, 2 p.m. ...	84°·6	84°·6	85°·5	83°·5	82°·6	81°·5	80°·8	80°·2	80°·8	81°·9	82°·6	83°·1	82°·6
,, ,, 9 p.m. ...	81°·1	80°·4	81°·1	77°·7	75°·9	73°·0	73°·0	73°·2	74°·1	75°·2	78°·1	79°·3	76°·8
Mean temperature[2]	81°·7	80°·6	81°·1	78°·6	76°·8	74°·3	73°·8	73°·8	74°·8	76°·6	79°·0	79°·9	77°·6
Mean maximum temperature	86°·4	86°·9	88°·0	86°·0	84°·2	82°·8	82°·4	82°·4	82°·6	83°·7	84°·2	84°·6	84°·5
Mean minimum temperature	77°·5	75°·9	74°·5	73°·2	71°·4	67°·5	66°·7	66°·4	67°·5	69°·8	72°·9	75°·7	71°·6
Mean of absolute monthly maxima ...	89°·2	89°·6	90°·3	91°·0	88°·9	85°·8	85°·6	84°·6	86°·2	87°·8	86°·2	87°·4	91°·8 (highest)
Mean of absolute monthly minima ...	73°·0	70°·9	71°·8	71°·1	67°·5	64°·6	63°·3	63°·5	64°·6	66°·9	68°·7	71°·8	62°·2 (lowest)
Mean relative humidity (per cent.) ...	80·7	80·3	84	87	85·3	83	84·7	83	83·3	83·7	84·6	85	83·7

[1] 1900—1904.

[2] The mean is derived from $\dfrac{7a + 2p + 9p}{3}$.

LINDI. Lat. 10° 2' S., Long. 39° 44' E. Alt. 268 feet.

Compiled from observations recorded during 9 years[1].

	Jan.	Feb.	March	April	May	June	July	Aug.	Sept.	Oct.	Nov.	Dec.	Year
Mean maximum temperature[2] (Fahr.°)	88°·9	89°·1	88°·9	88°·5	87°·8	88°·3	87°·8	85°·8	86°·2	86°·9	88°·9	90°·3	88°·1
Mean minimum temperature[3] ...	74°·5	73°·6	73°·9	73°·4	70°·0	65°·5	67°·6	67°·5	68°·7	70°·5	73°·2	74°·3	71°·1
Mean temperature[4] ...	81°·7	81°·3	81°·4	81°·0	78°·9	76°·9	77°·7	76°·6	77°·5	78°·7	81°·1	82°·3	79°·6
Mean daily range[5] ...	14°·4	15°·5	15°·0	15°·1	17°·8	22°·8	20°·2	18°·3	17°·5	16°·4	15°·7	16°·0	17°·0
Absolute maximum temperature	96°·6	96°·8	94°·1	96°·0	93°·4	92°·3	92°·1	93°·7	94°·5	93°·6	95°·9	99°·0	—
Absolute minimum temperature	68°·2	66°·9	70°·3	68°·7	61°·2	56°·7	63°·7	64°·0	63°·5	66°·4	68°·4	68°·7	—
Mean relative humidity[6] (per cent.) ...	84	84	85	84	78	69	74	76	78	79	78	81	79
Mean number of rainy days ...	18	13	18	16	5	1	2	3	4	4	7	14	105

[1] 1892, (1893 except March—June), (1894 December), (1895 except December), (1896 except March, April, August), 1897, 1898, 1899, 1900 (January, February, March, December), (1901 except January—March, and August).

[2] The maximum temperature is wanting for January and September 1896, March 1897.

[3] The minimum temperature is wanting for April 1897 and April 1898, May 1898, and June—August 1899.

[4] The mean temperature is derived from $\dfrac{\text{max.} + \text{min.}}{2}$.

[5] The daily range is max.—min.

[6] The relative humidity is wanting for June 1896, January 1897. The mean is derived from $\dfrac{7a + 2p + 9p}{3}$.

BAGAMOYO. Lat. 6° 27′ S., Long. 38° 53′ E. Alt. 26 feet.

From 2 years' observations[1]. Compiled from annual tables which appeared in *Mitteilungen aus den Deutschen Schutzgebieten* XIX. 1906, pp. 46, 47.

	Jan.	Feb.	March	April	May	June	July	Aug.	Sept.	Oct.	Nov.	Dec.	Year
Mean temperature at 7 a.m. (Fahr.°) ...	79°·3	79°·3	78°·6	78°·4	76°·3	71°·6	72°·1	71°·2	73°·8	74°·8	77°·9	79°·5	76°·1
,, ,, ,, 2 p.m. ...	83°·7	84°·7	86°·5	84°·0	82°·2	82°·6	81°·3	80°·6	81°·5	82°·8	84°·0	84°·9	83°·2
,, ,, ,, 9 p.m. ...	80°·6	80°·6	81°·1	78°·8	77°·2	74°·8	74°·1	74°·5	75°·9	76°·6	79°·5	80°·6	77°·9
Mean temperature[2] ...	81°·0	81°·3	81°·9	80°·1	78°·3	76°·1	75°·2	75°·2	76°·6	77°·7	80°·2	81°·3	78°·7
Mean maximum temperature ...	85°·6	85°·6	88°·0	85°·3	84°·4	83°·8	82°·9	82°·4	82°·6	84°·0	85°·1	86°·0	84°·6
Mean minimum temperature ...	77°·0	75°·2	74°·8	74°·8	73°·0	69°·1	69°·1	69°·3	69°·8	71°·2	74°·1	76°·1	72°·8
Mean relative humidity[3] (per cent.) ...	79	80	81	84	85	81·5	85·5	82·5	81·5	80·5	82·5	82	82

[1] April 1901 to December 1902.

[2] The mean temperature is derived from $\dfrac{7a + 2p + 9p}{3}$.

[3] The mean relative humidity is derived from $\dfrac{7a + 2p + 9p}{3}$.

KWAI. Lat. 4° 45′ S., Long. 38° 18′ E. Alt. 5280 feet.

From 3 years' observations. Compiled from annual tables which appeared in *Mitteilungen aus den Deutschen Schutzgebieten* XIX. 1906, pp. 70—72.

	Jan.	Feb.	March	April	May	June	July	Aug.	Sept.	Oct.	Nov.	Dec.	Year
Mean temperature at 7 a.m.[1] (Fahr.°)	64°·0	63°·5	61°·3	60°·6	59°·4	57°·9	55°·9	56°·3	58°·1	60°·4	62°·8	63°·0	60°·3
,, ,, ,, 2 p.m.[2]	73°·0	70°·2	70°·3	69°·8	65°·5	61°·7	60°·4	61°·9	64°·8	68°·7	70°·5	70°·2	67°·3
,, ,, ,, 8 p.m.[2]	63°·0	62°·6	61°·9	61°·5	59°·4	55°·4	54°·0	54°·7	56°·1	58°·5	61°·3	61°·7	59°·2
Mean temperature[3]	66°·7	65°·5	64°·6	64°·0	61°·3	58°·3	56°·7	57°·6	59°·7	62°·4	64°·9	64°·9	62°·2
Mean relative humidity (per cent.) at 7 a.m.[1]	77·6	84	86	86·6	86·6	82	81·3	76·6	73·3	75·3	75	83	80·6
Mean relative humidity (per cent.) at 2 p.m.[2]	54·5	72	71	70	74·5	72	71·5	67·5	65·5	60	61	69	67·4
Mean relative humidity (per cent.) at 8 p.m.[2]	83	89	88	88·5	88	88·5	84	84·5	84·5	82·5	84	90	85·9
Mean relative humidity[3] (per cent.) ...	71·7	80·7	81·6	81·7	83	79·8	77·9	76·2	74·4	72·6	73·3	80·6	71·1
Mean of absolute monthly minima of relative humidity[2]	28·5	54	48	47	51	55	55	48	49	39·5	41	45 (lowest in period)	23
Mean percentage of duration of sun-shine[4]	64·0	50·5	56·1	42·2	27·9	32·8	24·8	31·6	34·9	41·2	32·2	65·9	42

[1] 1900 (except February), 1901 (except November and December), 1902.

[2] 1900 (except February), 1901 (except November and December).

[3] The mean is derived from $\dfrac{7a + 2p + 8p}{3}$.

[4] 1900 (except October—December), 1901, 1902.

TABÓRA. Lat. 5° 3' S., Long. 32° 53' E. Alt. 4035 feet.
Compiled from 2—5 years' observations.

	Jan.	Feb.	March	April	May	June	July	Aug.	Sept.	Oct.	Nov.	Dec.	Year
Mean of temperature at 7 a.m.¹ (Fahr.°) (2 years)	66°·6	65°·1	67°·5	67°·8	67°·1	62°·4	63°·3	64°·8	67°·6	69°·6	69°·3	66°·6	66°·5
Mean of temperature at 2 p.m.¹ (2 years)	80°·2	81°·3	82°·2	82°·0	83°·8	83°·3	84°·0	86°·2	88°·3	89°·6	84°·7	80°·2	83°·8
" " 9 p.m.¹	68°·7	68°·7	69°·3	69°·1	70°·2	68°·9	70°·3	73°·0	75°·2	77°·0	72°·0	69°·4	71°·0
Mean temperature (5 years²) ...	71°·4	71°·8	71°·6	71°·4	71°·4	70°·2	70°·7	73°·8	76°·1	77°·9	75°·4	72°·3	72°·8
Mean maximum temperature (5 years²)	84°·7	85°·1	84°·4	84°·6	84°·7	84°·0	84°·9	87°·4	89°·4	92°·1	89°·1	86°·4	86°·4
Mean minimum temperature (2 years¹)	63°·1	62°·2	63°·9	63°·3	61°·7	58°·3	59°·2	60°·6	63°·7	65°·8	65°·1	63°·0	62°·5
Mean of absolute monthly maxima (2 years¹)	90°·5	88°·7	89°·1	88°·2	88°·5	87°·4	86°·7	90°·0	92°·1	94°·6	94°·6	89°·6	96°·8 (highest)
Mean of absolute monthly minima (2 years¹)	60°·3	59°·5	61°·5	58°·1	58°·3	53°·2	54°·9	56°·5	61°·5	61°·5	60°·6	59°·7	50°·7 (lowest)
Mean relative humidity (per cent.) 7 a.m.³	95	88	88	87	78·5	72·5	70	65·5	63·5	67·5	82·5	91·5	79
Mean relative humidity (per cent.) 2 p.m.	69	58	60·5	59	47·5	40	38	37	35	44·5	59·5	67·5	52·5
Mean relative humidity (per cent.) 9 p.m.	92	83	87	84·5	73	60·5	53·5	48	46·5	52	74	86·5	70
Mean relative humidity³ (per cent.) ...	85	76	78·5	77	66·5	57·5	53·5	50	48·5	54·5	72	83·5	66·5
Number of days with rain ...	23	17	17·5	14·5	4·5	1	·5	1	1	3	18·5	24·5	126
Number of days with rain of 0·2 inch and over	16	12	10	10	2	·5	0	0	·5	1	10	16	68

¹ 1903 and April to December 1904.

² This is the result of the mean temperature for 1894, 1895, and 1899 given by Captain Lyons in *The Physiography of the Nile and its Basin*, p. 21, and that derived from 1901 and 1902, which appeared in *Mitteilungen aus den Deutschen Schutzgebieten* xix. 1906, pp. 82, 83. The mean is derived from $\dfrac{7a+2p+9p}{3}$.

³ 1894, 1895, 1899, 1901, 1902. The mean is derived from $\dfrac{7a+2p+9p}{3}$.

MUANZA. Lat. 2° 31' S., Long. 32° 5' E. Alt. 3723 feet.

Compiled from 3—4 years' observations, which appeared from time to time in *Mitteilungen aus den Deutschen Schutzgebieten.*

	Jan.	Feb.	March	April	May	June	July	Aug.	Sept.	Oct.	Nov.	Dec.	Year
Mean maximum temperature[1] (Fahr.°)	82°·4	80°·4	82°·6	81°·7	81°·9	75°·6	82°·8	83°·7	84°·2	80°·8	81°·0	81°·9	81°·6
Mean minimum temperature[2] ...	61°·9	62°·6	61°·9	61°·9	62°·4	61°·0	60°·8	59°·0	60°·6	60°·1	60°·6	59°·2	61°·0
Mean temperature[3]	72°·1	71°·4	72°·1	71°·8	72°·1	68°·4	71°·8	71°·2	72°·5	70°·3	70°·9	70°·5	71°·0
Mean daily range[4] ...	20°·5	17°·8	20°·7	19°·8	19°·5	13°·6	22°·0	24°·7	23°·6	20°·7	20°·4	22°·7	20°·6
Mean relative humidity[5] (per cent.) ...	88	85	84	84	79	75	76	81	77	81	84	87	82
Mean number of days with rain[6] (5 years)	8	9	12	13	8	2	1	3	4	8	9	9	86
Mean number of days with rain of { 0·2 inch and over	5	5	7	9	5	2	0	1	2	4	6	7	53
Mean number of days with rain of { 1 inch and over[7]	0	1	2	2	1	1	0	1	1	1	1	1	12

[1] June—December 1903, 1904, 1905, 1906.
[2] 1903, January—July 1904, 1905, 1906.
[3] The mean temperature is derived from $\dfrac{max. + min.}{2}$.
[4] The daily range is max.—min.
[5] 1903—1906.
[6] 1902—1906. The records for June and July 1904 are wanting.
[7] The maximum record in 24 hours was 200 mm. (7·87 inches) in April 1903.

BUKOBA. Lat. 1° 20'·5 S., Long. 31° 52' E.   Alt. 3740 feet.

From 4 years' observations[1].

	Jan.	Feb.	March	April	May	June	July	Aug.	Sept.	Oct.	Nov.	Dec.	Year
Mean temperature[2] (Fahr.°) ...	68°·2	68°·7	69°·4	69°·4	68°·9	68°·0	66°·7	67°·1	68°·4	68°·2	68°·2	68°·0	68°·3
Mean maximum temperature	81°·1	79°·3	81°·3	83°·7	82°·2	78°·3	78°·1	77°·5	79°·9	81°·0	81°·3	81°·1	80°·4
Mean relative humidity (per cent.) (morning[3])	79	78	83	89	82	76	71	77	79	80	82	80	80
Mean relative humidity 2 p.m.	62	64	75	82	72	63	59	69	68	66	71	67	68
Mean number of days with rain (3 years[4])	9	8	17	21	14	3	3	7	7	10	16	13	128
Mean number of days with rain of 0·2 inch and over	5	6	10	17	2	1	2	4	3	6	12	7	85
Mean number of days with rain of 1 inch and over	0	2	3	6	4	0	0	0	1	2	2	2	22

[1] 1893, 1894, 1895, 1897.

[2] The mean is derived from $\dfrac{7a + 2p + 9p}{3}$.

[3] At 7 a.m.

[4] 1894—1897.

GERMAN EAST AFRICA. MEAN RAINFALL IN INCHES.
Inland.

	Lat.	Long.	Alt. in feet	No. of Years	Jan.	Feb.	March	April	May	June	July	Aug.	Sept.	Oct.	Nov.	Dec.	Year
Manow[1]	9°16'S.	33°53'E.	5183	(4)	10·98	7·09	14·92	26·26	16·30	2·13	3·98	1·22	0·67	1·22	2·48	9·49	96·7
Livale[2]	9°37'	37°38'	1640	(4)	9·25	7·95	5·00	4·13	0·20	0·03	0	0·02	0·47	0·55	1·50	6·42	35·5
Mahenge[3]	8°41'	36°43'	3360	(3)	13·58	12·17	14·76	10·35	5·12	0·75	0·43	0·95	0·91	0·59	3·54	10·35	73·5
Kisaki[4]	7°28'	37°40'	525	(5)	5·28	3·43	6·34	10·91	7·99	0·32	1·02	0·26	0·51	0·87	2·09	2·44	41·5
Mpwapwa[6]	6°21'	36°23'	3380	(6)	4·13	6·46	6·50	4·17	0·91	0	0·07	0	0	0	1·58	5·39	29·2
Msalabani-Magila[6]	5°49'	38°46'	755	(9)	1·85	3·43	4·13	9·92	13·19	1·81	3·90	2·60	1·97	3·66	7·64	4·53	58·6
Tabóra[7]	5°3'	32°53'	4035	(7)	4·53	4·17	4·69	3·15	0·79	0·16	0	0	0·28	0·59	3·15	5·08	26·6
Kondoa-Irangi*[8]	4°55'	35°57'	4660	(4)	4·31	4·53	3·15	4·02	0·79	0·04	0·01	0	0	0·12	0·91	6·14	23·6
Mombo[9]	4°53'	38°13'	1295	5	2·32	3·11	3·74	9·17	7·17	1·30	1·58	0·51	0·71	1·93	3·50	3·07	38·1
Ujiji[10]	4°52'	30°2'	2790	5	4·65	4·92	5·35	5·24	2·52	0·40	0	0·55	0·63	4·10	4·45	32·7
Kwai[11]	4°45'	38°18'	5280	3	2·84	2·72	4·76	3·31	4·13	0·87	1·69	0·32	0·43	1·65	3·62	4·49	30·8
Muanza[12]	2°31'	33°5'	3723	6	2·48	3·23	6·50	8·74	3·27	1·89	0·12	1·46	1·85	3·11	5·08	4·92	42·6
Bukoba[13]	1°20'	31°52'	3740	(10)	3·82	5·51	7·91	17·95	12·68	1·46	0·89	2·56	2·36	4·92	8·58	6·81	75·5
Shirati[14]	1°7'	33°59'		(4)	2·91	1·50	2·28	5·95	4·45	1·50	0·56	0·37	0·68	2·44	1·10	1·89	25·4

* The means for April and December are somewhat higher than a longer series of observations would probably give, in consequence of abnormal figures for the first month in 1905, and for the second in 1902.

[1] 1892—1895. [2] January, February, September—December 1902, 1903, 1904, January—May, October, December 1905. [3] September—December 1902, 1903, 1904, January—May 1905. [4] 1899, 1900, 1901, 1902, January—July 1903. [5] September—December 1898, January and May—July 1899, February, April, June—December 1900, January—March and June—December 1901, April—December 1904, 1905, January—April 1906. [6] March—December 1896, January—October 1897, November—December 1898, March—December 1899, 1900, 1901, 1902, 1903, 1904, January—August 1905. [7] 1894, January—July 1895, May—December 1903, January and April—December 1904, January—September 1905, 1906, 1907. [8] January—April, October—December 1902, 1903, 1904, January—July 1905. [9] 1901—1905. [10] 1896, 1897, 1903—1905. [11] 1900, 1901, 1902. [12] April—December 1894, January—May 1895, April—December 1901, 1902—1905. [13] August—December 1893, January, May—December 1894, 1895, January, and April—December 1896, January—March 1898, January, February 1901, January—August 1902, 1903, 1904 (only 21st—31st of May), 1905. [14] 1902, 1903, April—December 1904, January—August 1905.

(*M. D. S.* XIX. 1906, pp. 43—106, 172—180; XXI. 1908, pp. 44—104.)

GERMAN EAST AFRICA. MEAN RAINFALL IN INCHES.

Coast.

	Lat.	Long.	Alt. in feet	No. of Years	Jan.	Feb.	March	April	May	June	July	Aug.	Sept.	Oct.	Nov.	Dec.	Year
Pangani[1]	5° 26′ S.	38° 59′ E.	23	(6)	1·18	1·22	5·12	10·28	8·98	1·81	3·86	1·38	1·81	2·76	5·79	3·23	47·4
Tanga[2]	5° 4½′	39° 6′	80	(7)	2·17	1·65	4·41	9·33	12·05	1·61	1·81	2·95	2·44	3·82	14·96	3·07	60·3
Bagamoyo[3]	6° 27′	38° 53′	26	8	4·21	1·50	4·96	8·98	6·89	0·67	1·42	1·10	1·14	1·38	5·55	2·80	40·6
Dar-es-Salam[4]	6° 45′	39° 20′	46	(8)	4·37	2·52	3·66	11·34	8·70	0·79	1·73	1·06	1·22	2·13	5·00	3·11	45·6
Msalla[5]	7° 57′	39° 24′	16	(4)	3·11	1·69	3·23	17·99	5·04	1·54	1·93	0·16	0·79	0·79	1·30	5·71	43·3
Kilwa[6]	8° 44½′	39° 25′	36	(6)	5·83	3·27	5·95	8·90	4·10	0·07	0·39	0·70	0·75	0·32	2·84	2·95	36·1
Lindi[7]	10° 2′	39° 44′	268	(9)	6·06	4·06	7·52	4·69	0·98	0·06	0·30	0·53	0·59	0·65	1·77	4·84	32·1
Mikindani[8]	10° 17′	40° 8′	80	(4)	5·16	4·72	4·88	5·51	2·64	0·41	0·53	0·43	0·63	0·59	1·14	9·37	36·0

1 1899, January—June and September—December 1900, 1901, 1902, 1904, 1905.

2 1892, 1893, October—December 1894, January, December 1895, 1896, 1903, January, February, April—December 1904.

3 This result is obtained by combining the records for Kitopeni (Lat. 6° 26′·5 S., Long. 38° 52′·9 E.) with those for Bagamoyo (Lat. 6° 27′, Long. 38° 53′).

4 1897, 1898, January—October 1899, January, March—October, December 1900, 1901—1904.

5 1902, April, May, October—December 1903, January, February, August—December 1904, January, October 1905.

6 November, December 1891, 1892, 1893, July—December 1902, January—May 1903, 1904.

7 January, February, July—December 1893, December 1894, January—November 1895, January, February, May, June, and September—December 1896, 1897, 1898, February—November 1899, January—July and November, December 1900, 1902, 1903.

8 September—December 1901, 1902, 1903, 1905.

(*M. D. S.* XIX. 1906, pp. 43—106, 172—180; XXI. 1908, pp. 44—104.)

NORTHERN RHODESIA

NORTH-EASTERN RHODESIA.

North-eastern Rhodesia is a plateau land, with an average elevation of between 4000 and 6000 feet above sea level. This great plateau is cut throughout its entire length by the Luangwa valley. More properly speaking, perhaps, the region is a series of highlands. The Nyasa-Tanganyika plateau rises abruptly from Tanganyika (2624 ft.) to an elevation of over 6500 feet near Kambole. The altitude of Abercorn is 5550 feet, and north of this the elevation is over 6000 feet in German territory. This plateau continues at an average elevation of some 5000 feet to within 30 or 40 miles of Nyasa, and is continued in a south-westerly direction, forming the western flank of the Luangwa valley, and then trends westwards along the left bank of the Zambezi. Within these highlands is plateau land of an average elevation of over 4000 feet, cleft by the Chambezi, the parent stream of the Luapula, falling into Bangweulu (3700 ft.). Between the Luangwa valley and Nyasa the plateau reaches about 4000 feet, between the Luangwa and the Chambezi 5500 feet. On the south this plateau falls to the valley of the Zambezi. Between the Luapula, on the west, and Bangweulu, on the east, is more plateau land, with a mean altitude of 4500 feet, while to the south of the southern bend of the Luapula the altitude is also about 4500 feet. North of Bangweulu is the Chimpili plateau, with an altitude of 5500 feet, sending the Lovu river to Tanganyika, several tributaries to the Chambezi, smaller streams to Bangweulu direct, and the Loangwa (not to be confounded with the Luangwa) to Mweru (3000 ft.). The most abrupt fall in the land, except, of course, from the heights of Kambole to Tanganyika, is from the same heights (6500 ft.) to Mweru.

Nearly the whole of the highland area is covered with thin forest, and almost the entire country is more or less malarious, or it might be more correct to say that no part is absolutely free from malaria. But it should be noted that the residents, as a rule, enjoy good health in the uplands.

From April to October the South-East Trades are the prevailing winds. In the summer the direction is variable, but the wind most frequently experienced is from the north.

The Luangwa valley, shut in by hills on both sides, is very unpleasantly hot, and more typically tropical than the highlands, where the days are more like English summer days and the nights cool: but the valley is more tropical in its heat than in a constant luxuriance of vegetation, for though in the rainy season, with full, overflowing rivers and a steamy atmosphere, the vegetation would vie with that in any part of the world, yet the long dry season completely alters its appearance. The south-east wind, which blows across the valley, arrives desiccated by the ridge which it has crossed, which separates the valley from Nyasa[1]. Mr Money found that the wet season was heralded, on 6th October, by a premonitory shower, after which the gathered clouds dispersed, and hot dry weathei was experienced, the thermometer, in the screen at Kambwire, showing a temperature of 108° Fahr. (49°·2 C.). In the valley of the Rukusi, on which Kambwire lies, the wind got up every morning, during this period, about 9 a.m., a hot sickly wind, with a choking dust, which effectually dispersed whatever energy the cool of dawn had given. Often it took the form of a whirlwind, which was exceedingly unpleasant. On 24th November occurred a sudden and terrific thunderstorm, after which the rain became more and more regular. There was a cessation for a fortnight in mid January, and then rains occurred up to April, gradually becoming lighter[2].

In August, Mr Weatherley crossed the plateau between Chinama's, on the Luapula, and Bangweulu, and found it perfectly delightful, "though we were half frozen at night. On one memorable occasion the thermometer went down to 36° Fahr. (2°·2 C.). During the daytime, however, it was very pleasant[3]."

---

[1] Wallace, *Geographical Journal*, Vol. XXIX, 1907, p. 382.
[2] Money, *Geographical Journal*, Vol. X, 1897, p. 146.
[3] Weatherley, *Geographical Journal*, Vol. XII, 1898, p. 246.

The missionary, P. Foulon, writing in *Missions des Pères Blancs*, describes the climate of Kilubi or Kirui, at one time a peninsula, at others an island in Bangweulu, according to season, as relatively healthy, even for Europeans, though enormous marshes lie both to the east and south-east. The temperature varies between $42°\cdot8$ and $91°\cdot5$ Fahr. ($6°$ and $33°$ C.). The heat is moist, and therefore hard to bear. In the neighbourhood of the lake, storms are frequent and of great violence; they occur chiefly at night and are not of long duration, and appear not to bring so great a precipitation as might be expected.

At the western end of the Nyasa-Tanganyika plateau the rains begin in an undecided fashion about the end of October and last till April. There is generally a dry time of a fortnight or three weeks in December, and though it may rain every day after until April, the rain is not incessant, and almost every day there are a few hours of bright sunshine. Mr Wallace was at Abercorn, on and off, for a year, and estimates the rainfall there at 35 inches (889 mm.); on the other hand, Capt. Boileau, of the Boundary Commission, which was at work in a very rainy year, puts the amount of rainfall on the plateau at 70 inches (1778 mm.). The temperature at Abercorn does not vary much throughout the year, the clouds and rain having a cooling effect in summer. The maximum thermometer varies between $76°$ and $86°$ Fahr. ($24°\cdot5$ and $30°$ C.) and the minimum between $45°$ and $58°$ Fahr. ($7°\cdot2$ and $14°\cdot5$ C.). The nights are always cool on the plateau, and during June, July, and August, at least three blankets are necessary at night. From June to September a strong and bitterly cold S.E. wind blows day and night, and chills must be carefully guarded against, as they are the sure forerunners of fever. In swamps on the high plateau Mr Wallace has seen frost in the early morning, but at Abercorn the thermometer never went below $45°$ Fahr. ($7°\cdot2$ C.). Speaking generally, the climate is dry, and subject to no great fluctuations of temperature. Malaria unfortunately exists, but it is not virulent, and Europeans living on the high land look ruddy and well. Should the study of tropical diseases lead to such knowledge, that blackwater fever will have no terror for those who risk its attacks, then the Tanganyika plateau will be of value, not only as a residence for traders, but possibly also as a place suitable for colonisation.

Leaving Tanganyika and going westwards to Mweru, the country passed through is not so suitable for any attempt at settlement as the Nyasa-Tanganyika plateau. The elevation is lower, and the climate consequently warmer, and also, in Itawa, there is a longer dry season and a severer wet one, and the country for the most part consists of vast plains and swamps. Lunda (Kazembe's) is a comparatively hot country, but fertile and with a more luxuriant vegetation than the country north of it.

On the whole, it may be said that the highlands of North-Eastern Rhodesia, which constitute the greater part of the country, are relatively healthy and enjoy the benefits of a pleasant climate. The least healthy period of the year is that during which the rains are experienced and immediately after. This rainy season generally begins in November, towards the latter end of the month, and ends in April, the greatest precipitation occurring in the four intervening months. The lowlands do not enjoy a desirable climate. The range of temperature is not very large, that is to say, the temperature throughout the year does not vary much, the summer heat being usually tempered by rain and clouds. There can be little doubt that the hottest and unhealthiest district in this locality is the Luangwa valley, and the unhealthiest time of year just after the rains finish. Malarial fever is quite common, but it is not so general in the higher districts, and many of the white population have never suffered from it. Blackwater fever is uncommon, but by no means unknown. Dysentery is common and is usually caused by bad water. Generally speaking, all drinking water in the country requires boiling, though neglect of this precaution does not necessarily entail contraction of disease. During the winter months —June, July, and August—the temperature often falls very low, though it is stated that in the lower country frost is unknown. The tsetse fly is prevalent in all places where game abounds, but keeps to its accustomed belts of country.

The following is culled from the Report of the

Administrator of North-Eastern Rhodesia for the two years ending March 31, 1907, contained in the Report of the Directors of the South Africa Company.

"The public health of North-Eastern Rhodesia during this period has been most satisfactory. There have been no deaths among the officials of the Administration, the Staff of the African Transcontinental Telegraph Company, or among any of the farmers or genuine settlers, during the two years. This is the first time that such a healthy period has been recorded in North-Eastern Rhodesia. The past two years have witnessed an extraordinary decrease in the number of cases of malaria and its concomitant blackwater fever, to which alone I attribute the satisfactory record that has been established for this period. No better evidence than this can be cited in substantiation of the opinion expressed in previous reports, that once eliminate or control the malarial element as a factor in the causation of disease, then the climate of by far the greater portion of the country becomes as healthy as any in the world, and suitable to every class of European settler. It is perhaps necessary to state here that good healthy conditions cannot be maintained among Europeans in any country with such scanty natural resources as obtain in most parts of Africa. I refer especially to fresh fish and meat, vegetables, fruit and dairy produce, without which, in the best of climates, the health of Europeans will suffer. This is largely a matter of farming and cultivation, and it is noticeable how the health conditions have improved among the Europeans in small communities where these articles are more or less procurable. Milk, butter and fresh meat are now to be obtained at most of the Government Stations. One case of enteric fever has been admitted into the Hospital during this period and only two cases of blackwater fever. There have been a few cases of dysentery recorded with no mortality."

Meteorological observations for three years, 1900–1902, were published in the Official *Handbook of North-Eastern Rhodesia*, 1903, for Fort Jameson, consisting of temperature records taken at 9 a.m., noon, and 4 p.m. for one year and eight months, and at 8 a.m., 11 a.m., and 4 p.m. for the remaining sixteen months. Through the courtesy of the

British South Africa Company I have been furnished with the manuscript of the mean maximum and mean minimum temperatures at the same station for the two years 1905/6 and 1906/7. In both cases the rainfall is also given. I have mentioned the above details, because the table (p. 360) constructed from these observations is somewhat perplexing. In both of the two years 1900 and 1901 the noon temperature in April, and in 1902 the temperature at 11 a.m., are higher than the maximum temperature for each of the two years 1905/6 and 1906/7; and there are somewhat similar results for the months of May, July, and August. The actual figures for the mean noon temperature in 1900 and 1901 are $82°\cdot41$ and $85°\cdot50$ Fahr. ($28°$ and $29°\cdot7$ C.) and, for 11 a.m. in 1902, $84°\cdot37$ Fahr. ($29°\cdot1$ C.); whereas the maximum temperature in that month in 1905/6 was $80°\cdot5$ Fahr. ($26°\cdot9$ C.) and, in 1906/7, $80°\cdot8$ Fahr. ($27°\cdot1$ C.). I have given the means of both in the tables, and the only solution of the difficulty that I can conceive is that the thermometers were not, throughout, placed in the same situation. I had thought of making the mean temperature $\dfrac{\text{9 a.m.} + 12\ (\text{noon}) + 4\ \text{p.m.} + \text{max.} + \text{min.}}{5}$, but, eventually, came to the conclusion that not only would this be too high, but that probably the later observations were the more accurate, and, therefore, I have made the mean $\dfrac{\text{max.} + \text{min.}}{2}$ for two years.

At Magweru, near Fort Jameson, but standing 700 feet higher, the mean temperature, derived in the same manner, is a trifle lower, as might be expected, especially in May, June, September and October, and the daily range is considerably less, except towards the end of the year. The highest recorded temperature at Magweru is $98°$ Fahr. ($36°\cdot7$ C.) and the lowest $41°\cdot3$ Fahr. ($5°\cdot2$ C), and the mean morning temperature for the year $68°\cdot6$ Fahr. ($20°\cdot3$ C.). The mean annual rainfall, however, is three inches (76 mm.) more at Fort Jameson than at Magweru, and January is the rainiest month in both cases.

PRODUCTIONS, &c.

In North-Eastern Rhodesia cattle appear to do well, especially in the neighbourhood of Fort Jameson and in the Tanganyika plateau district. Horses also appear to thrive, for up to the present there has not been a single authenticated instance of horse-sickness. Cotton and tobacco are cultivated with great success in the low-lying Luangwa valley. At Kawimbe, twenty-five miles from Tanganyika, all sorts of European vegetables can be grown, as well as strawberries. The climate is also suitable for the cultivation of both wheat and oats, but irrigation would probably be necessary. Maize will certainly do well, and many varieties of fruits. Flax, ramie, and other fibre plants should also thrive. The Uwemba country is especially suited for the cultivation of coffee and wheat, while south of Mweru and on Kilwa Island the oil-palm flourishes. Such crops as ground-nuts and castor-oil berries might also be tried with every prospect of success. In the Choma division of Mweru district the crops consist of Indian corn, millet, manioc, sweet-potatoes, beans, oil-seeds, pumpkins, and tobacco. In the Luapula district the crops are manioc, malezi (a kind of millet) and sweet potatoes. In the forests are many useful kinds of timber, notably the mopani tree, which grows in low land, with a straight stem, and is exceedingly useful for building purposes ; the wood is dark and hard, and white ants will not touch it.

FORT JAMESON. Lat. 13° 38′ 12″ S, Long. 32° 40′ 15″ E. Alt. 3600 feet.

Compiled from observations taken during the 3 years 1900—1902, and during the 2 years 1905/6 and 1906/7.

	Jan.	Feb.	March	April	May	June	July	Aug.	Sept.	Oct.	Nov.	Dec.	Year
Mean morning temperature 8 a.m. for 2 years, 9 a.m. for 1 year (1900—1902) (Fahr.°)	72°·9	74°·2	75°·5	75°·9	73°·1	70°·1	67°·8	71°·5	73°·7	76°·1	75°·2	73°·5	73°·1
Mean midday temperature 12 noon and 11 a.m. (1900—1902)	77°·7	78°·9	80°·8	84°·1	80°·7[1]	76°·7	76°·1[1]	82°·2[1]	82°·6	84°·7	85°·0	78°·7	80°·7
Mean temperature at 4 p.m. (1900—1902)	76°·1	77°·0	78°·1	79°·5	77°·4	75°·0	74°·5	76°·8	81°·3	82°·8	83°·3	76°·2	78°·3
Mean maximum temperature (1905/6) and 1906/7)	83°·6	81°·1	81°·6	80°·6	79°·3	78°·7	74°·8	79°·5	87°·3	88°·3	83°·6	82°·5	81°·7
Mean minimum temperature (1905/6) and 1906/7)	64°·6	62°·5	61°·7	60°·5	58°·4	55°·5	52°·5	55°·3	62°·3	66°·4	63°·7	62°·9	60°·5
Mean temperature $\dfrac{\text{max.}+\text{min.}}{2}$ ...	76°·1	71°·8	71°·6	70°·6	68°·8	67°·1	63°·7	67°·4	74°·8	77°·3	73°·7	73°·7	71°·1
Mean daily range (max. – min.) ...	19°·0	18°·6	19°·9	20°·1	20°·9	23°·2	22°·3	24°·2	25°·0	21°·9	20°·9	19°·6	21°·2

[1] See above under Fort Jameson (pp. 357, 358).

MAGWERU. Lat. 13° 39′ 30″ S., Long. 33° 50′ E. Alt. 4300 feet.

Compiled from $2\frac{1}{3}$ years' observations[1].

	Jan.	Feb.	March	April	May	June	July	Aug.	Sept.	Oct.	Nov.	Dec.	Year
Mean maximum temperature (Fahr.°)	79°·2	79°·8	79°·8	79°·7	74°·6	71°·4	72°·2	76°·8	82°·7	84°·0	89°·3	85°·2	79°·3
Mean minimum temperature … …	63°·8	62°·9	62°·5	60°·2	55°·1	51°·6	52°·0	55°·3	61°·2	62°·6	67°·1	64°·8	60°·2
Mean temperature[2] … …	71°·5	71°·3	71°·2	69°·9	64°·9	61°·5	62°·1	66°·0	71°·9	73°·3	78°·2	75°·0	69°·8
Mean daily range[3] … …	13°·4	16°·9	17°·3	15°·5	19°·5	19°·8	20°·2	20°·5	21°·5	21°·4	22°·2	20°·4	19°·1
Mean amount of cloud (0—10) …	7·3	6·4	4·4	3·2	2·9	2·9	2·7	1·2	2·0	3·3	4·5	6·2	3·9
Mean number of rainy days …	22·3	16·4	12·3	5·0	0	2·0	0	0·5	2·5	2·0	5·5	11·0	79·5

[1] September—December 1905, 1906, 1907.

[2] The mean temperature is derived from $\dfrac{\text{max.} + \text{min.}}{2}$.

[3] The daily range is max. − min.

MEAN RAINFALL IN INCHES.

	Alt. in feet	No. of Years	Lat. S.	Lat. E.	Jan.	Feb.	March	April	May	June	July	Aug.	Sept.	Oct.	Nov.	Dec.	Year
Fife[1]	5400	3	9° 20′	33° 47′ 30″	5·64	8·02	6·29	2·94	0·35	0·03	0	0	0·02	0·53	3·59	6·99	34·40
Kilibula[2]		2	10° 1′	30° 57′	12·68	8·71	10·95	0·62	0·31	0	0	0	0	1·20	5·65	10·54	50·66
Fort Jameson[3]	3600	5	13° 38′12″	32° 40′15″	10·65	9·49	4·99	1·73	0·11	0·03	0·04	0·04	0	0·21	4·24	5·33	36·82
Magweru[4]	4300	3⅓	13° 39′30″	32° 50′	13·54	6·17	4·93	0·94	0·18	0·09	0·19	0·03	0·04	0·34	1·73	5·60	33·78

[1] 1900, 1901, 1902.　　[2] 1900, 1902.

[3] 1900—1902, 1905/6, 1906/7.　　[4] 1900—1902, January—April 1903.

NORTH-WESTERN RHODESIA.

In North-Western Rhodesia the Zambezi valley is the most prominent geographical feature, traversing its western districts, and curving round first to the east and then to the north-east to form the southern limit of the Protectorate. Within this great valley-curve lies another, of somewhat similar form, but of a smaller type, namely that of the Kafue river. Both streams have their sources in the Congo-Zambezi water-parting in the north of the territory. This parting is in many places unrecognisable as such, but the country in the neighbourhood stands at an elevation of between 4500 and 5000 feet. At the Zambezi source, in the extreme north-west, it has an altitude of 5000 feet, at the Kafue source in the north-east, of 4400 feet, and between these two points the general altitude is from 4500 to 4900 feet, though Mount Chafukuma rises to 5400 feet. The plateau sinks, speaking generally, from the water-parting southwards towards the Zambezi. At Lealui, on the Zambezi, the altitude is 3400 feet, and abreast of this at Kabulwebulwe, on the Kafue, it is 3500 feet. Within the Hook of the Kafue the plateau stands at an elevation of 3600 and 3700 feet, and there are similar altitudes to the south of the Hook. At Kazungula the altitude is 3200 feet. In the extreme east, to the east of the railway, are the Irumbi mountains in the north, with altitudes of 5000 feet and over, and the plateau traversed by the Lusenfwa river, with 4500 feet. From these heights the slope to the Zambezi is the steepest. The whole of the Marotse country was once a great lake, with the Zambezi flowing out of its southern extremity, which in part accounts for the curve in the river. The whole way from Kazungula to Lealui there is not a stone to be seen.

In the country north of the Kafue river and south of the Irumu mountains the climate, on the whole, is comparatively temperate, but in the valleys the temperature frequently rises to 108° Fahr. (42°·2 C.) in the shade, falling usually at night to about 85° Fahr. (29°·4 C.) though higher temperatures are often reached in November. The rainy season extends from December to March, that is to say, rains occur in all those months ; thunderstorms and showers frequently occur towards the latter end of October and in November. During the wet weather the temperature seldom rises above $100°$ Fahr. $(37°·8 C.)$ in the day or $90°$ Fahr. $(32°·2 C.)$ at night, but the atmosphere becomes "muggy," and the country is then less suited for Europeans. The most unhealthy time is immediately after the rains. Malaria is common in the low-lying districts. June, July and August are the winter months, and then the thermometer frequently falls very low. Dysentery should be guarded against by boiling all water which is to be used for drinking purposes.

In the Kafue valley the mean annual rainfall is about $30·5$ inches (775 mm.), and in the Zambezi valley to the south it is about the same, and by far the greater part, practically almost the whole, falls in January, February, and March, though, as mentioned above, there are rainy days over a longer period. In the plateau country the precipitation is said not to be so great. These rains are, of course, summer rains, the hottest season here being also the rainiest. At these times it is necessary to guard against the extreme heat, and, as in Nyasaland further east, a helmet should be worn. On the plateau, however, the heat is never oppressive. In June and July, and sometimes even in August and September, it is very cold at night and in the morning, hard frost being not at all uncommon, and blankets necessary. Clouds begin to appear in November, which is also characterised, as it is further north, by the first short thunderstorms. In the rainy season there is generally a break, usually towards the middle of December, which lasts between a fortnight and three weeks.

In the Kafue region the climate is good in the dry season, and during this period there is generally, in the mornings, for some three hours or so, a breeze from the east or south-east. At midday it is hot, but the heat is not excessive. The evenings are cool and pleasant, while in July and August it is actually cold at night. There is generally a heavy dew. The prevailing wind of the district is from the south-east.

At Monza, in the east of the Protectorate, in Long. 27° 30′ E. (approx.), a set of meteorological observations for a year is available. From these observations it appears that at this station, situated at an altitude of 4000 feet above sea level, the mean temperature was 69°·5 Fahr. (20°·8 C.), being higher than the mean in the four last months and the first three months of the year. The mean morning temperature (8 a.m.) was 4°·7 Fahr. (2°·6 C.) higher than the mean temperature. The mean maximum temperature was highest (90°·5 Fahr. or 32°·5 C.) in September, and the mean minimum lowest in June and July (47°·1 Fahr. or 8°·4 C.). The daily range was largest in June, July, August, and September. The absolute maximum temperature measured was 99° Fahr. (37°·2 C.) and the absolute minimum 41° Fahr. (5°C.). The total rainfall for the year 1900 was 24 inches (610 mm.), of which 6·50 inches (165 mm.) fell in January, 3·50 inches (89 mm.) in February, and 4·43 inches (112 mm.) in March. The relative humidity appears to vary between 81 per cent. in January and 31 per cent. in August.

The table which appears below, giving the temperature in the Marotse valley from Lealui to Kazungula, was compiled by M. A. Jalla[1], a missionary at Sefula, near the former place, from observations taken from 1890 to 1897. The absolute maximum, from 1889 to 1900, was 116°·6 Fahr. (47° C.), registered at Kazungula in the former year, and the absolute minimum was 17°·6 Fahr. (−8° C.), recorded at Nalisa on 10th July 1898, when there were 4·7 inches (12 cm.) of ice in a bucket. There was also a little ice at Lumbe in August 1892. The smallest daily variation of temperature, namely 7·2° Fahr. (4° C.), occurred on 4th December 1895 and 5th December 1896. The greatest daily variations occurred as follows :

52°·2 Fahr. (29° C.) on 5th October 1898, when the thermometer ranged between 98°·6 and 46°·4 Fahr. (37° and 8° C.).

48°·6 Fahr. (27° C.) on 1st August 1895.

47° Fahr. (26°·5 C.) on 1st July 1890 and 23rd September 1897.

The Marotse valley is, on the whole, very exposed to

---

[1] *Pionniers parmi les Ma-Rotse*, Florence, 1904.

wind, and the prevailing direction is from the north-east. The following is the mean number of windy days:

Jan.	6	April	7	July	15	Oct.	26
Feb.	8	May	8	Aug.	20	Nov.	20
March	8	June	10	Sept.	26	Dec.	8

These figures are the means for Lealui, and may be taken as approximately correct also for Nalolo. Kazungula is also very exposed to the wind, Sesheke a little less so; Senanga, Mabumbu and Mosi-wathunya still less, and Sefula the least of all. This difference is entirely due to the existence of woods. There is very seldom any rain in the months of May, June, and July, and only occasionally there may be slight showers in August, September, and October. The rainiest months are February and December, each with a mean rainfall of over 7.87 inches (200 mm.) The rains occasionally begin in August with a mean fall of 0.05 inches (1.29 mm.), September with a mean fall of 0.02 inches (0.52 mm.) or October with a mean fall of 0.8 inches (20.51 mm.), but the real rainy season lasts from December to March inclusive. It is a rare occurrence for a day to pass without the sun being seen. The absolute maximum rainfall in 24 hours was 4 inches (103 mm.).

The natives of the Marotse valley divide the year into four seasons, for which particulars are given below:

Season	Months	Rain (inches)	Windy days	Mean[1] temp. (Fahr.°)	Mean max. temp. (Fahr.°)	Mean min. temp. (Fahr.°)
Lelabula ("rains")	Dec.—Feb.	24·07	22	77°·4	88°·5	66°·0
Munda ("inundation")	March—May	6·38	23	75°·0	78°·9	61°·2
Maria (winter)	June—August	0·05	45	66°·0	87°·6	48°·9
Mbumbi ("dryness")	Sept.—Nov.	4·02	72	76°·6	94°·1	59°·9

In 1906/7 for the first time a considerable number of rain-stations were at work in North-Western Rhodesia, and, through the courtesy of the British South Africa Company, I am enabled to give (p. 368) a table of the rainfall, for that year, at an exceedingly representative set of stations, which shows, at a glance, the precipitation over the whole country for that year, but unfortunately the

---

[1] The mean temperature is $\dfrac{\text{max.} + \text{min.}}{2}$.

figures are available for one year only. Taking Kalomo, the old capital of the country, on the railway, as a starting-point, four of the stations lie roughly in a south to north line at considerable distances apart, and reaching to the northernmost confines of the territory. These are, in order northwards, Mumbwa, situated within the Hook of the Kafue river, Sable Antelope Mine, between Sitanda and the Lunga-Kafue junction, Kasempa, on the main route running from Kalomo, and Kansanshi, the terminus of that route. Again, south-west of Kalomo lies Livingstone, the present capital, on the Zambezi, near the Falls, and N.E. of Kalomo lie first Monza, and then Broken Hill in the extreme north-east of the Protectorate; while the remaining station, Lealui, lies in the west. From this table it will be seen that, speaking generally, the rainfall increases from south to north.

PRODUCTIONS, &c.

In North-Western Rhodesia the Matoka plateau and the country to the north, in the neighbourhood of the Kafue river, are especially suited for grazing cattle, sheep and goats. Many of the southern trees, such as the acacia, producing gum and bark, and the machabel, which gives a bark-fibre, extend as far north as the headwaters of the Congo river. The swamps of the Lukanga river are remarkable for the immense numbers of geese and ducks which make their home there. In the vicinity of the Zambezi, mealies and Kafir-corn do uncommonly well, deriving the necessary moisture from the river, and therefore being independent of rain. Other prominent crops are ground-nuts and sweet-potatoes. In the north very little maize is grown, but abundance of manioc, Kafir-corn and sweet-potatoes. There appears to be no reason why tobacco, and such fibre-plants as flax and ramie, should not do well, and also cotton. With respect to the last the climate is certainly suitable, while, as to soil, it is stated that the tributaries, *i.e.* the more northerly feeders, of the Kafue, "flow slowly in beds cut through deep rich black soil," which is most suitable for the production of this crop. In the clumps of virgin forests is a great wealth of rubber vines (*Landolphia*).

ZAMBEZI VALLEY FROM LEALUI TO KAZUNGULA.

Compiled from 8 years' observations[1].

	Jan.	Feb.	March	April	May	June	July	Aug.	Sept.	Oct.	Nov.	Dec.	Year
Mean maximum temperature (Fahr.°) ...	89°·6	89°·9	91°·9	88°·5	85°·5	78°·4	80°·1	86°·5	93°·0	96°·8	92°·3	86°·0	88°·2
Mean minimum temperature	67°·1	67°·6	68°·0	60°·3	55°·6	48°·0	47°·5	54°·9	56°·1	61°·7	61°·7	63°·5	59°·5
Mean temperature[2]	78°·3	78°·8	79°·9	74°·4	70°·5	63°·2	63°·8	70°·7	74°·6	79°·2	77°·0	74°·8	73°·8
Mean daily range[3]	22°·5	22°·3	23°·9	28°·2	29°·1	30°·4	32°·6	31°·6	36°·9	35°·1	30°·6	22°·5	28°·7
Maximum daily range	27°·9	30°·6	32°·4	39°·6	36°·9	37°·8	47°·7	48°·6	47°·7	52°·2	41°·4	37°·8	—
Minimum daily range	15°·3	14°·4	14°·4	23°·4	7°·2	14°·4	21°·6	23°·4	23°·4	21°·6	13°·5	7°·2	—
Mean of absolute monthly maxima ...	95°·0	102°·2	96°·8	97°·7	91°·4	84°·2	92°·3	96°·8	102°·2	109°·9	109°·4	104°·0	—
Mean of absolute monthly minima ...	61°·7	60°·8	66°·2	50°·0	42°·8	33°·8	26°·6	25°·1	26°·6	46°·4	50°·0	55°·4	—
Mean monthly range[4]	33°·3	41°·4	30°·6	47°·7	48°·6	50°·4	65°·7	71°·7	75°·6	63°·5	59°·4	48°·6	—
Mean rainfall (in inches) at Sefula } (4 years[5])	7·01	9·13	5·47	0·91	0	0	0	0·04	0·02	0·83	3·15	7·87	34·43

[1] 1890—1897. See also above.

[2] The mean temperature is derived from $\dfrac{\text{max.}+\text{min.}}{2}$.

[3] The daily range is max. – min.

[4] The monthly range is abs. max. - abs. min. for the month.

[5] From August 1895 to July 1899.

TABLE SHOWING RAINFALL (IN INCHES) AND NUMBER OF RAINY DAYS RECORDED AT STATIONS IN NORTH-WESTERN RHODESIA IN THE YEAR 1906/7.

	April 1906		May		June		July		Aug.		Sept.		Oct.	
	rain	days	rain	days	rain	days	rain	days	rain	days	rain	days	rain	days
Kalomo (mean of three stations: Police Camp, Hospital and Koch's Farm), alt. 5280 feet	Nil	Nil	Nil	Nil	Nil	Nil	Nil	Nil	Nil	Nil	·93	4·7	·69	2·7
Monza (Jesuit Mission)	·47	2	,,	,,	,,	,,	,,	,,	,,	,,	Nil	Nil	1·47	2
Livingstone (Victoria Falls)	·07	1	,,	,,	,,	,,	,,	,,	,,	,,	·59	4	1·8	4
Mumbwa (in Hook of Kafue)	·20	2	,,	,,	,,	,,	,,	,,	,,	,,	Nil	Nil	·58	2
Lealui (alt. 3400 feet)	·04	1	,,	,,	,,	,,	,,	,,	,,	,,	·65	3	·95	5
Kasempa	·23	2	,,	,,	,,	,,	,,	,,	,,	,,	·64	4	1·21	8
Kansanshi	3·63	8	,,	,,	,,	,,	,,	,,	,,	,,	·48	3	2·38	5
Broken Hill	·28	2	,,	,,	,,	,,	,,	,,	,,	,,	Nil	Nil	1·56	4
Sable Antelope Mine	·28	2	,,	,,	,,	,,	,,	,,	,,	,,	,,	,,	1·28	5

	Nov.		Dec.		Jan. 1907		Feb.		March		Year	
	rain	days	rain	days	rain	days	rain	days	rain	days	rain	days
Kalomo (mean of three stations: Police Camp, Hospital and Koch's Farm), alt. 5280 feet	2·48	10·7	5·51	11·3	7·45	16	9·19	20·3	3·81	11·7	29·68	77·3
Monza (Jesuit Mission)	·78	6	4·11	10	7·58	13	7·89	17	2·87	6	25·17	56
Livingstone (Victoria Falls)	2·50	7	1·40	8	4·80	12	14·72	14	2·39	8	27·65	58
Mumbwa (in Hook of Kafue)	2·89	10	3·53	8	11·37	18	11·48	17	2·61	4	32·66	61
Lealui (alt. 3400 feet)	1·85	8	4·99	14	7·39	19	7·38	17	3·76	10	27·01	77
Kasempa	4·46	12	3·26	15	12·05	23	9·01	24	5·54	15	36·40	103
Kansanshi	2·34	12	11·04	24	11·95	27	8·26	21	7·51	21	47·59	121
Broken Hill	3·46	9	2·89	11	5·32	17	6·48	18	2·12	7	22·11	68
Sable Antelope Mine	·96	4	4·16	18	6·65	18	11·37	23	2·16	10	26·86	80

NYASALAND.

In Nyasaland the elevated tableland preponderates over the low swampy country, and herein lies the secret of its comparatively healthy climate, though it is not altogether suited to Europeans. It has, in fact, two climates, one for the Shire valley and Nyasa, the other for the highlands. On the highlands near Blantyre, at an elevation of some 5000 feet (1525 m.), and in the Mlanje district, the heat is never oppressive, even in the hot season, when the thermometer registers about $96°$ or $97°$ Fahr. ($35°·5$ or $36°$ C.); and in the cold season, which is also the dry season, the mercury falls to about $40°$ Fahr. ($4°·4$ C.), and it is bracingly cold. There are some localities, the Nyika plateau for instance, where greater cold is experienced and there are frosts at night. At Blantyre the mean temperature is about $50°$ Fahr. ($10°$ C.) and the minimum $40°$ Fahr. ($4°·4$ C.), although it has been known to go so low as $30°$ Fahr. ($-1°·1$ C.), but this is only on exceptional occasions. The rainfall at Blantyre averages about 56 inches (1420 mm.).

In the low-lying plains the rainfall seldom exceeds 35 inches (889 mm.), except on the lake, where it is excessive; while at Mlanje, at an altitude of about 8000 feet (2440 m.), it is very wet, with a rainfall of 75 inches (1900 mm.), in fact the rainfall of the Shire highlands varies between 40 inches and 100 inches (1000 and 2500 mm.). In the low country are the unhealthy and malarious districts, but the nights are almost invariably cool, even in the extreme north, and from May to August they are quite cold, and the ground is frequently white in the morning. The temperature on the whole is moderate, but, before the rains, sometimes runs up as high as $115°$ Fahr. ($46°$ C.) in the shade. On the lake the midday temperature, in the hottest month of the year (November), is about $85°$ Fahr. ($29°·4$ C.), while the night temperature in the coolest months (May and June) is about $60°$ Fahr. ($15°·6$ C.). The extremes that have been measured are about $100°$ and $54°$ Fahr. ($37°·8$ and $12°·2$ C.).

The rainy season begins towards the end of November

K. A. 24

and lasts till the end of March or the beginning of April, the heaviest fall occurring during January and February; for the following six months the sky is, as a rule, perfectly clear, but mists are not infrequent. If it were not for the excessive precipitation the climate of Mlanje would be perfection. In the hill country above 3000 feet (915 m.) there are occasional showers in June, July, August, and September. It is said that European women stand the climate of Nyasaland better than men.

From March to October the South-West Monsoon prevails, and, coming from over the land, this wind brings no rain; from October to March the North-East Monsoon blows, and there are also light northerly winds and calms. At times a strong easterly wind blows for several days in succession from 3 a.m. to noon, when it dies away. Sudden heavy squalls, with thunder, lightning, and rain, are sometimes experienced coming from the south-west, but they are of short duration, seldom lasting more than an hour or two.

The daily range of the barometer on the lake is seldom more than 0·2 inch.

Sir H. H. Johnston says: "May is the unhealthiest month in Nyasaland, and June is not very good. The worst of the rains is over in March, and April is an agreeable month, cool, but not too chilly, bright sunshine, with occasional showers and without high winds. The unhealthiness of the middle and end of May is caused by strong cold winds from the southward, and the drying up of the marshes. The main cause of ill health in tropical Africa is catching cold. The cold winds of May produce fever by the sudden lowering of the temperature. The rainy season is not an unhealthy season, except for people who, travelling through a country without shelter, are constantly soaked through with rain, and unable to find a dry habitation."

The table on p. 371 is an extract from the meteorological register for the year ending 31st March 1902, and gives a fair general idea of the climatic conditions throughout the Protectorate, so far as temperature is concerned, the first three stations being in the Lower Shire District, Blantyre in the district of the same name, Fort Anderson in Mlanje, Zomba in the district of the same name, Liwonde and

Fort Johnston in Upper Shire, Nkata Bay in West Nyasaland, and Karonga in North Nyasaland.

The rainfall cannot be depended upon from year to year, nor from month to month, though the seasons are fairly regular. The mean fall for October, at Fort Johnston, for instance (see table below, p. 378), is almost entirely due to the abnormal fall of over 9 inches (230 mm.) of rain in that month in the year 1901. Omitting the fall for that year, the October mean would have been ·07 inch (2 mm.). Similarly, in the same month of the same year the fall at Fort Mlangeni was 5·30 inches (135 mm.), the usual record being not much above ·1 inch (2·5 mm.). The annual fall at Zomba varies between 63·5 and 38 inches (1613 and 965 mm.), at Lauderdale between 131·7 and 79 inches (3345 and 2000 mm.), at Nkata Bay between 90 and 60 inches (2286 and 1524 mm.). Similarly at Fort Mlangeni the fall in 1898/9 was 34·70 inches (881 mm.); in 1899/1900 41·21 inches (1046 mm.); in 1900/1 36·08 inches (935 mm.); in 1901/2 47·92 inches (1216 mm.); and in 1902/3 only 23·14 inches (599 mm.).

Station		Approx. alt.	Highest shade temp. Fahr.	Lowest shade temp. Fahr.	Mean annual temp. Fahr.	Rainfall (inches)
Port Herald	...	100 ft.	107° Nov. 6	54° July 23	Record broken	30·82 [2]
Chiromo	...	125 „	109° Nov. 6	52° July 22	78°·5	37·01 [3]
Chikwawa	...	135 „	106°[1] Nov.	52°[1] July	74°·4	31·09
Blantyre	...	3000 „	94° Nov. 1	50°[1] July	68°·4	46·40
Fort Anderson	...	3200 „	92° Nov. 10	51° July 27	69°·4	98·04
Zomba	...	2948 „	94°·8 Dec. 10	47°[1] July	69°·3	58·84
Liwonde	...	1190 „	98° Nov.	55° July 30	77°·2	36·89
Fort Johnston	...	1263 „	104° Oct. 9	48°·7 June 5	76°·3	37·17
Nkata Bay	...	1403 „	98° Nov. 8	56° June 12	75°·0	65 approx. record broken
Karonga	...	1260 „	94° Nov.	Record broken		

[1] Frequently.   [2] No record for March.   [3] No record for April.

Notwithstanding the usually heavy rainfall, the records show that even in the rainiest months there is a very considerable amount of sunshine. The results given in the table (p. 377) for Nkata Bay are derived from the daily estimated amount of sunshine; those, however, for Zomba (p. 375) are the results of records taken with a Whipple-Casella Universal Recorder. But these again are only approximate for the district, and are below the actual figures,

24—2

for, owing to the situation of the instrument, it cannot record the sunshine after 4.30 p.m. in the months of May, June, and July, when the sun approaches and recedes to and from the north, because a spur of Mount Zomba obscures the sun from the station at that time during these months; and similarly with regard to the east, the records seldom begin being burned before 7 a.m.

Mr (now Sir) Alfred Sharpe says that in the Shire highlands there is as good a climate as will be found in any of the elevated plateaux of Central Africa, but unfortunately what has been found there, as elsewhere, is that, no matter what height Europeans go to, they are still subject to periodical attacks of fever. Intermittent fever does no great harm, and to some extent one becomes inured to it. Remittent fever, however, is a far more troublesome type, and the worst of all is blackwater fever, which seems to make its appearance sooner or later in any part of Central Africa where Europeans settle.

And, speaking generally of Nyasaland, the same writer insists on the fact, which almost all writers in this country dilate upon, that the climate is not a hot one. That of the high levels is almost a European one, in which, for a great part of the year, the same clothes can be worn as would be used during an English summer, and cool breezes always blow. In fact the climate is a pleasant one.

On the plateau of central Angoniland the months of June and July are unpleasantly cold, and those of January and February unpleasantly damp. During the remainder of the year the climate may be truthfully described as delightful. This is the testimony of Mr Robert Codrington, Collector of the district.

At Karonga, on the north of Nyasa, the rains begin in November and last till about the end of April or the middle of May. During the other months it is absolutely rainless, but there are occasional cloudy days. The rainfall on the Nyasa-Tanganyika plateau is about 70 inches (1778 mm.), and below, at Karonga, about 50 inches[1] (1270 mm.), where

---

[1] These figures represent the estimate made by Capt. Boileau, during a very rainy season. Mr Wallace estimates the annual rainfall at Abercorn on the plateau at 35 inches (890 mm.). Estimates of rainfall, however, are not very trustworthy.

the maximum temperature is about $100°$ Fahr. ($37°·8$ C.), and it is a relief to mount to the plateau, where the temperature is only about $85°$ Fahr. ($29°·4$ C.). Here the nights are always cool, and in June, July, and August Capt. Boileau recommends three blankets. From June to September a strong and "bitterly cold" south-east wind blows day and night, giving rise to liability to chills and subsequent fever. A pith hat is desirable as a protection from the sun. The experiences of the Boundary Commission do not lead to the conclusion that the plateau is the healthiest of localities, as their party suffered from both ordinary malaria and blackwater fever. But this was during a very unhealthy season, owing to the extraordinary duration of the rains. Though in some parts of south Central Africa sleeping on the ground may be practised with impunity, this should never be attempted on the Nyasa-Tanganyika plateau, as fever is an almost certain result.

At Fort Mlangeni, in about $14° 41'$ S., and $34° 37'$ E., near the Portuguese boundary, the climate is described as exceedingly healthy for Europeans, on account of the height of the plateau. The temperature, throughout the year, appears to vary only slightly during the day, but the nights during the winter are decidedly cold. The prevailing wind is from the north-east, off the lake.

Mr Money says of the Hora plain, to the west of Nyasa, at an altitude of 4820 feet (1470 m.), that the climate is bracing, and free from the enervating winds, which sap one's energy on the lake. Years ago the Free Church of Scotland founded a mission station here, and the continued health of the present missionary, his wife, and their little son, points to the suitability of the land for the white man[1].

PRODUCTIONS, &c.

The dual character of the climate of Nyasaland renders it possible to produce very varied crops, and the comparative regularity of the heavy rains, combined with the sunny days, form a valuable asset in the cultivation of many tropical crops, while the lower temperature of the

[1] *Geographical Journal*, Vol. x, 1897, pp. 146, 148.

uplands is well suited to others. Rice grows well near the lake, and if rice thrives, then also should jute. The climate is well suited for the production of cotton, coffee, which is indigenous, and tobacco; certain localities are suitable for wheat; maize will grow almost anywhere. Rubber grows profusely, and chillies do well. Tea is now being produced, while, of the fibre plants, sisal-hemp, which is a native of East Africa, and ramie, are both certain of success, and one form of African hemp, namely, *Sanseviera ehrenbergii* is also a native of tropical East Africa, and therefore merely wants cultivation. This last is one of the class known as Bowstring Hemp, from the length of the fibre, which is used by natives for bowstrings. It is probable, too, that another variety, *Sanseviera cylindrica*, would do well, if tried, for its habitat is only a little south of Nyasaland. European fruits and vegetables are grown, especially in the Blantyre district.

With reference to special districts, the Shire highlands are said to be eminently suited for the growth of coffee, and for the rearing of cattle and horses. Tea, cacao, tobacco, and wheat also, will grow well, while rubber is indigenous. The principal agricultural produce of Central Angoniland consists of maize, sorghum, manioc, rice, millet, sweet-potatoes, varieties of beans and peas, cotton, oil seeds, and tobacco and coffee. The foot-hills of the Kirk range are also suitable for coffee. On the Hora plain, cattle, goats and sheep thrive; millet, maize, beans, peas, sweet-potatoes, ground-nuts and manioc yield good crops.

ZOMBA. Lat. 15° 23' S., Long. 30° 18' E. Alt. 2948 feet.

Compiled from the monthly numbers of the *British Central Africa (Nyasaland) Gazette* for 9 years[1].

	Jan.	Feb.	March	April	May	June	July	Aug.	Sept.	Oct.	Nov.	Dec.	Year
Mean maximum temperature (Fahr.°)	81°7	79°6	78°9	78°0	74°8	71°7	71°1	75°4	80°7	85°0	84°9	81°3	78°6
Mean minimum temperature ...	64°6	63°6	62°8	60°3	55°5	53°0	52°1	53°2	58°6	62°4	65°2	64°1	59°6
Mean temperature[2]	73°1	71°6	70°8	69°2	65°1	62°3	61°6	64°3	69°7	73°7	75°0	72°7	69°1
Mean daily range[3]	17°1	16°0	16°1	17°7	19°3	18°7	19°0	22°2	22°1	22°6	19°7	17°2	19°0
Mean relative humidity (per cent.) at 7 a.m.	90	93	93	91	86	87	88	84	80	75	80	82	86
Mean relative humidity (per cent.) at 2 p.m.	76	77	81	73	69	64	64	59	57	54	61	67	67
Mean relative humidity (per cent.) at 9 p.m.	85	88	90	89	84	84	81	76	69	67	75	82	81
Mean amount of cloud (0—10) at 7 a.m.	6·5	7·1	7·0	5·5	5·1	5·2	5·0	3·1	3·6	3·2	5·1	6·5	—
„ „ at 2 p.m.	6·7	7·7	6·7	6·0	5·2	5·8	5·8	3·8	3·8	4·2	5·9	6·9	—
„ „ at 9 p.m.	6·6	6·5	6·5	5·3	4·7	5·4	4·3	3·5	2·9	2·9	5·3	6·7	—
Mean number of hours' sunshine[4] ...	173	122	131	180	172	140	153	226	221	256	192	150	2116 hours

NOTE:—Zomba is fitted out as a first class meteorological station.

[1] 1899—1907.

[2] The mean temperature is derived from $\dfrac{\text{max.}+\text{min.}}{2}$.

[3] The daily range is max. – min.

[4] From 4 years' observations, 1904—1907.

LAUDERDALE (MLANJE). Lat. 16° 1′ S., Long. 35° 36′ E.　Alt. 2540 feet.

Compiled from 5 years' observations[1], recorded in the Reports of the British Association.

	Jan.	Feb.	March	April	May	June	July	Aug.	Sept.	Oct.	Nov.	Dec.	Year
Mean maximum temperature (Fahr.°)	81°.7	80°.4	79°.9	76°.8	74°.3	68°.4	70°.5	75°.6	82°.4	88°.6	86°.9	82°.0	78°.9
Mean minimum temperature	67°.3	64°.0	65°.8	63°.7	59°.2	55°.4	55°.3	56°.3	59°.7	64°.7	66°.2	66°.1	62°.0
Mean temperature[2]	74°.5	72°.2	72°.8	70°.3	66°.7	61°.9	62°.9	65°.9	70°.6	76°.6	76°.6	74°.0	70°.4
Mean daily range[3]	14°.4	16°.4	14°.1	13°.1	15°.1	13°.0	15°.2	19°.3	22°.7	23°.9	20°.7	15°.9	16°.9
Mean number of days on which rain was measured	23.0	23.2	20.2	18.0	13.4	12.2	9.4	6.2	3.8	4.4	11.8	20.8	166.4
Mean relative humidity (per cent.) at 6 a.m.[4]	90	87	91	91	85	83	82	82	74	73	84	90	84
Mean relative humidity (per cent.) at 2 p.m.[5]	77	83	80	81	74	75	72	72	52	55	62	83	72
Mean relative humidity (per cent.) at 9 p.m.[6]	83	87	87	85	81	82	76	67	61	61	71	82	77

NOTE:—The absolute maximum temperature recorded was 100°.4 in December; and the absolute minimum 47°.3 in July.

[1] 1896–1900.　　　[2] Derived from $\frac{\text{max.} + \text{min.}}{2}$.　　　[3] The daily range is max. − min.

[4] June, July and September, mean of 4 years, 1896, 1898–1900; August, mean of 3 years, 1898–1900.

[5] January—May, September, November and December, mean of 4 years, 1896—1899; June, July, October, mean of 3 years, 1896, 1898 and 1899; August, mean of 2 years, 1898, 1899.

[6] May and June, mean of 4 years, 1896, 1898–1900; August, mean of 3 years, 1898–1900.

FORT JOHNSTON. Lat. 14° 27′ S., Long. 35° 15′ E. Alt. 1590 feet.

Compiled from the supplements to the monthly numbers of the *British Central Africa (Nyasaland) Gazette* for 5 years[1].

	Jan.	Feb.	March	April	May	June	July	Aug.	Sept.	Oct.	Nov.	Dec.	Year
Mean maximum temperature (Fahr.°) ...	89°·6	88°·0	89°·8	87°·3	84°·0	80°·5	80°·6	84°·6	91°·5	96°·3	92°·6	92°·4	88°·1
Mean minimum temperature	70°·1	68°·1	68°·8	65°·3	60°·7	55°·6	57°·1	55°·0	62°·6	66°·6	72°·5	69°·1	64°·3
Mean temperature[2]	79°·8	78°·0	79°·3	76°·3	72°·4	68°·0	68°·8	69°·8	77°·0	81°·5	82°·5	80°·7	76°·2
Mean daily range[3]	19°·5	19°·9	21°·0	22°·0	23°·3	24°·9	23°·5	29°·6	28°·9	29°·7	20°·1	23°·3	23°·8

[1] 1903—1907.

[2] The mean temperature is derived from $\dfrac{max. + min.}{2}$.

[3] The daily range is max. − min.

NKATA BAY. Lat. 11° 36′ 30″ S., Long. 34° 18′ 30″ E. Alt. 1400 feet.

Compiled from 4 years' observations[1].

	Jan.	Feb.	March	April	May	June	July	Aug.	Sept.	Oct.	Nov.	Dec.	Year
Mean temperature at 7 a.m. (Fahr.°) ...	72°·1	71°·3	71°·7	69°·4	67°·6	62°·7	62°·0	63°·8	68°·0	72°·1	74°·9	72°·6	69°·0
,, ,, ,, 2 p.m.	84°·9	81°·2	84°·4	82°·6	80°·9	76°·4	75°·6	77°·6	82°·9	85°·7	87°·8	84°·0	82°·0
,, ,, ,, 9 p.m.	74°·7	74°·3	73°·8	73°·1	67°·4	64°·1	63°·1	64°·5	70°·0	73°·4	77°·2	77°·4	71°·1
Mean temperature $\dfrac{7a+2p+9p}{3}$ ...	77°·2	75°·6	76°·6	75°·0	72°·0	67°·7	66°·9	68°·6	73°·6	77°·1	80°·0	78°·0	74°·0
Mean number of hours' sunshine ...	207	146	212	163	277	218	209	292	242	274	266	203	2709 hours

NOTE:—Nkata bay is sheltered from all winds, except those from the East. The authority is the same as for the above.

[1] 1898—1900, March—December 1902. In 1899 the temperature observations are wanting for May, August and September. The sunshine records are wanting for May, August, and September for the same year, and for September 1900.

MEAN RAINFALL IN INCHES.

Station	Alt. in feet	No. of Years[1]	Jan.	Feb.	March	April	May	June	July	Aug.	Sept.	Oct.	Nov.	Dec.	Year
Chiromo[4] (Lower Shire)	125	5—7	4·56	6·51	5·01	1·08	·63	·51	·22	·30	·03	1·29	1·53	4·41	26·08
Chikwawa[5] „	127	5	8·71	6·40	5·27	·94	·54	1·27	·63	·10	·08	1·67	1·57	3·73	30·91
The Collectorate[6] (Blantyre)	3000	4	11·20	7·21	4·76	1·34	·20	·33	·15	·01	·26	·96	4·01	5·14	35·56
Myara[7] „	3000	3	9·47	7·39	5·97	1·51	·59	·57	·20	·05	·09	·79	3·95	9·45	40·03
Fort Anderson[8] (Mlanje)	3500	4—5	13·29	11·54	11·60	5·02	3·65	2·53	1·85	1·70	2·98	4·51	5·78	10·07	70·52
Lauderdale[9] „	2540	6—8	15·01	18·24	13·97	12·34	5·68	4·17	3·09	2·22	3·70	4·59	8·03	16·46	107·50
The Observatory[10] (Zomba)	2948	11	11·14	11·50	8·11	3·66	·71	·47	·32	·12	·32	1·73	5·39	10·95	54·42
Fort Johnston[11] (Upper Shire)	1590	6—8	8·49	7·00	4·04	2·86	·31	·13	·01	·07	·23	2·06[2]	1·88	6·42	33·50
Fort Mlangeni[12] (Angoniland)	3500	4—5	11·79	5·88	6·86	1·50	·12	0	·03	·01	·02	1·44[3]	1·74	5·05	33·44
Fort Hill[13] (North Nyasa)	3000	2—3	9·09	7·73	7·90	1·39	·17	0	0	0	0	·15	1·88	5·94	34·25
Nkata Bay[14] (West Nyasa)	1400	5—7	8·14	12·46	14·11	11·05	2·87	2·97	1·59	1·34	·33	·08	2·57	11·66	69·17

[1] Where two numbers are given, the larger number applies to the earlier months of the year, generally January, February and March.

[2] Due to an abnormal rainfall of 9·04 inches in 1901, the mean would otherwise be much less.

[3] Due to an abnormal rainfall of 5·30 inches in 1901, the mean would otherwise be much less.

[4] 1896, October—December 1898, January—March 1899, January—March 1900, 1901, 1902, 1903, January—March 1904.

[5] 1896, June, July 1898, November, December 1900, 1901, 1902, 1903, January—March 1904.

[6] October—December 1900, 1901—1903. [7] 1900—1902. [8] 1900—1903, January—March 1904.

[9] 1896—1902, January, February 1903. [10] November, December 1897, 1898—1907. [11] 1896, 1898, January—July, 1899, 1900, 1901, 1902, 1903, January—March 1904. [12] 1900—1903, January—March 1904. [13] 1901, 1902, January—July 1903. [14] 1898, January—April, June, July, November, December 1899, 1900, January—March 1901, March—December 1902, January—October, December 1903, January—March 1904.

MOZAMBIQUE.

The Zambezi river divides the Portuguese possessions in East Africa into Mozambique on the north and Gazaland on the south. Mozambique extends northwards from the Zambezi to German East Africa on the north, where the Ruvuma forms the boundary. Like the rest of Africa, the country rises by terraces from the sea to the interior. The low coast-belt is, in most places, fringed with mangrove swamps and is decidedly unhealthy. The most prominent feature in the interior is constituted by the Namuli mountains, which form a kind of *mont aux sources*, and govern the hydrography of the greater part of the region. From their slopes the Lurio flows northwards, the Ligonia south-eastwards and Likungo southwards, while from the little Lake Amaramba in the Nyasa plateau the Lujenda pursues a north-easterly course to join the Ruvuma. Pico Namuli rises to a height of 8000 feet (2440 m.), nearly rivalling Mount Mlanje to the south of Lake Shirwa. This peak is in the south of the Namuli heights, and Mount Inagu, near the northern extremity of the range, has an altitude of nearly 6500 feet (1982 m.). Mount Lichingo, further north, contributes feeders flowing east to the Lujenda, and is the origin of the river Luchiringo, which flows northwards to join the Ruvuma. From the general direction of the rivers it is easily seen that the slope of the country, with the Namuli heights, and the highlands lying between them and Mozambique, as governing elevations, is northeast in the north and south-east in the south. From October to March, when the sun is south of the equator, the prevailing winds on the Mozambique coast are from the north-east. On the other hand, from April to September, when the sun is north of the equator, south-westerly winds are chiefly experienced, though, as the season wears on, there appears to be a tendency for the wind to turn to the north.

The short dry season begins in December, on the coast, speaking generally of the whole stretch; for since the climatic factors follow the sun in its course, there is a very considerable divergence between the times of the seasons in the north and in the south. This short season lasts till mid January. The heavy rains then begin, when the sun is in the zenith on its way north, and last till the end of March or the beginning of April. Generally, for the whole coast, May and June are months of light rain and the remainder of the year constitutes the dry season. In the south, near the Zambezi, the seasons are earlier and the heavy rains begin in December, preceded by more moderate precipitation in November. In some localities, in some years, the dry season is very severe, and drought is experienced. In the northern parts of the interior the precipitation is much less than near the Zambezi delta.

It is generally said that the two most unhealthy months in the Zambezi and Shire valleys are April and May, in consequence of the action of the sun on the decaying vegetation after the rains. The height of the rainy season and the period of highest temperature are also considered unhealthy. The Zambezi delta and the neighbourhood of the Morabala and Elephant marshes are the worst localities. Even the region of the Upper Shire and Nyasa are very trying to Europeans. The *Africa Pilot* gives the following: "The report of H.M.S. *Herald*, 1891, on the health of various places is as follows:—Katunga, a very unhealthy locality; Chiromo, very fair; Vicenti, severe attacks of malarial fever at all seasons; Chinde river, no marked unhealthiness, but chills are dangerous; the Chinde was healthy, and free from fever in September and October 1890; maximum temperature 75° Fahr. (23°·9 C.), minimum 68° Fahr. (20° C.)."

On the Zambezi the rains come to an end towards the beginning of April, in July the crops are reaped, in August the grass begins to get scorched, and the country is exposed to the burning sun. The river-beds grow dry; even the Zambezi is changed from a mighty stream into a number of rivulets; the water-holes are empty. In September, and even more in October, the heat becomes sultry, and almost unbearable. The temperature rises to 113° Fahr. (45° C.) in the shade by day, and the nights, though cooler, often bring but little relief[1].

At Baroma, or Broma, on the Zambezi, in Lat. 16° S.

[1] Peters, *The El Dorado of the Ancients*, p. 77.

and Long. 33° 12' E., observations were taken from March 1891 to February 1892. There was no rain during the period May to October, and only ·02 inches (0·5 mm.) fell in April, while during the four months from November to February the total precipitation amounted to almost 30 inches (751 mm.).

At Mopea, situated on the Kwa-Kwa, an affluent of the Zambezi, 65 miles from the sea, in Lat. 17° 58' S., Long. 35° 44' E., at an altitude of 80 feet above sea level, the barometer is almost always highest in July and lowest in December or January: in 1895, however, it was highest in October. The pressure varies between 30·290 inches and 29·632 inches. The highest mean temperature for any month was 87°·1 Fahr. (30°·6 C.) for November 1897, and the lowest 63°·3 Fahr. (17°·4 C.) for June 1894. Of the mean maxima of temperature the two highest means were 102°·2 and 101°·5 Fahr. (39° and 38°·6 C.) for October and November of 1897 respectively, and the lowest mean maximum for any month was 77°·8 Fahr. (25°·4 C.) in June 1899. The lowest mean minimum for any month was 54°·1 Fahr. (12°·3 C.) in August 1892, and the highest mean minimum 75°·6 Fahr. (24°·2 C.) in January 1900. The highest temperature ever recorded was 112° Fahr. (44°·4 C.) on a day in November 1896, and the lowest 43° Fahr. (6°·1 C.) in August 1892 and 1898. The mean annual rainfall varies between 55·35 and 30·90 inches (1405 and 784·9 mm.), the heaviest fall for any one month being 16·91 inches (429·3 mm.) in December 1900, which fell on only 14 days, on one of which (24 hours) the precipitation was no less than 6·80 inches (173 mm.). The greatest number of rainy days recorded in any one month was 21 days in February 1899, the total precipitation being 14·70 inches (373 mm.) and the greatest fall in 24 hours in the month 2·45 inches (62 mm.). December is decidedly the rainiest month.

At Mozambique the mean annual rainfall is 33·96 inches (864 mm.), and that at Zumbo, on the Zambezi, is 35·51 inches (902 mm.).

The coastal districts of almost the whole of Mozambique are notoriously unhealthy. This is largely due to the swamps formed at the mouths of the many streams, and the action of a scorching sun on the vegetable matter brought down and deposited in their marshes; the unhealthiness is further added to by the heavy nightly dews and the large precipitation which follows the great heat. Mozambique town, says the *Africa Pilot*, is an unhealthy spot. The rainy season extends from November to March or April, but there are sometimes very considerable rains in May, and in 1897 there was a very marked example of this. During the period from October to April the wind is in the north, and for the rest of the year in the south. There is usually a sea-breeze, about 10 or 11 o'clock in the morning, from the south-east, which in the afternoon shifts round to the east: and this is replaced, as the early morning approaches, by a land-breeze. At rare intervals cyclones are experienced.

Fernão Veloso is a magnificent harbour, extending both north and south of its inlet, and enjoys regular land- and sea-breezes, which sweep the whole surface. There is here no trace of mangrove or swamp, and it therefore forms an exception to the ordinary run of coastal localities in Mozambique. The fact is that, though on the maps frequently called a river, no river enters the harbour, and it is therefore free from the influences mentioned above.

At Ibo harbour there is much rain, accompanied by thunder and lightning from about the middle of January to the middle of March, which is the unhealthy season, fevers being very prevalent.

At Kilimane, as elsewhere on this coast, the climate is not suited for Europeans. The early morning temperature in July has been registered as low as 62° Fahr. (16°·7 C.). The heaviest precipitation, usually accompanied by a vivid electric display, occurs in January and February, and there are light rains in November, and in May and June.

It appears strange that, in these days, when so many make what used to be called the Dark Continent their play-ground or hunting-ground, recourse must be had to almost historic records to learn anything of the climate of the interior of this vast tract of country, except in the regions bordering on the Zambezi. Two great trading companies have established themselves, the one in the north and the other in the south, but they have, neither of them, vouchsafed, so far, to grant any information of a reliable character with regard to the climate of the regions over which they hold sway.

Livingstone, however, tells us that, at a spot about 300 miles up the Ruvuma, at an altitude of some 800 feet (244 m.), cold winds from the highlands, situated further south, prevailed in mid June, and that the thermometer then registered only $55°$ Fahr. ($12°·8$ C.); and further, that to the south of this, at Muembe in the Yao country, which is situated about 2700 feet (823 m.) above sea level, on the hills, the mid July air was cool. In the mountain region to the west he experienced cold southerly breezes, and the sky was overcast every day after 10 a.m., so much so, in fact, that he was precluded from taking observations of the sun for latitude.

Bishop Maples speaks of the excellent climate of Muala district, which lies about 100 miles west of Lurio Bay, and of the keen fresh air, at an altitude of 2300 feet (700 m.) above sea level.

Consul O'Neill, who made many excursions through the country, in addition to surveying the coasts, was singularly reticent, in the many papers which he contributed to the *Proceedings of the Royal Geographical Society*, as to the climate of the country. He remarks, however, that the Makua country, in early September, and fairly near the coast, was "a country parched and tinted yellow by the burning sun"; and further he states that, when he was at some distance east of the Namuli heights, in early October, the southerly winds had been, for a week, drifting up masses of clouds and mists, and that these eventually descended, forming the first or early rains. Later on, during the same journey, when in the hill country, which lies to the immediate east of $38°$ E., on 25th October, he says, "in no part of the world have I felt the heat so intense as in passing through these hills," but it should be remembered that these hills were solid rock, devoid of vegetation, and that the radiation must therefore have considerably added to the discomfort on this occasion.

Major Spilsbury, writing lately in the *Journal of the African Society*, and speaking generally of the north, says that the rainfall is nearly 40 inches (1000 mm.), but admits that, when he arrived at the Lujenda river, there had been a prolonged drought. Abreast of Fort Maguire (Nyasa) and about 60 miles to the east, at Pandiuli, on the Luambala river, and at an altitude of 2825 feet (861 m.), he found the nights in mid October delightfully cool.

The Handbook of the "Companha do Nyassa" states, with regard to the northern districts, that "it appears that, in both the littoral and the medium zones, Europeans can reside for longer or shorter periods without risk of fever, while the interior zone is suitable for permanent residence and colonisation. In the two former zones, therefore, the resources of the country must be developed by native labour under European supervision; in the latter, European colonists can find a home." The Handbook quotes from a letter written from Mualia (about 100 miles inland from Pemba Bay): "During eight months of the year the evenings, nights, and mornings, are cool and delightful, the temperature varying from 65° to 70° Fahr. ($18°{\cdot}3$ to $21°{\cdot}1$ C.), and during the warmest portion of the days, from 11.30 a.m. to 3 p.m., varying from 80° to 85° Fahr. ($26°{\cdot}7$ to $29°{\cdot}4$ C.), and occasionally 90° Fahr. ($32°{\cdot}2$ C.) in the shade. During the hot season the thermometer ranges between 75° and 60° (?80°) Fahr. at night, and from 90° to 95° Fahr. ($32°{\cdot}2$ to 35° C.) in the shade between 11 a.m. and 3.30 p.m. The uplands are healthy at all seasons of the year." Another quotation runs thus: "The climate is well adapted to the southern European, and even the more northern races would bear it by careful avoidance of violent exertion and due regard for the rules of health, avoiding heavy wettings and the like." This was written from Metarika. It must be remembered that the above quotations are from persons interested in the country.

The following notes on the climate of the Province of Cabo Delgado, which embraces the whole of the northern portion of Mozambique, were kindly written, specially for this volume, by Mr L. H. L. Huddart, M.A. (Cantab.), A.R.S.M., &c., who spent nearly ten months in the country on a journey of observation.

1. *Heat.* The heat is fairly severe, especially near

the coast and in the valley of the Ruvuma river. The hottest months are September (latter half), October, and November. When the rains are late, December is also rather hot. Compared with other tropical continental countries the heat is not excessive, and is usually ameliorated by a pleasant breeze which gets up in the mornings and evenings. Shade temperatures are very seldom above 100° Fahr.

2. *Cold.* From a European point of view the weather can hardly ever be described as "cold," and frost is practically unknown, even on the highest ground bordering Lake Nyasa. On the other hand, the conditions are distinctly cool, and pleasantly so, in the rainy-season months of March to June, when temperatures under 60° Fahr. (15°·6 C.) are fairly common.

Upon the open high country, 100 to 200 miles west from the coast belt, the wind is frequently rather cool, making care necessary to avoid chills, but to anyone in reasonably good health these winds are delightful.

3. *Wind and rain.* Along the coast the effects of the N.E. and S.W. Monsoons are felt, and the weather conditions are consequently somewhat modified as compared with those further inland. The change of monsoon occurs with fair regularity at the end of October and in June, on each occasion giving rise to unsettled weather. The N.E. Monsoon brings rain and often high winds. Inland the effects of the changes are less felt, while in the extreme west of the Province the conditions are much the same as in the Shire Highlands.

The rains set in about November, although sometimes not until the end of January, or even later, causing great distress to the natives, who reap a poor harvest and are short of food for the next year. The natives begin to sow when the Pleiades, or the "Sowings" as the constellation is prettily called, appear above the horizon shortly after sunset.

The rains cease about June, when the millet and corn are harvested.

As a general rule the rainfall is greater on the coast belt and along the mountainous tract that borders Lake Nyasa, than in the intermediate country, some 150 miles broad, which runs north and south through the Province.

4. *Mist.* At certain times of the year, and not very often, thick mists are prevalent. They are naturally more common over the low-lying country near the valleys of the larger rivers, and upon the hill tops when the clouds hang low. In late February and March there is usually a good deal of mist inland.

The mists do not appear to be so dense nor to last so long as those of the Guinea coast on the other side of Africa.

5. *Dew.* The dew is usually fairly heavy and most refreshing in the early mornings of the dry season. Every trace of moisture disappears two hours after sunrise. With certain dry winds, especially towards the end of the dry season, no dew falls.

6. *Mornings and evenings.* The mornings during the dry season and at the end of the rains are delightful, fresh, and clear. Even in wet weather the mornings often break fine, the rain coming up later in the day.

In travelling, a halt is called about 10.30 or 11 a.m. in the dry season, as the ground gets so hot as to blister the feet of the porters.

The evenings are, *par excellence*, the best part of the day, a cool breeze springing up, and, in the fine weather, the stars come into view soon after sunset, and the clear night begins.

7. *Buoyancy.* The buoyancy of a climate is often proportional to the temperament of the man who describes it. To a healthy and hard-living man the climate of Cabo Delgado is distinctly buoyant in the mornings and evenings, except in the rains, when the air is humid. Humidity tells much more on a man than mere height of temperature. After long residence, say over five years, most Englishmen would feel the climate beginning to tell on them to some extent. In the mountainous country on the eastern side of Nyasa, the general climate cannot be very different from that of the Shire Highlands.

8. *Storms.* Cyclones occur, and a severe one did a good deal of damage around Pemba Bay a few years ago. The well-known severity of the storms in the Mozambique Channel is greatly modified by continental conditions in the inland zone.

9. *Health.* There are not many places where the *anopheles* mosquito is common, except in the north along the Ruvuma valley, and in some other low-lying tracts, as around Lake Amaramba, and consequently malaria is not prevalent in other districts. If a man lives with sense, there is no reason why he should not keep in good physical condition. Many come to the country with the fever in their system that has been contracted in other parts of Africa, *e.g.* the Zambezi valley, and blame the country for it.

10. *Clothes.* On the coast the usual white drill is mostly worn. For travelling and indeed all wear, nothing can beat the usual flannel shirt, with sleeves rolled up, and breeches with puttees or gaiters. A coat is not necessary except in the evenings. A mess jacket is useful.

11. *Habits.* These of course depend very much on the individual, but it is usual to have a siesta during the heat of the day, for "only dogs and Englishmen" go about in the sun in the middle of the day. Business hours, about 6 a.m. to 10 or 10.30, and 2.30 to 4 or 5 p.m. There is magnificent sport to be had, in fact some of the best in Africa.

PRODUCTIONS, &c.

Of the indigenous plants rubber is one of the most important, as it already constitutes one of the chief sources of native wealth. Coffee, sugar and cotton are all indigenous, and in the coastal regions, at Kilimane, the coffee shrub grows side by side with the cocoa-nut, but the districts especially suited to the cultivation of coffee are, of course, the highlands. Other crops which are, or might be, grown in this climate are rice, Indian corn, tobacco, sesame (semsem or benni-seed), manioc, ground-nuts, and sweet-potatoes. The Zambezi valley, on both banks, is especially suited for the cultivation of the sugar-cane. Wheat should do well, especially in the uplands of the Nyasa plateau. Near the source of the Luli, millet, manioc, tobacco, and sugar-cane were seen growing well. There is a cotton plantation to almost every village. European vegetables can be grown in the rainy season, and, with irrigation, all the year round.

25—2

MOPEA, ZAMBEZIA. Lat. 17° 58′ S, Long. 35° 44′ E. Alt. 80 feet.

Distance from the sea 65 miles, situated on the Kwa-Kwa, 3 miles from left bank of Zambezi.

Compiled from 10 years' observations[1], which appeared in *Meteorological Office: Climatological observations at Colonial and Foreign Stations. I. Tropical Africa.* 1904.

	Jan.	Feb.	March	April	May	June	July	Aug.	Sept.	Oct.	Nov.	Dec.	Year
Mean temperature 9 a.m. (Fahr.°) ...	81°·8	81°·8	81°·8	78°·9	73°·5	68°·3	68°·6	72°·0	77°·1	81°·6	83°·4	83°·4	76°·9
Mean daily range[2]	19°·5	19°·3	19°·6	20°·8	23°·6	24°·9	24°·8	27°·8	26°·2	29°·6	26°·7	20°·5	23°·6
Absolute maximum temperature ...	104°	101°	103°	99°	98°	93°	97°	102°	106°	110°	112°	109°	112° (period)
Absolute minimum temperature ...	64°	64°	63°	56°	50°	46°	47°	43°	50°	50°	57°	64°	43° (period)
Mean rainfall in inches	7·4	7·9	4·9	3·3	0·9	1·0	0·8	0·8	0·2	1·1	3·0	10·9	42·2
Mean number of rainy days ...	13·7	13·8	12·1	11·3	7·1	6·8	4·6	4·2	1·7	2·6	7·5	14·3	99·7
Mean relative humidity (per cent.)	78·5	78·5	77	78	81·5	83·5	84·5	80	76	70	70	76	77·8

NOTE.—The instruments in use were:—Max. Thermometer and Min. Thermometer, with numbers, but maker's name unknown. Dry and Wet Bulb Thermometers by G. Mason and Co., Glasgow. Barometer, maker's name unknown.

The mean of the barometric readings varies between 29·803 in January and 30·114 in July, the mean reading being 29·960.

The index errors of none of the instruments used is known.

The prevailing wind is from the south, and blows on 93 days in the year, compared with 58 from S.E. and 54 from S.W., 37 from E., 32 from N.W. and 31 from W. About 30 days are calm.

¹ 1891—1900.   ² The daily range is max.−min.

MOZAMBIQUE. Lat. 15° S., Long. 40° 44′ E. Alt. 6 feet (on the coast).

Compiled from 2 years' observations[1].

		Jan.	Feb.	March	April	May	June	July	Aug.	Sept.	Oct.	Nov.	Dec.	Year
Mean temperature[2] (Fahr.°)	...	81°.9	81°.7	83°.3	81°.1	78°.1	73°.8	73°.6	75°.0	76°.6	80°.1	82°.9	83°.8	79°.3
Mean maximum temperature	...	91°.7	89°.2	91°.8	88°.7	85°.6	81°.5	80°.8	82°.6	83°.3	86°.5	89°.8	90°.5	86°.8
Mean minimum temperature	...	79°.0	77°.4	78°.1	75°.6	73°.2	67°.8	68°.0	69°.8	72°.0	75°.4	77°.7	79°.5	74°.5
Mean daily range[3]	... ...	12°.7	11°.8	13°.7	13°.1	12°.4	13°.7	12°.8	12°.8	11°.3	11°.1	12°.1	11°.0	12°.3
Mean relative humidity (per cent.) 8 a.m.		85·5	88·0	86·5	85·0	86·5	82·5	85·0	82·0	82·5	79·5	78·0	80·0	83·4
″ ″ noon		71·0	75·0	66·5	67·5	68·0	65·5	68·5	64·0	67·5	67·5	67·0	69·0	68·1
″ ″ 8 p.m.		84·5	86·5	78·5	76·5	79·0	71·5	76·5	77·0	78·0	80·0	83·5	83·0	79·5
Mean rainfall (inches)	... ...	6·26	10·35	4·10	2·72	1·54	0·51	0·59	0·12	0·83	0	0·12	5·79	32·93

NOTE.—Mozambique is situated on a low coral island about a mile and a half long and some 400 yards broad. The country around is flat. The town is well kept. The island lies at the mouth of Mosuril Bay.

[1] 1900 and 1901 (except August).

[2] The figures from which the above table is compiled are given in the *Meteorologische Zeitschrift* (1904, p. 530) without any additional information; from internal evidence, however, it appears that the mean temperature is derived from $\dfrac{8a + 8p + \text{max.} + \text{min.}}{4}$. The mean for February 1900 is given as 37° C., but this is evidently an error and should be 27° C.

[3] The daily range is max. – min.

SOUTH AFRICA

Speaking generally of Southern Africa south of, say, 19° or 20° S., it may be said that the rainfall increases from west to east along any parallel, except in the extreme south of Cape Colony where irregularities exist, and in a portion of German South-West Africa, where there is a southerly intrusion of the higher rainfall area to the plateau in the neighbourhood of Windhoek, in which the precipitation is heavier than in the Kalahari. If a line be taken through Okahanja, in German South-West Africa, Bulawayo in Southern Rhodesia, and Beira in Portuguese East Africa (Gazaland), the records of the $9\frac{3}{4}$ years' mean annual rainfall for Okahanja, the 6 years' mean for Bulawayo and the 2 years' mean for Beira, are respectively 16, 22, and 65 inches (406·4, 558·8, and 1651 mm.). Again, on a line through Keetmanshoop (German S.W. Africa), Vryburg (Bechuanaland) and Lourenço Marques (Gazaland) there are means of 6, 8 and 10 years, giving the mean annual precipitation of 6·38, 22·38, and 28·23 inches (162, 568 and 717 mm.) respectively. Yet again, if three stations be taken still further south, the same conclusion is arrived at; at Port Nolloth the mean annual rainfall is 2·31 inches (59 mm.), at Kimberley it is 18·17 inches (461·5 mm.) and at Durban 42·46 inches (1080 mm.). This rule is true throughout any considerable areas, but is not so universally, the continuity of the gradations being, in many known instances, broken by the force of local circumstances. Thus it is true to say that, if three stations be taken at considerable distances apart, the rainfall will increase from west to east, but if any one station be taken, in all probability other stations may be found at no great distance to the east of it with lower annual rainfall and to the west with higher. In a less marked and less constant manner there is also an increase in rainfall from south to north; thus at Graaf Reinet, in Lat. 32° 16′ S., the mean annual rainfall is 15·29 inches (388 mm.), at Kimberley, in Lat. 28° 43′ S., it is 18·17 inches (462 mm.) and at Vryburg, in Lat.

$26°\ 55'$ S., it is $22\cdot38$ inches (568 mm.). Other examples may be found by consulting the rainfall tables given below for German S.W. Africa, Cape Colony, Orange Free State, Natal, the Transvaal, and Southern Rhodesia.

As with the rainfall, so with the temperature; it is found that there is a gradual increase along any parallel of Latitude, or in the neighbourhood of such parallel, from west to east. This is true in a general way, but the same limitation with regard to local circumstances also holds good. If a line be taken through Port Nolloth, O'okiep, Kimberley and Durban the following are the corresponding mean annual temperatures, namely, $57°\cdot5$, $63°$, $64°$ and $70°\cdot8$ Fahr. ($14°\cdot2$, $17°\cdot2$, $17°\cdot8$, and $21°\cdot5$ C.). There is also an increase of temperature from north to south along the West Coast, from west to east along the South Coast, and from south to north along the East Coast. The mean annual temperature at Walfisch Bay is $59°\cdot5$ Fahr. ($15°\cdot3$ C.), at Table Bay it is $61°\cdot3$ Fahr. ($16°\cdot3$ C.), but it is lower than either at Port Nolloth, viz. $57°\cdot5$ Fahr. ($14°\cdot2$ C.); at O'okiep it is $63°$ Fahr. ($17°\cdot2$ C.), further south, at Clanwilliam, $64°$ Fahr. ($17°\cdot8$ C.); and a comparison of the various tables given for South Africa will be found amply to bear out the other parts of the statement.

But from the peculiar terrace formation of Africa, the terraces rising one above the other, it so happens that almost throughout South Africa the mean temperature is nowhere far from $62°$ Fahr. ($16°\cdot7$ C.), the temperature, which would naturally increase with decrease in latitude, being modified by the corresponding rise in altitude. This, of course, is only true for the *mean* temperature; when the mean maximum and mean minimum temperatures are considered the result is widely different; for instance, the mean daily range at Cape Agulhas is only $10°\cdot4$ Fahr. ($5°\cdot8$ C.), while at Bloemfontein it is as much as $25°\cdot7$ Fahr. ($14°$ C.), and there are other stations where the range is still greater, the mean daily range for the whole of South Africa being $24°\cdot5$ Fahr. ($13°\cdot5$ C.).

The peculiarly marked difference between the conditions on the east and west coasts, alluded to above, has been accounted for by the warm Mozambique current washing the eastern shores in its southerly course, while the cold Benguela current in its northerly course performs an exactly contrary office for the western shores. The first gradually loses its warmth, while the second rises in temperature as it nears the equator.

Frozen water, in regions anywhere near the coast, is, in South Africa, an almost unknown phenomenon; but, in the highlands of the interior, frosts are not only of very frequent occurrence, as a reference to the absolute minima of temperature in the tables (below) will show, but have been known to occur in almost any month, the hottest months, December, January and February, not being entirely immune, though of course these frosts are of more frequent occurrence in the winter months, and especially in July, which is, almost throughout the country, the coldest month.

It often snows in localities situated at altitudes exceeding 3000 feet above sea-level, but snow is most frequently seen on the mountains, and, near the coast, even this is of infrequent occurrence. During a residence extending, off and on, over fifteen years, the writer only once saw snow on Table Mountain.

An examination of almost any one of the annual reports of the Meteorological Commission shows clearly that the greatest number of thunderstorms is reported in February, when, as a rule, the temperature is greatest; and it will be found that, speaking generally, the summer months are the thunderstorm months, while fewest storms are reported in June.

At present a pretty little controversy is raging as to the connection between the winds and the rains in Southern Africa. The old and long-accepted theory is receiving violent shocks and must either fall or, at any rate, receive considerable modifications[1]. This is so important a matter that a brief statement of the case appears to be desirable. The old theory is as follows: With the exception of the Cape Peninsula, which receives its rains with north-west winds, the remainder of South Africa is served in this respect by the south-east and south-west winds, acting

[1] *Science in South Africa.* Article on Meteorology by C. M. Stewart. Cape Town, 1905, pp. 36, 37.

as complements the one to the other. The rains borne by the south-east winds are at a minimum in August and are practically confined to the coast, while the rains brought by the south-west winds at this time of year are at a maximum. From August onwards there is a steady increase in the intensity of the south-east borne rains, and they advance further and further inland until January, when they attain their maximum, and the rains reach almost to the west coast. During this period the rains which come with the south-west winds have been gradually decreasing in intensity and receding before the advance of south-east rains, until in January they are at a minimum and confined practically to the extreme south coast. Again, from January the south-east borne rains decrease in intensity and recede towards the east until they reach a minimum in July and are confined to small areas near the south-east coast. During this period the rains which come with the south-west winds have been advancing towards the east and increasing in intensity until the maximum is again reached in August, so that the rains brought by the south-east and south-west winds are, so to speak, the converse of each other; as one advances the other recedes. And, further, it appears that the south-east rains fall in summer and the south-west rains in winter, so that the summer rains advance from the south-east, beginning about August, and culminating, as regards both scope and intensity, at midsummer, *i.e.* in January, when they steadily recede and decrease in intensity until August is again reached. With regard to the winter rains, the case is exactly the reverse. They advance from the south-west, beginning in January and increasing in intensity until mid-winter is reached in August, when they are at a maximum. There is a steady falling off until the minimum is again reached in January. Thus for Southern Africa the year may be divided as follows: the wettest six months for places served by the south-west rains are from April to October; and these places therefore have the wet season in winter. On the other hand the wettest six months for places served by the south-east rains are from November to May, these places having the wet season in summer.

That is the accepted theory in its most general form. But now, however, it appears to be established beyond question that, at Durban, the greater portion of the rain is brought by south-west winds, while south winds bring more than those from any of the remaining points of the compass. At East London, which was supposed to receive the greater portion of its rainfall from the S.E. Trades, it is now shown that the north-east winds perform this office. Port Elizabeth, which was also supposed to be favoured by these "trades," receives no rain with the wind in that direction, but derives almost the whole of its rain supply from west-south-west winds. At Cape Town the rain wind is not absolutely north-west, but north-north-west, while at Port Nolloth and O'okiep such rain as falls comes almost wholly with north and north-west winds; and further north, where the prevalent winds on the coast are, in both winter and summer, from the west and south-west, at Windhoek and in the interior of German S.W. Africa generally, the rain is invariably brought by winds in the east quadrant, mainly east, but also north-east and south-east[1].

These examples are all on the coast, but, if we go to the interior, similar evidence, contradictory of the old theory, is forthcoming. At Maclear, in the west of Griqualand East, the rains almost invariably come with an east wind. At Kimberley the heavy thunder-clouds and rain come from the west and north-west, while other rain-clouds and lighter storms come from the north-east and east, and no rain at all from the south-west. At Hopefontein, near Bulawayo, the prevailing wind is from the south-east, but changes to south-west before rain.

[1] *Science in South Africa*, pp. 36, 37.

SOUTHERN RHODESIA.

Southern Rhodesia, of course, forms part of the great South African plateau, and has an elevation of between 3500 and 5000 feet above sea-level. Across it extends a broad highland belt, forming the water-parting between the Zambezi, and the Limpopo and Sabi rivers, and containing such ranges as the Matoppo Hills. The whole country is a land of hills and dales and upland pastures, and lies well within the tropics.

It would be difficult to find a finer climate than that enjoyed by the highlands of Southern Rhodesia. The hottest time of the year is immediately before the rains, but the country reaps both the advantages and disadvantages of being in the region of the "summer rains." These rains moderate the heat, but are bad for many crops, which require early spring rains and a fine autumn. Though well within the tropics, the climate is anything but tropical, mainly, of course, owing to elevation, but also partly in consequence of this effect of the rains.

It is only in the low-lying regions that malaria is at all prevalent, and, with ordinary precautions, this need not be greatly dreaded, for the type of fever is not malignant.

There are but two seasons, summer and winter. The former lasts from October to March, and the latter, which is also the dry season, from April to September. In the latter season the nights are cold, and frosts are experienced in the coldest months, namely June, July and August, when warm clothes are appreciated, but the days are warm, bright, and sunny. No heavier frost than $2°$ Fahr. below freezing point has been experienced, and that in the low-lying regions. Snow is a thing unknown.

The mean annual rainfall varies from 15 inches (381 mm.) at Tuli and 22 inches (559 mm.) at Bulawayo to 28 inches (711 mm.) at Gwelo, 34 and 35 inches (863 and 889 mm.) at Salisbury and Umtali respectively, and over 43 inches (1092 mm.) at Melsetter, the average over the whole country being about 32 inches (813 mm.). Less winter rain falls in Southern Rhodesia than in any other part of South Africa, 93 per cent. of the precipitation occurring in summer.

The rains generally begin at the end of September or early in October and are followed by a spell of fine weather, to recommence in November and continue with occasional fine spells until the end of March or April. The average rainfall for the four years 1899–1902 in Mashonaland was 37·6 inches (955 mm.) and in Matabeleland 23·2 inches (589 mm.).

Reckoning from north to south, the mean annual temperature at Salisbury is $64°·4$, at Gwelo $64°·8$, at Bulawayo $67°·0$ and at Tuli $71°·3$ Fahr. ($18°$, $18°·2$, $19°·4$ and $21°·8$ C.). Similarly the mean temperature of the warmest month increases from $70°·2$ Fahr. ($21°·2$ C.) at Salisbury to $71°·3$ Fahr. ($21°·8$ C.) at Gwelo, $72°·5$ Fahr. ($22°·5$ C.) at Bulawayo and $80°·7$ Fahr. ($27°·1$ C.) at Tuli; and it is also found that the mean temperature of the coldest month, in the same way, increases from $56°·2$ Fahr. ($13°·4$ C.) at Salisbury to $58°·7$ Fahr. ($14°·8$ C.) at Tuli. The mean daily range of temperature is greater, throughout the country, in winter than in summer, but is less in the middle and north than in the extreme south.

The averages of the readings shown on the maximum and minimum thermometers in the shade at Salisbury, Bulawayo and Umtali for a period of nine years are given below:

		1897/8	1898/9	1899/0	1900/1	1901/2	1902/3	1903/4	1904/5	1905/6
Salisbury,	max.	77°·59	75°·47	75°·38	76°·3	74°·6	77°·5	76°·6	76°·0	78°·0
„	min.	53°·46	52°·57	53°·84	54°·7	52°·4	53°·8	53°·1	52°·0	53°·5
Bulawayo,	max.	79°·9	78°·66	86°·55	81°·7	78°·3	80°·0	78°·7	77°·7	79°·3
„	min.	55°·8	54°·96	52°·56	56°·1	54°·9	55°·2	54°·3	53°·8	54°·4
Umtali,	max.	72°·5	77°·71	76°·09	78°·2	76°·1	78°·3	77°·3	76°·9	78°·1
„	min.	50°·4	48°·22	49°·43	51°·5	48°·9	47°·0	53°·7	55°·7	55°·8

Comparing the extreme ranges of temperature, which will be mentioned below in the notes on the various stations, it is found that the difference between the highest and lowest recorded figures at the following places is

Empandeni	...	87° Fahr.	30°·5 C.
Tuli	...	84°	28°·9
Victoria	...	81°	27°·2
Gwelo	...	69°·3	20°·7
Hopefontein	...	65°·7	18°·7
Salisbury	...	58°	14°·4

The following is the comparative statement of the difference between the mean minima on grass and in the screen four feet above the ground at Salisbury and Bulawayo for 1906 and 1907:

Station	Year	Jan.	Feb.	March	April	May	June	July
Salisbury,	1906 ...	$1°·9$	$+1°·6$[1]	$0°·9$	$4°·5$	$6°·7$	$7°·6$	$7°·6$
„	1907 ...	$3°·0$	$1°·5$	$2°·7$	$3°·0$	$5°·1$	$6°·2$	$5°·4$
Bulawayo,	1906 ...	$2°·7$	$1°·3$	$2°·6$	$7°·8$	$8°·6$	$9°·3$	$7°·1$
„	1907 ...	$2°·3$	$1°·4$	$2°·5$	$2°·6$	$5°·6$	$5°·9$	$4°·4$

Station	Year	Aug.	Sept.	Oct.	Nov.	Dec.	Year
Salisbury,	1906 ...	$7°·1$	$4°·6$	$3°·2$	$2°·5$	$3°·2$	$4°·0$
„	1907 ...	$4°·8$	$4°·0$	$2°·6$	$1°·7$	$1°·5$	$3°·4$
Bulawayo,	1906 ...	$7°·3$	$6°·8$	$4°·1$	$3°·4$	$3°·1$	$5°·3$
„	1907 ...	$7°·7$	$4°·7$	$3°·4$	$2°·2$	$0°·2$	$3°·5$

The air temperature was invariably above the freezing point, but there were a few instances of ground frosts.

It would appear that the early rains and the late rains are due to different causes, for, while the former are general throughout the whole of Southern Rhodesia, the latter are far greater in the east than in the west, and are probably due to cyclones from the Indian Ocean.

---

[1] Sign + indicates excess of grass temperature over screen temperature.

MASHONALAND.

Salisbury, in Lat. $17° 48'$ S. and Long. $31° 5'$ E., is situated at an altitude of 4880 feet above sea-level, and the mean of the barometric readings varies between $25\cdot428$ inches and $25\cdot210$ inches. It is always highest in June or July and lowest in January or February. This result and those which follow are derived from observations taken during five successive years. The mean maximum temperature was highest in October in four of the years (in one of which the same maximum was reached in January) and in December in the remaining year; the mean readings were $84°\cdot4$, $82°\cdot8$, $84°\cdot8$, $78°\cdot7$ and $83°\cdot3$ Fahr. ($29°\cdot1$, $28°\cdot2$, $29°\cdot3$, $25°\cdot9$ and $28°\cdot5$ C.); the mean minimum temperature was always lowest in June or July and the corresponding figures were $40°\cdot4$, $45°\cdot2$, $45°\cdot1$, $42°\cdot4$ and $45°\cdot2$ Fahr. ($4°\cdot7$, $7°\cdot3$, $7°\cdot2$, $5°\cdot8$ and $7°\cdot3$ C.). The absolute maximum temperature was $88°\cdot4$ Fahr. ($31°\cdot3$ C.) (November), in 1901 and $90°\cdot3$ Fahr. ($32°\cdot4$ C.) (September) in 1902, but $92°\cdot6$ Fahr. ($33°\cdot6$ C.) has been registered; the absolute maximum was lowest in 1901 in June ($75°\cdot0$ Fahr. or $23°\cdot9$ C.) and in 1902, also in June ($77°\cdot5$ Fahr. or $25°\cdot3$ C.). The absolute minimum was lowest in both years in June, the thermometer readings being $34°\cdot8$ and $34°\cdot6$ Fahr. ($1°\cdot5$ and $1°\cdot4$ C.). The prevailing wind is sometimes between N. and N.W. in December, January and February, but otherwise it is invariably in the quadrant between N.E. and S.E., but chiefly from E.

The total rainfall is very uncertain. For the five consecutive years it was $21\cdot74$, $39\cdot60$, $42\cdot03$, $40\cdot92$ and $27\cdot01$ inches (552, 1006, 1067, 1039 and 686 mm.) respectively. The rainy season embraces the five months November to March, the heaviest fall is usually in January and February, and there are rains in April and October amounting on the average to an inch. July is rainless and August almost so. In September there were measured, in 1901, as much as $3\cdot08$ inches (78 mm.), but excepting that year, the month may be classed with June, July, and August. The mean temperature is fairly regular, but 1898 was a cool year and reduces all the means for the months throughout, as far as *mean* temperature is concerned. The first rain storm of the wet season generally occurs in October, but the heavy rains do not really begin till about the middle of November. The unpleasant hot winds, so frequent in other parts of South Africa, are unknown at Salisbury.

Umtali is situated in Lat. $18°\,45'$ S. and Long. $32°\,41'$ E., at an altitude of 3700 feet, or roughly 1100 feet lower than Salisbury, and the mean of the barometric readings varies between $26\cdot601$ inches in the middle of the year and $26\cdot377$ in the early part; at least that is the result derived from two years' observations. The figures may not therefore be quite those which would be given by a longer series of observations, but there seems to be no doubt that the barometer is highest in June and July and lowest in January and February.

The mean maximum temperature was lowest in June in each of the four years from which the results were obtained, and it was over $80°$ Fahr. ($26°\cdot7$ C.) from October to April. The mean minimum temperature was lowest in June in three of the years and in May in the fourth. The absolute maximum in 1902 was $95°$ Fahr. ($35°$ C.), and in 1901 $91°$ Fahr. ($32°\cdot8$ C.).

The greatest rainfall occurs in the period from November to March, but the precipitation is by no means regular. In 1899 no less than 21 inches (533 mm.) fell in February, and in the next year only $1\cdot47$ inches (37 mm.); in 1891 $8\cdot57$ inches (218 mm.) fell in December, and in the following year that month was completely rainless; so that the monthly means during the rainy season do not give a fair clue to what may be expected in any one month.

The prevailing winds are almost entirely from the east, though in the middle of the year they are frequently from the south-east or east-south-east.

Melsetter stands at an altitude of 5500 feet above sea-level, in Lat. $20°\,0'$ S. and Long. $32°\,44'$ E., and the mean of the barometric readings appears to vary between $24\cdot992$ inches and $25\cdot162$ inches. The mean annual temperature is $64°\cdot2$ Fahr. ($17°\cdot9$ C.). The warmest months are from September to March, November having the highest mean temperature, and also the highest mean maximum temperature. The highest absolute maximum temperature registered appears to be a little over $87°$ Fahr. ($30°·5$ C.). The dryest months are May, June, July and August, and the rainiest, November, December, January and February. The rainfall in November appears to be fairly regular, the mean being $4·67$ inches (118 mm.) for that month, and there is generally the heaviest precipitation in one of the other three months. In one year December had a fall of 10 inches (254 mm.) and in another only $1·03$ inches (26 mm.); in another year January had a fall of $12·50$ inches (318 mm.), with about 8 and 9 inches (200 and 230 mm.) in other years; and in yet another year nearly 17 inches (432 mm.) fell in February. The means, therefore, are not reliable for the actual months, though the season November to February generally is accountable for about three-quarters of the rainfall of the year. The total rainfall cannot be relied upon from year to year, there being as great a difference as 22 inches (559 mm.) between two consecutive years.

At Victoria, situated in Lat. $21°\ 5'$ S., Long. $31°\ 3'$ E., at an altitude of 3200 feet, the highest recorded temperature is $110°·4$ Fahr. ($43°·5$ C.), which was read in May 1904, and the lowest $29°$ Fahr. ($-1°·7$ C.) in June 1903 and July 1902, so that the extreme range amounts to $81°·4$ Fahr. ($45°·2$ C.) for the period for which the table (p. 405) is compiled; the thermometer, however, frequently reaches $98°$ and $99°$ Fahr. ($37°$ C.) in the summer months. The mean maximum temperature varies between $88°$ Fahr. or $31°$ C. (October 1904) and $69°$ Fahr. or $20°·5$ C. (June 1901); and the mean minimum between $37°·3$ Fahr. or $2°·9$ C. (July 1902) and $64°$ Fahr. or $17°·8$ C. (November 1904). The heaviest rains occur in November, December, January, February, and March, though not always in the same month in different years, $9·62$ inches (244 mm.) being recorded in December 1901, though only $2·15$ inches (55 mm.) fell in that month in the following year. Again, $5·76$ inches (146 mm.) were recorded in November 1902 and only $1·36$ inches (35 mm.) in 1904. January, however, appears to have a large total every year, ranging between $5·97$ and $8·02$ inches (151 and 204 mm.). In 1902 no rain fell in the five months April to August, and in July and August no rain was recorded in any of the years. The amount which falls on a rainy day varies greatly, for instance in June 1901 it rained on 7 days, but the total only amounted to ·33 inch (8 mm.), whereas, in January 1904, 8 inches (203 mm.) of rain fell in 18 days, and 7·23 inches (183 mm.) in 14 days in January 1903.

NOTE. All stations of the second class are provided with a barometer, a hygrometer (dry and wet bulb thermometers), a maximum and minimum thermometer, a rain gauge with measure, and a Stevenson screen.

SALISBURY. Lat. 17° 48′ S., Long. 31° 5′ E. Alt. 4880 feet.

Means compiled from 5 years' observations[1] as given in *I. Tropical Africa.* See note 1 under Umtali.

	Jan.	Feb.	March	April	May	June	July	Aug.	Sept.	Oct.	Nov.	Dec.	Year
Mean temperature at 9 a.m. (Fahr.°)	67°·7	66°·3	65°·6	64°·9	64°·9	56°·2	56°·7	60°·7	66°·8	70°·2	69°·5	69°·6	64°·9
Mean maximum temperature	78°·3	76°·1	76°·8	76°·3	72°·6	68°·3	69°·0	73°·5	78°·9	81°·9	79°·5	77°·7	75°·7
Mean minimum temperature	60°·3	59°·1	58°·9	54°·9	48°·2	44°·3	43°·7	46°·3	52°·5	56°·7	59°·9	59°·2	53°·7
Mean temperature $\frac{\text{max.}+\text{min.}}{2}$	69°·3	67°·6	67°·8	65°·6	60°·4	56°·3	56°·3	59°·9	65°·7	69°·3	69°·7	63°·5	64°·7
Mean daily range[2]	18°·0	17°·0	17°·9	21°·4	24°·4	24°·0	25°·3	27°·2	26°·4	25°·2	19°·6	18°·5	22°·0
Absolute maximum (1901[3])	84°·4	81°·0	82°·1	80°·6	79°·8	75°·0	75°·4	79°·6	82°·2	85°·4	88°·4	85°·0	—
,, (1902)	82°·4	85°·4	78°·8	84°·0	77°·5	75°·0	75°·8	84°·0	90°·3	86°·2	89°·8	—	—
,, (1907)	86°·5	78°·4	82°·5	80°·5	76°·6	79°·6	72°·6	80°·3	86°·2	88°·4	87°·4	82°·5	—
Absolute minimum (1901)	56°·2	53°·5	52°·8	47°·5	37°·8	34°·8	36°·5	39°·0	47°·2	46°·6	51°·6	54°·5	—
,, (1902)	57°·6	52°·2	57°·5	52°·0	42°·0	34°·6	39°·2	42°·4	41°·6	42°·5	55°·4	52°·1	—
,, (1907)	54°·0	57°·5	49°·5	45°·5	41°·2	35°·2	34°·0	34°·7	41°·5	46°·2	54°·0	57°·4	—
Mean relative humidity[4] (per cent.)	82	81	86	81	72	68	61	69	57	54	70	72	71

WIND.

	Jan.	Feb.	March	April	May	June	July	Aug.	Sept.	Oct.	Nov.	Dec.
1898	SE—NW	SE—NNE	E	E—SE	E	SE—NE	E	SE—E	E—SE	E	N—SE	NNW—SSE
1899	E	NE—SW	ENE—SSE	E	E	E	E	E	E	E	E	E
1900	E	E	E	NE	ESE	SE,E	NE	NE	ENE	ENE	NE,E	NE,E
1901	E	E	E	ESE	ESE	SE,E	ESE	E	E	NE	NE	NE
1902	E	ESE	NE	NE	SE	NE—E	SE—NE	NE—E	NE	NE	NE	NE

NOTE.—Salisbury is what is known as a Barometric Station in Rhodesia, *i.e.* it is supplied with a Barometer in addition to the usual thermometers.

[1] 1898—1902.   [2] The daily range is max.—min.

[3] The absolute maxima and minima are wanting for 1898—1900.

[4] The relative humidity in 1907, which was a very rainy year, was much below that represented by these means. In the first four months of the year it was only 62, 79, 70 and 63 per cent. respectively, while in August it was only 40 per cent., in October 46 and in June 49.

UMTALI. Lat. 18° 45′ S., Long. 32° 41′ E. Alt. 3700 feet.

Compiled from 4 years' observations[1].

	Jan.	Feb.	March	April	May	June	July	Aug.	Sept.	Oct.	Nov.	Dec.	Year
Mean temperature (8 a.m.) (Fahr.°) ...	70°·8	68°·8	69°·0	67°·1	61°·9	56°·6	56°·8	63°·6	64°·5	67°·9	69°·8	70°·6	65°·5
Mean maximum temperature ...	80°·2	80°·5	81°·8	81°·3	74°·5	68°·4	70°·0	73°·3	77°·6	80°·8	80°·5	80°·1	77°·4
Mean minimum temperature ...	58°·0	51°·9	52°·6	48°·4	43°·9	41°·7	42°·2	44°·4	47°·8	51°·2	53°·5	55°·3	49°·2
Mean temperature $\frac{max.+min.}{2}$ ...	69°·1	66°·2	67°·2	64°·9	59°·2	55°·1	56°·1	58°·8	62°·7	66°·0	67°·0	67°·7	63°·3
Mean daily range[2] ...	22°·2	28°·6	29°·2	38°·9	30°·6	26°·7	27°·8	28°·9	29°·8	29°·6	27°·0	24°·8	28°·2
Mean of absolute maxima of temperature[3]	89°	91°	86°·5	88°·5	86°·5	85°·5	79°	85°	90°·5	91°	91°	88°·5	87°·7
Mean number of rainy days[4] ...	18	14	15·5	5	2·5	4·6	1·6	0·3	5·6	5·3	12	14	98·4

NOTE.—Umtali is what is known in Rhodesia as a Barometric Station, *i.e.* it is supplied with a barometer in addition to the usual thermometers.

[1] 1899—1902. Compiled from *Meteorological Office: Climatological Observations at Colonial and Foreign Stations, I. Tropical Africa,* 1904. The observations are wanting for January, October and November 1899.
[2] The daily range is max. – min.
[3] For the years 1901 and 1902.
[4] Number of rainy days wanting for 1899 and the first 3 months of 1900. The prevailing wind throughout the year is from the East.

26—2

MELSETTER. Lat. 20° 0′ S., Long. 32° 44′ E. Alt. 5500 feet.

Compiled from 5 years' observations[1].

	Jan.	Feb.	March	April	May	June	July	Aug.	Sept.	Oct.	Nov.	Dec.	Year
Mean temperature at 8 a.m. (Fahr.°) ...	68°·6	67°·3	66°·5	65°·3	59°·8	54°·9	56°·1	60°·1	64°·2	67°·2	71°·1	70°·0	64°·2
Mean maximum temperature	74°·1	72°·2	71°·4	71°·3	66°·9	62°·5	63°·7	68°·1	71°·0	73°·4	75°·7	74°·2	70°·4
Mean of the absolute maxima of temperature[2]	81°·5	80°·7	76°·5	79°·5	74°·0	73°·1	73°·6	76°·0	83°·0	83°·5	82°·2	82°·3	78°·8
Mean relative humidity (per cent.) ...	79	83	80	66	64	66	61	52	54	53	63	65	66

NOTE.—Melsetter is supplied with the usual thermometers, as supplied in the Rhodesian meteorological service.

[1] 1900—1904.

[2] The absolute maxima for 1900 are wanting.

VICTORIA. Lat. 20° 21′ S., Long. 31° 3′ E. Alt. 3200 feet.

Compiled from 4 years' observations[1].

	Jan.	Feb.	March	April	May	June	July	Aug.	Sept.	Oct.	Nov.	Dec.	Year
Mean temperature at 8 a.m. and 8.30 a.m. combined[2] (Fahr.°)	73°·4	72°·8	71°·3	72°·7	64°·5	59°·7	58°·3	62°·8	66°·8	70°·4	72°·7	71°·7	68°·1
Mean maximum temperature ...	83°·1	82°·0	80°·9	82°·6	76°·7	73°·0	72°·4	78°·9	80°·8	84°·1	84°·2	82°·4	80°·1
Mean minimum temperature ...	62°·7	60°·3	59°·0	55°·1	45°·1	40°·5	39°·4	44°·3	52°·4	56°·0	62°·0	61°·9	53°·2
Mean temperature[3] ...	72°·9	71°·1	69°·9	68°·8	60°·9	56°·8	55°·9	61°·6	66°·6	70°·1	73°·1	72°·2	66°·6
Mean daily range[4]	20°·4	21°·7	21°·9	27°·5	31°·6	32°·5	33°·0	34°·6	28°·4	28°·1	22°·2	20°·5	26°·9
Mean of absolute maxima[5]	93°·3	92°·0	90°·0	92°·2	96°·3	86°·0	85°·2	88°·3	95°·0	99°·0	96°·5	97°·0	—
Mean of absolute minima[5]	55°·3	52°·7	48°·3	45°·3	34°·0	31°·5	30°·0	35°·5	43°·0	42°·0	56°·0	49°·0	—
Average monthly range[6] ...	38°·0	40°·3	41°·7	46°·9	62°·3	54°·5	55°·2	52°·8	52°·0	57°·0	40°·5	48°·0	—
Relative humidity (per cent.) (2 years, 6 months[7])	69	64	70	60	59	64	67	61	59	61	71	60	64
Mean number of days on which rain was measured	16	4	11	3	1	2	0	0	1	4	11	12	65

NOTE.—Victoria is one of the meteorological stations in the Rhodesian service and is supplied with the usual thermometers.

[1] 1901—1904.
[2] Derived from 1902 at 8 a.m. and 1903 and 1904 at 8.30 a.m.
[3] The mean temperature is derived from $\dfrac{\text{max.} + \text{min.}}{2}$.
[4] The daily range is max. − min.
[5] The absolute maxima and minima are wanting for the year 1901.
[6] The monthly range is absolute maximum − absolute minimum for the month.
[7] For 1902, 1903 and 6 months of 1904.

MATABELELAND.

At Gwelo, situated in Lat. $19°27'$ S. and Long. $29°49'$ E., at an altitude of 4645 feet above sea-level, the morning temperature in mid-winter is a couple of degrees higher than at almost any other station. This mean morning temperature varies between $77°$ Fahr. recorded for October 1900, and $55°·1$ Fahr. for June 1901 ($25°$ and $12°·8$ C.). The mean maximum temperature was highest in November 1903, viz. $83°·9$ Fahr. ($28°·8$ C.) ($83°·8$ Fahr. or $28°·77$ C. in December the previous year), while the lowest recorded was $67°$ Fahr. ($19°·4$ C.) for June 1901. The lowest mean minimum recorded was $39°·2$ Fahr. ($4°$ C.) in July 1901, and the highest $61°·7$ Fahr. ($16°·5$ C.) in December the same year. The absolute maximum reading generally occurs in October or November, the highest recorded, in the period for which the table below was compiled, was $97°·4$ Fahr. ($36°·3$ C.) in October 1904, but in 1902 the highest reading occurred in September, when $93°$ Fahr. ($33°·8$ C.) was recorded. The lowest temperature recorded was $27°·7$ Fahr. ($-2°·3$ C.) in June 1901, and the thermometer also touched the freezing point in May and July; while in 1904 the lowest readings were $31°·5$ and $31°·0$ Fahr. ($-0°·28$ and $-0°·56$ C.) in June and July respectively; in the other years the freezing point was not reached. The extreme range of temperature is, therefore, $69°·3$ Fahr. ($38°·5$ C.).

The mean annual rainfall at Gwelo varies considerably from year to year, and the monthly means are by no means regular. A little over 19 inches (482 mm.) was the total precipitation in 1903, while over $40·5$ inches (1029 mm.) were recorded in 1901. The greatest fall in any one month was $15·58$ inches (396 mm.) in December 1901, while in the same month in 1902 only $2·35$ inches (60 mm.) fell; $10·14$ and $9·56$ inches (258 and 243 mm.) were recorded for January 1904 and 1900 respectively. The $15·58$ inches (396 mm.), mentioned above, fell in 24 days. The number of rainy days in a year is between 50 and 70, but, in 1901, 104 were recorded. The prevailing wind at Gwelo is south-east almost throughout the year, but N.N.E. is recorded as the prevailing direction for September and November 1902, and E.S.E. for November 1901.

At Bulawayo, situated in Lat. 20° 9′ S. and Long. 28° 35′ E., at an altitude of 4469 feet above sea-level, the hottest months are from October to February, the most oppressive heat occurring before the rains set in. In October the mean temperature varies between 76° and 68°·5 Fahr. (24°·5 and 20°·3 C.), and in November between 74° Fahr. (23°·4 C.) and the same lower limit. In December and January the mean temperature has almost the same extremes. In some years September is also hot, notably in 1902, when the highest mean annual temperature (68°·3 Fahr. or 20°·2 C.) was recorded, and September was the hottest month of the year. The lowest mean annual temperature, during the period for which the table below was compiled, occurred in 1904, viz. 65°·2 Fahr. (18°·5 C.). September, however, as a rule, and April, may be regarded as transition months. The winter is the dry season, and June and July are both the coldest months and the dryest months, in fact they are rainless, and May, August, and September almost so, though, in some years, there may be slight showers. Frost is a thing unknown. The dryest year was 1899, when only 17·91 inches (455 mm.) of rain fell, while in 1901 the total precipitation for the year amounted to 29·43 inches (748 mm.), the greatest fall known at Bulawayo. The heaviest rainfall for any one month in the period of seven years, for which the table below is compiled, was 9·69 inches (246 mm.) in December 1901, followed by 8·30 inches (211 mm.) in January 1898, and 8·15 inches (207 mm.) in December 1900.

During the dry season the prevailing winds are from E., E.N.E., or E.S.E.; in the rainy season they either back to north or come from the west. If the wind backs from east to north or north-west, the air always becomes oppressive, the barometer at once sinks, and one or two days afterwards the weather is disturbed, and one of two results is sure to follow. Either a dust storm rises, or the sky becomes cloudy and it rains, or threatens to rain. It is a curious and noticeable fact that, during certain hours in the morning, no rain has been recorded at Bulawayo throughout the whole period of nine years, during which observations have been made.

At Hopefontein, Lat. 20° 12' S. and Long. 28° 48' E., the mean temperatures for November, December, January, and February are almost identical, being a trifle over 71° Fahr. (21°·6 C.). The mean maximum temperature is highest just before the rains in October, and the mean minimum is lowest in July. The absolute maximum is registered, as a rule, in either October or November, and the absolute minimum in June or July. The extreme range of temperature, during the four years' period for which the table below is compiled, was 65°·7 Fahr. (36°·5 C.), being the difference between the extreme maximum (97°·7 Fahr. or 36°·5 C.) read in November 1904, and the extreme minimum (32° Fahr. or 0° C.) recorded in June 1901, when the thermometer just touched the freezing point. The mean annual rainfall, as deduced from the figures for fifteen years is 27·86 inches (708 mm.), which is lower by one inch than that given by the first ten years of the period, namely, 28·86 inches (733 mm.). This is brought about by the exceedingly small precipitation recorded for the year 1905, namely, 16·81 inches (427 mm.). In fact the last three years of the fifteen-year period, viz. 1902–4–5 (the figures for 1903 being wanting) were all poor in rainfall, and considerably below the general average. In December 1902 only ·36 inch (9 mm.) was recorded, as against the average of 4·79 inches (122 mm.), and in January 1905 only 2·56 inches (65 mm.), the average for the month being 8·05 inches (205 mm.). The greatest fall in any one month, during the last five years of the period, was 11·24 inches (286 mm.) in January 1900. As a rule the heavy rains are preceded by a thunderstorm, with a sudden downpour, succeeded by a gentle wind. In January and February there is sometimes a continuous downpour for several days without thunder. The general direction of the wind is S.E., changing to S.W. before rain.

Empandeni, in Lat. 20° 39' S. and Long. 27° 45' E., is situated at an altitude of 4500 feet above sea-level, or at about the same elevation as Bulawayo and Gwelo. The mean temperatures for November, December, January, and

February are all over 72° Fahr. (22°·2 C.), the first being the hottest month, with a mean maximum of 85°·4 Fahr. (29°·7 C.). The absolute maxima, however, in the five-year period, for which the table below is compiled, occur once in September, twice in October (once with a similar reading in February), once in March, and once, the extreme maximum recorded, viz. 110° Fahr. (43°·3 C.) in April. The absolute minima are usually recorded in June, though in one year (1904) there was a similar reading in September. All these absolute minima are below freezing point, and the thermometer sinks below that point every year, the extreme minimum being 23° Fahr. (−5° C.) or 9° Fahr. of frost. The extreme range for the period is thus very large, namely, 87° Fahr. (48°·3 C.), *i.e.* the difference between 110° Fahr. recorded in April 1900, and 23° Fahr., the reading in June 1901. The winter of 1900 was, on the whole, severe, and the mean minimum for June was only just over the freezing point, viz. 32°·2 Fahr. (0°·11 C.). The mean monthly range for this month is larger than for any other month, viz. 56° Fahr. (31°·1 C.), and the mean daily range is 33°·3 Fahr. (18°·5 C.), only exceeded by the months of July and August. The mean annual rainfall, in the period under consideration, varies between 19·5 inches and 23 inches (495 and 584 mm.), but the monthly records are not so consistent. In January 1902, no less than 10·20 inches (259 mm.) of rain were recorded, while the total for the same month in 1901 was only 2·37 inches (60 mm.), the former figure being the highest record for any one month. In some years the period from April to September is absolutely rainless.

Tuli, in Lat. 21° 52' S. and Long. 29° 12' E., is situated on the Shashi river, about 24 miles above its junction with the Limpopo, and is therefore in the valley of the latter. The altitude is only 1750 feet above sea-level, and Tuli therefore is by far the lowest of the stations we have been considering. Indeed it has less than half the altitude of Umtali and only a little more than half that of Victoria, while it lies nearly 3000 feet below Hopefontein and more than 3000 feet below Salisbury. The climate is therefore more tropical. The absolute maxima of temperature from September to March are, most frequently, considerably over

100° Fahr. (37°·8 C.). In September 1902 the thermometer reached 106° Fahr. (41°·1 C.), and, in 1904, 99° Fahr. (37°·2 C.) was recorded; in October the maxima for the four years 1901–1904 were 98°, 102°, 106°, 106° Fahr. (36°·7, 38°·9, 41°·1, 41°·1 C.); in November they were 103°, 106°, 101°, 105° Fahr. (39°·4, 41°·1, 38°·3, 40°·5 C.), and in December 100°, 107°, 108°, 102° Fahr. (37°·8, 41°·7, 42°·2, 38°·9 C.). In January 106° Fahr. (41°·1 C.) was registered in 1901, and 105° Fahr. (40°·5 C.) in 1903; in February the highest readings were 103° Fahr. in 1901, 102° Fahr. in 1902, and 104° Fahr. in 1903 (39°·4, 38°·9, 40° C.); in March 1903 the record showed 104° Fahr. (40° C.), and finally 100° Fahr. (37°·8 C.) was the maximum reading for April in both 1902 and 1903. The monthly range throughout the year was never less than 44° Fahr. (24°·4 C.) (the range for January), and was as much as 58°·5 Fahr. (32°·5 C.) in May, and in the same month the daily range was 38°·6 Fahr. (21°·4 C.). The annual rainfall varied between 10·43 inches in 1903 and 18·94 in 1901 (265 and 481 mm.). The greatest precipitation in any one month was 6·21 inches (158 mm.) in January 1901.

PRODUCTIONS, &c.

The climate of Southern Rhodesia—and, it may be stated, the soil as well—is well suited for the growth of wheat, oats, and barley, as well as rye and lucerne, and also for all European vegetables. Mealies, Kafir corn, and other native grains thrive almost anywhere. It is a suitable climate for tobacco and also for cotton, the temperature being sufficiently high. Southern Rhodesia is not without the limits of rubber growth, except perhaps in the south-west, and many fibre plants, such as ramie, sisal-hemp, flax and jute should do well. Such fruits as figs, oranges, peaches, almonds, walnuts, lemons, bananas, quinces, apples, apricots and pomegranates have already been tested and found to produce good results.

GWELO. Lat. 19° 27′ S., Long. 29° 49′ E. Alt. 4645 feet.

Compiled from 5 years' observations[1].

	Jan.	Feb.	March	April	May	June	July	Aug.	Sept.	Oct.	Nov.	Dec.	Year
Mean morning temperature (3 years at 8 a.m.[2], 2 years at 8.30 a.m.[3]) (Fahr.[3])	69°·6	69°·1	68°·2	68°·2	62°·7	58°·2	58°·4	63°·0	66°·4	71°·1	72°·0	69°·6	66°·4
Mean maximum temperature	79°·2	79°·1	78°·2	79°·0	74°·7	70°·7	70°·4	76°·0	80°·3	81°·5	83°·0	80°·0	77°·7
Mean minimum temperature	59°·8	57°·1	57°·2	52°·6	45°·6	41°·0	40°·2	45°·3	49°·9	55°·4	59°·5	59°·0	51°·9
Mean temperature[4]	69°·5	68°·1	67°·7	65°·8	60°·2	55°·8	55°·3	60°·7	65°·1	68°·4	71°·3	69°·5	64°·8
Mean daily range[5]	19°·4	22°·0	21°·0	27°·4	29°·1	29°·7	30°·2	30°·7	30°·4	26°·1	23°·5	21°·0	25°·8
Mean of absolute monthly maxima[6] ...	89°·8	87°·4	84°·6	87°·0	83°·1	79°·9	78°·5	86°·0	89°·3	93°·1	93°·0	88°·8	—
Mean of absolute monthly minima[6] ...	52°·1	48°·1	49°·3	46°·3	34°·8	31°·5	32°·9	34°·4	41°·7	44°·1	50°·9	50°·3	—
Average monthly range[7]	37°·7	39°·3	35°·3	40°·7	48°·3	48°·4	45°·6	51°·6	48°·6	49°·0	42°·1	38°·5	—
Mean relative humidity (per cent.) ...	71	71	70	60	55	57	55	47	43	45	56	69	58
Mean number of days on which rain was measured	17·2	9·5	10·2	4·5	0·8	1·2	0	0·2	0·6	4·8	12·6	14·4	76

NOTE.—Gwelo is supplied with all the thermometers usual in the Rhodesian service (a second order station in connection with the Cape Meteorological Commission).

[1] 1900—1904.   [2] 1900—1902.   [3] 1903, 1904.

[4] The mean temperature is derived from $\dfrac{\text{max.} + \text{min.}}{2}$.

[5] The daily range is max. – min.

[6] The absolute maxima and minima are wanting for the year 1900.

[7] The monthly range is absolute maximum – absolute minimum for the month.

BULAWAYO. Lat. 20° 9′ S., Long. 28° 35′ E. Alt. 4469 feet.

Compiled from 7 years' observations[1].

	Jan.	Feb.	March	April	May	June	July	Aug.	Sept.	Oct.	Nov.	Dec.	Year
Mean temperature at 8 a.m. (Fahr.°) ...	71°5	69°6	69°6	67°8	62°5	57°6	58°0	62°4	68°8	72°5	72°5	71°6	67°0
Mean maximum temperature ...	83°1	80°9	81°2	82°1	78°5	71°4	73°0	77°7	83°1	88°1	85°1	82°4	80°3
Mean minimum temperature ...	62°3	60°2	59°8	55°3	48°1	44°7	43°8	47°3	52°9	58°8	61°2	61°1	54°7
Mean temperature $\frac{max. + min.}{2}$ ...	72°7	70°6	70°5	68°7	63°3	58°1	58°4	62°5	68°0	73°5	73°1	71°7	67°5
Mean daily range[2] ...	20°8	20°7	21°4	26°8	30°4	26°7	29°2	30°4	30°2	29°3	25°9	21°3	25°6
Mean of absolute monthly maxima[3] ...	91°8	90°6	88°0	88°6	86°2	79°0	78°2	87°7	92°2	94°7	94°4	90°4	—
Mean relative humidity (per cent.) ...	70	69	70	62	55	56	54	48	40	46	59	65	58
Extremes of rainfall in {highest (inches)	8·30	4·40	3·57	1·80	1·02	0·15	0	0·47	0·71	2·22	6·81	9·69	29·42
7 years {lowest „	2·25	0·12	1·82	0	0	0	0	0	0	0·16	2·36	0·27	17·91

[1] 1898—1904.
[2] The daily range is max.− min.
[3] The absolute maxima are wanting for the years 1898, 1899, 1900.

Note.—Bulawayo is supplied with all the thermometers usual in the Rhodesian service (a second order station in connection with the Cape Meteorological Commission) about 390 miles distant from the nearest point on the coast.

HOPEFONTEIN. Lat. 20° 12′ S., Long. 28° 48′ E. Alt. 4700 feet.

Compiled from 4 years' observations[1].

	Jan.	Feb.	March	April	May	June	July	Aug.	Sept.	Oct.	Nov.	Dec.	Year
Mean temperature at 8 a.m. (Fahr.°) ...	68°9	67°3	64°7	64°9	59°0	55°7	53°4	59°9	65°6	70°4	72°2	77°1	64°9
Mean maximum temperature ...	81°6	81°4	77°8	78°8	73°7	70°1	68°6	75°7	80°4	84°9	83°2	82°0	78°2
Mean minimum temperature ...	61°1	59°2	58°1	54°1	47°4	44°2	43°3	46°7	52°9	58°0	60°6	60°8	53°9
Mean temperature[2] ...	71°3	70°3	67°9	66°5	60°6	57°1	55°9	61°2	66°6	71°4	71°9	71°4	66°0
Mean daily range[3] ...	20°5	22°2	19°7	24°7	26°3	25°9	25°3	29°0	27°5	26°9	22°6	21°2	24°3
Mean of absolute monthly maxima ...	89°5	89°3	84°9	85°6	83°0	79°3	76°6	85°5	92°6	96°2	95°0	91°7	—
Mean of absolute monthly minima ...	53°1	53°8	50°2	47°9	40°0	34°7	36°8	39°8	43°4	46°7	52°9	53°2	—
Average monthly range[4] ...	36°4	35°5	34°7	37°7	43°0	44°6	39°8	46°4	49°2	49°5	42°1	38°5	—
Mean relative humidity (per cent.) ...	74	80	81	72	70	66	70	65	67	52	59	64	68

[1] 1901—1904.

[2] The mean temperature is derived from $\dfrac{max. + min.}{2}$.

[3] The daily range is max. − min.

[4] The monthly range is absolute maximum − absolute minimum for the month.

NOTE.—Hopefontein (now frequently known as Hope Fountain) is a meteorological station of the second order, in connection with the Cape Meteorological Commission, distant about 370 miles from the nearest point on the coast.

EMPANDENI. Lat. 20° 39' S., Long. 27° 45' E. Alt. 4500 feet.

Compiled from 5 years' observations[1].

	Jan.	Feb.	March	April	May	June	July	Aug.	Sept.	Oct.	Nov.	Dec.	Year
Mean morning temperature (4 years at 8 a.m., 1 year at 8.30 a.m.[2]) (Fahr.°)	71°·7	71°·1	68°·8	69°·3	63°·1	56°·9	53°·4	60°·5	64°·2	68°·5	72°·5	70°·9	65°·9
Mean maximum temperature	82°·7	83°·6	80°·4	82°·8	78°·0	72°·3	72°·7	79°·9	80°·7	83°·8	85°·4	83°·3	80°·5
Mean minimum temperature	62°·6	60°·4	59°·3	54°·7	45°·1	39°·0	37°·7	43°·8	51°·1	58°·2	62°·6	62°·5	53°·1
Mean temperature[3]	72°·7	72°·0	69°·8	68°·8	61°·6	55°·7	55°·2	61°·8	65°·9	71°·0	73°·5	72°·9	66°·8
Mean daily range[4]	20°·1	23°·2	21°·1	28°·1	32°·9	33°·3	35°·0	36°·1	29°·6	25°·6	22°·8	21°·8	27°·4
Mean of absolute monthly maxima	91°·7	93°·6	91°·0	94°·3	89°·9	85°·0	80°·0	88°·0	93°·0	97°·0	95°·2	94°·2	—
Mean of absolute monthly minima	53°·7	50°·8	50°·0	45°·3	37°·2	29°·0	31°·6	35°·1	38°·6	46°·1	55°·8	43°·4	—
Average monthly range[5]	38°·0	42°·8	41°·0	49°·0	52°·7	56°·0	48°·4	52°·9	54°·4	50°·9	39°·4	50°·8	—
Mean relative humidity (per cent.) ...	70	72	75	66	63	54	65	65	61	61	62	68	65
Mean number of days on which rain was measured	12·2	3·7	8·5	3·2	0·7	0	0	0	0·5	4·0	9·5	7·0	49·3

NOTE.—All the usual thermometers are in use here, Empandeni being a second order station in connection with the Cape Meteorological Commission.

[1] 1901, 1902, 1903, 1904, 1905. [2] 1905.

[3] The mean temperature is derived from $\dfrac{max.+min.}{2}$.

[4] The daily range is max. – min.

[5] The monthly range is difference between absolute maximum and absolute minimum for the month.

TULI. Lat. 21° 52′ S., Long. 29° 12′ E. Alt. 1750 feet.

Compiled from 4 years' observations.[1]

	Jan.	Feb.	March	April	May	June	July	Aug.	Sept.	Oct.	Nov.	Dec.	Year
Mean temperature at 8 a.m. (Fahr.°)	78°·5	78°·2	74°·6	73°·1	68°·2	63°·1	56°·7	62°·0	68°·4	74°·1	75°·3	77°·3	70°·8
Mean maximum temperature	92°·3	91°·1	86°·9	87°·7	85°·8	80°·7	76°·2	83°·0	86°·3	88°·0	90°·6	91°·5	86°·7
Mean minimum temperature	69°·1	63°·5	60°·5	55°·7	47°·2	43°·9	41°·2	47°·6	55°·7	59°·5	63°·4	64°·8	56°·0
Mean temperature[2]	80°·7	77°·3	73°·7	71°·7	66°·5	62°·3	58°·7	65°·3	71°·0	73°·7	77°·0	78°·1	71°·3
Mean daily range[3]	23°·2	27°·6	26°·4	32°·0	38°·6	36°·8	35°·0	35°·4	30°·6	28°·5	26°·2	26°·7	30°·7
Mean of absolute monthly maxima	103°	101°	98°	97°	95°·5	91°	83°	93°	99°·5	103°	104°	101°	—
Mean of absolute monthly minima	59°	52°·5	52°·5	47°	37°	38°	35°	39°·5	44°	45°	50°	55°	—
Average monthly range[4]	44°	48°·5	45°·5	50°	58°·5	53°	48°	53°·5	55°·5	58°	54°	46°	—
Mean relative humidity (per cent.)	68·7	72·2	79·2	70·5	66·7	65·7	76·0	78·5	64·0	67·2	70·5	71·5	70·9
Mean number of days on which rain was measured	6·7	2·5	3·2	2·0	0	0·7	0	0·2	0·5	3·5	7·7	4·7	31·7

[1] 1901, 1902, 1903, 1904.

[2] The mean temperature is derived from $\dfrac{max. + min.}{2}$.

[3] The daily range is max. − min.

[4] The monthly range is absolute maximum − absolute minimum.

NOTE.—All the usual thermometers are in use here (a third order or climatological station in connection with the Meteorological Commission of the Cape).

SOUTHERN RHODESIA. MEAN RAINFALL IN INCHES.

Mashonaland.

	Lat. S.	Long. E	Alt. in feet	No. of Years	Jan.	Feb.	March	April	May	June	July	Aug.	Sept.	Oct.	Nov.	Dec.	Year
Salisbury[1]	17° 48′	31° 5′	4880	5	7·59	9·06	5·48	1·10	0·20	0·07	0	0·07	0·64	0·74	4·40	5·02	34·37
Umtali[2]	18° 45′	32° 41′	3700	4	9·64	4·38	5·25	1·91	1·51	1·25	0·24	0·03	0·84	0·78	4·39	5·26	35·48
Melsetter[3]	20° 0′	32° 44′	5500	6	9·89	10·34	6·28	1·63	0·51	0·73	0·47	0·04	1·44	1·43	4·67	6·23	43·66
Victoria[4]	20° 21′	31° 3′	3200	4	7·16	2·44	4·25	0·64	0·13	0·08	0	0	0·38	1·53	3·29	5·38	25·28

Matabeleland.

	Lat. S.	Long. E	Alt. in feet	No. of Years	Jan.	Feb.	March	April	May	June	July	Aug.	Sept.	Oct.	Nov.	Dec.	Year
Gwelo[5]	19° 27′	29° 49′	4500	5	7·80	3·58	3·35	0·42	0·32	0·09	0	0·36	0·14	0·87	4·50	6·87	28·30
Bulawayo[6]	20° 9′	28° 35′	4469	7	5·92	1·78	2·62	0·86	0·31	0·05	0	0·07	0·13	0·88	4·91	5·08	22·61
Hopefontein[7]	20° 12′	28° 48′	4700	15	8·05	5·37	3·51	1·96	0·13	0·14	0·03	0·07	0·12	0·71	3·01	4·79	27·89
Empandeni[8]	20° 39′	27° 45′	4500	4	7·33	1·62	2·52	1·01	0·16	0	0	0	0·03	1·40	3·82	2·95	20·84
Tuli[8]	21° 52′	29° 12′	1750	4	4·63	0·91	2·38	0·84	0	0·04	0	0·01	0·27	0·72	3·33	1·79	14·92

[1] 1898—1902. [2] February—September 1899, 1900, 1901, 1902. [3] 1900—1905.
[4] 1901, 1902, 1904, 1905. [5] 1900—1902, 1904, 1905. [6] 1898—1902, 1904, 1905.
[7] January—April, October—December 1890, 1891, 1892, January—May 1893, 1894, 1895, 1897, 1900—1907.
[8] 1901—1904.

The following table shows the mean annual rainfall and the mean number of rainy days at all stations in Southern Rhodesia, where the rainfall has been recorded for three years and upwards.

Name of Station	Mean Rainfall		No. of Years during which observations have been recorded
	Amount (inches)	Days	
Mashonaland			
Ayrshire Mine	30·89	83	6
Borrowdale	38·80	86	3
Charter	32·15	79	4
Chishawasha	41·09	94	7
Experimental Farm	32·94	75	3
Enkeldoorn	32·77	70	3
Gutu	30·43	77	3
Hartley	35·70	75	5
Hartmann Hill	29·66	58	7
Hospital, Salisbury	32·74	89	10
Macheke	34·03	74	3
Marandellas	38·05	79	7
Melsetter	45·77	105	9
Mount Darwin	26·79	63	6
M'Rewas	34·84	73	3
Progress Farm	37·50	81	4
Public Gardens, Salisbury	31·57	80	5
Rusapi	35·96	83	3
Sinoia	30·46	73	4
Sipolilos	24·29	66	3
Umtali	33·02	93	8
Utopia, Umtali	33·99	105	7
Victoria	27·28	61	9
Westridge, Salisbury	30·66	82	6
Matabeleland			
Bulawayo	23·53	79	10
Empandeni	23·38	49	8
Filabusi	21·67	57	4
Fort Rixon	20·48	60	3
Government House, Bulawayo	23·88	56	3
Gwanda	19·68	57	3
Gwelo	25·01	75	7
Hopefontein	26·70	80	10
Inyati	23·58	57	4
Matoppo Mission	22·99	67	3
Sebungwe	26·52	55	3
Selukwe	41·01	75	4
Shiloh	22·12	53	3
Tegwani	19·98	49	6
Tuli	14·97	34	8
Westacre	23·18	45	5

THE TRANSVAAL.

It is necessary to have some idea of the relief to appreciate the differences in climate in different areas in the Transvaal. The Transvaal is made up of the northern prolongation of the great South African Plateau, the Banken or slopes from the Plateau to the lower ground, and this lower ground or Low Veld. The Plateau is bounded on the north and east by a line which, starting from about Pitsani and following the crest of the Witwaters Rand escarpment, passes just north of Johannesburg and then turns northwards to a point some 20 miles east of Pretoria, and, bending east and north-east, curves sharply round Lydenburg and Pilgrim's Rest, joining the eastern backbone of the continent, the Drakensberg, and then runs almost due south to about 20 miles east of Wakkerstroom.

This plateau area embraces the following territorial divisions and parts of divisions :

1. The S.E. portion of Lydenburg district, including the towns of Lydenburg, Belfast and Pilgrim's Rest.

2. A very small portion of Barberton district in which Kaapsche Hoop is situated.

3. Carolina district.

4. The whole of Ermelo except the extreme easterly portion.

5. Wakkerstroom district.

6. All the southern portion of Middelburg district, *i.e.* exclusive of the northern extension.

7. Bethel district.

8. Standerton district.

9. The extreme south-east of Pretoria district (not including the town of Pretoria).

10. Heidelburg district.

11. Witwaters Rand district, except the portion north of 26° S.

12. Potchefstroom district.

13. Lichtenburg district.

14. Bloemhof district.

15. Wolmaranstad district.

16. The extreme south of Marico district.

17. The extreme south of Rustenburg district.

This plateau land has an elevation of about 6000 feet in the east, and slopes down to about 4000 on the west. A line joining the intersection of 29° E. and 27° S. with the intersection of 28°31′ E. and 26° S., *i.e.* a line running from a point 20 miles west of Standerton to a point about 15 miles east of Johannesburg, divides this plateau into the High Veld on the east and the Middle Veld on the west. The climate of the former is "absolutely healthy, and is one of the most invigorating in the world." The temperature varies from an extreme summer heat of 90° Fahr. in the shade to several degrees of frost in the cold winter nights.

The altitude of the High Veld varies from 5000 feet on the west to 6000 feet on the east, and that of the Middle Veld from 4000 feet on the west to 5000 feet on the east, though the Witwaters Rand is somewhat higher. The rainfall diminishes from east to west. On the eastern slopes of the Drakensberg the annual precipitation amounts to from 40 to 55 inches (1016 to 1397 mm.), the atmosphere is charged with moisture and mists are the rule and not the exception. This narrow belt is known as the "mist belt," and on either side the rainfall grows rapidly less. On the east, the precipitation at the Lebompo Mountains is 28 inches (711 mm.). On the west the rate of decrease is more gradual; east of a Heidelburg-Witbank line the mean annual rainfall varies from 30 to 35 inches (762 to 989 mm.); in the Witwaters Rand and Pretoria districts the precipitation amounts to from 25 to 30 inches (635 to 762 mm.), and west of Klerksdorp, as the Kalahari desert is neared, the rainfall only amounts to from 15 to 25 inches (381 to 635 mm.). The rainfall, however, is uncertain after the High Veld is left, and varies greatly from year to year. At Rietfontein, near Johannesburg, the total fall was about 20 inches (508 mm.) in 1888 and over 31 inches (788 mm.) in 1890, while at Rustenburg in 1895 the fall was 40 inches (1016 mm.). On the High Veld a good proportion of the rain falls steadily on cloudy days, but heavy tropical thunderstorms are frequent; in the Middle Veld, however, on going west, the steady rains are more rare and the tropical thunderstorms increase. On the Middle Veld the winter is shorter than on the High Veld. The whole of the plateau, including both High Veld and Middle Veld, is free from malaria, except a few places on the latter.

The plateau is surrounded on the north and east by the Banken or band of slopes to the lower ground below. On the Bechuanaland border this band extends from Pitsani northwards for some 20 miles to 25° S., and gradually increases in width till at Pretoria it is 35 miles wide. It is then narrowed down by the northerly protrusion of the High Veld, but expands again in the region of the Olifants river, to fine down to a width of only 10 miles as it curves round the plateau at Pilgrim's Rest, while on the east it has an average width of 45 miles.

The Banken area embraces the following portions of the territorial divisions:

1. The middle section of Marico district, including the town of Zeerust.

2. A band of Rustenburg district, including the town of that name.

3. The portion of the Witwaters Rand district lying north of 26° S.

4. The middle section of Pretoria district, including Pretoria and the Blands river district.

5. The greater part of north Middelburg district (except the extreme north).

6. A band of Lydenburg district curving round the High Veld.

7. West Barberton, including the town of Barberton.

8. The extreme eastern portion of Ermelo district.

9. All Piet Retief district except the extreme easterly portion.

10. Western Swaziland, including the town of Bremersdorp.

The High Veld consists of rolling grass-clad treeless downs; in the Middle Veld the lines of hills are covered with scrub and the lower river valleys are adorned with trees, though the eastern portion strongly resembles the High Veld: but in the Banken zone the whole country is beautifully watered with running brooks and streams, whose valley bottoms are well wooded, and the slopes, especially on the east, are clothed with patches of forest. The Banken form the most attractive part of the scenery of the Transvaal, and are perfectly healthy for man, except in the lower valleys, where malaria is rife and has been known to drive out the whole population of some villages.

The remainder of the country constitutes the Low Veld, with the exception of two Middle-Veld outliers of the plateau, each with its banken. One of these is situated in the Zoutpansberg district around Pietersburg and Marabastad, and extending N.E. to beyond 22° S., and the other in the Waterberg district to the immediate north of Nylstroom. The climate of these two outliers partakes of the nature of that of the Middle Veld, but, lying further north, the heat is greater, and even the higher regions are malarious.

The altitude of the Low Veld varies from 1000 to 4000 feet, the lower regions being on the east. The Low Veld is flat, and covered with scrub bush, and water is very scarce. The climate, owing to lower elevation, and, for the most part, more northerly situation, is more tropical than that of the other zones. The summer heat rises to 113° Fahr. (45° C.) in the shade, but the nights are cool, though frosts are only experienced along the river bottoms. In the neighbourhood of the Banken the rainfall may amount to 30 inches (762 mm.), but in districts removed from the influence of the mountains the rainfall grows less and less and is also uncertain. In the low parts of Zoutpansberg the mean annual rainfall is between 10 and 12 inches (254 and 305 mm.). The whole of this country is a hot-bed of malaria, though it is more prevalent and of a worse type in some districts than in others. In this fever-stricken region are included :

Zoutpansberg and Waterberg (except the High Veld).
The northern parts of Marico, Rustenburg and Pretoria.
The extreme north of Middelburg.
The north-west, north, and east of Lydenburg.
The eastern parts of Barberton and Swaziland.
The extreme east of Piet Retief.

In the Transvaal the winter is the dry season, the country lying in the region of summer rains. On the plateau the summer is hot, but, as a rule, not oppressive, and the thunderstorms tend to lower the temperature. In the winter, the nights and mornings are very cold, but the days are bright and the air clear and bracing, in fact the winter climate is most enjoyable. Ermelo has been re-commended as a health resort for those afflicted with consumption. The Standerton district is said to be trying in winter because of the very cold west and south-west winds, while the Wolmaranstad winter has the reputation of being milder than that of most of the Transvaal districts, but the summer is hot and droughts frequently occur.

In the extreme north, in the Limpopo valley, the mean annual temperature is about $70°$ or $71°$ Fahr. ($21°·1$ or $21°·7$ C.). Nearly every year extreme temperatures of $106°$ to $108°$ Fahr. ($41°·1$ to $42°·2$ C.) are experienced between October and March, and the mean maximum temperature for November to February is over $90°$ Fahr. ($32°·2$ C.). It is cold in the winter months and sometimes there are frosts on the coldest winter nights, though not as a rule. The lowest temperature recorded is $22°·9$ Fahr. ($-5°$ C.), or $9°$ Fahr. of frost in May 1904. The monthly range of temperature is large, varying from $44°$ to $46°$ Fahr. ($24°·4$ to $25°·5$ C.) in summer to $56°$ to $58°$ Fahr. ($31°·1$ to $32°·2$ C.) in the winter months, while the daily range is between $35°$ and $40°$ Fahr. ($19°·4$ and $22°·2$ C.) in winter, and from $24°$ to $26°$ Fahr. ($13°·2$ to $14°·4$ C.) in summer. The mean annual rainfall is about 15 inches (380 mm.), experienced on some 34 days in the year, almost entirely in the period from November to February.

At Johannesburg the mean of the barometric readings varies between 24·746 and 24·468 inches, the lowest readings being taken in January and the highest in June, July, and August. May, June, July, and August are the dryest months in the year and the coolest. The rains generally begin in September, but sometimes not till October; and November, December, January, and February are the wettest and hottest months. The daily range of temperature is large, as might be expected at a distance of 300 miles from the nearest point on the sea coast, and is highest in September, October, and November. The ther-mometer frequently falls below freezing point in the colder months, and the absolute maximum is generally measured in November and December; the greatest range of temperature for the period during which the observations were taken, which form the basis of the table below, is 75° Fahr. (41°·6 C.), being the difference between 96° Fahr. (35°·5 C.) measured in November 1897, and 21° Fahr. (−6°·1 C.) recorded in July 1898. On the whole Johannesburg enjoys a magnificent climate, being on the plateau, and the only drawback consists in frequent sandstorms which are very disagreeable.

At Pretoria the winters are agreeable, and milder than at Johannesburg, but the summer heat is somewhat greater, as the town is on the Banken. The number of rainy days at Pretoria is almost the same as at Johannesburg and is very similarly distributed throughout the year. In both December and January it rains, on an average, on 15 days, in March and November on 12, on 10 in February, while October is credited with 8 days, April with 6, September with 4, and there are 7 rainy days in the period May to August, making 89 in the year.

For the southern, more elevated, healthier, and more thickly populated portion of the Transvaal, some very interesting particulars have lately been published with regard to the rainfall, in the report of the Commissioner of Lands for the year ending 30th June 1906, but unfortunately they are only for one year, viz. 1904/5, which was a dry year. These particulars refer to the rainfall in the catchment areas of various rivers :—(1) upper Crocodile river (proper), in the divisions of Pretoria, Witwaters Rand, Rustenburg and Krugersdorp, and the catchment areas of its tributaries, Magalies river, Hex river, Plat river, and Pienaars river; (2) the upper Vaal, in the divisions of Standerton, Ermelo and Wakkerstroom, and its tributaries, the Mooi river in Potchefstroom, Krugersdorp and Rustenburg divisions, the Schoonspruit in Potchefstroom and Lichtenburg; (3) the upper Great Olifants river, in Middelburg and Standerton, with the Bronkhorstspruit in Pretoria and the White river, an affluent of the Crocodile river of the Komati basin, in the Barberton division.

In the upper Crocodile catchment area 17 rain-gauge stations were at work, and the average precipitation for the year amounted to 20·20 inches (513 mm.), the fall at the various stations ranging from 24·75 inches (629 mm.) at Joubert Park, Johannesburg, to 16·87 inches (428 mm.) at Kromdraai; while the average monthly maximum was 4·70 inches (119 mm.) in November. In the Magalies river area, in the divisions of Pretoria, Rustenburg, and Krugersdorp, the average annual rainfall was slightly less, viz. 19·42 inches (493 mm.), and the difference between the extremes was not so great, the maximum fall for the year being 22·01 inches (500 mm.) recorded at Hekpoort, and the minimum 16·60 inches (411 mm.) at Kaalfontein. In the smaller areas of Hex river, in Rustenburg, and Plat river, in Waterberg, the rainfall was still less, averaging 18·01 and 17·66 inches (458 and 448 mm.) respectively; the maximum in the former was 23·51 inches (597 mm.) at Sandfontein, and in the latter area 18·13 inches (460 mm.) at Rietfontein, and the corresponding minima 14·92 inches (379 mm.) at Vlakfontein and 17·22 inches (438 mm.) at Elandsfontein. In the Pienaars river catchment area in Pretoria division the average of 10 rain-gauge stations gave a fall of 21·95 inches (558 mm.), the greatest annual fall, 24·71 inches (628 mm.) occurring at Bryntirion, and the smallest, 11·88 inches (302 mm.), at Wagensdrift. The heaviest fall occurred in November, which also witnessed the greatest monthly fall in the Magalies river area, while that for the Hex river and Plat river areas was in December and February respectively.

In the areas of the upper Vaal river, the Mooi river, and Schoonspruit the heaviest fall was in January, and measured a little over 7 inches (178 mm.) in the first, a little more than 4·5 inches (114 mm.) in the second, and a little less than 4 inches (102 mm.) in the last-named area. In the upper Vaal area there were 20 rain-gauge stations working, and the average annual precipitation, deduced from the results obtained at these stations, was found to be 29·63 inches (752 mm.), the station with the highest figure, 38·25 inches (971 mm.), being Standerton, and that with the lowest, 23·11 inches (587 mm.), Bethel. In the Mooi river area there were 12 stations at work, and the results gave an average of only 17·32 inches (440 mm.) for the average annual fall; the Schoonspruit area with 10 stations gave the least fall of any of the areas, viz. 16·1 inches (409 mm.). In the Mooi river area 23·51 inches (597 mm.) was the greatest fall, and was recorded at Zandfontein; 14·62 inches (371 mm.) at Klerkskraal was the least; the corresponding figures for the Schoonspruit area were 18·71 inches (475 mm.) at Rhenosterfontein, and 14·12 inches (359 mm.) at Putfontein.

In the catchment area of the upper Great Olifants river, in the divisions of Middelburg and Standerton, the average of 13 stations for the year was 21·56 inches (547 mm.), with a maximum of 26·76 inches (680 mm.) at Trichardsfontein, and a minimum of 15·13 inches (385 mm.) at De Wittekranz, the maximum fall for any one month being 6·65 inches (169 mm.) in November. The Bronkhorstspruit area in the Pretoria division had an average of 20·78 inches (527 mm.) for the year. Eight stations contributed to the result, and at Witklip the maximum, 22·94 inches (583 mm.), was recorded, and the minimum, 18·85 inches (479 mm.), at Waarkraal. Finally, the most easterly area for which measurements were recorded was that of the catchment area of the White river, a feeder of the Crocodile river, which joins the Komati river. This area is situated in Barberton division and there were two recording stations, viz. Spitzkop and White River Settlement. The fall for the year at the former was 38·12 inches (968 mm.), and at the latter 28·99 inches (736 mm.). The heaviest fall was in February, when 7·77 inches (197 mm.) were recorded.

The following table gives the mean temperature and the maximum and minimum temperatures, based on observations extending from July 1904 to June 1906 inclusive.

Place	Mean temperature	Maximum	Minimum
Amsterdam (Fahr.°) ...	62°·2	97°·5	24°·1
Bloemhof	63°·3	100°·5	20°·0
Komati Poort ...	74°·5	110°·9	35°·0
Krabbefontein ...	67°·0	103°·1	33°·9
Pietersburg	64°·3	99°·1	27°·0
Pretoria (Arcadia) ...	63°·8	97°·4	25°·2
Vereeniging	61°·0	96°·0	17°·7
Volksrust	57°·0	91°·1	15°·0
Zeerust	65°·0	103°·7	23°·2

PRODUCTIONS, &c.

The Transvaal, apart from its mining industries, is, like the Orange Free State, essentially a stock-rearing country, and also to a large extent a fruit-growing area. Such fruits as apricots, peaches, nectarines, pears, plums, and figs do very well indeed in the districts of Marico, Rustenburg, Pretoria, Waterberg, Piet Retief, Barberton, Lydenburg, and Zoutpansberg, that is, in the north and east. Wheat, oats, and barley do well, and mealies are a staple crop. Tobacco-growing is quite an industry in some localities, especially in Lichtenburg, Marico, and Potchefstroom. Potatoes thrive. Coffee and cotton may be grown in some districts, notably in Marico, where rye also does well. European vegetables may be grown in most localities, though Rustenburg, up to the present, appears to have been most successful in this respect. In Swaziland cotton is indigenous and does especially well in the eastern parts.

JOHANNESBURG. Lat. 26° 12′ S., Long. 28° 3′ E. Alt. 5735 feet.

Compiled from 5—13 years' observations [1].

	Jan.	Feb.	March	April	May	June	July	Aug.	Sept.	Oct.	Nov.	Dec.	Year
Mean temperature at 8 a.m. (Fahr.°) ...	67°.9	68°.2	65°.1	61°.7	52°.1	44°.9	45°.4	53°.7	61°.4	63°.5	64°.9	64°.3	59°.4
Mean maximum temperature	80°.0	81°.4	77°.7	74°.7	67°.9	62°.3	59°.5	66°.4	75°.9	80°.9	82°.3	83°.5	74°.4
Mean minimum temperature	56°.9	55°.5	54°.0	49°.3	40°.4	36°.3	35°.3	39°.6	46°.2	51°.1	53°.5	57°.2	47°.9
Mean temperature [2]	68°.5	68°.5	65°.8	62°.0	54°.2	49°.3	47°.4	53°.0	61°.0	66°.0	67°.9	70°.3	61°.2
Mean daily range (max.−min.)	23°.1	25°.9	23°.7	25°.4	27°.5	26°.0	24°.2	26°.8	29°.7	29°.8	28°.8	26°.3	26°.5
Mean of absolute monthly maxima	90°.1	90°.0	87°.8	82°.6	75°.9	71°.5	66°.7	71°.0	88°.0	91°.0	91°.7	93°.3	—
Mean of absolute monthly minima	47°.2	48°.6	48°.2	39°.8	28°.6	27°.7	26°.2	27°.7	34°.0	39°.3	40°.0	50°.0	—
Average monthly range [3] ...	42°.9	41°.4	39°.6	42°.8	47°.3	43°.8	40°.5	43°.3	54°.0	51°.7	51°.7	43°.3	—
Mean relative humidity (per cent.) ...	68	66	73	75	71	69	67	57	49	54	58	63	64
Mean number of days on which rain was measured (13 years)	14.0	9.7	10.5	7.2	3.5	1.2	0.7	1.4	3.7	7.5	11.8	13.5	84.7
Extreme monthly rainfall during 13 years { greatest	10.21	8.09	7.02	5.16	2.79	0.71	0.75	1.52	2.76	5.43	7.94	7.63	40.85 (year)
{ least	1.65	2.91	1.10	0.44	0	0	0	0	0	0.06	2.19	1.49	19.85 (year)
Mean annual rainfall (13 years) ...	6.63	4.77	3.63	2.04	0.74	0.18	0.16	0.34	1.07	1.95	4.05	4.80	30.36

[1] From temperature and humidity observations recorded in 1897, 1898, 1904 (July—December), 1905, 1906 (January—June). The years for rainfall are given on p. 428.

[2] The mean temperature is derived from $\dfrac{\text{max.}+\text{min.}}{2}$.

[3] The monthly range is abs. max.−abs. min. for the month.

TRANSVAAL MEAN ANNUAL RAINFALL IN INCHES.

	Lat. S.	Long. E.	Alt. in feet	No. of Years	Jan.	Feb.	March	April	May	June	July	Aug.	Sept.	Oct.	Nov.	Dec.	Year
Johannesburg[1]	26° 12'	28° 3'	5735	13	6·63	4·77	3·63	2·04	0·74	0·18	0·16	0·34	1·07	1·95	4·05	4·80	30·36
Witwaters Rand[2]	...	...	...	4	6·32	3·77	3·28	2·01	0·41	0·16	0·24	0·55	0·74	2·77	4·34	3·82	28·41
Pretoria[3] ...	...	...	4471	10	6·77	3·69	3·11	0·91	0·56	0·19	0·07	0·19	1·18	1·93	3·99	4·35	26·94
Rietfontein[4] ...	...	...	...	2	5·36	3·12	1·70	2·06	0·64	0·23	0	0	0·45	2·98	4·83	4·28	25·65
Bremersdorp[5] (Swaziland)	26° 42'	31° 19'	2100	3	12·07	1·37	5·34	3·42	2·16	0·25	0	0·25	0·99	3·47	5·16	3·29	37·77
Birthday Camp[6] (Klein Letaba R.)				$1\frac{5}{12}$	7·6	8·4	0·6	0·4	0·06	0	0·03	0·06	1·7	1·7	2·9	6·2	29·9

[1] September—November 1893, 1894—1898, January—August 1899, 1900, 1903—1907.
[2] This result is the average of 15 stations on the Rand.
[3] 1894—1898, 1903—1907.
[4] 1888, 1890.
[5] 1905—1907.
[6] August—December 1892, 1893.

ORANGE FREE STATE.

The climate of the Orange Free State is on the whole healthy. Fever (malaria) is not prevalent, and such cases as occur are not of a malignant type. The area may be divided into three zones, viz. a westerly very dry zone, a central transition zone, and an eastern zone, which is better watered and has a moist climate. In the western zone, at Boshof, the rainfall is very small, at Jacobsdal and at Philippolis it is only between 19 and 20 inches (482 and 500 mm.) annually, while at Bloemfontein it is between 25 and 26 inches (635 and 660 mm.), at Wepener 28·5 inches (724 mm.), and at Ladibrand over 38 inches (965 mm.); and speaking generally the rainfall increases from west to east.

In the summer months—December, January, and February—the heat is excessive, especially in the western and central zones, and the thermometer often registers 108° and 109° Fahr. (42°·5 C.) in the shade. On the other hand, the cold in winter is frequently severe, and the changes in temperature are very trying. The mean maximum temperature varies between 87° and 64° Fahr. (30°·5 and 17°·8 C.), and the mean minimum between 58° or 59° Fahr. (15° C.) and freezing point. In June and July the daily range is very great, for while the mean maximum has reached its lowest point and is increasing, the mean minimum is still on the downward grade to its lowest point. The mean temperature of the warmest month is between 70° and 73° Fahr. (21°·1 and 22°·8 C.) and of the coldest about 46° or 47° Fahr. (8° C.); the absolute minimum registered is 16° Fahr. (−8°·9 C.) or 16° Fahr. of frost. The mean annual rainfall for the whole area is nearly 24 inches (610 mm.) falling on 55 days, by far the greater part in the summer months. The average amount of cloud is between 26 and 27 per cent.

The rainfall cannot be depended on. This is true not only for the Orange Free State, but for most other parts of our southern and eastern possessions in Africa, notably in the East Africa Protectorate. The mean annual rainfall of 19·88 inches or 505 mm., as given in the table below for Jacobsdal, is derived from such divergent figures as 30 inches in one year and only 15 inches in the next (762 and 381 mm.). At Philippolis there was a drought in 1897, and only 8·6 inches (218 mm.) were recorded, whereas, in 1891, the total annual precipitation amounted to over 31 inches (787 mm.). Similarly, at Jagersfontein, there is a difference of no less than 22 inches (560 mm.) between the records of two of the years, equal, in point of fact, to the total mean for the four years. At Bethulie and Bloemfontein there is an almost equivalent difference. At Ladibrand nearly 14 inches (355 mm.) of rain fell in January of one year, and only 3 inches (76 mm.) in the same month three years before; and at Kroonstad, in the same month, nearly 12·5 inches (317 mm.) were recorded in 1892, and only 1·75 inches (45 mm.) in 1896. These figures are enough to show the small amount of reliance that can be placed in the precipitation from time to time, and the constant anxiety that dogs the steps of agriculturists.

At Bloemfontein, situated at an altitude of 4518 feet above sea-level, the mean atmospheric pressure appears to vary between 25·790 inches and 25·325 inches, the barometer being highest in June, July, and August, and lowest in November, December, and January, the minimum being generally recorded in the last mentioned month. The mean morning temperature in each month is fairly regular from year to year, the greatest difference between the six years' observations, from which the table below is compiled, being not more than between 4° and 5° Fahr. (2°·5 C.) except in October, when there is a difference of 7°·3 Fahr. (4° C.) between the mean for 1898 (59°·5 Fahr.) and that for 1897 (66°·8 Fahr.), and in December, when the mean in 1897 is 73°·9 Fahr. and that in 1904 is 67°·8 Fahr., giving a difference of 6°·1 Fahr. (3°·4 C.). The mean maximum temperature is not quite so steady for each month, and the greatest divergencies occur in April and May; in 1897 those months were hotter than usual, and in 1899 colder than the average; and hence we get a difference of 12°·3 Fahr. (6°·8 C.) in April, and 13°·1 Fahr. (7°·3 C.) in May, between the means, the actual figures in those years being $82°·3$ and $70°$ Fahr. ($27°·9$ and $21°·1$ C.) for April and $73°$ and $59°·9$ Fahr. ($22°·8$ and $15°·5$ C.) for May. In the mean maxima the greatest difference for any one month, throughout the six years, is $7°$ Fahr. ($3°·9$ C.).

The highest absolute maximum in any year is generally measured in November, December, or January, though in 1898 the April maximum was the highest. The highest reading of the thermometer in the six-year period was $101°·7$ Fahr. ($38°·7$ C.) in January 1903, and the lowest $22°$ Fahr. ($-5°·6$ C.) in June 1905, and thus the range of temperature for the whole period was $79°·7$ Fahr. ($44°·3$ C.). The lowest absolute minimum is always measured in May, June, or July. The mean relative humidity varies greatly from year to year in almost every month, the average variation being about 16 per cent., and December being the most constant. The amount of cloud is very small and bears testimony to the propriety of the name of the "land of sunshine."

The rainfall varies considerably from year to year and also from month to month. In 1897 the precipitation amounted to only 15 inches (381 mm.) and in 1891 it was as much as $34·5$ inches (876 mm.). Again, in November 1891 the monthly fall was $7·5$ inches (165 mm.), in 1897 that month was absolutely dry, and in 1892 only one inch of rain (25 mm.) was recorded. Similarly, in January 1897 the rainfall measured $7·25$ inches (184 mm.), while less than three-quarters of the amount was measured in the preceding year. The dryest season is from June to September.

At Philippolis, situated in Lat. $30°\,15'$ S. and Long. $25°\,18'$ E., at an altitude of 4700 feet above sea-level, in the south of the Orange Free State, the mean atmospheric pressure varies between $25·772$ inches and $25·481$ inches, the barometer usually standing highest in June, July, and August, and lowest in November, December, and January, the lowest reading generally being taken in the last-mentioned month. The mean temperature for the year is $59°·2$ Fahr. ($15°·1$ C.), December, January, and February being the hottest months, and June and July the coldest; but during May, June, July and August frosts are frequent, though the absolute minimum temperature is generally measured in June or July, $16°$ Fahr. ($-8°·9$ C.) being the lowest recorded temperature. The extreme range of temperature for the three-year period, during which the observations were taken, from which the table below is compiled, was 83° Fahr. (46°·2 C.), being the difference between 99°·2 Fahr. (37°·3 C.) recorded in January 1899, and 16°·2 Fahr. (−8°·9 C.) recorded in July 1898. The daily range of temperature is greatest in August and September. The mean amount of cloud is very small, and bright days are the rule. The mean annual rainfall is 19·65 inches, or 495 mm., though this is probably too low, as in 1897, which was a year of drought, only 8·6 inches (218 mm.) were measured. Omitting this year the total precipitation varies between 15 and 30 inches (381 and 762 mm.), and falls almost entirely in the summer months.

PRODUCTIONS, &c.

The Orange Free State is essentially a stock-rearing area, which is more the creature of climate in its suitability for various agricultural productions than many a larger tract of country. It has been pointed out that the rainfall, such as it is, diminishes from east to west, and it follows that, though stock may be raised throughout the whole area, and though the soil is almost equally fertile over the whole country, yet the west is almost entirely unsuited for agriculture. In the east, however, where the rainfall is comparatively much greater, the climate is suitable for the production of wheat, oats, and barley, the first-named especially in the districts of Ficksburg and Ladibrand. In the few western localities where irrigation can be practised, these will do equally well. Kafir corn and maize or mealies will grow almost anywhere. Pumpkins and potatoes and nearly all European vegetables do well in the east. In some places root-crops thrive, especially in the neighbourhood of Bethlehem. Tobacco is grown, and produces good results, especially near Vrede, and fruits, such as peaches, apricots, plums, &c., will grow in most localities. The climate of Bloemfontein is said to be suitable for the modern treatment of phthisis.

BLOEMFONTEIN. Lat. 28° 56′ S., Long. 26° 18′ E. Alt. 4518 feet.

Compiled from 6 years' observations[1].

	Jan.	Feb.	March	April	May	June	July	Aug.	Sept.	Oct.	Nov.	Dec.	Year
Mean temperature at 8 a.m. and 7.30 a.m. combined[2] (Fahr.°)	69°·9	68°·1	64°·3	57°·4	48°·1	41°·2	41°·8	47°·1	54°·4	63°·3	68°·2	71°·2	57°·9
Mean maximum temperature ...	85°·1	84°·3	79°·7	75°·3	66°·6	61°·4	63°·0	67°·6	73°·9	78°·8	82°·6	87°·2	75°·4
Mean minimum temperature ...	60°·6	59°·4	55°·6	48°·9	40°·4	33°·8	34°·3	38°·6	45°·3	50°·4	54°·5	59°·5	48°·4
Mean temperature[3] ...	72°·8	71°·8	67°·6	62°·1	53°·5	47°·6	48°·6	53°·1	59°·6	64°·6	68°·5	73°·3	61°·9
Mean daily range[4] ...	24°·5	24°·9	24°·1	26°·4	26°·2	27°·6	28°·7	29°·0	28°·6	28°·4	28°·1	27°·7	26°·5
Mean of absolute monthly maxima ...	95°·7	93°·3	89°·0	85°·8	76°·6	69°·7	71°·2	77°·9	88°·1	90°·9	93°·7	96°·1	—
Mean of absolute monthly minima ...	50°·0	49°·9	45°·0	38°·0	28°·9	25°·6	25°·3	27°·8	29°·7	36°·4	40°·0	48°·6	—
Average monthly range[5] ...	45°·7	43°·4	44°·0	47°·8	47°·7	44°·1	45°·9	50°·1	58°·4	54°·5	53°·7	47°·5	—
Mean relative humidity (per cent.) ...	65	68	72	70	70	70	65	57	51	53	53	55	62
Mean amount of cloud (0—10) ...	3·4	3·3	2·9	2·8	2·7	1·8	1·9	2·4	2·0	2·9	2·5	2·9	2·6
Mean number of days on which rain was measured	8·7	6·9	6·9	5·7	2·8	1·8	1·2	0·9	1·8	2·8	4·7	5·2	49·4

[1] 1897—1899 and 1903—1905.

[2] 8 a.m. from 1903—1905 and 7.30 a.m. 1897—1899.

[3] The mean temperature is derived from $\dfrac{\text{max.} + \text{min.}}{2}$.

[4] The daily range is max.—min.

[5] The monthly range is abs. max.—abs. min. for the month.

NOTE.—Bloemfontein is a meteorological station of the second order, distant about 260 miles from the coast.

K. A.

28

PHILIPPOLIS. Lat. 30° 12′ S., Long. 25° 16′ E. Alt. 4700 feet.

Compiled from 3 years' observations[1].

	Jan.	Feb.	March	April	May	June	July	Aug.	Sept.	Oct.	Nov.	Dec.	Year
Mean temperature at 8 a.m. (Fahr.°) ...	70°·4	68°·9	63°·0	55°·6	46°·2	41°·5	40°·4	47°·9	55°·5	57°·6	67°·1	70°·9	57°·1
Mean maximum temperature	83°·8	83°·2	76°·2	71°·7	63°·3	60°·7	60°·4	67°·3	73°·4	74°·3	78°·6	84°·3	73°·1
Mean minimum temperature	57°·1	56°·0	52°·7	47°·0	37°·9	31°·3	35°·1	36°·1	41°·6	45°·5	50°·1	56°·6	45°·5
Mean temperature[2] ...	70°·5	69°·6	64°·5	59°·3	50°·1	46°·0	47°·7	51°·7	57°·5	59°·9	64°·3	70°·5	59°·3
Mean daily range[3]	26°·7	27°·2	23°·5	24°·7	25°·4	29°·4	25°·3	31°·2	31°·8	28°·8	28°·5	27°·7	27°·6
Mean of absolute monthly maxima ...	96°·0	92°·0	83°·5	76°·8	77°·5	67°·2	68°·9	77°·5	84°·8	83°·6	88°·9	93°·4	—
Mean of absolute monthly minima ...	44°·2	43°·7	39°·7	38°·2	23°·4	21°·7	20°·7	24°·8	26°·4	31°·9	35°·5	44°·8	—
Average monthly range[4] ...	51°·8	48°·3	43°·8	38°·6	54°·1	45°·5	48°·2	52°·7	58°·4	51°·7	53°·4	48°·6	—
Mean relative humidity (per cent.) ...	66	65	75	75	74	71	73	66	54	57	55	57	63
Mean amount of cloud (0—10) ...	3·6	3·2	2·7	2·4	2·1	1·6	1·5	1·2	1·6	2·1	1·9	2·5	2·2
Mean number of days on which rain was measured	9	7	7·3	7·7	4·3	0·7	2·6	1	1	3·3	2·3	2·6	49

[1] 1897—1899, when the station was one of the second order.

[2] The mean temperature is derived from $\dfrac{max. + min.}{2}$.

[3] The daily range is max. − min.

[4] The monthly range is abs. max. − abs. min. for the month.

ORANGE FREE STATE. MEAN RAINFALL IN INCHES.

Stations from W. to E.	No. of Years	Lat. S.	Long. E.	Jan.	Feb.	March	April	May	June	July	Aug.	Sept.	Oct.	Nov.	Dec.	Year
Jacobsdal[1] ...	8	29° 10'	24° 45'	3·08	2·66	4·01	1·85	1·05	0·82	0·14	0·35	0·25	0·69	2·10	2·88	19·88
Philippolis[2] ...	8	30° 12'	25° 16'	3·20	2·39	3·27	1·97	1·08	1·06	0·27	0·67	0·48	1·07	2·20	2·02	19·68
Jagersfontein[3] (near Fauresmith)	4	29° 41'	25° 26'	4·61	2·62	4·68	1·62	0·93	0·70	0·57	0·95	0·30	1·21	2·67	1·81	22·67
Bethulie[4]	8	30° 30'	25° 59'	3·15	2·96	3·70	1·32	1·05	0·81	0·46	0·73	0·74	0·96	1·92	3·42	21·22
Bloemfontein[5] ...	8	28° 56'	26° 18'	4·31	3·36	3·74	2·19	1·46	0·89	0·31	0·24	0·68	1·74	3·43	3·30	25·65
Smithfield[6] ...	7	30° 10'	26° 27'	4·13	3·18	3·31	2·01	1·21	0·78	0·67	0·96	0·79	1·26	1·65	4·19	24·14
Wepener[7] ...	7	29° 41'	27° 1'	5·90	3·01	4·52	1·56	1·12	0·86	0·46	1·12	0·59	1·70	2·09	5·63	28·56
Ladibrand[8] ...	4	29° 15'	27° 47'	8·42	5·04	3·97	2·78	1·08	1·74	0·73	1·16	1·23	3·20	5·36	3·73	38·44
Kroonstad[9] ...	8	27° 29'	27° 54'	5·95	3·23	3·04	1·91	1·31	0·46	0·17	0·42	0·93	1·68	3·28	3·23	25·61
Harrismith[10] ...	4	28° 20'	29° 9'	6·31	4·13	3·28	1·49	0·84	0·48	0·36	0·60	1·62	2·18	6·31	2·72	30·32

[1] 1890—1897. May 1893 is wanting.

[2] 1890—1897. April, June, and December 1896 are wanting.

[3] July, September—December 1890, 1891, 1893, 1897.

[4] 1890—1897. May and December 1892, May—July 1894 are wanting.

[5] 1890—1897. May 1893, July—December 1896 are wanting.

[6] 1891, January, July—December 1892, 1893, January—March, and May—December 1894, 1895—1897.

[7] 1891, January—November 1892, 1893, January—June, September—December 1894, January—July, September, November, December 1895, February—June, August—December 1896, 1897.

[8] 1890—1893.

[9] 1890—1893, January—November 1894, 1895—1897.

[10] January—November 1890, 1891, January—May, August—December 1892, January—March, May, September—December 1893.

28—2

NATAL.

The climate of Natal is decidedly healthy, and, owing to the terrace formation being confined here within narrow limits, the country rises by steps from the sea-level to some 6000 feet in about a hundred miles, and not only is the climate salubrious but there is great variety, in fact there are several climatic zones, ranging from warm, through temperate, to bracing.

Comparing the temperature of Durban with that of other towns along the coast, it will be found that the mean annual temperature at Cape Town (Observatory) is $62°$ Fahr. ($16°\cdot7$ C.), at Port Elizabeth $63°\cdot9$ Fahr. ($17°\cdot7$ C.), at East London $64°\cdot8$ Fahr. ($18°\cdot2$ C.), at Port St John (Kaffraria) $66°\cdot9$ Fahr. ($19°\cdot4$ C.), while that at Durban is $70°\cdot7$ Fahr. ($21°\cdot5$ C.), showing a gradual increase along the coast northwards, as might be expected. The mean of the warmest month at Durban is $77°$ Fahr. ($25°$ C.), and of the coolest $64°\cdot2$ Fahr. ($17°\cdot9$ C.), while the absolute maximum and minimum are $110°\cdot6$ and $41°\cdot8$ Fahr. ($43°\cdot7$ and $5°\cdot4$ C.) respectively, but during the warmest month the thermometer frequently reaches $90°$ Fahr. ($32°\cdot2$ C.). June, July and August are the coolest months, though the thermometer seldom falls below $50°$ Fahr. ($10°$ C.). November, December, January, February, March are the hot months, and the registered temperature is rarely below $60°$ Fahr. ($15°\cdot6$ C.).

In the more elevated districts of the interior the temperature is lower; at Pietermaritzburg, which stands at an altitude of 2,225 feet above sea-level, the mean annual temperature is about $65°\cdot5$ Fahr. ($18°\cdot6$ C.) or $5°\cdot2$ Fahr. lower than at Durban and the thermometer at times not only touches the freezing point, but reaches as low a level as $28°$ Fahr. ($-2°\cdot2$ C.); while, on the other hand, the temperature may rise for a few days in the summer to $103°$ Fahr. ($39°\cdot4$ C.). In the highlands the climate is more bracing; at Howick, whose elevation above sea-level is 3,439 feet, the mean temperature is only $63°$ Fahr. ($17°\cdot2$ C.), or just one degree higher than at Cape Town, the difference in latitude being $4°21'$. The warmest month has a mean of $70°\cdot5$ Fahr. ($21°\cdot4$ C.), as against $69°\cdot7$ Fahr.

(20°·95 C.) at Cape Town and 70°·3 Fahr. (21°·3 C.) at Port Elizabeth, and for the coolest month the mean is 52°·2 Fahr. (11°·2 C.); the absolute maximum is 102° Fahr. (38°·9 C.) and the absolute minimum 21° Fahr. (-6°·1 C.). The daily range on the coast is about 20° Fahr. (11°·1 C.) and this is pretty constant throughout the year. Here the extreme sun-heat is tempered by the presence of clouds—the average amount at Durban being 45·6 per cent.—by thunderstorms, and rain.

The average rainfall over the whole of Natal is 34·09 inches (866 mm.) and the average number of days on which rain is measured 117, 77 per cent. in the summer (October—March) and 23 per cent. in the winter. At Durban the mean annual rainfall is 41 inches or 1041 mm., at Pietermaritzburg it is 33·55 inches (852 mm.) and the number of days on which rain is registered is 135; at Howick the fall is only 29·45 inches (748 mm.), and, generally speaking, it rains more on the coast than on the interior uplands. The number of real rainy days, omitting those on which there are passing showers, is at Durban 61 and at Pietermaritzburg 58. Roughly about 5 inches (125 mm.) fall each summer month, and 2 inches (50 mm.) each winter month. The Natal winter is dry and almost always bright. May, June, and July are the finest months at Durban and in the vicinity, a light breeze coming in from seaward during the day, to be replaced by a land breeze at night; but strong east and west gales are not infrequent in these months, though not so frequent as in the boisterous months of August, September, and October.

At Durban the rainfall varies considerably from year to year. In 1893 the total precipitation amounted to no less than 71·3 inches (1811 mm.), and in 1895 to 51·46 inches (1307 mm.), while in 1899 the fall was only 28·75 inches (730 mm.), and in 1900 only 27·23 inches (692 mm.). The monthly rainfall, in the same manner, cannot be depended on. The following table shows the greatest and least amounts (in inches) recorded for each month during the ten-year period 1890—1899.

	Jan.	Feb.	March	April	May	June	July	Aug.	Sept.	Oct.	Nov.	Dec.
Greatest	6·94	10·17	11·10	8·00	3·88	3·80	2·60	3·39	13·84	13·65	8·33	11·23
Least	1·10	1·35	2·01	0·64	0·30	0·08	0·02	0·23	0·84	2·56	1·93	3·62

The winds at Durban which blow most frequently come from directions lying in the two quadrants between north-north-east and east-south-east, and between west-south-west and south-south-east, in fairly equal proportions, and alternating throughout the year in short periods of a few days. The only other winds which are of any account are those from north and north-west. The former occur throughout the year distributed in about equal proportions in all months, with a slight predominance for March, and occurring on some 31 days in the year in all; the latter blowing on about 25 days, 17 of which are in the period May to August.

The following table shows the number of days on which the wind blows from various directions in each month, compiled from 19 years' observations:

		N.	N.E.	E.	S.E.	S.	S.W.	W.	N.W.	Calm
January	...	3	6	4	6	5	3	0	1	3
February	...	3	6	4	5	4	2	0	1	3
March	...	4	5	3	5	5	3	0	1	5
April	...	2	4	3	4	4	4	1	2	6
May	...	2	4	2	3	3	3	3	4	7
June	...	2	3	2	4	3	2	3	5	6
July	...	2	3	3	3	3	3	3	5	6
August	...	3	4	3	4	4	3	1	3	6
September	...	3	5	3	5	4	4	1	1	4
October	...	3	6	4	6	5	4	0	0	3
November	...	2	5	4	7	5	3	0	1	3
December	...	2	6	5	7	5	3	0	1	2
Year	...	31	57	40	59	50	37	12	25	54

The coastal parts of Zululand, unlike those of Natal, are unhealthy and malarious, owing chiefly to the existence of lagoons and swamps; but the higher lands are much healthier, and in the dry winter season, March to September, the air is very cold and bleak.

A trying hot wind from the north-west is sometimes experienced in the interior, but the prevailing wind is from the south-east. In the hot summer months, there are heavy rains accompanied by thunder and lightning. January is usually the rainiest month. Qudeni has an average fall of 13·25 inches (336 mm.) in that month, considerably more than the fall for the whole year in many parts of the Karoo.

The figures from which the table for Maritzburg (below) is constructed show that, especially so far as the mean maximum and mean minimum temperatures are concerned, the means of the table are very good means, and very reliable. The monthly records are very regular from year to year and differ very little from the mean of the table; for instance the greatest difference from the mean maximum temperature throughout the whole seven years, for which the table is constructed, is $3°·8$ Fahr. ($2°·1$ C.), in January 1902, when the mean maximum for that month was $81°·7$ Fahr. ($27°·6$ C.), whereas the resultant mean for the period is $85°·5$ Fahr. ($29°·7$ C.). The greatest corresponding difference in the mean minimum is $3°·9$, in June 1907, when the mean minimum temperature for that month was $36°·7$ Fahr. ($2°·6$ C.), as against $40°·6$ Fahr. ($4°·7$ C.), the mean minimum for the whole period. The mean temperature of the table, being the result of $\frac{\text{max.} + \text{min.}}{2}$, should therefore be reliable. The absolute maximum for the whole period was $109°$ Fahr. ($42°·8$ C.) in January 1906, and the absolute minimum $28°$ Fahr. ($-2°·2$ C.) in July 1907 and August 1906, and therefore the extreme range of temperature during the seven years was $81°$ Fahr. ($45°$ C). In every one of the years the thermometer reached over $100°$ Fahr. ($37°·8$ C.) in one or more of the months September to February. On the other hand the thermometer touched freezing point or went below it, in addition to the two occasions just mentioned, in May, June, and August 1907, and in June 1903. The rainfall cannot be depended upon from year to year. Of the yearly totals from which the mean of $35·96$ inches (913 mm.) is derived, one was as much as $51·64$ inches (1312 mm.) and another only $26·68$ inches (678 mm.). Similarly the monthly rainfall cannot be depended upon: for instance, in February 1907, no less than $12·92$ inches (328 mm.) of rain fell, while the record for the same month in 1903 was only $2·55$ inches (65 mm.). The most regular months appear to be November and December.

PRODUCTIONS, &c.

So varied is the climate of Natal, that a habitat can be found for almost any class of plant. It is a kind of floral meeting ground, where the products of tropical regions and those of the temperate zone meet. A climate can there be found which is suitable for growing sugar cane and tea, coffee and cotton, maize and Kafir corn, wheat and barley, tobacco and cayenne pepper, oats and lucerne. Aloe fibre is produced and rubber will grow. The climate is also suitable for horse-breeding and stock-raising. All ordinary vegetables will thrive near the coast, and tropical and other fruits both near the coast and in the midland districts. In the uplands, where the wheat thrives, black-wattle grows well, especially in the Umvoti country. With a little care in the selection of the locality, anything, in fact, will grow and, not only grow, but produce fairly good results.

Natal is the southernmost country where rubber will grow, and it does well in Tongaland.

In addition to all this, there are certain parts of the country which experience has shown to be well suited for the modern treatment of the *early* stages of consumption.

DURBAN. CAPE NATAL LIGHTHOUSE. Lat. 29° 52′ 40″ S., Long. 31° 3′ 50″ E.

19 years' observations[1].

	Jan.	Feb.	March	April	May	June	July	Aug.	Sept.	Oct.	Nov.	Dec.	Year
Mean temperature (Fahr.°) ...	76°·6	77°·0	75°·5	72°·2	67°·9	65°·2	64°·5	64°·2	67°·7	69°·7	72°·7	75°·0	70°·7
Mean maximum temperature ...	85°·5	86°·0	84°·3	81°·6	77°·9	76°·1	75°·4	72°·1	76°·9	78°·2	81°·7	83°·8	79°·9
Mean minimum temperature ...	67°·7	68°·0	66°·7	62°·8	57°·9	54°·4	53°·7	56°·2	58°·6	61°·3	63°·8	66°·2	61°·4
Mean daily range (max. − min.)	17°·8	18°·0	17°·6	18°·8	20°·0	21°·7	21°·7	15°·9	18°·3	16°·9	17°·9	17°·6	18°·5
Absolute maximum temperature ...	105°·2	98°·8	97°·1	98°·1	91°·3	92°·8	93°·6	102°·9	110°·6	107°·8	104°·7	101°·7	110°·6 (period)
Absolute minimum temperature ...	56°·4	56°·4	51°·9	50°·0	45°·0	43°·2	41°·8	45°·6	46°·5	48°·4	52°·2	55°·3	41°·8 (period)
Number of rainy days	17	13	15	10	10	5	5	9	12	17	19	18	150
Mean amount of cloud (0—10) ...	6	6	5	4	3	2	3	4	5	6	6	6	5

NOTE.—The mean barometric records vary between 29·99 in January and 30·24 in June and July, the extreme range being least in February, viz. 0·80, and greatest (1·29) in September.

[1] From the *Africa Pilot*. No intimation is given as to how the means were obtained, nor as to what years are included in the 19-year period.

PIETERMARITZBURG. Lat. 29° 34' S., Long. 30° 22' E. (approx.). Alt. 2225 feet.

Compiled from 7 years' observations[1].

	Jan.	Feb.	March	April	May	June	July	Aug.	Sept.	Oct.	Nov.	Dec.	Year
Mean maximum temperature (Fahr.°)	85°.5	84°.0	80°.7	78°.6	76°.4	74°.2	73°.9	76°.1	77°.5	78°.7	79°.8	82°.3	78°.9
Mean minimum temperature	61°.5	60°.8	58°.0	54°.0	46°.8	40°.6	40°.4	45°.0	50°.9	53°.3	56°.6	59°.1	52°.2
Mean temperature[2]	73°.5	72°.4	69°.3	66°.3	62°.6	57°.4	57°.1	60°.6	64°.2	66°.0	68°.2	70°.7	65°.5
Mean daily range[3]	24°.0	23°.2	22°.7	24°.6	29°.6	33°.6	33°.5	31°.1	26°.6	25°.4	23°.2	23°.2	26°.7
Mean of absolute monthly maxima ...	102°.0	99°.8	94°.4	90°.0	87°.8	85°.4	85°.6	89°.7	97°.6	101°.0	100°.0	99°.4	—
Mean of absolute monthly minima ...	53°.7	52°.0	51°.3	44°.0	35°.0	33°.1	33°.1	33°.7	40°.0	43°.3	46°.7	50°.0	—
Mean monthly range[4]	48°.3	47°.8	43°.1	46°.0	52°.8	52°.3	52°.5	56°.0	57°.6	58°.7	57°.3	49°.4	—
Mean rainfall (inches)	5.11	6.19	5.10	2.57	1.06	0.28	0.12	0.83	1.81	2.50	5.34	5.05	35.96
Mean number of days on which rain is measured	18	16	16	8	4	2	2	3	8	14	19	18	128

NOTE.—Pietermaritzburg is a Meteorological Station under the Government Astronomer of Natal, and the above table is based on his Reports.

[1] 1901—1907.

[2] The mean temperature is derived from $\dfrac{\text{max.} + \text{min.}}{2}$.

[3] The daily range is max. – min.

[4] The monthly range is abs. max. – abs. min. for the month.

TEMPERATURE AND TOTAL ANNUAL RAINFALL.

Place	No. of years	Mean		Mean temperature	Mean of absolute yearly		Absolute (for whole period)		Range			Rainfall	
		max.	min.		max.	min.	max.	min.	Mean daily	Mean monthly	Extreme (for period)	Mean annual (inches)	Mean No. of days
Stanger	14	78°·4	59°·3	68°·8	110°	42°	116°	30°	19°·1	68°	86°	39·51	166
Verulam	14	82°·1	60°·3	71°·2	106°	41°	115°	33°	21°·8	65°	82°	36·61	110
Port Shepstone ...	7	78°·2	60°·3	68°·2	96°	45°	107°	40°	17°·9	51°	67°	48·76	129
Richmond	12	76°·3	52°·2	64°·3	102°	30°	108°	28°	24°·1	72°	80°	39·18	134
Pietermaritzburg ...	11	79°·7	52°·9	66°·3	106°	32°	109°	27°	26°·8	74°	82°	33·55	132
Howick	10	77°·2	47°·7	62°·4	100°	24°	102°	19°	29°·5	76°	83°	30·09	134
Greytown	10	79°·1	54°·9	67°·0	96°	33°	101°	24°	24°·2	63°	77°	33·71	126
Weenen	10	84°·2	49°·0	66°·6	107°	23°	109°	18°	35°·2	84°	91°	26·92	88
Estcourt	12	78°·8	48°·2	63°·5	101°	26°	109°	16°	30°·6	75°	93°	29·18	90
Newcastle	10	79°·2	57°·1	68°·1	102°	28°	105°	23°	22°·1	74°	82°	32·97	90
Umzinto	9	81°·5	56°·6	69°·0	101°	44°	106°	41°	24°·9	57°	65°	36·62	90

NOTE.—These places are Meteorological Stations under the Government Astronomer of Natal, and the above figures are based on his Reports.

MEAN RAINFALL IN INCHES.

NATAL.

	No. of Years	Jan.	Feb.	March	April	May	June	July	Aug.	Sept.	Oct.	Nov.	Dec.	Year
Durban[1] ...	10	3·89	4·93	5·12	3·53	1·87	0·65	0·73	1·22	4·62	5·76	4·19	5·70	42·21
	19	4·50	4·47	4·99	3·14	1·90	0·86	0·76	1·51	4·01	4·88	4·30	4·39	39·71
Ovington[2] ...	5	5·55	5·43	3·47	1·81	0·47	0·20	0·08	0·59	2·48	3·23	4·92	5·20	33·43
Ottawa[3] ...	13	4·57	4·33	4·06	3·82	1·42	0·98	1·46	2·01	2·13	3·27	5·91	5·20	39·16
Mount Edgecombe[3] ...	7	4·49	4·57	4·06	2·28	2·76	0·39	0·91	1·65	3·43	5·59	3·70	3·86	37·69
Pietermaritzburg[4] ...	28	4·57	4·76	4·49	2·24	0·83	0·36	0·91	1·26	1·65	3·43	5·67	5·08	34·58

[1] The 10-year figures are for the 10 years ending 1899.
[2] In the West, under the Drakensberg.
[3] On the coastal railway south of Verulam. For period ending 1907.
[4] For the period ending 1907.

ZULULAND.

	Lat.	Long.	Alt. in feet	No. of Years	Jan.	Feb.	March	April	May	June	July	Aug.	Sept.	Oct.	Nov.	Dec.	Year
Melmoth	28° 35'	31° 24'		4½	9·08	2·95	6·22	1·51	1·72	0·66	0·16	0·87	1·91	3·25	4·86	5·35	38·54
Nkandhla	28° 32'	31° 0'	3600	4½	6·72	3·96	6·32	1·76	1·46	0·85	0·19	5·74	2·08	2·87	5·22	5·94	43·11
Eshowe	28° 54'	31° 27'	1400	6½	9·17	4·53	8·23	3·36	2·26	1·17	0·66	1·72	4·68	6·86	6·12	8·23	56·99
Nongoma	...	...	2650	4½	8·90	3·32	7·28	0·79	2·08	0·62	0·60	0·36	1·58	2·94	3·80	4·64	36·91
Nondweni	...	...		4½	5·17	3·33	4·30	1·82	1·10	0·28	0·01	0·92	0·94	2·70	3·27	3·82	27·66
Qudeni	28° 31'	30° 31'	5800	2¾	13·25	4·43	7·70	2·54	3·54	1·06	0·47	0·58	2·16	6·34	6·82	7·49	56·38

NOTE.—The above table for Zululand is compiled from the Reports of the Cape Meteorological Commission.

CAPE COLONY[1].

If it be borne in mind that Africa rises from the sea to the interior by terraces, little need be said with reference to the relief. The coast plateau on the west is bounded by the mountains of Namaqualand and by the Drakenstein mountains and the southern portion is bounded on the north by the Zonderende plateau, the Langebergen and Lange Kloof mountains. In the east the coastal plateau is almost lost, the mountains coming right down to the sea. The Southern Karoo lies north of the south portion of the coastal plateau, and between the coastal range and the Zwaartebergen. The Great or Central Karoo lies north of this again and is limited northwards by the Nieuwveld and Sneeuwberg ranges. Roughly this area lies between 2000 and 3000 feet above sea-level. North of the Central Karoo are the Great Plains, with an altitude of between 3000 and 4000 feet in the southern portion; and these plains are continued northward into Bechuanaland. In the east the uplands are greatly accidented and broken, and vary in altitude from 1800 to 3000 feet and over. Basutoland has been not inaptly called the Switzerland of the Cape.

Over a large area like this, the climate naturally varies greatly, but, speaking generally it is decidedly healthy, the air is, as a rule, clear and buoyant, and the humidity low. Fogs are unknown; bright sunny days, blue skies, a soft, yet bracing, atmosphere, and starry nights are the rule. Where the two currents, the Mozambique and Benguela, the one hot, the other cold, meet, in the neighbourhood of Cape Agulhas, there is, as might naturally be expected, the greatest amount of cloud. But it is seldom that the sun does not shine throughout a whole day. There are only three stations in the Colony, at present, where the sunshine is recorded, viz. at the Royal Observatory, near Cape Town, at Kimberley and at Stutterheim. The records for Cape Town and Kimberley are almost identical for the year, namely 7·86 hours and 7·96 hours respectively for each day. Stutterheim is rather badly placed in this

---

[1] Now called the Cape of Good Hope.

respect, being shut in by mountains on the west, and therefore, though the record commences with the rising of the sun, it is lost in the latter part of the day, when the sunshine is shut out by the mountains a long time before it reaches what would be the western horizon ; and the record here is consequently much less, viz. 6·61 hours on an average, throughout the year. As a rule July is, throughout the Colony, the clearest and brightest month.

In point of climate the eastern districts are, by the majority of people, considered preferable to the western. In the former the summer is the rainy season, and in the west the winter. Taking the whole area, January and February are the hottest months, and July is the coldest. There are occasions when, and localities at which, the sun is fiercely hot, but the atmosphere being correspondingly dry, the intensity of the sun's rays is not overpowering, but is, on the contrary, quite endurable. The coastal climate is, naturally, from the proximity of the sea, equable, but in the interior a climate of a more continental type is experienced, the daily range being, of course, much greater than on the coast.

As a general rule, the rainfall throughout the Colony is light and moderate in intensity. In fact except along the south and east coasts there is a marked deficiency. It is only on exceptional occasions and in few localities that the fall takes the form of a tropical downpour. Some exceptionally heavy downpours have, however, been recorded, *e.g.* at the Kowie (Port Alfred) nearly 9·5 inches (240 mm.) of rain fell in $15\frac{1}{2}$ hours in September 1897 ; 4·5 inches (114 mm.) at Newlands, near Cape Town, in May 1899 ; 7·75 inches (197 mm.) near Port Elizabeth in November 1903, and 10·3 inches (262 mm.) near Mossel Bay in April 1905, but these stations are all on or near the coast. Thunderstorms, as already mentioned in the remarks on South Africa, are most frequent in the summer months, and reach their maximum of both frequency and intensity simultaneously with the occurrence of the greatest heat, and when, in the eastern districts, the heaviest rainfall is close at hand.

The Colony might conveniently, so far as climate is concerned, be divided into (1) a south-westerly region, the land of the winter rains, the land of European cereals, the land of the vine and the orchard, with a hot dry summer; (2) a temperate easterly region, the land of summer rains and thunderstorms, the land of maize and millet, of tobacco and of fruit trees, with a dry winter, and (3) a desert region. Or again it might be divided into (1) an area where the larger proportion of rain falls in winter, (2) a region which receives the greater part of its rain in summer, and (3) a small area which receives equal attention, in the matter of precipitation, in summer and winter. The Meteorological Commission, however, for record purposes, divides the Colony into a number of areas, which will now be considered in detail.

I. In the *Cape Peninsula*, as a whole, February is the warmest month, with a mean temperature of $69°·7$ Fahr. ($20°·9$ C.) at the Royal Observatory and $72°·0$ Fahr. ($22°·2$ C.) at Simon's Town, while the coldest months are June and July, the mean temperature being between $54°$ and $58°·4$ Fahr. ($12°·2$ and $14°·7$ C.). The mean annual temperature for the whole Peninsula is $62°·0$ Fahr. ($16°·7$ C.). The means of the extremes of temperature are about $80°$ and $47°$ Fahr. ($26°·7$ and $8°·3$ C.), and the absolute maximum and minimum $101°·3$ and $34°·1$ Fahr. ($38°·5$ and $1°·2$ C.) at the Royal Observatory and $97°$ and $47°$ Fahr. ($36°·1$ and $8°·3$ C.) respectively at Simon's Town. The daily range of temperature is less than in the neighbouring districts, and is fairly regular throughout the year, though slightly smaller in winter than in summer.

The rainfall in the Peninsula averages $38·24$ inches (969 mm.) in the year, 77 per cent. of this total falling in winter, and 23 per cent. from October to March, the whole precipitation occurring on some 92 days. But the total fall in various localities differs greatly, even though their situation may be comparatively close; thus the mean annual rainfall at the Town House, Cape Town, is $22·54$ inches (573 mm.), at the Royal Observatory $24·52$ inches (623 mm.), and at Simon's Town $27·80$ inches (706 mm.), but at Wynberg and Bishop's Court, in close proximity to the mountain, the mean annual precipitation amounts to $40·60$ and $52·46$ inches (1031 and 1334 mm.) respectively. The Peninsula receives a greater rainfall than any other of the Cape districts, Basutoland coming next with about 34 inches (860 mm.). The heaviest rain occurs in July, when the average is 6·39 inches (162 mm.) on 12 days. The winter rains are brought by the north-west winds. The average amount of cloud on the Peninsula is 46 per cent., but on top of Table Mountain it is as much as 53·3 per cent.

The following table[1] shows the number of days on which the wind blows from various directions in each month, compiled from 18 years' observations:

	N.	N.E.	E	S.E.	S.	S.W.	W.	N.W.	Calm
January	1	0	0	2	21	1	2	4	—
February	1	0	0	2	18	1	2	4	—
March	1	0	1	2	17	1	3	6	—
April	2	0	0	3	14	2	3	6	—
May	3	0	0	2	13	1	3	9	—
June	5	0	0	1	9	3	4	8	—
July	5	0	0	1	12	2	4	7	—
August	3	0	0	2	11	2	5	8	—
September	2	0	0	2	12	2	5	7	—
October	2	0	0	1	14	2	6	6	—
November	2	0	0	2	17	1	3	5	—
December	1	0	0	3	20	1	3	3	—
Year	28	0	1	23	178	19	43	73	—

[1] Extracted from the *Africa Pilot* (see p. 449).

TABLE BAY[1].

From observations for a long series of years varying from 18 to 30.

	Jan.	Feb.	March	April	May	June	July	Aug.	Sept.	Oct.	Nov.	Dec.	Year
Mean temperature (Fahr.°) ...	$69°.2$	$68°.3$	$68°.0$	$61°.7$	$57°.7$	$55°.1$	$54°.2$	$55°.7$	$57°.1$	$60°.2$	$63°.3$	$67°.1$	$61°.3$
Mean daily range	$11°.7$	$12°.5$	$12°.7$	$11°.8$	$9°.4$	$8°.6$	$9°.4$	$9°.1$	$9°.5$	$10°.7$	$10°.8$	$11°.7$	$10°.6$
Maximum range on any day ...	$26°.2$	$28°.1$	$28°.7$	$27°.9$	$25°.2$	$23°.8$	$22°.3$	$23°.3$	$23°.1$	$26°.2$	$25°.5$	$26°.4$	$28°.7$
Minimum range on any day ...	$8°.8$	$8°.1$	$7°.9$	$5°.3$	$4°.9$	$4°.7$	$5°.2$	$5°.3$	$4°.9$	$7°.1$	$8°.4$	$9°.4$	$4°.7$
Mean relative humidity (per cent.) ...	67	68	69	74	80	81	80	80	77	73	69	67	74
Mean daily range of relative humidity...	22	22	22	22	16	13	14	15	16	20	20	22	19
Mean rainfall (inches)	0·72	0·67	0·78	1·74	3·64	4·65	3·47	3·06	2·13	1·58	1·06	0·60	24·10
Mean number of rainy days	6	5	3	7	10	11	12	14	8	12	8	5	103
Mean amount of cloud (0—10)	3	3	3	4	5	5	5	5	5	4	4	3	4

NOTE.—The mean of the barometric readings varies between 29·96 in February and 30·20 in July.

(Extracted from the Admiralty publication *Africa Pilot*, Part iii, 1905, p. 566.)

[1] The authorities from which the data were taken were Buchan's *"Challenger" Report* and Symon's *Meteorological Magazine*.

WEST DIVISION.

II. The *South-West*, that is, roughly, the country in the neighbourhood of the Paarl, Ceres, Worcester, Caledon, Stellenbosch, Tulbagh, and Montagu. In this area the mean temperature varies from about 59° Fahr. at Ceres to 62° Fahr. at Worcester (15° to 16°·7 C.). The warmest month is February and the coolest July; in the former the mean is 69° Fahr. at the first place and 72° Fahr. at the second (20°·6 and 22°·2 C.), and the corresponding figures for July are 48° and 52° Fahr. (8°·9 and 11°·1 C.), the absolute maximum for Ceres being 100°·6 Fahr. (38°·1 C.), recorded in March 1902, and for Worcester 105°·3 Fahr. (40°·7 C.), recorded in the same month of the same year, while the absolute minima are 26°·5 Fahr. for Ceres and 29°·7 Fahr. for Worcester (−3°·0 and −1°·3 C.), the first having been measured in July 1903 and the second in June 1904.

The means of the extremes of daily temperature are about 85° and 34° Fahr. (29°·4 and 1°·1 C.), and the daily range varies between 30° and 23° Fahr. (16°·6 and 12°·7 C.). The heaviest rains are usually experienced in June, July and August; May is a rainy month, and September and October in a smaller degree. The annual amount of rain varies considerably from place to place. At Ceres it is 40 inches (1016 mm.), at the Paarl over 32 inches (812 mm.), and at Tulbagh only 17·5 inches (445 mm.). The percentage of cloud is less than in the Cape Peninsula. In this district as a whole the total annual rainfall is 23·12 inches (587 mm.) on 62 days, distributed as follows: June 3·09 inches (78·5 mm.) on 7 days, July 3·23 inches (82 mm.) on 8 days; May is very similar to July; August 3·10 inches (79 mm.) on 8 days; then follow September with 2·18 inches (55 mm.) on 6 days, April and October with 1·98 inches (50 mm.) on 5 and 6 days respectively, January and November with a trifle over an inch (25 mm.) and December and February with less than an inch; March is very like December. The mean relative humidity is much higher at Ceres than at Worcester according to the meteorological returns.

With regard to the tables (below), it is noticeable, from

the figures from which they are compiled, that, with regard to Ceres the absolute maximum temperature in the seven years was recorded three times in March, and three times in December, twice in February and once in November. The absolute minimum always occurred in May, June, and July, and in 1899 the same figure was recorded in all three months ($29°\cdot4$ Fahr. or $-1°\cdot5$ C.), this being the only occasion, however, when the absolute minimum for the year was recorded in May. The number of rainy days in the year varied from 65 in 1899 to 90 in 1902, and in two years it did not rain at all in February, and in one year in November. At both Ceres and Worcester the figures for the morning temperature and the mean maxima and minima are good means, *i.e.* fairly constant throughout the months from year to year. The absolute maximum, at Worcester, was recorded, during eight years, three times in February, twice each in December and March, and once in January; while the absolute minimum was recorded four times in June, twice in July, and once each in May and August. Only on two occasions, at Worcester, was the record below freezing point, while at Ceres it nearly always freezes on occasions in June, July, and August. In some years it does not rain at all in January, February and December at Worcester.

III. *West Coast*, including the western parts of Namaqualand and Clanwilliam, with Piquetberg and Malmesbury. This area is proverbially dry and hot—there being less and less rain going northwards along the coast districts—with from 17 to 19 inches (432 to 483 mm.) of rain annually round Piquetberg and Malmesbury, and only a little over 8 inches (203 mm.) at Clanwilliam, and no more than about 2 inches (50 mm.) at Port Nolloth. The mean annual temperature at Clanwilliam, one of the hottest places in the Colony, is about $65°$ Fahr. ($18°\cdot3$ C.), and at Port Nolloth $57°\cdot5$ Fahr. ($14°\cdot2$ C.). February is the warmest month, and the mean for that month is $75°\cdot3$ Fahr. ($23°\cdot1$ C.) at the former place, $59°\cdot9$ Fahr. ($15°\cdot5$ C.) at the latter. The coldest time is between June and July, when the mean is about $55°$ Fahr. ($12°\cdot8$ C.). At this time, though it is small throughout the year, there is the greatest daily range at Port Nolloth, and the least at Clanwilliam.

29—2

This daily range represents the difference between the greatest day temperature and the lowest night temperature. At Clanwilliam the average amount of cloud is 30 out of 100. At this place the absolute maximum temperature appears to be measured in any of the four months December to March, and the absolute minimum usually in June or July, though it was recorded in August in 1905 and in October in 1900. In June and July it sometimes freezes, while the absolute minimum for the period of 10 years was the August 1905 record, though it does not usually freeze in that month. The absolute maximum for the period was $110°·9$ Fahr. ($43°·8$ C.), recorded in December 1899. The figures for the relative humidity are not very good means for the months of May and July, but otherwise they are fairly good results.

Along the west coast hot, strong, dry, east winds are experienced, for two or three days at a time, from the end of November to the middle of April. These winds usually back northward, and when they have reached north-east clouds appear, but do not bring rain. The weather preceding the occurrence of these easterly winds is oppressive, but, when they have passed away, it is cool and pleasant. The barometer always rises abnormally before the arrival of these hot winds, which occasionally appear to bring about an almost total inversion of the seasons[1].

---

[1] *Science in South Africa*, 1905, p. 40.

CERES. Lat. 33° 22' S., Long. 19° 20' E. Alt. 1493 feet.

Compiled from 7 years' observations.

	Jan.	Feb.	March	April	May	June	July	Aug.	Sept.	Oct.	Nov.	Dec.	Year
Mean temperature at 8 a.m. (Fahr.°) ...	63°·6	63°·5	57°·8	48°·6	43°·6	39°·1	39°·8	42°·7	47°·6	53°·4	58°·9	64°·3	52°·0
Mean maximum temperature[1] ...	80°·9	83°·7	80°·6	73°·8	68°·2	62°·7	60°·9	63°·6	67°·9	71°·2	75°·4	81°·8	72°·6
Mean minimum temperature[1] ...	53°·6	54°·6	51°·2	44°·6	39°·1	34°·6	35°·1	37°·6	40°·9	45°·2	48°·6	52°·9	44°·9
Mean daily range[2]	27°·3	29°·1	29°·4	29°·2	29°·1	28°·1	25°·8	26°·0	27°·0	26°·0	26°·8	28°·9	27°·7
Mean temperature[3]	67°·4	69°·1	65°·9	59°·2	53°·6	48°·6	48°·0	50°·6	54°·4	58°·2	62°·0	67°·4	58°·7
Mean of absolute monthly maxima	93°·6	94°·9	94°·5	90°·4	81°·8	73°·3	72°·9	81°·6	85°·3	86°·4	90°·7	95°·4	—
Mean of absolute monthly minima	44°·5	43°·1	40°·8	35°·5	33°·2	28°·8	29°·1	31°·5	33°·1	34°·5	40°·1	43°·6	—
Mean monthly range	49°·1	51°·8	43°·7	54°·9	48°·6	44°·5	43°·8	50°·1	52°·2	51°·9	50°·6	51°·8	—
Mean relative humidity[4] (per cent.) ...	88	91	90	94	94	95	95	94	94	89	85	85	91
Mean number of rainy days ...	47	2·4	4·0	7·1	8·7	7·6	10·3	8·1	7·1	7·4	5·1	3·4	76

NOTE.—Ceres is a Meteorological Station of the second order, distant about 55 miles from the sea. All second class stations are provided with a Barometer, a Hygrometer (Dry and Wet Bulb Thermometers), a maximum and minimum Thermometer, a Rain Gauge with measure, and a Stevenson Screen.

[1] The mean maximum and minimum temperatures for 21 years are available, but the figures differ so little from the above that it has been thought better to retain the whole series for 7 years intact. The greatest differences occur in January, May and November. The 21-year mean maximum for January is 84° Fahr., but only on one occasion, viz. in 1897, has the mean maximum reached this figure during the period 1897—1903, for which the above table is compiled. Similar remarks apply to the differences between the means for the longer and shorter periods for the months of May and November, but the differences are not so great.

[2] The mean daily range is the mean of max.−min.

[3] The mean temperature is derived from $\dfrac{max.+min.}{2}$.

[4] The mean relative humidity should be accepted with the greatest caution, as it is apparently much too high.

WORCESTER. Lat. 33° 40′ S., Long. 19° 27′ E. Alt. 771 feet.

Compiled from 8 years' observations[1].

	Jan.	Feb.	March	April	May	June	July	Aug.	Sept.	Oct.	Nov.	Dec.	Year
Mean temperature at 8 a.m. (Fahr.°) ...	68°·1	67°·5	64°·8	58°·3	51°·9	47°·5	46°·9	50°·4	54°·4	59°·8	62°·7	67°·9	58°·3
Mean maximum temperature[2] ...	84°·4	85°·2	82°·9	75°·7	70°·1	65°·4	64°·1	66°·3	70°·1	73°·6	77°·8	84°·1	75°·0
Mean minimum temperature[2] ...	57°·0	59°·1	56°·6	51°·0	46°·1	42°·1	40°·7	43°·2	46°·0	49°·3	51°·0	55°·0	49°·8
Mean daily range[3] ...	27°·4	26°·1	26°·3	24°·7	30°·0	23°·3	23°·4	23°·1	24°·1	24°·3	26°·8	29°·1	25°·2
Mean temperature[4] ...	70°·7	72°·1	69°·7	63°·3	58°·1	53°·7	52°·1	54°·7	58°·1	61°·4	64°·4	69°·5	62°·4
Mean of absolute monthly maxima	97°·8	99°·4	99°·0	92°·1	82°·9	76°·8	77°·9	84°·2	86°·0	89°·0	92°·7	97°·4	—
Mean of absolute monthly minima	47°·3	47°·6	44°·8	39°·8	37°·9	32°·4	33°·3	34°·9	36°·8	38°·5	42°·6	46°·9	—
Mean monthly range ...	50°·5	50°·8	44°·2	52°·3	45°·0	44°·4	44°·6	49°·3	49°·2	50°·5	46°·1	40°·5	—
Mean relative humidity (per cent.) ...	65	71	74	80	84	82	84	84	77	71	67	65	75
Mean amount of cloud (0—10) ...	1·7	2·6	2·2	2·9	2·9	2·3	2·6	2·8	2·9	3·1	2·8	1·2	2·0
Mean number of rainy days ...	1·5	1·6	2·9	4·5	4·4	4·0	5·8	4·2	5·4	3·5	2·1	1·2	41

NOTE.—Worcester is a Meteorological Station of the second order, distant 50 miles from the nearest point on the coast. See note on p. 453.

1 1897—1904.

2 The mean maximum and minimum temperatures for 18 years are available, but the figures differ so little from the above that it has been thought better to retain the whole series for 8 years intact. The greatest difference between the mean maxima for the longer and shorter periods being 1°·5 Fahr. in any one month, and for the mean minima considerably less.

3 The mean daily range is the mean of max.−min.; similarly the mean monthly range is the mean of abs. max.−abs. min. for each month.

4 The mean temperature is derived from $\dfrac{\text{max.}+\text{min.}}{2}$.

CLANWILLIAM. Lat. 32° 10' S., Long. 18° 53' E. Alt. 245 feet.

Compiled from 10 years' observations[1]

	Jan.	Feb.	March	April	May	June	July	Aug.	Sept.	Oct.	Nov.	Dec.	Year
Mean temperature at 8 a.m. (Fahr.°) ...	70°4	69°5	66°1	60°1	52°7	48°3	48°1	49°0	53°9	60°2	65°4	69°4	59°4
Mean maximum temperature[2] ...	88°5	91°4	88°9	83°1	74°9	69°2	68°4	71°3	75°2	78°5	82°9	88°1	80°0
Mean minimum temperature[2] ...	58°2	59°2	57°3	51°5	45°9	41°2	41°5	41°6	44°6	48°1	51°8	56°3	49°8
Mean daily range[3] ...	30°3	32°2	31°6	31°6	29°0	28°0	26°9	29°7	30°6	30°4	31°1	31°8	30°2
Mean temperature[4] ...	73°3	75°3	73°1	67°3	60°4	55°2	54°9	56°4	59°9	63°3	67°3	72°2	64°9
Mean of absolute monthly maxima	103°2	105°2	103°0	98°3	91°2	80°4	84°5	88°2	93°0	95°1	99°2	102°8	—
Mean of absolute monthly minima	48°3	49°3	46°2	39°6	37°4	33°7	33°7	34°8	36°2	36°6	43°2	47°7	—
Mean monthly range[3] ...	54°9	55°9	56°8	58°7	53°8	46°7	50°8	53°4	55°8	58°5	56°0	55°1	—
Mean relative humidity (per cent.) ...	67	67	73	80	85	88	86	89	81	73	65	61	76
Mean amount of cloud (0—10) ...	2·1	1·8	2·5	3·1	3·4	3·3	3·4	3·4	3·8	3·7	2·8	2·2	2·9
Mean number of rainy days ...	2·6	1·1	1·6	3·5	7·2	6·5	6·9	5·9	4·1	5·2	2·0	0·8	47·4

NOTE.—Clanwilliam is a Meteorological Station of the second order, distant 30 miles from the nearest point on the coast. See note on p. 453.

[1] 1897—1906.

[2] The mean maximum and minimum temperatures for eight of the months for 21 years are available, but the figures differ so little from the above that it has been thought better to retain the whole series for 10 years intact. Only on one occasion, viz. in 1897, did the mean maximum, during the period 1897—1906, reach the mean maximum for the 21-year period in January. In this month the greatest difference between the long and short periods occurs.

[3] The mean daily range is the mean of max. – min., the mean monthly range the mean of abs. max. – abs. min. in each month.

[4] The mean temperature is derived from $\dfrac{\text{max.} + \text{min.}}{2}$.

PORT NOLLOTH. Lat. 29° 15′ S., Long. 16° 52′ E. Alt. 16 feet.

	Jan.	Feb.	March	April	May	June	July	Aug.	Sept.	Oct.	Nov.	Dec.	Year
Mean temperature (Fahr.°) ...	59°·5	59°·9	59°·3	57°·7	56°·8	55°·4	55°·2	53°·8	55°·0	58°·1	59°·0	60°·3	57°·6
Mean daily range	12°·3	12°·0	14°·0	15°·8	16°·2	17°·5	20°·0	17°·8	17°·6	17°·6	12°·9	13°·9	15°·7
Mean of absolute monthly maxima ...	70°·9	72°·0	82°·9	95°·4	88°·4	82°·0	83°·6	85°·5	100°·0	95°·0	89°·5	94°·0	100°
Mean of absolute monthly minima ...	44°·5	45°·3	46°·5	43°·0	38°·5	42°·5	37°·5	36°·5	36°·5	41°·5	45°·0	44°·5	36°·5
Mean relative humidity (per cent.) ...	92·5	91·5	92·0	83·5	88·0	87·0	79·0	81·5	82·0	77·5	85·5	89·0	85·7

NOTE.—These results are the means for 2 years, 1890 and 1891. The absolute maximum temperature recorded was 102°·9 Fahr. and the absolute minimum 35°·1 Fahr. These figures are taken from the *Africa Pilot*, Part II, 1901, p. 357, and it is not there stated how the mean temperature was arrived at.

SOUTH COAST.

IV. The *South Coast* division includes a narrow coast belt extending from False Bay to Port Elizabeth. The western half of this division receives over 50 per cent. of its rain in winter and the eastern half in summer. The whole area enjoys an equable climate owing to its proximity to the sea, and it is comparatively moist and warm. The range between the mean daily maximum and the mean daily minimum temperatures is never much more than $14°$ Fahr. ($7°·8$ C.), except in June, July and August, when it is slightly more. The mean daily temperature is never higher than $76°$ Fahr. ($24°·4$ C.) or lower than $51°$ Fahr. ($10°·6$ C.), while the absolute maximum in this division varies between $105°$ and $112°$ Fahr. ($40°·5$ and $44°·4$ C.), and the absolute minimum between $39°$ and $25°$ Fahr. ($3°·9$ and $-3°·9$ C.). The percentage of cloud is between 46 and 50. The mean annual rainfall at Port Elizabeth is $19·74$ inches (502 mm.), at Mossel Bay about $16·7$ inches (424 mm.), and at the Knysna some 28 inches (711 mm.). The average rainfall for the whole South Coast division has been calculated to be $24·87$ inches (632 mm.) on 79 days.

With regard to the tables (below) it is noticeable that the absolute maximum recorded at Port Elizabeth is $105°$ Fahr. ($40°·6$ C.), and that this figure has been recorded both in December and March, and $103°$ Fahr. ($39°·4$ C.) in April. The absolute minimum at the same place is $37°$ Fahr. ($2°·8$ C.) recorded in July. The mean relative humidity is almost constant throughout the year, ranging between 72 and 79 per cent. At Mossel Bay the absolute maximum temperature recorded during the eight years for which the table is compiled, is $97°·2$ Fahr. ($36°·2$ C.) measured in April 1905, and the absolute maximum for the year varies between this figure and $81°·2$ Fahr. ($27°·3$ C.), and has been recorded in all months except June, August, September, October, and November; in February and April 1898 it was $94°$ Fahr. ($34°·4$ C.), and in April 1901 the recorded figure was $94°·2$ Fahr. ($34°·5$ C.). The absolute minimum for the year, at Mossel Bay, is always recorded in June, July,

August, or September. The lowest figure for the eight years is 40°·05 Fahr. (4°·4 C.), which was measured in June, August, and September 1905. The relative humidity, though fairly even, is not so constant as at Port Elizabeth, ranging between 81 and 69 per cent.

On the south coast hot winds, called Berg winds, are experienced from north by west, between May and August, and also in April and September at times, similar to the hot easterly coast winds described under the Western Division. They last about 24 hours and occur from four to six times in these months. A Berg wind is usually preceded by a bright orange glow in the sky, lasting for about ten minutes after sunset and before sunrise.

The following table shows the number of days on which the wind blows from various directions in each month, compiled from four years' observations :

	N.	N.E.	E.	S.E.	S.	S.W.	W.	N.W.	Calm
January	0	1	5	5	1	5	11	1	2
February	0	1	5	4	1	3	9	2	3
March	1	1	3	5	1	4	10	2	4
April	1	3	4	2	2	2	9	3	4
May	2	7	4	2	1	5	5	4	1
June	2	7	2	2	0	2	8	4	3
July	2	8	3	3	0	2	6	4	3
August	3	4	3	3	1	3	9	4	1
September	2	2	4	4	1	3	10	3	1
October	1	1	4	6	1	2	11	4	1
November	1	2	3	5	1	5	10	2	1
December	1	1	4	4	1	3	12	2	3
Year	16	38	44	45	11	39	110	35	27

MOSSEL BAY. Lat. 34° 11' S., Long. 22° 9' E. Alt. 105 feet.

Compiled from 8 years' observations.[1]

	Jan.	Feb.	March	April	May	June	July	Aug.	Sept.	Oct.	Nov.	Dec.	Year
Mean morning temperature (4 years at 7.30 a.m., 2 years at 8 a.m. and 2 years at 8.30 a.m.) (Fahr°.)	68°·3	68°·6	66°·2	62°·8	57°·9	55°·2	51°·2	55°·0	56°·9	60°·7	63°·5	66°·9	61°·1
Greatest difference in above in the 8 years	3°·5	1°·5	4°·7	3°·4	3°·6	2°·2	3°·8	5°·0	2°·1	2°·1	5°·0	4°·3	1°·8
Mean maximum temperature[2]	74·8	75·3	73·7	71·4	69·5	66·3	66·4	65·3	65·1	67·4	69·8	73·5	69·9
Mean minimum temperature[2]	62·7	64·6	61·7	58·4	55·1	51·2	49·8	50·6	51·8	54·8	56·8	61·0	56·5
Mean daily range[3]	12·1	10·7	12·0	13·0	14·4	15·1	16·6	14·7	13·3	12·6	13·0	12·5	13·4
Mean temperature[4]	68·7	69·9	67·7	64·9	62·3	58·7	58·1	57·9	58·4	61·1	63·3	67·3	63·2
Mean of absolute monthly maxima	82·6	83·5	87·6	89·0	84·2	78·0	80·6	80·6	77·0	77·9	78·0	82·2	—
Mean of absolute monthly minima	55·6	56·4	54·4	50·6	48·2	46·8	44·8	44·2	44·6	47·2	49·9	54·7	—
Mean monthly range[3]	27·0	27·1	33·2	38·4	36·0	31·2	35·8	36·4	32·4	30·7	28·1	27·5	—
Mean relative humidity (per cent.)	69	76	78	81	77	74	74	78	79	75	71	74	75
Mean amount of cloud (0—10)	4·7	5·7	5·3	4·5	4·4	3·6	3·4	4·4	5·4	5·4	4·7	4·9	4·7
Mean number of rainy days	7·5	8·5	7·9	7·6	6·7	4·7	6·7	8·3	9·5	8·3	5·6	5·7	87

NOTE.—Mossel Bay is a Meteorological Station of the second order. See note on p. 453.

[1] 1897—1904.

[2] The mean maximum and minimum temperatures for 22 years are available, but the figures differ so little from the above that it has been thought better to retain the whole series for 8 years intact. The greatest difference between the means for the long and short periods occurs in June, the mean maximum for the 22-year period being 68°·8 Fahr., but in no year of the period 1897—1904 did the mean reach that figure.

[3] The daily range is max.−min. The monthly range is abs. max. − abs. min. for each month.

[4] The mean temperature is derived from $\dfrac{\text{max.}+\text{min.}}{2}$.

PORT ELIZABETH. Lat. 33° 58′ S., Long. 25° 37′ E. Alt. 181 feet (approx.).

Compiled from 20 years' observations.

	Jan.	Feb.	March	April	May	June	July	Aug.	Sept.	Oct.	Nov.	Dec.	Year
Mean maximum temperature (Fahr.°)[1]	75°·7	76°·2	74°·7	72°·1	69°·6	68°·1	66°·8	66°·6	66°·7	68°·2	71°·0	73°·8	70°·8
Mean minimum temperature[1] ...	64°·0	64°·3	61°·9	58°·1	54°·0	51°·3	49°·3	50°·9	53°·2	55°·7	58°·4	61°·4	56°·9
Mean temperature[1] ...	69°·9	70°·3	68°·3	65°·1	61°·8	59°·7	58°·1	58°·8	60°·0	62°·0	64°·7	67°·6	63°·9
Mean daily range (max.−min.) ...	11°·7	11°·9	12°·8	14°·0	15°·6	16°·8	17°·5	15°·7	13°·5	12°·5	12°·6	12°·4	13°·9
Absolute maximum temperature ...	96°	98°	105°	103°	97°	88°	89°	94°	92°	99°	99°	105°	105°·2
Absolute minimum temperature ...	53°	52°	47°	46°	40°	40°	37°	41°	40°	45°	44°	52°	37°·2
Mean relative humidity (per cent.) ...	76	77	79	78	77	75	75	78	77	75	73	72	76
Mean number of rainy days ...	5	5	6	7	6	5	6	7	7	7	7	6	74
Mean amount of cloud (0—10) ...	5	6	5	5	5	4	4	5	5	6	6	5	5

NOTE.—The mean of the barometric readings varies between 29·96 in January and 30·18 in July.

(From the *Africa Pilot*.)

[1] The mean maximum and minimum temperatures for 25 years are available, but the figures differ so little from the above that it has been thought better to retain the whole series for 20 years intact. It is not stated in the *Africa Pilot* which the precise years were, nor how the means were obtained, but the figures appear to show that the mean temperature was derived from $\frac{\text{max.}+\text{min.}}{2}$.

[2] These are the absolute maximum and minimum for the period.

KAROO.

V. The *Southern Karoo* division is a narrow band lying to the north of the South Coast belt and includes such neighbourhoods as those around Ladismith, Amalienstein, Oudtshoorn, Glenconnor, and Uniondale. In this division February is the warmest month and July the coldest. The mean temperature of the former month is (for the whole division) about $75°\cdot5$ Fahr. ($24°\cdot2$ C.) and of the latter month a trifle over $51°$ Fahr. ($10°\cdot6$ C.), the mean temperature for the year being $63°\cdot8$ Fahr. ($17°\cdot7$ C.). The highest temperature recorded is $112°$ Fahr. ($44°\cdot4$ C.) and the lowest $18°$ Fahr. ($-7°\cdot8$ C.), *i.e.* $14°$ Fahr. below freezing point, a marked difference between this inland area and the southern seaboard. The average rainfall is only $10\cdot56$ inches (268 mm.), measured on 35 days; in fact throughout the whole of the Karoo, from Tulbagh on the west to Cradock on the east, the climate is dry and healthy; in winter the nights are cold and a fire is by no means unwelcome. The daily range of temperature is large and is fairly regular, except in June and July, for in those months the maximum temperature having reached its lowest point, begins to rise, while the minimum temperature continues to sink lower and lower. The average amount of cloud is about 36 per cent.

VI. The *West Central Karoo* division is another band, narrow on the west, but expanding northwards on the east, lying north of the Southern Karoo band, and containing the towns of Prince Albert and Beaufort West, and part of the district of Graaf Reinet. Here the rainfall sinks to only 9 inches (228 mm.) per annum recorded on some 30 days, there being somewhat over an inch (25 mm.) in each of January, February, and March, the rainiest months. The elevation being higher than in the Southern Karoo, the mean temperature is only $59°\cdot8$ Fahr. ($15°\cdot4$ C.). The period from November to January has the highest mean temperature, varying from $58°$ to $59°\cdot6$ Fahr. ($14°\cdot4$ to $15°\cdot3$ C.). June and July are the coldest months, the mean temperature being about $50°$ Fahr. ($10°$ C.). The highest temperature recorded in this division is $105°$ Fahr. ($40°·6$ C.) and the lowest $28°$ Fahr. ($-2°·2$ C.). The daily range is largest in January and July, and the highest and lowest temperatures are measured in those months respectively. The division on the whole has a dry healthy climate (*see* remarks under Section V). The heaviest rainfall occurs in March, at any rate in the west portion of this area, the least in July when, on the average, there are two days on which rain is measured with a total of $0·34$ inch (8 mm.).

VII. The *East Central Karoo* section is the area which includes the neighbourhoods of Aberdeen, Graaf Reinet, Bloemhof, Somerset East, &c. The mean temperature in this section is about $63°$ Fahr. ($17°·2$ C.). June and July are the coldest months and February the warmest. The mean morning temperature is about $59°·5$ Fahr. ($15°·3$ C.). The mean maximum temperature is a trifle higher in the north than in the south, being $78°·7$ Fahr. ($26°$ C.) at Graaf Reinet and $76°·2$ Fahr. ($24°·5$ C.) at Somerset East, the mean minimum being about the same at each, and therefore the mean daily range is greater in the north than in the south. Somerset East enjoys a greater rainfall than Graaf Reinet and, as might be expected, the percentage of cloud is greater, being 42 per cent. at the former and only 28 at the latter. The figures, from which the tables (below) are derived, show that the absolute maximum is always measured in December, January, or February, and the absolute minimum in the period from May to September, it having been recorded, in various years, in all these months. The absolute maximum recorded at Graaf Reinet, during the six years for which the table is calculated, is $108°$ Fahr. ($42°·2$ C.) in January 1898, the next figure being $105°$ Fahr. ($40°·6$ C.) in February 1897. At Somerset East, during the ten years for which the table is calculated, the absolute maximum is $106°·8$ Fahr. ($41°·5$ C.) recorded in February 1901. The absolute minimum at Graaf Reinet ($29°·7$ Fahr. or $-1°·3$ C.) was recorded in both May and June 1901, and that at Somerset East ($26°·8$ Fahr. or $-2°·9$ C.) in September 1898. At the latter place the thermometer always falls below freezing point in July, almost invariably in June and August, and not infrequently in May and September. The relative humidity varies between 70·9 per cent. and 58·4 per cent. at Somerset East, and between 72·6 per cent. and 56·5 per cent. at Graaf Reinet (*see* remarks under section V).

VIII. The *Northern Karoo* section extends from a little to the east of the town of Clanwilliam eastwards, including the neighbourhoods of Sutherland, Fraserburg, Victoria West, Richmond, Hanover, Steynsberg, and Cradock. The western portion of this area has only a very small rainfall; in the centre, at Wagenaar's Kraal, the precipitation, though larger, only amounts to 11 inches (280 mm.), while further east at Hanover the corresponding figure is 13·5 inches (343 mm.), and in the south-east at Cradock 16 inches (406 mm.). It is hotter at Cradock than at either Hanover or Wagenaar's Kraal, the mean annual temperature being 62° Fahr. (16°·7 C.), as against 57° and 59° Fahr. (13°·9 and 15° C.) at the other places respectively; and, again, the hottest month at Cradock, February, has a mean temperature of 72°·3 Fahr. (22°·4 C.), as against 69°·7 Fahr. (20°·9 C.) for January, the hottest month at Hanover, and 70°·7 Fahr. (21°·5 C.) for February, the hottest month at Wagenaar's Kraal; and further, the coldest months, June and July, have a mean temperature of 50° Fahr. (10° C.) at Cradock, against 42°·3 and 46°·2 Fahr. (5°·67 and 7°·9 C.) at the other two places respectively. The absolute maximum for the year at Hanover, to judge from the figures for eight years, is always measured in either December, January, or February, but most frequently in December. The absolute maximum recorded during the period is 97° Fahr. (36°·1 C.) in December 1898 and 1899, February 1900 and January 1904. The absolute maximum recorded at Cradock is 110° Fahr. (43°·4 C.) and at Wagenaar's Kraal 101° Fahr. (38°·3 C.). The absolute minimum at Hanover is always measured in June or July in any year. From March to October almost invariably, and sometimes even in November on cold nights, the thermometer falls below freezing point. The absolute minimum recorded is 10° Fahr. (−12°·2 C.) in June and July 1898, the corresponding figures for Wagenaar's Kraal and Cradock being 18° and 19° Fahr. (−7°·8 and −7°·2 C.) respectively (*see* remarks under section V).

IX. *Northern Border.* This section includes the

eastern parts of Namaqualand and Clanwilliam Divisions, the north of the Division of Fraserburg, Hope Town, and Griqualand West with the towns of Pella, Upington, Prieska, and Kimberley. The western portion of this large area comes under what is sometimes called the Upper Karoo, and the rainfall here is very small, droughts being not infrequent. Kenhardt has a mean annual rainfall of 5·41 inches (138 mm.). The summer is very hot, the mean temperature of the warmest month being $78°·5$ Fahr. ($25°·8$ C.), and the thermometer sometimes reaches $112°$ Fahr. ($44°·4$ C.). On the other hand the nights are always cool, and in the middle of winter $12°$ Fahr. of frost have been recorded, though the mean temperature of the coldest month is $51°$ Fahr. ($10°·6$ C.). The winter is dry and nearly always bright and bracing, and, as alluded to above, there are sharp frosts at night. The eastern portion of this area includes Griqualand West, with Kimberley, and here, speaking generally, the climate is healthy, though very hot in summer; the thermometer has been known to register $108°$ Fahr. ($42°·2$ C.), though the mean temperature of the warmest month, December, is only $78°$ Fahr. ($25°·6$ C.), and of the coolest $51°·4$ Fahr. ($10°·8$ C.). The nights in winter are cold and $20°$ Fahr. ($-6°·7$ C.) has been registered. The mean annual rainfall here is 19 inches (482 mm.); most rain falls in March, followed by January, December, and February. From May to October is the dry season. At Kimberley, for which the table (below) is calculated from the figures for the ten-year period 1897–1906, it is noticeable that the absolute maximum for any year is always recorded in the period from November to February, and that the highest figures recorded are $104°·3$ Fahr. ($40°·1$ C.) in December 1898, and $104°·1$ Fahr. ($40°·0$ C.) in November 1901 and January 1903. The absolute minimum temperature is usually recorded in June or July, but sometimes also in May or August. The lowest temperature record, in the period 1897–1906, shows $21°·5$ Fahr. ($-5°·8$ C.) in July 1898. The thermometer always sinks below freezing point in May, June, July, and August, and not infrequently in September. The mean relative humidity is comparatively low and ranges between 46 and 66 per cent.

GRAAF REINET. Lat. 32° 16' S, Long. 24° 32' E. Alt. 2500 feet.

Compiled from 6 years' observations, 1897, 1898, and 1900—1903.

	Jan.	Feb.	March	April	May	June	July	Aug.	Sept.	Oct.	Nov.	Dec.	Year
Mean morning temperature (5 years at 8 a.m. and 1 year at 8.30 a.m.) (Fahr.°)	68°6	66°8	62°8	58°6	55°1	50°3	51°6	50°9	55°2	59°6	63°4	68°6	59°3
Mean maximum temperature[1]	88°9	88°6	83°6	78°8	73°3	68°1	69°3	70°6	75°2	78°4	82°0	88°1	78°7
Mean minimum temperature[1]	55°2	59°0	54°2	50°8	46°8	42°9	43°2	42°7	43°9	48°2	50°5	57°7	49°6
Mean daily range (max. − min.)	33°7	29°6	29°4	28°0	26°5	25°2	26°1	27°9	31°3	30°2	31°5	30°4	29°1
Mean temperature[2]	72°0	73°8	68°9	64°8	61°0	55°9	56°2	56°6	59°6	63°3	66°2	72°9	64°2
Mean of absolute monthly maxima	103°7	102°5	99°3	92°1	87°7	79°0	81°4	84°7	90°7	95°8	94°6	95°5	—
Mean of absolute monthly minima	45°4	49°5	41°7	39°9	37°7	32°3	31°4	31°5	35°8	35°5	39°5	44°3	—
Mean monthly range	58°3	53°0	57°6	52°2	50°0	47°7	50°0	53°2	54°9	60°3	55°1	51°2	—
Mean relative humidity (per cent.)	64	73	69	69	62	64	57	62	60	68	65	63	65
Mean amount of cloud (0—10)	2·4	3·2	2·4	2·3	2·4	1·8	1·6	2·5	3·1	3·1	2·6	2·7	2·5
Mean number of rainy days	4·8	7·0	6·2	5·2	4·0	2·3	2·3	3·0	2·8	5·8	3·7	4·3	51·4

NOTE.—Graaf Reinet is a Meteorological Station of the second order distant about 126 miles from the sea. See note on p. 453.

[1] The mean maximum and minimum for 18 years are available, but the figures differ so little from the above that it has been thought better to retain the whole series for 6 years intact. It is noticeable, however, that the means for the shorter period are good means, that is to say, no year differs very largely from the mean except in January and February, and that in these months also occur the greatest differences between the means for the long and short periods.

[2] The mean temperature is derived from $\dfrac{\text{max.} + \text{min.}}{2}$.

K. A.

30

SOMERSET EAST. Lat. 32° 44′ S., Long. 25° 35′ E. Alt. 2400 feet.

Compiled from 10 years' observations, 1897—1906.

	Jan.	Feb.	March	April	May	June	July	Aug.	Sept.	Oct.	Nov.	Dec.	Year
Mean morning temperature (3 years at 7.48 a.m., 3 years at 8 a.m., 3 years at 8.30 a.m., 1 year at 8.50 a.m.) (Fahr.°)	67°·2	67°·6	64°·6	60°·1	54°·5	50°·4	50°·3	53°·0	56°·3	59°·2	63°·3	68°·0	59°·5
Mean maximum temperature[1] ...	82°·2	85°·1	81°·0	77°·7	71°·5	66°·5	67°·6	69°·9	72°·8	75°·3	79°·6	84°·1	76°·1
Mean minimum temperature[1] ...	57°·2	58°·3	55°·2	51°·5	46°·0	42°·2	40°·7	43°·4	46°·0	47°·9	51°·0	54°·9	49°·5
Mean daily range (max. − min.) ...	25°·0	26°·8	25°·8	26°·2	25°·5	24°·3	26°·9	26°·5	26°·8	27°·4	28°·6	29°·2	26°·6
Mean temperature[2]	69°·7	71°·7	68°·1	64°·6	58°·7	54°·3	54°·1	56°·6	59°·4	61°·6	65°·3	69°·5	62°·8
Mean of absolute monthly maxima	103°·0	101°·4	96°·1	92°·0	85°·3	78°·6	79°·0	85°·1	91°·0	95°·4	98°·9	100°·8	—
Mean of absolute monthly minima	47°·1	47°·1	43°·5	38°·9	35°·7	31°·3	30°·4	32°·0	33°·4	34°·4	40°·0	44°·1	—
Mean monthly range[3]	55°·9	54°·3	52°·6	53°·1	49°·6	47°·3	48°·6	53°·1	57°·6	61°·0	58°·9	56°·7	—
Mean relative humidity (per cent.)	64	68	71	68	64	60	61	59	61	62	60	58	63
Mean amount of cloud (0—10) ...	4·5	8·1	4·9	4·2	3·3	3·0	2·6	3·2	4·3	4·8	4·3	4·0	4·0
Mean number of rainy days ...	9·8	10·9	11·1	7·9	5·4	4·3	4·2	7·5	8·7	8·7	6·2	7·5	91·9

NOTE.—Somerset East is a Meteorological Station of the second order, distant about 70 miles from the nearest point on the coast. See note on p. 453.

[1] The mean maximum and minimum temperatures for 24 years are available, but the figures differ so little from the above that it has been thought better to retain the whole series for 10 years intact. The only large differences that occur are in January and April. In the former month the maximum for the longer period is 2°·5 higher and in the latter 2°·4 lower than for the shorter period, and in neither case are the means good, i.e. the years vary considerably.

[2] The mean temperature is derived from $\dfrac{\text{max.} + \text{min.}}{2}$.

[3] The monthly range is absolute maximum − absolute minimum for the month.

HANOVER. Lat. 31° 3′ S., Long. 24° 26′ E. Alt. 4500 feet.

Compiled from 8 years' observations, 1898—1905.

	Jan.	Feb.	March	April	May	June	July	Aug.	Sept.	Oct.	Nov.	Dec.	Year
Mean temperature at 8 a.m. (Fahr.°)	69°0	68°1	60°5	53°3	44°2	35°9	35°4	41°3	50°3	56°1	60°3	68°2	53°5
Mean maximum temperature	85°0	83°9	78°1	72°7	65°3	58°2	58°1	63°3	68°7	72°8	79°6	85°4	72°6
Mean minimum temperature	54°4	53°9	48°7	41°6	34°4	26°9	26°2	29°6	36°8	41°6	47°0	52°8	41°2
Mean daily range[1]	33°1	30°0	29°4	31°1	30°9	31°3	31°9	33°7	31°9	31°2	32°6	32°6	31°4
Mean temperature[2]	69°7	68°9	63°4	57°1	49°8	42°5	42°1	46°4	52°7	57°2	63°3	69°1	56°9
Mean of absolute monthly maxima	93°7	94°0	89°0	82°1	76°9	66°7	66°1	74°6	80°7	86°1	91°5	94°4	—
Mean of absolute monthly minima	46°1	43°1	37°0	29°0	22°9	17°0	16°4	20°9	22°7	28°7	34°0	41°4	—
Mean monthly range[1]	47°6	50°9	52°0	53°1	54°0	49°7	49°7	53°7	58°0	57°4	57°5	53°0	—
Mean relative humidity (per cent.)	72	72	75	75	77	71	71	71	71	71	67	67	72
Mean number of rainy days	4·2	4·5	5·5	4·1	2·2	2·0	1·2	1·1	2·7	3·0	1·5	2·9	35

NOTE.—Hanover is a Meteorological Station of the third order, or, as it is called, a Climatological Station, distant about 100 miles from the nearest point on the coast.

[1] The daily range is max.−min.; the mean monthly range the difference between the mean of absolute maxima and the mean of absolute minima.

[2] The mean temperature is derived from $\dfrac{max.+min}{2}$.

KIMBERLEY. Lat 28° 43′ S., Long. 24° 46′ E. Alt. 4042 feet.

Compiled from 10 years' observations, 1897—1906.

	Jan.	Feb.	March	April	May	June	July	Aug.	Sept.	Oct.	Nov.	Dec.	Year
Mean morning temperature (6 years at 8 a.m., 3 years at 8.30 a.m., and 1 year at 9 a.m.) (Fahr.°)	77°·2	73°·7	68°·0	61°·8	52°·8	46°·2	46°·2	51°·3	59°·6	66°·8	73°·1	76°·7	62°·8
Mean maximum temperature[1]	92°·3	91°·7	86°·1	80°·5	73°·0	66°·5	68°·3	72°·9	79°·0	84°·7	89°·6	93°·3	81°·5
Mean minimum temperature[1]	61°·8	61°·9	57°·3	51°·1	42°·6	36°·3	36°·2	39°·8	45°·6	51°·1	56°·1	62°·5	50°·2
Mean daily range[2]	30°·5	29°·8	28°·8	29°·4	30°·4	30°·2	32°·1	33°·1	33°·4	33°·6	33°·5	30°·8	31°·3
Mean temperature[3]	76°·0	76°·8	71°·7	67°·8	57°·8	51°·4	52°·2	56°·3	62°·3	67°·9	72°·8	77°·9	65°·8
Mean of absolute monthly maxima	101°·6	101°·7	94°·9	88°·3	84°·2	76°·4	77°·6	83°·5	92°·8	96°·9	101°·2	100°·8	—
Mean of absolute monthly minima	54°·5	54°·0	48°·3	39°·6	29°·9	27°·3	26°·4	28°·1	32°·5	39°·0	43°·8	51°·5	—
Mean monthly range[2]	47°·1	47°·7	46°·6	48°·7	54°·3	49°·9	51°·2	55°·4	60°·3	57°·9	57°·4	49°·3	—
Mean relative humidity (per cent.)	47	55	62	66	64	66	63	56	52	47	46	49	56
Mean amount of cloud (0—10)	4·7	4·8	4·2	3·8	3·3	2·7	1·9	2·7	3·5	3·8	2·7	4·2	3·6
Mean number of rainy days	8·8	10·5	9·3	7·0	3·1	1·8	1·5	1·0	2·8	4·0	5·0	8·1	62·9

NOTE.—Kimberley is a Meteorological Station of the second order, distant about 340 miles from the nearest point on the coast. See note on p. 453.

[1] The mean maximum and minimum for eight of the months for 22 years are available, but the figures differ so little from the above that it has been thought better to retain the whole series for 10 years intact.

[2] The daily range is max.−min.: the mean monthly range is the difference between the mean of absolute maxima and the mean of absolute minima.

[3] The mean temperature is derived from $\dfrac{\text{max.}+\text{min.}}{2}$.

X. *South-East.* This section includes the country round the towns Grahamstown, King William's Town, East London, Fort Beaufort, Alexandria, and Alice, and generally speaking, the climate is more bracing inland than on the coast. The absolute minimum temperature recorded at Grahamstown is $29°$ Fahr. $(-1°·7$ C.); at King William's Town $28°$ Fahr. $(-2°·2$ C.) was recorded in July 1900, and at Lovedale $27°$ Fahr. $(-2°·8$ C.) in July 1898; while at East London the absolute minimum is $40°$ Fahr. $(4°·4$ C.), and this was recorded in July 1898, August 1900, June 1903, and August 1905. The mean temperature of the coldest month at Grahamstown is $54°·6$ Fahr. $(12°·5$ C.), at King William's Town $55°·7$ Fahr. $(13°·2$ C.), at Lovedale $54°·4$ Fahr. $(12°·5$ C.), but at East London $60°$ Fahr. $(15°·6$ C.). The mean annual temperature at the same places is $62°·3$, $65°·1$, $64°·1$, and $64°·4$ Fahr. $(16°·8$, $18°·4$, $17°·8$, and $18°$ C.) respectively. The highest recorded temperature at Grahamstown is $101°$ Fahr. $(38°·3$ C.). During the six years for which the Lovedale table (below) is constructed, the absolute maximum there recorded was $113°·5$ Fahr. $(45°·3$ C.) in December 1899. During the ten years 1897—1906 the highest temperature recorded at King William's Town was $115°$ Fahr. $(46°·1$ C.) in December 1899, while during the same period at East London the highest record was $98°$ Fahr. $(36°·7$ C.) in August 1901. The rainfall in this section varies considerably. At King William's Town it is about 26·25 inches (667 mm.) per annum, at Grahamstown it is nearer 30 inches (762 mm.) and at East London only 23·4 inches (594 mm.), while at Stutterheim, in the same section, the mean annual precipitation amounts to 31·5 inches (800 mm.). The average for the section, taken as a whole, is 26 inches (660 mm.), measured on 74 days. December, January, February, and March are the rainiest months with a little over 3 inches (76 mm.) each on eight or nine days, and July is the dryest month with a little less than half an inch on the average. The average amount of cloud at Lovedale and King William's Town is 3·5 out of 10, while at East London it is over 5 out of 10.

LOVEDALE. Lat. 32° 46′ S., Long. 26° 51′ E. Alt. 1720 feet.

Compiled from 6 years' observations, 1897—1900, 1903, 1904.

	Jan.	Feb.	March	April	May	June	July	Aug.	Sept.	Oct.	Nov.	Dec.	Year
Mean morning temperature (4 years at 8 a.m. and 2 years at 8.30 a.m.) (Fahr.°)	68°·5	67°·6	63°·9	60°·1	53°·1	49°·1	49°·9	52°·5	56°·6	60°·3	64°·6	68°·9	59°·6
Mean maximum temperature[1] ...	84°·0	84°·9	82°·0	78°·8	72°·6	67°·9	68°·8	72°·5	76°·1	77°·1	80°·0	84°·4	77°·4
Mean minimum temperature[1] ...	60°·5	60°·2	56°·9	52°·9	46°·1	40°·9	40°·3	43°·1	47°·2	50°·8	53°·8	58°·4	50°·9
Mean daily range[2]	23°·5	24°·7	26°·1	25°·9	26°·5	27°·0	28°·5	29°·4	29°·0	26°·3	26°·2	26°·0	26°·5
Mean temperature[3]	72°·2	72°·5	69°·5	65°·8	59°·3	54°·4	54°·5	57°·8	61°·7	63°·9	66°·9	71°·4	64°·1
Mean of absolute monthly maxima	101°·2	101°·5	98°·9	94°·7	86°·2	80°·6	81°·8	87°·5	94°·4	98°·5	102°·3	103°·7	—
Mean of absolute monthly minima	50°·5	48°·7	45°·3	39°·8	35°·8	31°·3	29°·6	33°·2	34°·2	37°·3	44°·3	48°·4	—
Mean monthly range[2]	50°·7	52°·8	53°·6	54°·9	50°·4	49°·3	52°·2	54°·6	60°·2	61°·2	58°·0	55°·3	—
Mean relative humidity (per cent.)	76	79	83	83	82	75	77	73	75	72	74	71	77
Mean amount of cloud (0—10) ...	4·7	3·9	3·3	3·5	2·5	1·9	2·2	2·6	2·8	3·6	3·2	3·2	3·1
Mean number of rainy days ...	9·7	8·2	8·7	6·3	4·7	3·0	3·7	4·9	5·0	7·7	6·0	6·1	74

NOTE.—Lovedale is a Meteorological Station of the second order, distant about 50 miles from the sea. See note on p. 453.

[1] The mean maximum and minimum temperatures for eight of the months for 19 years are available, but the figures differ so little from the above that it has been thought better to retain the whole series for 6 years intact.

[2] The daily range is max.−min.; the mean monthly range is the difference between the mean of absolute maxima and the mean of absolute minima.

[3] The mean temperature is derived from $\dfrac{max.+min.}{2}$.

KING WILLIAM'S TOWN. Lat. 32° 52' S., Long. 27° 23' E. Alt. 1314 feet.

Compiled from 10 years' observations, 1897—1906.

	Jan.	Feb.	March	April	May	June	July	Aug.	Sept.	Oct.	Nov.	Dec.	Year
Mean morning temperature (6 years at 8 a.m., and 4 years at 8.30 a.m.) (Fahr.°)	71°·5	71°·3	67°·6	64°·5	58°·3	52°·8	52°·7	57°·3	61°·5	63°·8	67°·4	71°·4	63°·3
Mean maximum temperature[1] ...	84°·8	85°·4	82°·7	80°·7	76°·0	70°·7	72°·8	73°·2	75°·4	76°·9	79°·6	83°·9	78°·5
Mean minimum temperature[1] ...	60°·2	61°·4	57°·7	54°·0	46°·5	41°·7	41°·5	44°·4	48°·5	51°·5	54°·2	59°·3	51°·7
Mean daily range[2] ...	24°·6	24°·0	25°·0	26°·7	29°·5	29°·0	31°·3	28°·8	26°·9	25°·4	25°·4	24°·6	26°·8
Mean temperature[3] ...	72°·0	73°·4	70°·2	67°·3	61°·2	55°·7	57°·1	58°·8	61°·9	64°·2	66°·9	72°·6	65°·1
Mean of absolute monthly maxima	102°·1	101°·9	102°·2	96°·5	92°·3	84°·7	86°·1	92°·9	96°·8	98°·6	101°·7	106°·0	—
Mean of absolute monthly minima	49°·9	50°·3	48°·4	41°·9	35°·1	33°·7	31°·4	33°·3	36°·5	38°·9	44°·0	48°·0	—
Mean monthly range[2] ...	52°·2	51°·6	53°·8	54°·6	57°·2	51°·0	54°·7	59°·6	60°·3	59°·7	57°·7	58°·0	—
Mean relative humidity (per cent.) ...	65	70	72	73	68	70	70	65	67	68	61	61	67
Mean amount of cloud (0—10) ...	4·3	4·2	4·0	3·3	3·1	2·5	2·6	3·0	4·0	4·5	3·7	3·7	3·6
Mean number of rainy days ...	11·1	10·2	10·7	8·1	5·5	4·6	2·9	5·3	7·7	10·1	8·6	8·7	93·5

NOTE.—King William's Town is a Meteorological Station of the second order, distant about 25 miles from the coast. See note on p. 453.

[1] The mean maximum and minimum temperatures for nine of the months for 23 years are available, but the figures differ so little from the above that it has been thought advisable to retain the whole series for 10 years intact.

[2] The daily range is max.−min.; the monthly range is abs. max.−abs. min.

[3] The mean temperature is derived from $\dfrac{\text{max.}+\text{min.}}{2}$.

EAST LONDON. Lat. 33° 2' S., Long. 27° 55' E. Alt. 33 feet.

Compiled from 10 years' observations, 1897—1906.

	Jan.	Feb.	March	April	May	June	July	Aug.	Sept.	Oct.	Nov.	Dec.	Year
Mean morning temperature (6 years at 8 a.m. and 4 years at 8.30 a.m.) (Fahr.°)	70°·5	69°·4	66°·9	64°·6	59°·5	55°·3	55°·2	57°·4	61°·2	63°·7	66°·7	69°·0	63°·3
Mean maximum temperature¹ ...	75·8	74·7	73·0	72·3	71·3	70·4	69·9	69·5	68·4	69·9	71·6	73·9	71·7
Mean minimum temperature¹ ...	64·6	64·0	61·7	59·6	53·9	49·9	49·7	51·5	54·9	57·0	59·4	63·0	57·4
Mean daily range... ...	11·2	10·7	11·3	12·7	17·4	20·5	20·2	18·0	13·5	12·9	12·2	10·9	14·3
Mean temperature ...	70°·2	69°·3	67°·3	65°·9	62°·6	60°·1	59°·8	60°·5	61°·6	63°·5	65°·5	68°·5	64°·5
Mean of absolute monthly maxima	83°·0	81°·0	82°·0	83°·0	88°·0	85°·0	85°·0	91°·0	83°·0	84°·0	80°·0	80°·0	—
Mean of absolute monthly minima	57°·0	57°·0	54°·0	51°·0	45°·0	43°·0	42°·0	43°·0	46°·0	46°·0	50°·0	55°·0	—
Mean monthly range ...	26°·0	24°·0	28°·0	32°·0	43°·0	42°·0	43°·0	48°·0	37°·0	38°·0	30°·0	25°·0	—
Mean relative humidity (per cent.) ...	80	81	83	82	77	74	75	76	77	75	76	77	78
Mean amount of cloud (0—10) ...	5·7	5·7	5·3	4·8	4·3	4·1	3·8	4·5	5·4	6·0	5·5	5·6	5·1
Mean number of rainy days ...	8·7	8·2	9·6	6·0	4·6	4·0	3·6	5·2	6·6	9·0	7·4	7·9	80·8

NOTE.—East London is a Meteorological Station of the second order on the coast. See note on p. 453.

¹ The mean maximum and minimum temperatures for 25 years are available, but the figures differ so little from the above that it has been thought better to retain the whole series for 10 years intact. The greatest difference between the long and short series is 1°·2 in March, when the 25-year period gives a mean maximum temperature of 74°·4. In only one year, however, of the 10-year period (1897—1906) was this figure reached, namely in 1902, when the mean was 74°·7. For method of obtaining the means, *see* under King William's Town.

XI. The *North-East* section embraces the area around Aliwal North and Queen's Town. The mean annual rainfall in this area varies between about $24\cdot5$ and 26 inches (622 and 660 mm.), the rainiest months being January, March, and December, July being the dryest month, taking the section as a whole. The number of days on which rain is measured appears to vary between 60 and 70 in the year. The warmest months are November, December, January, and February, the last-named being the hottest, with a mean temperature of about 70° Fahr. (21° C.). The mean temperature of the coldest month varies between $46^{\circ}\cdot5$ and 51° Fahr. ($8^{\circ}\cdot1$ and $10^{\circ}\cdot6$ C.). The tables (below) for Queen's Town and Aliwal North are calculated for the ten-year period 1897—1906, and it appears from these figures that the absolute maximum temperature may be measured, in any one year, in any of the four months from November to February. The highest recorded temperature for Aliwal North is 97° Fahr. ($36^{\circ}\cdot1$ C.) in January 1899, and for Queen's Town 104° Fahr. (40° C.) in January 1903, and 103° Fahr. ($39^{\circ}\cdot4$ C.) has been recorded in both December and February, the former in 1902, the latter in 1901. The absolute minimum at Aliwal North is always recorded in either June or July, but at Queen's Town it may occur in any of the four months from May to August. The lowest recorded temperature at the former is 15° Fahr. ($-9^{\circ}\cdot4$ C.), measured in July 1898, and at the latter 20° Fahr. ($-6^{\circ}\cdot7$ C.) recorded in 1898. The mean annual temperature for the whole area is a little below the average of 62° Fahr. ($16^{\circ}\cdot7$ C.), that at Aliwal North being 60° Fahr. ($15^{\circ}\cdot6$ C.), and that at Queen's Town 62° Fahr. At Aliwal North the thermometer invariably sinks below freezing point some time in May, June, July, August, and September, and not infrequently in April and October; while at Queen's Town it never freezes in April.

ALIWAL NORTH. Lat. 30° 41' S., Long. 26° 40' E. Alt. 4330 feet.

Compiled from 10 years' observations, 1897—1906.

	Jan.	Feb.	March	April	May	June	July	Aug.	Sept.	Oct.	Nov.	Dec.	Year
Mean morning temperature (6 years at 8 a.m., and 4 years at 8.30 a.m.) (Fahr.°)	69°·6	68°·2	62°·8	55°·5	44°·7	37°·7	35°·5	44°·8	55°·0	61°·6	66°·8	71°·0	56°·1
Mean maximum temperature[1] ...	84°·2	82°·6	77°·8	74°·2	68°·6	63°·4	64°·8	67°·7	72°·6	75°·7	79°·8	84°·6	74°·6
Mean minimum temperature[1] ...	57°·3	56°·7	53°·4	45°·7	35°·5	29°·6	29°·6	34°·0	41°·0	46°·1	50°·1	55°·8	44°·5
Mean daily range[2]	26°·9	25°·9	24°·4	28°·5	33°·1	33°·8	35°·2	33°·7	31°·6	29°·6	29°·7	28°·8	30°·1
Mean temperature[2] ...	70°·7	69°·6	65°·6	59°·9	52°·0	46°·5	47°·2	50°·8	56°·8	60°·9	64°·9	70°·2	59°·6
Mean of absolute monthly maxima	92°·9	93°·0	86°·6	81°·9	78°·0	71°·6	72°·0	78°·7	85°·2	87·7	92°·6	94°·6	—
Mean of absolute monthly minima	47°·6	47°·4	41°·3	35°·0	25°·5	20°·1	20°·9	22°·7	27°·9	31°·5	37°·5	45°·0	—
Mean monthly range[2]	45°·3	45°·6	45°·3	46°·9	52°·5	51°·5	51°·1	56°·0	57°·3	56°·2	55°·1	49°·6	—
Mean relative humidity (per cent.)	60	66	76	78	77	79	79	65	58	56	53	53	66
Mean amount of cloud (0—10) ...	3·3	3·5	3·3	3·0	3·0	3·0	2·4	2·7	4·0	3·8	2·8	2·6	3·1
Mean number of rainy days ...	9·4	11·2	10·1	7·3	5·4	2·8	1·7	3·2	4·4	4·9	3·8	8·2	72·9

NOTE.—Aliwal North is a Meteorological Station of the second order, distant 165 miles from the nearest point on the coast. See note on p. 453.

[1] The mean maximum and minimum temperatures for 25 years are available, but the figures differ so little from the above that it has been thought better to retain the whole series for 10 years intact. The mean maxima for the longer period are lower than for the shorter 10-year (1897—1906) period in May and July by 2°·0 and 2°·3 respectively, but only in one year of the shorter period, viz. 1899, did the means for these months sink so low as the long-period means.

[2] The daily range is max. − min. ; the monthly range is abs. max. − abs. min.; the mean temperature $\dfrac{\text{max.} + \text{min.}}{2}$.

QUEEN'S TOWN. Lat. 31° 54′ S., Long. 26° 52′ E. Alt. 3500 feet.

Compiled from 10 years' observations, 1897—1906.

	Jan.	Feb.	March	April	May	June	July	Aug.	Sept.	Oct.	Nov.	Dec.	Year
Mean morning temperature (6 years at 8 a.m., and 4 years at 8.30 a.m.) (Fahr.°)	66°·6	66°·0	62°·2	57°·0	49°·5	45°·3	46°·7	49°·7	55°·7	59°·3	62°·6	66°·0	57°·2
Mean maximum temperature[1] ...	85°·1	83°·7	79°·5	77°·0	70°·8	65°·5	66°·9	69°·0	73°·4	77°·9	81°·7	85°·5	76°·3
Mean minimum temperature[1] ...	58°·1	57°·5	54°·7	49°·2	40°·6	36°·6	36°·6	39°·2	44°·4	48°·4	51°·3	56°·4	47°·8
Mean daily range[2]	27°·0	26°·2	24°·8	27°·8	30°·2	28°·9	30°·3	29°·8	29°·0	29°·5	30°·4	29°·1	28°·5
Mean temperature[2]	71°·6	70°·6	67°·1	63°·1	55°·7	51°·0	51°·7	54°·1	58°·9	63°·1	66°·5	70°·9	62°·0
Mean of absolute monthly maxima	99°·5	98°·0	91°·9	87°·6	82°·6	74°·5	75°·1	81°·4	88°·9	94°·0	97°·5	98°·5	—
Mean of absolute monthly minima	48°·4	49°·9	44°·0	37°·4	29°·0	24°·7	24°·3	25°·6	29°·2	33°·2	38°·5	44°·4	—
Mean monthly range[2]	51°·1	48°·1	47°·9	50°·2	53°·6	49°·8	50°·8	55°·8	59°·5	60°·8	59°·0	54°·1	—
Mean relative humidity (per cent.)	69	74	77	78	71	66	63	62	62	65	64	66	68
Mean amount of cloud (0—10) ...	4·2	4·5	4·7	3·8	3·3	2·4	2·2	3·1	4·2	3·9	3·5	3·7	3·6
Mean number of rainy days ...	10·0	11·1	9·8	7·4	3·9	3·4	1·9	4·1	6·4	6·4	5·7	8·6	78·7

NOTE.—Queen's Town is a Meteorological Station of the second order, distant about 100 miles from the coast. See note on p. 453.

[1] The mean maximum and minimum temperatures for 25 years are available, but the figures differ so little from the above that it has been thought better to retain the whole series for 10 years intact. The mean maxima for the longer period are slightly lower throughout, the greatest differences occurring in April and December, and only in 2 years of the 10-year (1897—1906) period did the means sink so low as those for the longer period in these months.

[2] See note under Aliwal North.

XII. *Kaffraria*. The rainfall on the coast at Port St John's is heavy, amounting to over 42 inches (1067 mm.) in the year. A little inland at Umtata it is only 23 inches (584 mm.), though directly north of Port St John's, at an elevation of 6850 feet above sea-level, in the Kokstadt neighbourhood, it is over 31 inches (787 mm.). Some 74 per cent. of this rain falls during the summer months, and is distributed over between 80 and 100 days in the year. Every month has its rain, but May, June, July, and August are the dryest months, while at Umtata the months of June and July, in the year 1898, were absolutely rainless. In the table at p. 479, the mean precipitation for January, April, and July at Umtata is probably above the average, in consequence of a fall of 9·5 inches (241 mm.) in January 1898, of over 5 inches (127 mm.) in April 1903, and of over 7 inches (178 mm.) in July 1902, all of which considerably swell the totals from which the means are derived. The annual rainfall at this station varies between 17·34 inches in 1899 and 29·71 inches (440 and 755 mm.) in 1898. At Port St John's, the total annual fall is very irregular and varies between 23·75 inches in 1899 and 54·07 inches in 1897 (603 and 1372 mm.). The monthly fall, too, cannot be depended on. The mean for March is 4 inches (100 mm.), but over 11 inches (280 mm.) fell in that month in 1897. Again the mean for April is 2·94 inches (75 mm.), but 13·25 inches (336 mm.) fell in that month in 1903; and similarly there was a total fall of over 12 inches (305 mm.) in June 1902, though the mean for that month is 2·86 inches (73 mm.). The absolute maximum temperature recorded at Port St John's, during the ten years for which the table is compiled, was 104° Fahr. (40° C.) in October 1898, while 100° Fahr. (37°·8 C.) was recorded in 1901, and 101° Fahr. (38°·3 C.) both in September and October 1904, these being the only instances when 100° Fahr. and over was recorded during this period; and the figure 99° Fahr. (37°·2 C.) only occurs twice, viz. in August 1899 and October 1905. At Umtata 100° Fahr. (37°·8 C.) and over was recorded every year, almost always in two months, and in some years in three. The highest recorded temperature during the ten-year period 1897—1906 was 110° Fahr. (43°·4 C.) in November 1897, while 108° Fahr. (42°·2 C.) was measured in February 1900, and in the first-mentioned year there were recorded in October 101°, and in December 100° Fahr. (38°·3 and 37°·8 C.). The lowest temperature at Port St John's was 40° Fahr. (4°·4 C.), recorded in October 1898 and 1901, and at Umtata 21° Fahr. ($-6°·1$ C.) in July 1898. Thus the extreme range of temperature at the first place was 64° Fahr. (35°·6 C.), and being on the coast this is of course much less than the extreme range at Umtata, viz. 89° Fahr. (49°·4 C.). Taking Kaffraria as a whole, the mean for the hottest month varies between 72·5° Fahr. in the low-lying districts, and 61° Fahr. in the higher parts (22°·5 and 16°·1 C.), while the mean for the coldest month is 47° Fahr. (8°·3 C.), that at Port St John's being 61° Fahr. (16°·1 C.). The amount of cloud is high, especially during the summer or rainy months.

PORT ST JOHN'S. Lat. 31° 38′ S., Long. 29° 35′ E. Alt. 22 feet.

Compiled from 10 years' observations, 1897—1906.

	Jan.	Feb.	March	April	May	June	July	Aug.	Sept.	Oct.	Nov.	Dec.	Year
Mean morning temperature observed partly at 8 a.m, and partly at 8.30 a.m. (Fahr.°)	73°·5	74°·2	71°·7	69°·5	65°·2	61°·3	61°·4	62°·4	64°·9	67°·3	70°·1	72°·7	67°·8
Mean maximum temperature	79°·4	79°·8	78°·1	77°·6	74°·5	71°·8	71°·8	72°·0	72°·6	73°·7	75°·2	78°·6	75°·4
Mean minimum temperature	65°·4	64°·8	62°·3	60°·9	55°·6	51°·9	50°·4	50°·5	54°·7	56°·5	60°·2	63°·7	58°·1
Mean temperature[1]	72°·4	72°·3	70°·2	69°·2	65°·0	61°·8	61°·1	61°·3	63°·7	65°·1	67°·7	71°·1	66°·8
Mean daily range[2]	14°·0	15°·0	15°·8	16°·7	18°·9	19°·9	21°·4	21°·5	17°·9	17°·2	15°·0	14°·9	17°·3
Mean of absolute monthly maxima ...	86°·6	85°·9	85°·6	85°·5	85°·2	81°·3	82°·9	89°·9	89°·3	89°·0	85°·0	85°·0	—
Mean of absolute monthly minima ...	57°·0	58°·7	53°·3	52°·1	49°·8	46°·2	43°·3	44°·7	43°·2	45°·8	52°·4	55°·2	—
Mean monthly range[3]	29°·6	27°·2	32°·3	31°·4	35°·4	35°·1	39°·6	45°·2	46°·1	43°·2	32°·6	29°·8	—
Mean relative humidity (per cent.) ...	80	79	79	77	66	67	65	67	72	73	76	76	73
Mean rainfall in inches	5·53	4·10	4·00	2·94	1·85	2·86	·75	1·48	3·51	5·26	5·05	4·73	42·06
Mean number of rainy days	16·1	10·1	7·8	7·3	4·9	3·4	3·3	5·8	6·7	10·9	11·0	10·5	97·8
Mean amount of cloud (0—10)	6·7	6·1	5·9	5·2	4·6	5·0	4·7	4·8	5·9	6·6	6·1	6·7	57

[1] The mean temperature is derived from $\dfrac{\text{max.}+\text{min.}}{2}$.

[2] The daily range is max. − min.

[3] The monthly range is abs. max. − abs. min.

NOTE.—Port St John's is a Meteorological Station of the second order. See note on p. 453.

UMTATA. Lat. 31° 35′ S., Long. 28° 46′ E. Alt. 2400 feet.

Compiled from 10 years' observations, 1897—1906.

	Jan.	Feb.	March	April	May	June	July	Aug.	Sept.	Oct.	Nov.	Dec.	Year
Mean morning temperature (Fahr.°) observed partly at 7.30 a.m., partly at 8 a.m., and partly at 8.30 a.m.	58°·9	68°·6	65°·0	61°·0	52°·3	45°·2	45°·7	52°·1	58°·3	62°·5	66°·3	69°·4	59°·6
Mean maximum temperature ...	80°·8	81°·2	78°·2	78°·1	74°·6	69°·6	71°·9	72°·3	74°·4	75°·2	78°·0	81°·2	76°·3
Mean minimum temperature ...	60°·3	60°·6	58°·1	52°·1	43°·1	36°·7	36°·8	41°·3	47°·0	51°·0	54°·4	58°·5	50°·0
Mean temperature[1] ...	70°·5	70°·9	68°·2	65°·1	58°·8	53°·2	54°·3	56°·8	60°·7	63°·1	66°·2	69°·8	63°·1
Mean daily range[2] ...	20°·5	20°·6	20°·1	26°·0	31°·5	32°·9	35°·1	31°·0	27°·4	24°·2	23°·6	22°·7	26°·3
Mean of absolute monthly maxima ...	99°·0	98°·7	93°·3	91°·6	87°·4	82°·9	83°·2	89°·3	94°·7	95°·8	99°·7	99°·2	—
Mean of absolute monthly minima ...	50°·2	52°·4	49°·4	39°·8	30°·7	28°·3	28°·0	29°·6	33°·6	35°·9	42°·6	48°·8	—
Mean monthly range[3] ...	48°·8	46°·3	43°·9	51°·8	56°·7	54°·6	55°·2	59°·7	61°·1	59°·9	57°·1	50°·4	—
Mean relative humidity (per cent.) ...	74	75	81	81	79	77	77	73	73	70	68	69	75
Mean rainfall in inches ...	3·50	3·58	3·06	1·39	·92	1·67	·33	·62	1·25	2·03	2·17	2·97	23·49
Mean number of rainy days ...	15	12	13	7	4	3	3	4	7	11	11	12	102
Mean amount of cloud (0—10) ...	6·4	6·0	6·0	4·6	3·6	3·2	3·0	3·8	4·9	5·8	5·4	5·6	4·9

Note.—Umtata is a Meteorological Station of the second order, distant about 35 miles from the nearest point on the coast. See note on p. 453.

[1] The mean temperature is derived from $\frac{\text{max.}+\text{min.}}{2}$.

[2] The daily range is max. − min.

[3] The monthly range is abs. max. − abs. min.

In *Basutoland*, frequently alluded to as the Switzerland of the Cape, the rainfall varies between 37 and 19 inches (940 and 482 mm.), 77 per cent. of which falls in the summer months. The rainfall, moreover, is not dependable from month to month. In January 1900 no less than $9·08$ inches (229 mm.) fell at Mohalie's Hoek and only $1·23$ inches (31 mm.) in the same month in 1903. Similarly at Teyateyaneng, $9·61$ inches (244 mm.) fell in March 1902, and only $0·32$ inch (8 mm.) in the same month in 1903. One usually, sometimes two or three, of the months May, June, July, and August are absolutely rainless, but there is sure to be *some* rain in the remaining month or months, though very little. The number of rainy days in the year varies between 46 and 81. In the tables given below, which cover a period of six years, the relative humidity means are fairly good means, but that for the August at Teyateyaneng is not dependable, being derived from such divergent figures as 93, 44, 90, 57, &c. The mean annual temperature in Basutoland is about $58°$ Fahr. ($14°·4$ C.), or some four degrees below the general South African mean, this being due to the elevated nature of the country. The absolute maximum temperature for Teyateyaneng is $100°$ Fahr. ($37°·8$ C.) recorded in January 1904, though the thermometer not infrequently touches $97°$, $98°$ and $99°$ Fahr. ($36°·1$, $36°·6$ and $37°·2$ C.) towards the end of the year. The absolute minimum is usually recorded in June, though in 1900 August held the record for the year. The lowest figure yet recorded is $18°$ Fahr. ($-7°·8$ C.) in 1903, so that the extreme range of temperature at this station is $82°$ Fahr. ($45°·6$ C.). At Mohalie's Hoek the extreme range is $79°$ Fahr. ($43°·9$ C.), the absolute maximum being $95°$ Fahr. ($35°$ C.), measured in January and December 1902, and the absolute minimum $16°$ Fahr. ($-8°·9$ C.) recorded in June 1903. Frosts, at this station, are sometimes recorded as early as March and as late as November, *e.g.* $30°$ Fahr. ($-1°·1$ C.) was the minimum recorded for March 1903, and $31°$ Fahr. ($-0°·6$ C.) for November 1902. The mean daily range for both stations, though large, is fairly regular, varying only between $32°·9$ and $27°·6$ Fahr. ($18°·2$ and $15°·3$ C.).

TEYATEYANENG. Lat. 29° 8′ S., Long. 27° 45′ E. Alt. 5690 feet.

Compiled from 6 years' observations[1].

	Jan.	Feb.	March	April	May	June	July	Aug.	Sept.	Oct.	Nov.	Dec.	Year
Mean morning temperature, partly at 8 a.m., and partly at 8.30 a.m. (Fahr.°)	68°5	67°7	62°9	60°1	50°0	44°2	46°5	52°3	57°9	62°2	66°9	69°8	59°1
Mean maximum temperature ...	82°8	81°4	76°6	73°0	67°3	60°1	62°4	66°7	71°8	77°7	80°2	83°7	73°6
Mean minimum temperature ...	54°0	53°1	49°0	43°9	36°4	30°6	30°7	34°8	39°7	44°8	47°4	52°1	43°0
Mean temperature[2] ...	68°4	67°2	62°8	58°5	51°8	45°3	46°5	50°7	55°8	61°3	63°8	67°9	58°3
Mean daily range of temperature[3] ...	28°8	28°3	27°6	29°1	30°9	29°5	31°7	31°9	32°1	32°9	32°8	31°6	30°6
Mean of absolute monthly maxima of temperature	94°	91°	84°	80°	77°	69°	68°	75°	82°	89°	92°	94°	—
Mean of absolute monthly minima ...	45°	48°	41°	35°	26°	21°	24°	25°	27°	32°	37°	44°	—
Mean monthly range[4]	49°	43°	43°	45°	51°	48°	44°	50°	55°	57°	55°	50°	—
Mean relative humidity (per cent.) ...	59	61	69	59	59	64	58	66	59	54	46	54	59
Mean number of rainy days ...	11	10	7	5	1	2	1	1·5	3	5	6	8	60·5

[1] 1900—1905.
[2] The mean temperature is derived from $\frac{\text{max.} + \text{min.}}{2}$.
[3] The daily range is max. – min.
[4] The monthly range is abs. max. – abs. min.

NOTE.—Teyateyaneng is a Meteorological Station of the third order, or Climatological Station, about 195 miles from the nearest point on the coast.

K. A.

31

MOHALIE'S HOEK. Lat. 30° 8′ S., Long. 27° 28′ E.

Compiled from 6 years' observations[1].

	Jan.	Feb.	March	April	May	June	July	Aug.	Sept.	Oct.	Nov.	Dec.	Year
Mean morning temperature, partly at 8 a.m., and partly at 8.30 a.m. (Fahr.°)	63°·1	62°·7	57°·6	53°·6	49°·2	44°·8	46°·9	49°·5	53°·5	57°·8	66°·2	61°·7	55°·1
Mean maximum temperature ...	83°·4	82°·9	76°·8	73°·1	67°·0	59°·9	61°·9	65°·7	70°·1	75°·9	79°·3	82°·8	73°·2
Mean minimum temperature ...	53°·5	52°·3	49°·0	43°·3	34°·6	29°·8	30°·2	33°·4	40°·8	44°·8	47°·6	52°·5	42°·6
Mean temperature[2] ...	68°·4	67°·6	62°·9	58°·2	50°·8	44°·8	46°·1	49°·6	55°·5	60°·3	62°·9	67°·7	57°·9
Mean daily range[3]	29°·9	30°·6	27°·8	29°·8	32°·4	30°·1	31°·7	32°·3	29°·3	31°·1	31°·7	30°·3	30°·6
Mean of absolute monthly maxima	93°	91°	85°	82°	76°	68°	68°	75°	82°	88°	90°	92°	—
Mean of absolute monthly minima	42°	46°	40°	32°	24°	20°	21°	21°	25°	30°	35°	42°	—
Mean monthly range[4]	51°	45°	45°	50°	52°	48°	47°	54°	57°	58°	55°	50°	—
Mean relative humidity (per cent.) ...	66	68	74	71	62	58	56	54	54	55	55	61	61
Mean number of rainy days ...	8·0	11·5	9·2	6·2	3·0	2·7	1·2	2·3	3·7	5·0	6·2	9·3	68·3

NOTE.—Mohalie's Hoek is a Meteorological Station of the third order, or Climatological Station, distant about 150 miles from the coast.

[1] 1900—1904 and March, April and June—December 1905.

[2] The mean temperature is derived from $\dfrac{\text{max.} + \text{min.}}{2}$.

[3] The daily range is max. − min.

[4] The monthly range is abs. max. − abs. min.

Bechuanaland is a northerly prolongation of the Karoo, and gradually merges into the Kalahari desert. The climate on the whole is healthy for a good part of the year. The winter months extend from April to September, the remaining months forming the summer. The extremes of heat and cold are experienced, the situation being continental, *i.e.* being at a considerable distance from the sea in all directions. This is especially the case in the winter, when long droughts occur. At Vryburg the thermometer sometimes rises to $107°$ or $108°$ Fahr. $(41°·6$ or $42°·2$ C.$)$, while the lowest readings reach $20°$ Fahr. $(-6°·7$ C.$)$, the absolute minimum, however, at Kuruman was about $25°$ Fahr. $(-3°·9$ C.$)$ for the year 1905, the only year for which a complete set of observations is available. The annual rainfall varies between about 19 and 25 inches (482 and 635 mm.), which almost entirely falls in the period November/April, the average number of rainy days being about 56 in the year. The morning temperature at Kuruman varies between $70°$ and $73°$ Fahr. $(21°·1$ and $22°·8$ C.$)$ in November, December, and January, and $34°$ and $35°$ Fahr. $(1°·1$ and $1°·7$ C.$)$ in June and July. The mean daily range is about $30°$ Fahr. $(16°·6$ C.$)$ and the monthly range nearly $48°$ Fahr. $(26°·6$ C.$)$. The mean rainfall at Mafeking, for 15 years, is 25·5 inches (648 mm.), of which 5·7 inches (145 mm.) fell on nine days in January, 4·11 inches (104 mm.) on nine days in February, 5·76 inches (146 mm.) on nine days in March, and 3·74 inches (95 mm.) on eight days in December, the fall in November being 2·65 inches (67 mm.) on eight days. The months May, June, July, and August were absolutely dry in the years 1903—1905.

PRODUCTIONS, &c.

The great deficiency of rain in the Cape Colony, regarded as a whole, and its uncertainty, especially in the inland districts, militate against the production of most crops, but this intense dryness renders many places in the colony, which are also well suited by their altitude, excellent spots for the modern treatment of consumption. Wherever water is available, however, the soil is more productive than would be expected from its nature, and the corn-grasses of temperate regions, wheat, barley, and oats, do well, and so, too, does maize. Tobacco is produced in some localities and lucerne thrives. The hot dry autumn, in the south-west, is eminently favourable for viticulture, and of this many avail themselves. Orchard culture, too, is practised with success in both east and west. But the main industry of the colony consists in the breeding of sheep, horses, mules, cattle, goats—and also ostriches, for which the climate is excellently suited—though the lack of water and fodder is at times severely felt, and it is surprising how the animals adapt themselves to circumstances, drink filthy water and eat sour grass.

The climate of Basutoland is suitable for the production of wheat, oats, mealies, and Kafir corn.

CAPE COLONY, &c. MEAN ANNUAL RAINFALL (IN INCHES).

Compiled from the Annual Reports of the Meteorological Commission for periods varying from 7 to 10 years, between the years 1897 and 1906.

Cape Peninsula.

Station	Lat. S.	Long. E.	Alt. in feet	Jan.	Feb.	March	April	May	June	July	Aug.	Sept.	Oct.	Nov.	Dec.	Year
Simonstown ...	34° 12'	18° 26'	12	0·54	0·82	1·49	2·16	3·77	3·96	4·36	3·46	3·00	1·91	1·39	0·94	27·80
Wynberg ...	34° 0'	18° 28'	235	0·48	0·53	1·75	2·93	5·52	6·47	7·70	6·67	4·31	2·03	1·26	0·95	40·60
Bishop's Court ...	33° 58'	18° 28'	250	0·66	0·69	2·01	3·89	6·26	9·20	9·24	8·52	5·17	3·17	2·00	1·65	52·46
Royal Observatory	33° 56'	18° 29'	40	0·42	0·64	0·84	1·84	3·65	3·89	4·14	3·40	2·32	1·35	0·96	0·67	24·12
Town House (Cape Town)	33° 55'	18° 25'	94	0·45	0·61	0·88	1·67	3·59	3·48	3·59	3·09	2·37	1·29	0·90	0·63	22·55

West.

SOUTH-WEST.

Station	Lat. S.	Long. E.	Alt. in feet	Jan.	Feb.	March	April	May	June	July	Aug.	Sept.	Oct.	Nov.	Dec.	Year
Paarl ...	33° 45'	18° 57'	500	0·53	0·69	0·93	2·72	4·15	6·01	5·17	4·72	3·11	2·23	1·23	0·86	32·35
Ceres ...	33° 22'	19° 19'	1493	0·51	0·58	0·97	3·27	5·74	6·82	4·74	4·78	5·18	3·19	2·27	1·90	39·95
Tulbagh ...	33° 18'	19° 8'	490	0·21	0·59	0·58	1·74	2·54	2·11	1·87	2·03	2·30	1·66	1·07	0·70	17·40

WEST COAST.

Station	Lat. S.	Long. E.	Alt. in feet	Jan.	Feb.	March	April	May	June	July	Aug.	Sept.	Oct.	Nov.	Dec.	Year
Malmesbury ...	33° 28'	18° 43'	460	0·29	0·49	0·68	1·52	2·45	2·82	2·43	2·49	1·87	1·17	0·89	0·48	17·58
Piquetberg ...	32° 54'	18° 43'	700	0·12	0·42	0·50	1·65	2·76	3·25	2·29	2·40	2·33	1·79	1·23	0·66	19·40
Port Nolloth ...	29° 14'	16° 51'	40	0·05	0·10	0·16	0·20	0·42	0·33	0·16	0·44	0·16	0·03	0·17	0·09	2·31
Klipfontein ...	29° 12'	17° 39'	3104	0·14	0·35	0·57	0·75	1·15	0·97	0·52	0·86	0·33	0·34	0·34	0·20	6·52

MEAN ANNUAL RAINFALL (IN INCHES):—*continued.*

South Coast.

Station	Lat. S.	Long. E.	Alt. in feet	Jan.	Feb.	March	April	May	June	July	Aug.	Sept.	Oct.	Nov.	Dec.	Year
Mossel Bay ...	34° 11'	22° 9'	105	1·02	1·11	1·33	2·25	1·66	1·72	0·84	1·03	1·47	1·60	1·68	1·06	16·77
Swellendam ...	34° 4'	20° 27'	500	2·62	2·25	3·16	2·83	3·75	2·65	2·11	2·73	2·70	3·07	3·01	1·74	32·62
Knysna	34° 2'	23° 2'	950	2·03	1·83	2·01	1·76	2·09	2·66	1·89	2·43	2·88	2·61	3·37	2·47	28·03
Port Elizabeth ...	33° 57'	25° 37'	180	0·78	0·92	0·87	1·37	2·62	1·78	1·15	1·87	2·77	1·86	2·16	1·60	19·75
Uitenhage ...	33° 47'	25° 24'	192	0·95	0·93	1·85	1·77	1·39	1·57	0·57	1·49	1·87	1·58	2·79	1·19	17·95

Karoo.

SOUTH KAROO.

Station	Lat. S.	Long. E.	Alt. in feet	Jan.	Feb.	March	April	May	June	July	Aug.	Sept.	Oct.	Nov.	Dec.	Year
Oudtshoorn ...	33° 36'	22° 13'	1085	0·52	0·85	0·81	0·56	1·10	0·67	0·35	0·59	0·96	0·87	0·96	0·35	8·59
Amalienstein ...	33° 27'	21° 24'	1545	1·03	0·99	1·32	1·04	1·52	1·10	0·64	0·65	0·92	1·39	1·24	0·58	12·42
Glenconnor ...	32° 25'	25° 5'	730	1·75	0·79	1·54	1·00	0·65	1·03	0·11	0·78	1·24	0·76	2·50	0·68	12·83

W. CENTRAL KAROO.

Station	Lat. S.	Long. E.	Alt. in feet	Jan.	Feb.	March	April	May	June	July	Aug.	Sept.	Oct.	Nov.	Dec.	Year
Matjesfontein ...	33° 14'	20° 40'	2593	0·28	0·69	0·44	0·63	1·10	0·54	0·57	0·53	0·36	0·58	0·60	0·19	6·51
Prince Albert ...	33° 11'	22° 2'	2120	1·67	1·68	1·98	1·55	2·24	0·44	0·45	1·09	1·54	1·23	0·68	0·71	15·26
Nel's Poort ...	32° 14'	23° 4'	3125	2·13	1·34	2·13	0·91	0·71	0·30	0·17	0·27	0·74	0·53	1·66	1·02	11·91

E. CENTRAL KAROO.

Station	Lat. S.	Long. E.	Alt. in feet	Jan.	Feb.	March	April	May	June	July	Aug.	Sept.	Oct.	Nov.	Dec.	Year
Graaf Reinet ...	32° 16'	24° 34'	2500	1·72	1·39	2·73	0·87	1·21	0·41	0·17	0·73	1·29	0·88	2·47	1·42	15·29
Rietfontein ...	32° 52'	25° 5'	2271	1·29	0·73	1·74	0·95	0·73	0·53	0·11	0·34	0·63	0·66	1·34	0·33	9·38
Somerset East ...	32° 44'	25° 35'	2400	4·20	3·00	3·19	2·85	1·44	0·50	2·05	1·13	1·25	2·42	2·57	4·25	28·85

MEAN ANNUAL RAINFALL (IN INCHES):—*continued.*

N. KAROO.

Station	Lat. S.	Long. E.	Alt. in feet	Jan.	Feb.	March	April	May	June	July	Aug.	Sept.	Oct.	Nov.	Dec.	Year
Sutherland ...	32° 25'	20° 42'	4776	0·53	0·78	0·77	0·83	1·25	0·98	0·69	0·71	0·49	0·81	1·28	0·16	9·26
Colesberg ...	30° 43'	25° 7'	4470	2·86	2·13	3·32	1·77	1·26	1·08	0·19	0·48	0·23	0·95	2·00	1·99	18·26
Cradock ...	32° 11'	25° 38'	2856	1·51	2·66	2·08	2·54	1·26	0·36	0·94	0·93	0·68	0·79	2·00	2·04	17·79

N. BORDER.

Station	Lat. S.	Long. E.	Alt. in feet	Jan.	Feb.	March	April	May	June	July	Aug.	Sept.	Oct.	Nov.	Dec.	Year
Pella ...	29° 2'	19° 9'	1800	0·14	0·40	0·59	0·33	0·31	0·11	0	0·28	0·02	0·46	0·09	0·23	2·96
Kimberley ...	28° 43'	24° 46'	4042	2·78	2·98	2·84	1·96	0·71	0·51	0·10	0·26	0·25	0·79	2·10	2·89	18·17
Upington ...	28° 36'	21° 15'	2800	1·40	1·02	2·13	1·79	0·65	0·12	0	0·07	0·07	1·14	0·68	1·74	10·81

South-East.

Station	Lat. S.	Long. E.	Alt. in feet	Jan.	Feb.	March	April	May	June	July	Aug.	Sept.	Oct.	Nov.	Dec.	Year
Alicedale ...	33° 20'	26° 4'	1906	1·62	1·24	1·48	1·39	1·39	1·07	0·36	0·62	1·76	1·39	2·80	1·18	16·30
Lovedale ...	32° 46'	26° 51'	1716	3·23	2·92	3·40	2·31	1·17	0·93	0·62	1·16	1·93	1·60	3·98	2·55	25·80
Katberg (Lower)	32° 29'	26° 40'	3380	6·23	4·54	5·39	3·83	1·70	1·07	0·62	1·70	3·19	4·18	5·57	6·12	44·14
Grahamstown ...	32° 20'	26° 33'	1800	2·87	2·63	3·21	2·70	2·08	1·62	0·67	1·08	3·28	3·10	4·55	2·16	29·95
King William's Town	32° 51'	27° 22'	1310	2·79	2·50	2·98	3·07	1·53	1·49	0·47	1·28	2·25	2·13	3·71	2·08	26·28

North-East.

Station	Lat. S.	Long. E.	Alt. in feet	Jan.	Feb.	March	April	May	June	July	Aug.	Sept.	Oct.	Nov.	Dec.	Year
Queenstown ...	31° 51'	26° 51'	3500	5·20	2·79	3·61	2·04	0·81	0·77	0·52	0·93	1·01	1·57	3·16	3·33	25·74
Burghersdorp ...	31° 0'	26° 21'	4480	2·88	2·54	4·00	1·91	1·11	0·87	0·42	1·22	0·63	1·13	2·48	2·85	22·04
Aliwal North ...	30° 43'	26° 43'	4330	3·29	3·03	3·71	1·90	1·43	0·84	0·52	1·25	0·90	1·28	2·40	3·93	24·48

MEAN ANNUAL RAINFALL (IN INCHES):—*continued.*

Station	Lat. S.	Long. E.	Alt. in feet	Jan.	Feb.	March	April	May	June	July	Aug.	Sept.	Oct.	Nov.	Dec.	Year
Kaffraria.																
Tsomo[1] ...	32° 2′	27° 51′	51	3·65	3·82	3·19	2·24	0·55	0·35	0·39	1·23	0·98	1·98	2·99	2·97	24·34
Kokstad[2] ...	30° 33′	29° 26′	4284	4·56	3·72	4·25	2·00	0·84	0·95	0·23	0·89	1·77	2·15	3·96	5·55	30·87
Basutoland.																
Quthing[3] ...	30° 23′	27° 45′	6000	5·78	4·12	6·18	2·87	1·60	1·04	0·34	1·51	1·30	2·48	4·06	6·45	37·73
Maseru[4] ...	29° 17′	27° 30′	5065	5·35	5·40	3·97	1·98	1·32	1·55	0·63	0·95	1·48	2·63	5·11	3·99	34·36
Teyateyaneng[5] ...	29° 8′	27° 45′	5690	5·54	2·99	4·63	2·42	1·55	1·14	0·23	1·42	0·82	2·73	3·00	5·36	31·83
Mohalie's Hoek[6]	30° 8′	27° 28′		3·7	4·1	3·3	1·9	1·1	0·9	0·2	0·6	1·7	1·9	2·0	4·3	25·7
Leribe[7] ...	28° 54′	28° 12′	5210	5·54	4·06	4·73	2·06	1·12	0·82	0·27	0·96	1·14	2·61	2·87	5·50	31·68
Bechuanaland.																
Vryburg[8] ...	26° 55′	24° 43′	4286	5·69	3·80	3·73	1·84	0·79	0·34	0·04	0·42	0·18	0·91	1·03	3·61	22·38

[1] 1890—1896.   [2] 1890—1897.

[3] 1890—1897. The mean for December is probably too high owing to a fall of over 13 inches in 1896.

[4] 1890—1895.   [5] 1891—1897.   [6] 1890—1895.   [7] 1890—1897.

[8] 1891—1897. The means for January and February are probably too high, in consequence of a fall of over 10 inches in the first month in 1893, and of over 6 and over 7 inches in the second month in 1892 and 1894 respectively. The total fall in 1897 was as low as 11 inches and in 1894 was over 28 inches.

GERMAN SOUTH-WEST AFRICA

German South-West Africa affords no exception to the general terrace formation of Africa. On the contrary it is here typically developed. The narrow coastal belt—consisting mainly of sand dunes—varies in width from five to twenty or thirty miles, and is followed by bands of higher altitude until at about 110 miles east of the coast (at Walfisch Bay) the great Nama-Dama tableland is reached, with a general elevation of from 4000 to 6000 feet, and with rugged ridges and outstanding peaks rising to over 8000 feet. Abreast of Windhoek this great tableland extends into Bechuanaland, but fines down in breadth till it reaches 20° S. in the north, and is continued southwards until, when about 24° S. is reached, it is cut into by a depression extending east and west for 120 miles; then suddenly nearing the coast, and following its general direction, it is continued southwards, and terminates in the Huib plateau, with outliers in the Little and Great Karas Mountains further east.

In German South-West Africa there are two well-marked seasons; a cold dry season which lasts from May to September, and a hot season from October to April. The rainfall increases from south to north and from west to east, as will be presently shown. The rains, however, are not regular, some years being very dry, others more moist.

The climate in summer is hot but healthy. The winter is moderate throughout, and, in the interior, night frosts are not infrequent; the coast is cool, and has only mists for some 30 miles inland. The winds are chiefly from the south and west. In the warmer half of the year, *i.e.* from October to March, there are winds from the north, and these are the cause of the chief wet season, from January to March. In the northern parts, in normal years, the rains begin in November and end in March. The two most dangerous months, in regard to fever, are April and May. The climate of Ovampoland, in the north, is tropical and is considered to be healthy for Europeans. In Hereroland, in spite of the sharp variations of temperature, the climate is very healthy and favourable to colonisation.

From the peculiar formation of the ground in South Africa, with its raised shores, one would expect, as a matter of course, that the coast lands would be rich in rainfall and the inland portions comparatively poor; and so far as the south and east coasts are concerned this is true ; and it is true also of the inland Karoo region. But this does not hold good for the west coast. Here the prevailing winds on the coast are west and south-west throughout the year, and at the same time the coasts are washed by the cold Benguela current and the still colder tides, and thus it occurs that the winds are deprived of the greater part of their moisture before reaching the warm coast lands. Such moisture as is condensed is in the form of constant thick mists and heavy falls of dew, and these form the only regular precipitation on the coast lands in winter.

In the summer the westerly high pressure area has approached nearer to the coast, and there is therefore less opportunity for the winds to gather moisture, and, during this season also, the Namieb desert region has increased in temperature, so that there is little or no opportunity for condensation, and the mists are consequently not so common, though dews are fairly copious.

In the inland areas, during winter, the winds are the same as on the coast, viz. west and south-west, and the winter rainfall accompanying them only reaches the interior in quite isolated and extraordinary cases, for as soon as the sea-winds, already robbed by the cold sea of most of their moisture, reach the warmer country, their capacity for condensation is at once decreased, and even the ascent to the higher land does not appear to be able to produce precipitation. In the summer, the prevailing winds in the interior are from the eastern quadrant, and the easterly rains reach as far as the highlands of German South-West Africa (see p. 393), but are expended before reaching the coast lands.

So great is the rate of absorption and evaporation that Andersson found that pools of from 40 to 50 feet long and several feet deep would dry up in the course of a week[1]. This was in December, in the interior of Damaraland.

---

[1] Andersson, *The Okovango River*, p. 146.

The rainfall, as mentioned above, increases from west to east, and from south to north. This is borne out by the following facts : if a series of stations be taken in the southern part of the region, it is found that the mean annual rainfall at Lüderitz Bay is 0·79 inch (20 mm.) only, at Kubub, farther east, it is 2·87 inches (73 mm.) ; and, still farther, at Keetmanshoop it is 6·3 inches (162 mm.), and at Hasur 6·85 inches (174 mm.) ; or if better-known places, though not in such a direct west-east line, be taken, the result is

Lüderitz Bay	...	...	...	0·79 inch	or	20 mm.
Warmbad	...	...	...	4·25 inches	„	108 „
Upington	...	...	...	8·39 „	„	213 „

Similarly, a west-east line may be taken farther north, *e.g.*

Zesfontein	...	...	...	2·91 inches	or	74 mm.
Outyo ...	...	...	...	17·48 „	„	444 „
Grootfontein	...	...	...	23·39 „	„	594 „

and a line between these two, and parallel with them, gives

Swakopmund	...	...	...	0·83 inch	or	21 mm.
Tsoabis ...	...	...	...	6·46 inches	„	164 „
Windhoek	...	...	...	15·04 „	„	382 „
Gobabis ...	...	...	...	16·46 „	„	418 „

Similarly, a south-north line gives the following result

Warmbad	...	...	...	4·25 inches	or	108 mm.
Keetmanshoop	...	...	...	6·38 „	„	162 „
Hoakhanas	...	...	...	7·87 „	„	200 „
Windhoek	...	...	...	15·04 „	„	382 „
Waterberg	...	...	...	20·51 „	„	521 „
Grootfontein	...	...	...	23·39 „	„	594 „

The heaviest rainfall therefore occurs in the north-east portion of the territory, and the least on the coast, where it is very small indeed, as is shown by the following figures, the names of the stations being given from north to south:— Zesfontein 2·91 inches (74 mm.), Swakopmund 0·83 inch (21 mm.), Walfisch Bay 0·32 inch (8 mm.), and Lüderitz Bay 0·79 inch (20 mm.).

In German South-West Africa the beginning and end of the rainy season are not invariable, but differ from year to year ; observations recorded during the nine-year period 1891—1899 show that at Waterberg the season may begin any time between October 15th and November 4th, and end any time between April 6th and May 21st. The corresponding periods for the beginning and ending of the season at Rehoboth are from October 2nd to November 26th, and from April 3rd to May 9th.

With regard to the number of rainy days, at Waterberg the average number of days, based on the figures for four years, on which ·008 inch (0·2 mm.) of rain was measured, was 51, and the number on which more than ·04 inch (1 mm.) fell was 42, so that, the mean annual rainfall being 20·47 inches (520 mm.), it follows that the average fall on a rainy day, omitting the days of less than 1 mm., is about ·48 inch (12 mm.). At Otavi the corresponding number of days on which rain falls is 59 (·008 inch and over), 50 (·04 inch and over), and for Grootfontein 55 and 47 respectively; and, the mean annual rainfall at these two places being 598 mm. or 23·54 inches and 594 mm. or 23·39 inches, the average precipitation on a rainy day is about 12 and 13 millimetres or about half an inch. These three places are in the region of the territory's greatest rainfall, Windhoek and Rehoboth are fairly central, while Keetmanshoop and Warmbad lie in the south; and the corresponding figures for these four places are

		No. of days with ·008 inch of rain and more	No. of days with ·04 inch of rain	Annual rainfall in inches	Average fall in a rainy day
Windhoek	...	45	36	15·00	·42
Rehoboth	...	27	23	10·43	·45
Keetmanshoop	...	19	15	6·38	·42
Warmbad	...	15	11	4·29	·39

The night fogs of the Namieb, or tableland east of the sand dunes, form a peculiar feature of the South-West Africa climate. Pearson[1], in his notes on a journey from Walfisch Bay to Windhoek, after describing the mirage, says, "The night fogs, in so arid a region, are hardly less remarkable. About 8 o'clock in the evening of January 24, as we crossed the dry bed of the Tubas river, a cloud appeared in the west. The stars were gradually obliterated, and by 10 o'clock we were shrouded in a cold 'Scotch mist.' After sleeping on the ground from midnight until 4 o'clock on the following morning, I was able to wring the water out of my top-covering, a woollen rug. As

[1] *Some notes on a journey from Walfish Bay to Windhuk* by H. H. W. Pearson. 1907. (Kew Bulletin.)

soon as daylight came, the ground was seen to be discoloured by the moisture, and the plants were copiously sprinkled with dew. At 7 a.m. on January 30 the water was dripping from the branches of the Tamarisks in the Khan valley at Haikamchab. . That so regular and abundant a deposit of moisture should not be able to support a greater number of shallow-rooted plants than we found on the Namieb seems at first remarkable."

Windhoek is situated in Lat. $22°34'$ S. and Long. $17°5'$ E. at an altitude of 5512 feet or 1680 metres. The mean annual temperature is $66°·2$ Fahr. ($19°$ C.). The coldest months are June, July, August, and September, but the thermometer also falls very low in October. November, December, January, and February are the warmest months, and sometimes it is very hot in May. The extremes of temperature recorded are $98°·3$ Fahr. ($36°·8$ C.) and $26°·5$ Fahr. ($-2°·9$ C.). Throughout the year at Windhoek it is coolest between five and six o'clock in the morning, and warmest between two and three o'clock in the afternoon. The mean temperature, and the mean maximum and minimum temperatures, are fairly regular for the various months from year to year, and are sometimes almost identical; for instance, the mean minimum temperature for March in four consecutive years was $58°·5$, $57°·7$, $58°·6$ and $58°·3$ Fahr., or $14°·7$, $14°·3$, $14°·8$ and $14°·6$ C., respectively, and the mean maximum for May for the same years was $73°·6$, $72°·9$, $73°·9$ and $74°·3$ Fahr., or $23°·1$, $22°·7$, $23°·3$ and $23°·5$ C.

The greatest rainfall occurs in the summer, the average, calculated from observations for over 13 years, being $9·3$ inches (236 mm.) for the period from January to March. The winter months from July to September are practically rainless, and so also is June; in the spring (October—December) December is the rainiest month, with 2 inches out of $3·3$ inches (53 out of 84 mm.) for the whole period.

The table below, taken from a paper in the *Mitteilungen aus den Deutschen Schutzgebieten*[1], shows the mean annual temperature at the successive hours of the day in degrees Fahr.

Throughout the year the barometer stands highest at 9 or 10 a.m., and lowest at 3 or 4 p.m. It is highest in

[1] *M.D.S.*, 1907, p. 103.

April and May, and in July and August, and lowest in December, January, and February, and October.

a.m. time	1	2	3	4	5	6	7	8	9	10	11	noon
Temp. F.°	59°·4	58°·6	57°·9	57°·4	56°·7	57°·0	60°·1	64°·2	68°·7	71°·2	73°·4	75°·0
p.m. time	1	2	3	4	5	6	7	8	9	10	11	mid-night
Temp. F.°	76°·3	76°·8	77°·0	76°·3	74°·8	71°·6	68°·2	66°·2	64°·2	63°·1	61°·5	60°·4

In the paper quoted above, the writer compares the days, at Windhoek, during June and July with those in mid-Germany during the second half of September, but the nights are cold.

At Omaruru in 1893 there were 16 frosty nights during June—July—August, the greatest cold being experienced in the two last mentioned, while the absolute minimum for the winter was $15°·8$ Fahr. $(-9°$ C.), recorded in the middle of July. The first frost in that year occurred on May 16th, when $30°·5$ Fahr. $(-1°·2$ C.) was registered. Below is given the number of frosty nights at two stations.

		Otyiseva alt. $\begin{cases}5585 \text{ feet} \\ 1550 \text{ m.}\end{cases}$	Rehoboth alt. $\begin{cases}4593 \text{ feet} \\ 1400 \text{ m.}\end{cases}$
May	...	0	1
June	...	8	17
July	...	3	23
August	...	0	3
September	...	0	1
Totals	...	11	45

The period January—February—March is essentially the rainy season, or rather such rain as falls is usually registered in those months. At Tsoabis 84·2 per cent. of the total annual precipitation is recorded during that period, at Rehoboth 67·9 per cent., at Omaruru 63·7 per cent., and at Olukonda 66·9 per cent.

The following is the record of the days on which rain was registered at Windhoek and Rehoboth during the years 1892 and 1893.

January 1892. Windhoek 8th—17th, 23rd, 27th—31st; Rehoboth 7th—20th, with short pauses, 22nd, 23rd, 27th—31st.

February 1892. Windhoek 4th—6th, 14th—18th; Rehoboth 1st—4th, 14th—16th, 23rd.

March 1892. Windhoek 4th—6th, 11th—19th, 24th, 25th, 28th—1st April; Rehoboth 11th, 12th, 17th—19th, 24th.

January 1893. Windhoek 1st, 8th—31st; Rehoboth 1st, 7th—29th.

March 1893. Windhoek 5th, 8th—21st, 26th, 27th; Rehoboth 1st—4th, 10th, 11th, 18th—26th.

Combining the results given by Dove[1] with those contained in the *Mitteilungen a. d. Deutschen Schutzgebieten*[2], the following table, showing the mean temperature, in Fahr. degrees, at stations in Hereroland and Namaqualand is obtained.

		Omaruru $21\frac{1}{2}°$ 3806 ft.	Otyiseva $22\frac{1}{3}°$ 5084 ft.	Rehoboth $23\frac{1}{3}°$ 4560 ft.	Hoakhanas $23\frac{5}{8}°$ 4593 ft.
	Lat. Alt.				
January	...	$75°\cdot2$	—	$77°\cdot5$	$78°\cdot8$
February	...	$75°\cdot6$	$73°\cdot6$	$76°\cdot3$	$78°\cdot3$
March	...	$72°\cdot1$	$71°\cdot2$	$74°\cdot3$	$75°\cdot2$
April	...	$69°\cdot4$	$67°\cdot3$	$68°\cdot5$	$68°\cdot0$
May	...	$60°\cdot4$	$58°\cdot3$	$60°\cdot1$	$59°\cdot0$
June	...	$55°\cdot0$	$50°\cdot2$	$50°\cdot7$	$53°\cdot1$
July	...	$54°\cdot7$	$50°\cdot7$	$51°\cdot3$	$50°\cdot9$
August	...	$55°\cdot9$	$56°\cdot8$	$56°\cdot8$	$53°\cdot9$
September	...	$67°\cdot5$	$65°\cdot7$	$64°\cdot2$	$65°\cdot7$
October	...	$70°\cdot9$	$71°\cdot4$	$70°\cdot3$	$72°\cdot3$
November	...	$75°\cdot7$	$76°\cdot3$	$76°\cdot1$	$75°\cdot9$
December	...	$78°\cdot3$	$76°\cdot5$	$73°\cdot0$	$79°\cdot5$

Of Angra Pequena (Lüderitz Bay) the *Africa Pilot* says that it is a healthy place with a good climate. The prevailing winds are from the west and south-west, and they are generally fresh. The maximum temperature is about 75° Fahr. (24° C.).

At Swakopmund the mean daily temperature at 7 o'clock in the morning appears to vary between $62°\cdot6$ Fahr. (17° C.), and $49°\cdot3$ Fahr. ($9°\cdot6$ C.) in the course of the year; at 2 p.m. between 69° Fahr. ($20°\cdot5$ C.) and $56°\cdot5$ Fahr. ($13°\cdot6$ C.); and at 9 o'clock in the evening between about $64°\cdot4$ Fahr. (18° C.) and 52° Fahr. ($11°\cdot2$ C.).

---

[1] Dove, *Deutsch-Südwestafrika*, Berlin, 1902, p. 81.
[2] *M.D.S.*, 1907, p. 103.

In the table given below the temperature means are probably too high for the months of March and May in consequence of the abnormal heat in these months in the year 1902. In that year the mean maximum temperature for March was $72°·7$ Fahr. ($22°·6$ C.) and for May $75°·2$ Fahr. ($24°$ C.), while the absolute maxima for those months were $100°$ and $101°$ Fahr. ($37°·8$ and $38°·2$ C.) respectively. The lowest mean maximum for any month in the seven-year period, from which the means are calculated, was $59°·5$ Fahr. ($15°·3$ C.). The mean minimum temperature varied between $60°·8$ Fahr. ($16°$ C.) and $45°·5$ Fahr. ($7°·5$ C.), while the absolute minimum in the period was $36°·5$ Fahr. ($2°·5$ C.) and the absolute maximum $105°·1$ Fahr. ($40°·6$ C.), giving a maximum range, for the whole seven-year period, of $68°·6$ Fahr. ($38°·1$ C.). The amount of cloud was always much greater in the morning than at any other time, and varied between 8·9 (December) and 5·4 (May and June) out of 10; and it was slightly greater in the evening than at 2 p.m.; and similar remarks apply to the relative humidity, though there was a greater difference between that at 2 p.m. and that at 9 p.m. The mean of the barometric readings varied between 29·9 inches (759·6 mm.) in August and 29·69 inches (784·1 mm.) in April.

At Hoakhanas in the northern part of Great Namaqualand the rain came with winds in the following directions and proportionate frequency.

S	S.W	W	N.W	N	N.E	E	S.E
2	2	12	4	16	14	36	14

This is the result of observing 50 rainfalls on the 44 days on which it rained, at 7 a.m., 2 p.m., and 9 p.m.

PRODUCTIONS, &c.

German South-West Africa, though said to be rich in minerals, is no paradise for the agriculturist. Namaqualand is nearly entirely an arid waste. The grassy plains and slopes of Damaraland, however, afford excellent opportunities for stock-raising. In the north, in Ovampoland, are many large areas well adapted, both by soil and climate, for raising crops of maize, barley, oats, and lucerne. The north-east, it will be remembered, is the most favoured so far as rainfall is concerned, and there appears to be no reason why other crops, such as wheat, Kafir corn, &c., should not be grown. In the extreme north-east are gummiferous trees and also the *mopani* tree, which grows very straight and is good for building purposes, especially as white ants will not touch it.

With reference to the stations for which meteorological tables are given below—

Swakopmund is situated on the coast at a distance of about a mile from the mouth of the Swakop river on the north bank. The instruments in use at this station were a dry-bulb thermometer, a wet-bulb thermometer, a maximum thermometer, and a minimum thermometer.

Windhoek lies about 140 miles inland, close to the tropic. The station is provided with all the usual thermometers, and a barometer, as well as a Richards Barograph and Thermograph. These are placed in a specially constructed screen.

SWAKOPMUND. Lat. 22° 42′ S., Long. 14° 33′·5 E.

Compiled from 7 years' observations[1], which appeared in *Mitteilungen aus den Deutschen Schutzgebieten.*

	Jan.	Feb.	March	April	May	June	July	Aug.	Sept.	Oct.	Nov.	Dec.	Year
Mean temperature at 7 a.m. (Fahr.°)	61°·3	61°·5	61°·7	57°·4	57°·2	54°·7	53°·1	51°·4	54°·0	56°·7	56°·8	60°·3	57°·2
,, ,, 2 p.m.	64°·9	65°·7	65°·8	62°·8	65°·3	63°·7	61°·0	58°·8	59°·0	60°·8	61°·2	63°·7	62°·7
,, ,, 9 p.m.	61°·7	62°·2	62°·6	59°·4	59°·4	56°·8	55°·2	54°·3	55°·4	57°·0	57°·6	60°·4	58°·5
Mean temperature[2]	62°·6	63°·1	63°·3	59°·9	60°·6	58°·5	56°·5	54°·9	56°·1	58°·1	58°·6	61°·5	59°·5
Mean maximum temperature	69°·4	69°·8	70°·7	68°·2	72°·0	69°·6	67°·6	64°·8	64°·0	66°·0	66°·0	68°·5	68°·1
Mean minimum temperature	58°·6	59°·2	58°·8	55°·2	53°·2	50°·2	49°·3	47°·7	50°·4	50°·9	53°·2	57°·6	53°·7
Mean daily range[3]	10°·8	10°·6	11°·9	13°·0	18°·8	19°·4	18°·3	17°·1	13°·6	15°·1	12°·8	11°·3	14°·4
Absolute maximum temperature	78°·3	81°·9	100°·4	98°·6	103°·6	97°·7	105°·1	96°·4	89°·8	99°·3	84°·0	85°·1	105°·1 (highest)
Absolute minimum temperature	51°·3	50°·7	50°·5	46°·2	40°·1	37°·8	38°·1	36°·5	40°·5	39°·7	45°·0	47°·3	36°·5 (lowest)
Mean relative humidity[4] (per cent.) at 7 a.m.	85	88	86	90	70	67	77	87	83	84	85	85	83
Mean relative humidity (per cent.) at 2 p.m.	77	78	77	78	61	59	69	75	75	75	78	77	74
Mean relative humidity (per cent.) at 9 p.m.	85	86	84	86	70	71	78	85	84	84	83	85	82
Mean relative humidity (daily)	82	84	82	84	67	66	75	82	81	81	81	82	80
Mean amount of cloud (0—10)	6·4	6·7	6·3	5·8	4·1	3·9	4·8	5·1	5·6	5·7	6·0	7·0	5·6

NOTE.—Swakopmund is situated on the coast about one mile north of the mouth of the Swakop river.

[1] 1899—1905.

[2] The mean temperature is derived from $\dfrac{7a+2p+9p}{3}$.

[3] The daily range is max.−min.

[4] Humidity records are wanting from 12th September 1899 to 12th May 1900. The mean is derived from $\dfrac{7a+2p+9p}{3}$.

WINDHOEK. Lat. 22° 34' S., Long. 17° 5' E. Alt. 5494 feet.

Compiled from observations for (5) years[1], which appeared in *Mitteilungen aus den Deutschen Schutzgebieten.*

	Jan.	Feb.	March	April	May	June	July	Aug.	Sept.	Oct.	Nov.	Dec.	Year
Mean maximum temperature ...	86°·5	83°·5	82°·8	79°·3	73°·6	69°·3	69°·1	73°·2	78°·6	82°·6	86°·2	85°·8	79°·2
Mean minimum temperature ...	63°·1	61°·3	58°·3	53°·8	48°·6	43°·2	44°·4	46°·8	51°·1	55°·0	59°·0	61°·9	53°·9
Mean temperature[2] (Fahr.°)	74°·8	72°·3	70°·5	66°·6	61°·0	56°·3	55°·9	59°·9	64°·9	68°·9	72°·7	73°·9	66°·5
Mean daily range[3]	23°·4	22°·2	24°·5	25°·5	25°·0	26°·1	24°·7	26°·4	27°·5	27°·6	27°·2	23°·9	25°·3
Mean of absolute monthly maxima[4] ...	91°·9	90°·7	87°·4	84°·9	85°·5	74°·8	75°·0	78°·4	86°·0	90°·7	91°·6	92°·5	95°·4 (highest)
Mean of absolute monthly minima[5] ...	56°·7	53°·8	51°·6	46°·0	37°·9	52°·9	36°·5	29°·8	38°·3	40°·8	48°·4	55°·8	29°·8 (lowest)

NOTE.—Windhoek lies about 140 miles inland, close to the Tropic of Capricorn.

[1] October—December 1901, 1902, 1903, March—December 1904, January—June 1905.

[2] The mean is derived from $\frac{\text{max.} + \text{min.}}{2}$. Observations at 7 a.m., 2 p.m., and 9 p.m. were only commenced in July 1905: previous to this the records were only taken at 8 a.m.

[3] The daily range is max. – min.

[4] The absolute maximum recorded was 98° Fahr.

[5] The absolute minimum recorded was 26°·5 Fahr.

32—2

WALFISCH BAY.

The table, which appears below, is compiled from observations taken during four years, and the figures show that the mean temperature at 8 o'clock in the morning is very regular for almost every month during the four years, except perhaps April, June, and October, when the greatest divergencies from the mean occur. In the first month the mean temperature at 8 a.m. was $60°$ Fahr. ($15°·6$ C.) in one year and $54°·8$ Fahr. ($12°·7$ C.) in another; and in the second month it was $57°·9$ Fahr. in one year and $49°·6$ Fahr. in another ($14°·4$ and $9°·8$ C.); and in the third it was $65°·8$ Fahr. ($18°·8$ C.) in one year and $55°·9$ Fahr. ($13°·3$ C.) in another. In all the other months the figures differ but little from year to year.

With regard to the mean maximum temperature, the greatest divergencies from the mean occur in April, May, and June, amounting to nearly $4°$, $3°·5$ and $3°$ Fahr. ($2°·2$, $2°$, and $1°·7$ C.) respectively, but on the whole the means are very good; the means of the absolute maxima are derived from almost identical figures in December, January, and February, but they vary considerably in the other months. The extreme range of temperature during the whole four years was $59°·9$ Fahr. ($33°·2$ C.), between $94°·8$ Fahr. ($34°·8$ C.) measured in one April and $34°·9$ Fahr. ($1°·6$ C.) measured in July. In some years the temperature does not fall much below $40°$ Fahr. ($4°·4$ C.), and never exceeds $95°$ Fahr. ($35°$ C.).

Practically Walfisch Bay is rainless; in one of the four years absolutely no rain was measured, and the total precipitation in the 48 months was only $2·53$ inches (64 mm.). It is less cloudy in the middle of the year than at either the beginning or end.

The climate at Walfisch Bay is generally healthy and there is no malaria. For its position the place is comparatively cool while the south-west wind is blowing, and this, it should be noticed, is the prevalent wind; but in clear still weather it is hot. The thermometer frequently runs riot and there are sudden changes of temperature. The nights are generally damp and chilly, as is usual on this part of the coast, where the heavy dews and thick fogs and mists prevail. These mists frequently last well into the forenoon, when the weather becomes clear and generally remains so from 10 a.m. to 2 p.m., but so soon as the south-west wind freshens the mists reappear.

WALFISCH BAY.

Percentage table of wind frequency (from 8 years' observations).

	N.	N.E.	E.	S.E.	S.	S.W.	W.	N.W.	Calm
Spring (October—December)	23	2	3	1	2	39	9	4	17
Summer (January—March)	19	2	2	1	2	32	18	6	18
Autumn (April—June) ...	7	2	8	2	9	38	10	2	22
Winter (July—September)	18	3	9	1	6	35	8	3	17
Year	17	3	5	1	5	36	11	4	18

WALFISCH BAY. Lat. 22° 57′ S., Long. 14° 26′ 30″ E. Alt. 7 feet.

Compiled from 4 years' observations[1].

	Jan.	Feb.	March	April	May	June	July	Aug.	Sept.	Oct.	Nov.	Dec.	Year
Mean temperature at 8 a.m.[2] (Fahr.°) ...	62°8	61°5	61°9	58°1	56°1	54°0	52°9	51°0	55°7	59°1	57°9	61°3	57°7
Mean maximum temperature[2] ...	69°1	68°6	69°7	67°2	67°9	64°4	62°9	62°1	63°0	64°7	63°6	67°7	65°9
Mean minimum temperature[2] ...	57°5	56°6	57°9	54°0	51°4	47°4	53°9	46°2	50°7	52°0	53°8	57°6	53°2
Mean temperature[3] ...	63°3	62°6	61°8	60°6	59°6	55°9	58°4	54°1	56°8	58°3	58°7	62°6	59°5
Mean daily range (max. − min.) ...	11°6	12°0	11°8	13°2	16°5	17°0	9°0	15°9	12°3	12°7	9°8	10°1	12°7
Mean of absolute monthly maxima[2] ...	73°4	72°7	82°5	82°3	81°7	83°3	78°7	72°4	88°4	73°6	73°1	73°2	77°9
Mean of absolute monthly minima[2] ...	53°8	55°2	50°4	49°4	43°2	39°1	35°9	38°2	43°5	43°9	47°9	51°7	46°0
Monthly extremes {highest abs. max. in 4 years	73°7	72°9	88°6	94°8	93°5	88°6	90°6	76°9	93°8	83°8	83°8	73°9	94°8
{lowest ” ”	73°1	72°4	71°8	73°4	83°8	73°1	70°8	66°5	76°9	66°9	66°0	72°9	66°0
Monthly extremes {highest abs. min. in 4 years	54°5	57°0	53°5	51°5	46°5	41°0	40°0	39°2	46°2	45°2	50°1	56°0	57°0
{lowest ” ”	53°0	52°0	46°0	46°0	39°0	38°0	34°9	36°3	40°0	43°0	46°0	49°0	34°9
Mean relative humidity[4] (per cent.) ...	80	88	87	86	83	77	84	91	87	85	84	86	85
Mean amount of cloud[5] (0—10) ...	8·0	8·1	7·0	6·4	5·9	5·3	7·1	4·9	7·5	6·8	7·6	9·1	6·9
Mean number of days on which rain was measured[6]	0	·5	·25	·5	·25	·25	0	0	·5	·5	0	·25	3
Mean rainfall (in inches)[6] ...	0	·085	·002	·005	·002	·035	0	0	·225	·027	0	·250	·631
Total rainfall in 4 years (in inches)[6]	0	·34	·01	·02	·01	·14	0	0	·90	·11	0	1·00	2·53

[1] 1900, 1903, 1904, 1905.
[2] Temperature records are wanting for January and May 1903, and for July 1905.
[3] The mean temperature is derived from $\dfrac{\text{max.} + \text{min.}}{2}$.
[4] Humidity records are wanting for January and May 1903, and for July 1905.
[5] The cloud records are only for 3 years 1903—1905.
[6] Rain records are wanting for May 1905 and July 1905. See table on p. 503.

GERMAN SOUTH-WEST AFRICA. MEAN RAINFALL IN INCHES.

	S. Lat.	E. Long.	Alt. in feet	No. of Years	Jan.	Feb.	March	April	May	June	July	Aug.	Sept.	Oct.	Nov.	Dec.	Year
Olukonda[1] ...	17° 57′	16° 18′	3509	9—10	3·47	5·28	3·78	1·54	0·04	0	0	0	0·12	0·55	1·77	2·87	19·42
Grootfontein[2] ...	19° 40′	18° 2′	5003	5¼	4·76	3·56	4·45	1·89	0·36	0	0	0	0·12	0·83	1·61	5·55	23·13
Waterberg[3] ...	20° 35′			10	4·29	4·37	3·50	2·17	0·39	0	0	0·04	0·04	0·32	1·42	4·02	20·56
Okahanja[4] ...	21° 59′	16° 57′	4349	11½	2·95	3·19	3·47	1·42	0·39	0	0·04	0·29	0·03	0·79	0·75	2·68	15·96
Gobabis[5] ...	22° 21′	19° 1′	4634	8	4·69	2·95	2·36	1·54	0·21	0	0·07	0·05	0·08	0·55	0·98	2·95	16·43
Otyimbinge[6] ...	22° 20′	16° 10′	3074	5½	1·46	1·50	0·91	0·43	0·02	0	0	0	0·08	0·08	0·21	1·38	6·07
Windhoek[7] ...	22° 34′	17° 5′	5494	14½	3·50	2·64	3·15	1·93	0·20	0	0·10	0·16	0·04	0·32	0·95	2·09	15·08
Schaaprivier[8] ...	22° 45′	17° 22′	5723	11⅚	3·66	2·36	2·84	1·38	0·24	0·03	0·08	0·16	0	0·20	0·67	2·32	13·94
Rehoboth[9] ...	23° 19′	17° 3′	4774	7½	2·80	2·21	2·01	1·22	0·24	0·03	0	0·14	0·02	0·26	0·54	1·02	10·49
Hoakhanas[10] ...	23° 57′	17° 56′	4120	7	1·61	1·87	1·50	1·65	0·22	0·06	0·02	0	0	0·04	0·18	0·71	7·86
Nomtsas[11] ...	24° 28′	16° 46′		6½	1·64	1·24	1·54	1·34	0·12	0	0	0	0	0·10	0·12	1·16	7·26
Gibeon[12] ...	25° 7′	17° 46′	3695	7	2·36	0·91	1·22	0·75	0·02	0·03	0·02	0	0	0·28	0·28	0·51	6·38
Keetmanshoop[13]	26° 32′	18° 2′	3362	6¾	1·14	1·18	1·65	1·10	0·08	0·04	0·12	0	0	0·24	0·24	0·51	6·30
Bethanien[14] ...	26° 30′	16° 52′	3057	6⅚	1·02	0·83	1·46	0·95	0·02	0·16	0	0·08	0·04	0·28	0·08	0·55	5·47
Kubub[15] ...	26° 42′	16° 10′	4676	3⅙	0·85	0·92	1·47	0·63	0·43	0·59	0	0·14	0·13	0·02	0	0·87	6·05
Warmbad[16] ...	28° 27′	18° 42′	2354	5¾	0·59	1·06	0·79	1·06	0·08	0·16	0·12	0·06	0·16	0·08	0·04	0·08	4·28
Zesfontein[17] ...	18° 35′	13° 36′		3⅙	0·39	0·53	0·51	0·32	0·24	0	0	0	0·03	0·10	0·04	0·78	2·94
Swakopmund[18] ...	22° 42′	14° 33½′		6½	0·05	0·10	0·14	0·04	0·08	0	0·02	0·02	0·02	0·08	0	0·24	0·79
Walfisch Bay[19] ...	22° 57′	14° 26½′	7	13¾	0·03	0·04	0·06	0·04	0·02	0	0	0·02	0·02	0·04	0·02	0	0·29
Lüderitzbucht[20]	26° 36′	15° 15′		3 1/12	0	0·22	0·01	0·04	0·15	0·06	0	0·20	0·11	0	0·31	0·06	1·16
Omaruru[21] ...	21° 35′	16° 13′	3793	8	2·95	2·44	1·46	0·98	0·12	0	0	0	0·06	0·21	0·27	1·46	9·95

[1] March—December 1896, 1897—1900, June—December 1901, January—October 1902, November, December 1903, January—March 1904, February, March 1905. [2] April—December 1899, January 1900—June 1905. [3] 1893, 1894, June—December 1895, 1896, 1897, June—December 1898, 1899—1903. [4] June—December 1891, 1892, 1893, January—June 1894, 1895, January—June 1896, November, December, 1897, July—December 1898, 1899—1903, April—December 1904, January—June 1905. [5] July—December 1897, January 1898—June 1905. [6] October—December 1892, January—March 1893, December 1896, January—June 1897, July—December 1899, July 1900—December 1902, July—December 1903, July—December 1904, January—June 1905. [7] January 1891—June 1905. [8] July—December 1892, 1893—1903, January, March, May, November 1904. [9] July—December 1883, 1884, 1885, July—December 1888, October 1900, 1901—1904, January—June 1905. [10] July 1898—June 1905. [11] July 1898—December 1903, January—August, December 1904, January—May 1905. [12] July 1898—June 1905. [13] February—April, November 1898, January 1899—June 1905. [14] January—June, August, November 1893, May—December 1899, 1900—1904, January—June 1905. [15] August—December 1892, January—September 1893, July—December 1903, 1904, January—June 1905. [16] 1898—1902, June—October 1903, June—September 1904. [17] May—December 1902, 1903, 1904, January—June 1905. [18] 1899—1904, January—June 1905. [19] 1886—1896, April—December 1899, 1900, 1901. [20] August—December 1892, January—September 1893, 1894, January—November 1895. [21] 1883—1885, 1900—1904. (*M.D.S.* 1907, p. 54.)

GAZALAND.

The whole of the northern part of Gazaland consists of the right-bank portion of the Zambezi basin, bounded roughly on the south by a line drawn from the point where the 36th meridian east of Greenwich cuts the coast, through the intersection of the 34th meridian with the 18th parallel of latitude. The general slope of this portion of the country is north-eastwards sinking down to the river. The middle of the region has a general easterly slope, and the southern portion a south-easterly. The country rises gradually, by terraces, from the ocean, the westerly limit, except in the Zambezi basin, being the highlands of the Manika plateau in the north, and the Lebompo range in the south, in the gap between which both the Sabi and Limpopo find their way to the sea. Abreast of the Zambezi delta are the Nhamonga and Gorongoza mountains, the birth-place of many streams which flow, some to the Zambezi, some to the Pungwe, and some directly to the sea.

There is the same lack of precise information as to the climate of Gazaland as there is in the case of Mozambique. Almost all that can be done, therefore, is to glean what is possible from general statements. The coastal plain, as is the case with Mozambique, has an evil reputation ; the rivers bring down masses of vegetation, which a warm sun soon helps to decay in the swamps where they collect. The result may be easily imagined. All the coastal towns, which, as a rule, are located at the river mouths, are malarious. The comparatively heavy rainfall and nightly dews tend also to add to the unhealthiness of the coast.

At Delagoa Bay the rains generally last from September to March or April. North of this, as far as the Zambezi, the rains, following the sun in its course northwards, begin later, about November, and continue till April. Not very long ago Lourenço Marques used to be surrounded by swamps, but a large proportion of the land has been reclaimed, and the swamps filled up; the roads, too, have been improved and macadamised ; and the town, which was previously very unhealthy has consequently improved greatly in this respect. The Governor registered meteorological observations for 21 months, from October 1876 to June 1878, and from these it may be gathered that the mean temperature for those months was $75°\cdot3$ Fahr. ($24°\cdot1$ C.). The mean maximum was $89°\cdot2$ Fahr. ($31°\cdot7$ C.), and the mean minimum $65°\cdot1$ Fahr. ($18°\cdot4$ C.); while the highest recorded temperature was $100°\cdot4$ Fahr. ($38°$ C.), and the absolute minimum $49°\cdot1$ Fahr. ($9°\cdot5$ C.), the latter recorded in June, and the former in January. The mean annual rainfall, compiled from the records for ten years, is $28\cdot23$ inches or 717 mm. In the year 1878 $8\cdot11$ inches (206 mm.) fell in January. The coolest period is from May to August, when there is least rain but heavy dew.

Some 16 miles north of Lourenço Marques is the Swiss Mission Station of Rikatla. Meteorological observations were taken here during the period 1889—1891, but unfortunately I have not been able to secure the results in full. The following general results, however, are available. The mean morning temperature (8 a.m.) is $67°\cdot46$ Fahr. ($19°\cdot7$ C.), the mean maximum $93°\cdot5$ Fahr. ($34°\cdot1$ C.), which is considerably higher than at Lourenço Marques, and the mean minimum $64°\cdot8$ Fahr. ($18°\cdot2$ C.), or almost the same as at that place. The highest temperatures are registered in January, February, and March. The absolute maximum, for the three years, was $112°$ Fahr. ($44°\cdot5$ C.), recorded in January, and the absolute minimum $48°\cdot2$ Fahr. ($9°$ C.). May, June, July, and August are the coolest months. Westerly winds are seldom experienced, though the general direction is fairly distributed between north, south, and east. The north winds are dry and hot, and are rarely accompanied by rain, and during the prevalence of these winds fever is rife. The south wind varies much both in force and humidity. This is the wind which brings the rain, and when the direction is due south or south-west, the precipitation is greatest. The easterly winds are rather sea-breezes, and begin, generally, towards midday, lasting till about 3 p.m., to be replaced by either the northerly or southerly wind. These sea-breezes are merely local, but are healthy. The rains at this station usually last from October to March.

For Beira a table of meteorological observations, compiled from the results for two years, is given below, and this speaks for itself. It may be well, however, to draw attention to the fact that the greatest range of temperature occurs during the cooler months; and to call attention to the heavy rainfall in December and February. During May, June, July, August, and September the mean temperature is below the mean for the year. During the remaining months it is above that figure.

Inhambane is considered to be the least unhealthy of the coast towns. The hottest months are February and March, when the mean temperature varies between $82°\cdot4$ and $84°\cdot2$ Fahr. ($28°$ and $29°$ C.). In July the temperature is as low as $62°$ Fahr. ($16°\cdot7$ C.), but during the period from November to May fever is prevalent.

Chinde is reported to be comparatively healthy, and this is borne out by the fact that it is sometimes made use of as a kind of sanatorium, a place to go to for a change, those who have suffered from malarial fever in the interior finding a short stay here beneficial, owing mainly to the bracing sea-breezes.

To leave the coast lands and enquire concerning the interior. Mr St V. Erskine, from experience in the early seventies, remarks that the climate in the highlands of the west, between Manika and the Sabi river, is eminently healthy and cool, and suited for stock raising, and other agricultural pursuits, especially the cultivation of wheat. The rain sets in here about mid-October, with driving mists from the east and south-south-east. These mists are followed by thunder and lightning, and then the rain falls for two or three days, but either north or west winds restore the fine weather. In December the thermometer, he says, frequently marked $102°$—$104°$ Fahr. ($38°\cdot8$—$40°$ C.), and sometimes even $105°$ Fahr. ($40°\cdot5$ C.), while at the end of that month the night temperature was $74°$ Fahr. ($23°\cdot3$ C.). Lower down on the Sabi river the heat was intense in mid-September.

The Barue district is reported to be extremely healthy, and mosquitoes are but rarely seen. In the low-lying localities, however, which are not numerous, there may be danger of fever. The cold at certain seasons is considerable, and at an altitude of 1500 feet above the sea-level, the thermometer has been known to register 42° Fahr. (5°·6 C.), the midday temperature rarely exceeding 70° Fahr. (21°·1 C.) at the same period. Col. Arnold[1] writes as follows of the same district (Barue):—"The climate as a whole is healthy. On the plateau it is dry, cool, and invigorating, and consistently temperate throughout the year. Even on the lower plains colonization by Europeans is certainly possible —much more so than in the valleys of Manika. Even in the rains two or more dry sunny days intervene, as a rule, between the downpours, while the prevailing winds from the Indian Ocean are cooling and pleasant."

At Boroma or Broma, on the Zambezi, in Lat. 16° S., Long. 33° 12′ E., observations were taken from March 1891 to February 1892. During the period from May to October there was no rain, and only ·02 inch (0·5 mm.) in April, but during the period November to February the total precipitation amounted to 29·57 inches (751 mm.).

Tete has a very hot climate. In February 103° Fahr. (39°·4 C.) is not unfrequently registered in the shade, while in November 84° Fahr. (28°·9 C.) is an ordinary figure to register, and in July, which is the coolest month, 72° Fahr. (22°·2 C.) is about the mean of the midday temperature. Between this and the coast the midday temperature is about 98° Fahr. (36°·6 C.), sinking to 80° Fahr. (26°·7 C.) at night during the hotter months. The mean annual rainfall is 33·6 inches or 853 mm.

In the Inyanga district Dr Peters[2] says it was bitterly cold at night in August, and water froze.

The interior of Gazaland north of the Limpopo river up to the 20th parallel of latitude is an almost rainless district, and the tracts in the neighbourhood of the Limpopo bend are malarious.

Mr R. L. Reid[3] says that, in the Urema valley, when he was there, August to December 1902, the Gorongoza country was suffering from drought.

For information on the climate on the Zambezi *see* under Mozambique at page 380.

M. G. Vasse, who hunted in the country for three years,

[1] *Geographical Journal*, Vol. XVIII, 1901, p. 515.
[2] *The El Dorado of the Ancients.*
[3] *Geographical Journal*, Vol. XXV, 1905, p. 64.

divides the seasons, in Manika district, thus:—a springtime of storms consisting of October, November, and December, revivifying nature; a summer of heavy rain extending over January, February, and March, with great heat; an autumn season with cool nights occupying the next three months, when the waters retire from the inundated lands to their proper river-courses; this is the worst time for fever; July, August, and September form the dry winter months, when the natives burn the grass[1].

PRODUCTIONS, &c.

As to the vegetable products of Gazaland, Col. Arnold says that they might be made to include tea, coffee, cotton, indigo, cacao, tobacco, and vanilla. Rice grows well on the lower Sabi river; in the Manika district wheat might easily be grown, and there, too, cotton, sugar, and coffee thrive in the deeper and warmer valleys. Inhambane is especially noted for the production of sugar, and ground-nuts. Rubber might be successfully cultivated in the north, as it is indigenous. Mangrove bark is obtainable almost anywhere along the coasts. Sisal hemp might profitably be cultivated near Inhambane and Lourenço Marques, as well as the *Agave americana*.

[1] Vasse, *Tour du Monde*, 1898, p. 226.

BEIRA. Lat. 19° 50′ S., Long. 34° 50′ E.

Compiled from 2 years' observations[1].

	Jan.	Feb.	March	April	May	June	July	Aug.	Sept.	Oct.	Nov.	Dec.	Year
Mean temperature[2] (Fahr.°) ...	79°·8	80°·2	80°·2	77°·2	73°·5	69°·2	69°·0	70°·7	73°·5	76°·0	79°·0	80°·8	75°·8
Maximum temperature	89°·8	88°·8	88°·0	86°·0	86°·0	81°·5	83°·0	83°·8	83°·0	87°·2	90°·2	90°·5	—
Minimum temperature	72°·0	74°·0	72°·3	68°·3	62°·8	59°·3	58°·0	59°·8	61°·8	64°·3	69°·8	72°·8	—
Mean rainfall in inches[3]	9·2	18·8	9·2	5·0	1·9	1·4	·1	·2	·6	2·1	1·7	15·0	65·2
Maximum fall in 24 hours	3·9	4·3	1·9	2·0	·7	·3	·1	·1	·7	1·0	·8	4·6	4·6

The mean of the barometric readings varies between 29·91 in January and 30·21 in June, the greatest range being ·55 in September and the least ·24 in March. The most prevalent directions of the wind are :—

January	...	S.E.	S.S.E.	E.	E.S.E.	July	...	E.S.E.	S.E. S.
February	...	E.	S.E.			August	...	S.S.E.	E.S.E. E.
March	...	S.E.	E.S.E.	S.S.E.		September	...	S.E.	E.S.E. E.
April	...	S.E.	E.S.E.			October	...	E.	E.S.E.
May	...	S.	S.E.			November	...	E.	E.N.E. S.E.
June	...	S.W.	N.W.	S.S.E.	E.S.E.	December	...	E.	E.S.E. S.E.

1 1895, 1896.

2 There is no statement in the *Africa Pilot*, from which this Table is extracted, as to how the means were obtained.

3 The rainfall in 1895 was 95 inches, in 1896 it was only 36 inches.

LOURENÇO MARQUES. Lat. 25° 59′ N., Long. 32° 35′ E.

Compiled from observations for 21 months. See p. 505.

		Jan.	Feb.	March	April	May	June	July	Aug.	Sept.	Oct.	Nov.	Dec.	Year
Barometric pressure (inches)	...	29·83	29·82	29·90	30·12	30·16	30·32	30·19	30·12	30·01	30·04	29·88	29·81	30·02
Extreme range (inches)	...	·80	·66	·70	·84	·96	·74	·55	·45	·82	1·14	·99	·81	1·34[1]
Mean temperature (mean of 8 a.m., noon, and 8 p.m.) Fahr.°	81°·2	82°·8	80°·6	76°·7	74°·1	68°·1	68°·0	70°·9	70°·3	73°·4	77°·1	80°·1	75°·3
Absolute maximum	...	100°·4	96°·8	94°·1	86°·5	85°·3	79°·7	81°·1	84°·2	89°·6	88°·2	92°·1	95°·5	100°·4[1]
Mean rainfall (inches)	...	5·83	6·50	3·97	·90	·54	·24	·32	·09	1·58	1·97	1·02	3·82	26·78[2]
Mean number of rainy days	...	12	12	10	6	3	3	6	2	6	7	10	11	88

The observations for January to June are for the years 1877, 1878, July, August, and September for 1877, October, November, and December for 1876 and 1877.

The mean temperature is above the true mean, the means for January to June 1878 being from 2° to 2¼° above the mean of the maximum and minimum observations.

[1] Extreme for the period.

[2] 680 mm., as compared with the 10-years mean 717 mm. (28·23 inches).

APPENDIX I.

EXPLANATORY GLOSSARY OF THE PRINCIPAL VEGETABLE PRODUCTIONS EXCEPT TIMBER-TREES.

[By Miss MARY S. KNOX.]

Acacias, of which there are many species indigenous in Africa, are shrubs or small trees of the *Mimosa* tribe. Gum arabic (chiefly from *Acacia vera*) and gum senegal are exudations from several species of Acacia, some of which are native on the Moroccan Atlantic coast and in Senegal, others in the Atlas, in the sub-desertic regions, and elsewhere. Bark (Black-Wattle) for tanning is furnished by some species in Natal and other localities, and pods for the same purpose are gathered from other species in Egypt, Nubia, etc.

Agave, see under **Hemp.**

Aloe, see under **Hemp.**

Alpha, Halfa or **Halpha** (*Macrochloa tenacissima*) is one of the xerophil grasses, from which fibre is extracted, growing in the sub-desertic regions of the Mediterranean and the Sudan, that is, both north and south of the Sahara. The fibre, known commercially as Esparto, is used in the manufacture of paper.

Amendoim, see **Ground-nuts.**

Bamboo, see **Raphia vinifera.**

Banana, see under **Plantain.**

Bark, see **Acacia, Mangrove,** and **Uganda Bark.**

Barley (*Hordeum*), one of the corn-grasses, is used for malt and produces, when deprived of the husk, pearl-barley; it has a wider range of climate than any other cereal but prefers a fairly warm and temperate climate; it thrives best on light friable loam or calcareous soil.

Benni seed, see **Sesame.**

Black-Wattle, see **Acacia.**

Cacao or **Cocoa** is produced from the seeds of *Theobroma cacao*, a small tree, seldom more than 16 or 17 feet high, and is supposed to be, or should be, the basis of chocolate. Cocoa requires a very damp and very warm climate.

Carapa is a fat or vegetable oil, produced from a species of *Carapa* indigenous in the West Coast. The Carapa is a tree, the fruit of which contains oily seeds, from which the fat is made.

Cassava, see **Manioc.**

Chillies, the ripe and dried fruits of several species of *Capsicum*, e.g., *baccalium, frutescens, annuum*, are ground to powder and form cayenne pepper. Some species are natives of Egypt.

Cocoa-nut Palm (*Cocos nucifera*) is put to so many uses that it would be almost impossible to name them. But so far as Europeans are concerned the chief products are the fibre, which is obtained from the outside coating of the nuts, the nuts themselves, and cocoa-nut oil, obtained from the white kernel inside the nut. This oil is a constituent of margarine. Copra is the kernel of the nut, properly after the oil has been pressed out, and is used as a spice. These Palms are cultivated throughout the tropics, but they prefer the vicinity of the sea, though found growing at considerable distances inland.

Coffee (*Coffea arabica*). This shrub or small tree, when allowed to grow freely, will reach a height of twenty feet with a stem three or four inches thick, but, as cultivated, is kept down to about ten feet. The fruit looks like a small cherry, and each contains two seeds, which are ground into coffee. The plant grows best with a heavy rainfall, moderate elevation, and a warm but temperate climate. For the best results the rainfall should not fall short of 75 inches, and should be well distributed ; but coffee can be, and is, grown in drier regions, though smaller crops are then obtained. It is a native of South-West Abyssinia, in the district of Kaffa, hence the name. It is also indigenous along the whole of the West Coast, from Sierra Leone to Angola.

Copal gum, so far as Africa is concerned, is obtained from leguminous trees which are indigenous in Zanzibar and also in the Congo. It is now produced in many parts of Africa.

Copra, see **Cocoa-nut Palm.**

Cork-Oak is *Quercus suber*, the bark of which is cork.

Cornflour, see **Maize.**

Cotton is the hairy covering of the seeds of various species of *Gossypium*. It is mainly a tropical or sub-tropical crop. The various species may be grouped under two heads, viz., the late-ripening and the early-ripening. The former take a longer time to come to maturity, the latter ripening sooner. The late-ripening species require a prolonged rainy season, which is not needed in the case of the early-ripening sorts ; but both are distinctly warm-climate crops.

Cotton-Silk, see **Kapok.**

Cumin is a dwarf, fennel-like plant (*Cuminum cyminum*) which produces hot, aromatic fruits, very much like those of the caraway ; but now they are little used compared with the latter.

Date Palm (*Phœnix dactylifera*) grows to a great height. In addition to the fruit being used for food, a valuable fibre is obtained from the bases of the stalks. The leaves are put to many uses by the natives—building huts, &c. ; the heart of the young leaves is also used as a vegetable. Date Palms are chiefly cultivated in the north of Africa.

Dhurra, see **Millet.**

Earth-nuts, see **Ground-nuts.**

Flax (*Linum usitatissimum*) is an annual, mainly grown in temperate regions. The seed—Linseed—contains a large amount of oil and is imported into Britain in considerable quantities. From the residue, after the expression of the oil, is formed the oil-cake, greatly valued for fattening cattle. The finer kinds of linen are manufactured from the woody fibre of the stems of *Linum*.

Ginger is furnished by the rhizomes, or races as they are called, of *Zingiber officinale* after they have been dried in the sun and, in some cases, scraped when green. The plant, which is herbaceous, will grow in most tropical countries.

Ground-nuts (*Arachis hypogœa*), also called Amendoim, Mandobim, and Mancarra, are found native in West Africa, Mozambique, and German East Africa. The plant, which prefers light, sandy soil, is used as herbage food for cattle, and is remarkable for the way it thrusts its fruit into the ground after the flower has fallen ; this fruit, the ground-nut, or earth-nut as it is sometimes called, is an edible pod containing two seeds about the size of peas, which are eaten as food. But the chief mercantile product of the plant is the oil which is expressed from the seeds and used as an illuminant, or for food, and is said to be equal in quality to olive oil.

Guinea Corn, see **Millet.**

Guinea Wheat, see **Maize.**

Gum, see **Acacia, Copal.**

Halfa Grass, see **Alpha.**

Hemp, or what is known as Common Hemp (*Cannabis sativa*), is in the main a fibre plant, but it also has an oleaginous seed, and in hot, and preferably dry places, produces a resin. It is an annual, usually growing to a height of from 4 to 10 feet. The plants are grown close together so as to produce straight stems and no branches, as the fibre is contained in the thin bark of the stem. Bowstring Hemp—so-called because the fibre, from the thick fleshy leaves some 4 feet in length and 3 or 4 inches in breadth, is used as bowstrings by the African natives—includes four different African species. *Sanseviera guineensis*, the best known, is indigenous in the West Coast of Africa and Hinterland, and is also found in Abyssinia. *S. cylindrica* is indigenous in South Africa, *S. longiflora* in equatorial Africa, and *S. ehrenbergii* in East Africa. They are all stemless perennials with runners. Sisal Hemp (*Agave Sisalana*), native in both British and German East Africa, wants a strong dry soil, preferably limestone. *Agave americana* (commonly called aloe) produces fibre which is called "pita" (native Mexican name). These species of *Agave* have a large spreading tuft of hard, spiny leaves 3 to 6 feet long. The fibre, from both roots and leaves, is used for twine, rope, and other purposes such as paper-making, and the juice of the leaves yields a substitute for soap, which lathers in salt water as well as in fresh. The fibre of any of these hemps is separated from the roots, stems, or leaves, as the case may be, by bruising them, steeping them in water, and afterwards beating them.

Henna is the Persian name, and El Khanna the Arabic (though sometimes in Egypt called Khenna) of *Lawsonia alba* or *inermis*, a dwarf shrub, 8 to 10 feet high, grown in Egypt and North Africa, of which the dried leaves when powdered are used as a cosmetic and, also, for dyeing skins, Morocco leather, etc. a reddish-yellow colour.

Indian Corn, see **Maize.**

Indigo is mainly yielded by several species of the tropical and sub-tropical genus, *Indigofera*, and by *Tephrosia apollinea*, a native of Egypt and Nubia, all of which are leguminous plants. The dye is prepared in somewhat the following manner :—the leaves and stems are soaked in water, which becomes yellow, and is then drawn off, and kept in motion. By oxidation a blue precipitate is produced, which is subsequently boiled and dried.

Jute is the very valuable and important fibre obtained from the stems of *Corchorus capsularis* and *Corchorus olitorius*, both herbaceous annuals, seldom found far beyond the tropics. They require dampness throughout the whole period of their growth, and therefore a rice climate will also suit jute. The soft fibre is often 12 feet in length, and is used in the manufacture of carpets, some kinds of cloth, gunny-bags, &c., the coarser and stronger kinds for cordage, while paper is made from the refuse. The young shoots of both plants are used as a pot-herb, that of *C. olitorius*, largely grown in Egypt and Syria, being called Jew's Mallow.

Kafir Corn, see **Millet.**

Kapok (a species of *Bombax*). Cotton silk or raw silk is obtained from the silky hairs attached to the pea-like seeds, of which there are numbers in the large fruits of this tree, a native of West Africa.

K. A. 33

Khanna } , see **Henna.**
Khenna }

Kola-nuts are bitterish, white or red seeds, about the size of walnuts, eaten or used as a condiment by the natives in Western and Central tropical Africa. They are the production of several species of *Sterculia* and *Cola*, fairly high trees, which prefer the coastal districts.

Landolphia, see under **Rubber.**

Linseed, see **Flax.**

Lucerne (*Medicago sativa*) is a clover-like, perennial, fodder plant yielding annually two or more crops of green food much relished by horses and cattle.

Madder is the common name of the genus *Rubia*, and especially *R. tinctorum*, trailing or climbing herbs of which the roots, when dry, furnish a red dye.

Maize (*Zea Mays*), *i.e.* Indian corn or mealies, sometimes called Guinea wheat or Turkey wheat, is an important cereal grass, the produce of which is largely imported into Britain. There are 300 varieties, all tall plants with large leaves, which are used as fodder for cattle and horses, as also are the steeped seeds sometimes. The immature cobs are used as a vegetable, and the dried seeds ground to a fine flour, maizena or oswego, which we call cornflour. It is probably the grain which is most largely consumed by human beings, next to rice, and is used greatly as food in many parts of Africa. It grows in very different soils and in dissimilar states of moisture and dryness, for instance it will grow almost equally well in the wet regions of the equatorial belt and in dry South Africa. It is sown in rows 2 to 4 feet apart and very frequently hoed ; the crop is easily saved and the grain easily preserved. It is often attacked by a disease called Ergot.

Mancarra, see **Ground-nuts.**

Mandioc, see **Manioc.**

Mandobim, see **Ground-nuts.**

Mangroves (*Rhizophora* and *Conocarpus*) are trees which grow in the muddy, coastal swamps of the tropics and are valuable as producing tannin and dyes which are obtained from the bark.

Manihot, see **Manioc.**

Manioc, Mandioc or Manihot, the product of which is called Cassava, is a tropical plant, and there are two different kinds, which, though botanically closely allied and both woody or shrubby plants 6 to 8 feet high, growing from fleshy tuberous roots, differ largely in other respects.

(1) Sweet Cassava (*Manihot Aipi*),
(2) Bitter Cassava (*Manihot utilissima*).

Sweet Cassava has a reddish tuber which, when boiled, makes an excellent vegetable food and when dried is ground into meal. Bitter, or yellow, Cassava is characterised by a virulently poisonous juice. The poison is mainly got rid of by grating the tubers to a pulp and expressing the juice, and the residual mass is pounded into a coarse meal, resembling bread crumbs, which is made into thin cakes, or cooked in various ways, the heat dissipating any remaining poison. The poisonous expressed juice, if allowed to settle, deposits a large quantity of starch, known as Tapioca meal, which is subsequently prepared, by heating, for trade. Both kinds of Cassava are largely used as food by the natives in tropical climates.

Mealies, see **Maize.**

Millets are large grasses, and the term is commonly used for various species of small seed corn, the seeds being generally smaller than wheat, and, when ground, used for human food. They are also used for poultry. Those chiefly grown in Africa are (1) Italian millet (*Setaria italica*, sometimes included under the genus *Panicum*) in the North and other parts. (2) Common millet (*Panicum miliaceum*). This is found in many parts and needs little rain. It can be cultivated where wheat and maize fail on account of poor or sandy soil. (3) Guinea corn or Sorghum (*Andropogon Sorghum*) is a native of South Africa and called there Kafir corn. It also grows in the Sudan and Egypt, there called Dhurra. There is a very large number of varieties of the species, which are all tall plants, and grow best in the rainy season, where there is a rainfall of from 30 to 40 inches, but some of them require a rainfall late in the season. The three chief varieties are (*a*) red Kafir corn, 5 to 6 feet high, (*b*) white Kafir corn, smaller, (*c*) African millet, 5 to 6 feet. The Sorghum makes fodder for cattle, but care should be taken that it is not too young, as the young plants contain a certain amount of prussic acid. One variety is a fibre plant. (4) Pearl millet, *i.e.* Spiked or Bullrush millet (*Pennisetum typhoideum*), is about 6 feet high and grows in Egypt and in many parts of tropical Africa, which is its true home.

Niger Seed is the small black seed of *Guizotia oleifera*, an annual composite herb, indigenous in Egypt and Abyssinia. A sweet edible oil, also used as a luminant, is expressed from the seeds.

Oats (*Avena sativa*—several varieties) grow on almost any soil, on high land, cold mountain valleys, or marshy ground. They are used as grain for cattle, meal for human food, and straw, which also forms food for cattle. This crop requires a cool summer, but differs from barley in requiring rain in spring and dry weather before harvest.

Oil Palm (*Elæis guineensis*) is a native of tropical West Africa from Sierra Leone eastwards, and along the shores of the Central African lakes. It is from 20 to 30 feet high and wants moist soil, flourishing in warm, damp valleys, where it grows in extensive forests. It produces two different oils, viz.:

(1) Palm-oil made from the fleshy outer covering of the fruit,
(2) Palm-kernel oil from the seeds.

These are used in the manufacture of candles and as lubricants and, in Africa, eaten as butter.

Olive (*Olea europæa var. sativa*) grows anywhere in the Mediterranean climatic province. The tree is of slow growth and attains a great age, but seldom exceeds 20 feet in height. The fruits vary in shape and size ; the flesh is very oily, the edible oil, of which there are three qualities, according to the method of extraction, being extensively exported as well as used for home consumption. The unripe fruits, after preparation, are used for pickling.

Palm, see **Oil Palm, Cocoa-nut Palm.**

Papyrus antiquorum is the Paper Reed, which was early worked in Lower Egypt, and grown on river banks in Abyssinia ; it is also the chief plant of the Nile Sudd, having underground stems which send up ordinary stems, 8 to 10 feet in height, parts of which are above the water. The old Egyptians made paper from slices of these stems cut vertically.

Pepper—African—is furnished by the dried fruits of various species of *Habzelia*, either tall shrubs or trees, natives of West Africa. The *Habzelia* flourishes best in a tropical climate with moderate extremes of heat and cold and a very heavy rainfall.

Pepper—Cayenne, see **Chillies.**

Pine-apple Fibre is obtained from *Ananassa sativa*, the pine-apple plant, so-called because the fruit resembles a pine or fir cone. The plant is a biennial, not unlike an aloe in habit, having hard, but thinner, leaves from which the fibre is obtained.

33—2

Plantain and **Banana** (*Musa paradisiaca* and *Musa sapientum* respectively, with several varieties) differ very little from each other, being both tropical or sub-tropical trees, growing as high as 20 feet or more, which have the appearance of palms and, with very little attention, yield fruit in enormous quantities, which is eaten either raw or cooked, and forms the chief food of a large part of the human race. Fibre, used in paper-making, &c., is obtained from the leaf-stalks, and the large leaves are employed by the natives for thatching their dwellings. Plantain meal is also made by powdering the dried fruit. The flower-stalk of *Musa Ensete*, a native of Abyssinia, is, when cooked, used as food.

Potatoes (innumerable varieties of *Solanum tuberosum*). Almost the only requirement for potatoes is a well-drained soil. They have been grown with success in many parts of Africa.

Ramie, see **Rhea.**

Rape Seed is the seed of *Brassica campestris* or *B. napus*, cabbage-like fodder plants. A yellow oil, used in the manufacture of soap and as a lubricator, is expressed from the seed, and the residue is formed into cake, largely used as a fertilizer and, also, as food for cattle and sheep.

Raphia vinifera or Bamboo is indigenous in the low swampy parts of tropical Africa, especially in the west, and is put to endless uses, *e.g.*, the natives build houses with the woody stalks, baskets are made from thin strips of the hard outer part, as well as from the undeveloped leaves, which are, also, made into cloth. This is the plant which produces the palm-wine, whence its name.

Raw Silk, see **Kapok.**

Rhea or Ramie (*Böhmeria nivea*) is a small, shrubby plant, 3 to 4 feet high, belonging to the nettle tribe, but having no stinging hairs. It requires a more or less tropical climate for its best development. Fibre is obtained from the inner bark of the stems and is manufactured into grass-cloth.

Rice (*Oryza sativa*) is a grass, whose grain supplies food to human beings to a greater extent than any other known plant, and it constitutes a large part of the cereal crop of Africa. It is confined to damp tropical or semi-tropical climates, and requires a temperature of from 60° to 80° Fahr. to ripen it. The best results are obtained where, during the growing season, there is a moderate amount of sunshine, and a damp, warm atmosphere; it therefore grows best in such localities as have a heavy and certain rainfall. Where irrigation is practised, a portion of the necessary rainfall can, in a measure, be dispensed with. There are 13 other species of the genus *Oryza*.

Rubber is obtained from three sources :—trees, climbers, and grassy or shrubby plants.

Lago Silk Rubber (*Funtumia elastica*) is a medium-sized tree, sometimes, in error, called *Kickxia elastica*. It is a native of Liberia, Gold Coast, Southern Nigeria, the Cameroons, Congo, Sierra Leone, and possibly Uganda. Para Rubber (*Hevea brasiliensis*) is a large tree, which has been introduced into the West coast of Africa and thrives well, the necessary conditions for its successful growth being there satisfied, namely—about 100 inches of rainfall, a mean annual temperature of about 80° Fahr. with a minimum of 60° Fahr. Ceãra Rubber or Manicoba (*Manihot Glaziovii*) closely resembles the Cassava plant (see under **Manioc**). It thrives in places quite unsuited to most cultivated plants, *e.g.* rocky or stony soils of poor quality, and arid districts. A native of Brazil, it has been introduced into West Africa, Uganda, Natal, and Zanzibar.

The Landolphias are woody climbers growing in the warm parts of Africa. The species which grows in the more Northern parts of the West Coast (Senegal, Gambia, Sierra Leone) is *Landolphia Heudelotii*. From Sierra Leone southwards to Angola is found *L. owariensis*, and many other species in the

Congo region. *L. Kirkii* is an East Coast species found from German East Africa to Natal; a variety of this has been found in Uganda and also occurs in the Cameroons.

The grass or root rubber, which is found in the neighbourhood of Stanley Pool and the Kasai region, is chiefly obtained from *Landolphia Tholloni*, a little shrub not more than 15 to 30 centimetres in height, and with habit like the *Aveille Myrtille* of Europe, but with very long clotted rhizomes, from which the rubber is extracted.

Safflower or Bastard Saffron (*Carthamus tinctorius*) is an annual composite plant, from 2 to 3 feet high, cultivated in Egypt. The flowers, containing yellow and red colouring matter, of which the latter is most valued, are made up into small discs and largely imported into Britain, where they are chiefly used for dyeing silk. From the seeds of one species of *Carthamus* oil for burning or culinary purposes is extracted.

Saffron is employed for flavouring and colouring, and largely used for painting and dyeing. African Saffron (*Lyperia crocea*), the flowers of which have been imported into Britain, is a native of Cape Colony and other parts, but the saffron of the shops is chiefly obtained from the orange-coloured style and stigmas of *Crocus sativa*.

Sanseviera, see under **Hemp.**

Semsem, see **Sesame.**

Sesame (*S. orientale* and *S. indicum*), also called Sesamum, Semsem, and Benni-seed, is an annual herb indigenous in Egypt. It has edible seeds which make excellent oil, said, by some, to be as good as olive oil. In its crude state the oil is used in soap-making and for burning in lamps.

Shaddock (*Citrus decumana*) is somewhat larger than an orange or lemon tree. The rounded fruit, also called Shaddock, has a smooth pale yellow skin and white or reddish sub-acid pulp. It sometimes weighs 10 to 20 lbs., and then is known as Pompelmousse; when small it goes by the name of Forbidden Fruit and a smaller clustered variety is called Grape Fruit.

Shea Butter is a solid white fat, extracted by boiling from the seeds, of which there are not more than three or four in a fruit, of *Bassia Parkii*, a tree, native in North tropical Africa. It keeps well and tastes not unlike ordinary butter.

Sisal, see **Hemp.**

Sorghum, see **Millet.**

Sugar Cane (many varieties of *Saccharum officinarum*) requires a tropical or sub-tropical climate. The thick, juicy kinds require a great deal of moisture; the thin, hard sorts, a well-distributed rainfall. Speaking generally, sugar requires an average temperature of about $80°$ F. and a rainfall of 60 inches if irrigation be not possible, but it can be grown with lower requirements.

Sweet Potato, see under **Yam.**

Tamarinds are the pods produced by an evergreen tree (*Tamarindus indica*) growing to 80 feet in height. The pods contain an acid pulp and are preserved in syrup as a confection, the seeds yield oil, and flowers, fruit, and leaves are used in dyeing.

Tapioca, see **Manioc.**

Tea is composed of the dried leaves of a species of *Thea*, an evergreen shrub or small tree, which is a native of Upper India, China, and Japan, but has been grown successfully in several localities in Africa, notably Nyasaland and Natal. It requires a heavy rainfall, well distributed, and a moist, warm atmosphere during the growing period; hot, dry winds are distinctly injurious.

Tobacco (various species of *Nicotiana*) is an herbaceous plant usually standing from 3 to 5 feet high. It is grown in most parts of Africa. The finer varieties, when properly cultivated, are generally restricted to twelve leaves, measuring about two feet and a half by one foot, which, when prepared, make half a pound of tobacco. The dried leaves when stretched in the hand are as soft as the finest kid.

Turkey Wheat, see **Maize.**

Uganda Bark cloth is woven from the fibre, which is obtained from the bark of a species of *Ficus.*

Vanilla, the well-known flavouring, is obtained from the thin, pod-like fruit of *Vanilla planifolia* and other species, all climbing orchids, natives of the tropics.

Vine (*Vitis*) requires a temperate climate and a moderate amount of moisture. For the fruit to ripen properly a warm dry autumn is requisite. The plants should have a generous supply of warmth and light, but should not be liable to be scorched by the sun. The vine is cultivated in the extreme south-west of Africa and in the extreme north.

Wheat (*Triticum vulgare*) withstands a greater range of temperature than most other crops. It is a crop of the warmer and drier parts of the temperate zones, but so varied are the habits of the numerous varieties that some will be found to grow well in almost any climate. The young seedlings will stand frost, but cold weather when the crop is growing towards maturity is decidedly detrimental, *i.e.* a temperature below 55° F. for three months or so ; on the other hand, wheat will not thrive in warm summer rain.

Yam (*Dioscorea*). Many people confound this with **Sweet Potato** (*Batatas edulis* and *B. paniculata*). As a matter of fact, they have little in common. The former is a monocotyledon, the latter a dicotyledon. The yam grows very much larger than the sweet potato, and really the only point in which they are similar is that they are both creeping or twining plants. From the seeds of *Batatas paniculata* is obtained Natal cotton.

APPENDIX II.

BRUSSELS.

Compiled from 67 years' observations.

	Jan.	Feb.	March	April	May	June	July	Aug.	Sept.	Oct.	Nov.	Dec.	Year
Mean temperature (C.°)	2°·28	3°·77	5°·73	9°·50	13°·28	17°·00	18°·39	17°·89	15°·11	10°·78	6°·22	3°·22	10°·22
" " (Fahr.°)	36°·1	38°·8	42°·3	49°·1	55°·9	62°·6	65°·1	64°·2	59°·2	51°·4	43°·2	37°·8	50°·4
Mean maximum temperature (C.°)	4°·61	6°·55	9°·11	13°·77	18°·22	21°·89	23°·11	22°·39	19°·11	14°·00	8°·77	5°·50	13°·89
" " (Fahr.°)	40°·3	43°·8	48°·4	56°·8	64°·8	71°·4	73°·6	72°·3	66°·4	57°·2	47°·8	41°·9	57°·0
Mean minimum temperature (C.°)	0°·0	1°·11	2°·22	5°·22	8°·36	12°·11	13°·72	13°·50	11°·11	7°·50	3°·72	1°·00	6°·61
" " (Fahr.°)	32°·0	34°·0	36°·0	41°·4	47°·1	53°·8	56°·7	56°·3	52°·0	45°·5	38°·7	33°·8	43°·9
Absolute maximum temperature (C.°)	13°·89	18°·22	21°·00	25°·78	30°·73	34°·72	35°·22	34°·61	31°·11	25°·00	19°·11	15°·28	35°·22
" " (Fahr.°)	57°·0	64°·8	69°·8	78°·4	87°·3	94°·5	95°·4	94°·3	88°·0	77°·0	66°·4	59°·5	95°·4
Absolute minimum temperature (C.°)	−19°·78	−16°·61	−13°·00	−4°·11	−0°·78	+4°·00	6°·99	8°·99	2°·78	−2°·56	−11°·23	−16°·89	−19°·78
" " (Fahr.°)	−3°·6	+2°·0	8°·6	24°·6	30°·6	39°·2	44°·5	48°·1	37°·0	27°·5	11°·8	1°·6	−3°·6
Mean relative humidity (per cent.)	86·3	82·0	72·7	64·1	61·9	63·9	65·2	66·3	71·7	78·3	83·6	87·6	73·6
Mean rainfall (in mm.)	54	47	49	47	56	65	74	75	65	71	62	60	72·5
" " (in inches)	2·13	1·85	1·93	1·85	2·20	2·56	2·91	2·95	2·56	2·80	2·44	2·36	28·54
Mean amount of cloud (0—10)	7·5	7·2	6·9	6·3	6·4	6·5	6·5	6·3	6·0	6·7	7·3	7·5	6·8
Mean number of days of — Rain	14·4	13·2	14·8	15·5	16·4	16·2	16·8	16·6	15·7	17·9	17·2	15·4	190
Snow	6·2	5·9	6·1	2·0	0·4	0·0	0·0	0·0	0·0	0·3	3·1	4·9	29
Hail	0·7	0·8	1·6	2·0	1·4	0·7	0·4	0·3	0·5	0·9	0·7	0·7	11
Thunder	0·2	0·2	0·6	1·2	2·7	3·8	4·1	3·8	1·8	0·6	0·1	0·1	16
Fog	9·6	6·6	5·7	3·1	2·2	1·5	1·3	2·8	5·7	8·3	9·5	10·7	67

LISBON. Alt. 194 feet.

Results of Meteorological Observations made at the Observatory "Infante D. Luiz," 1856—1905.

	Jan.	Feb.	March	April	May	June	July	Aug.	Sept.	Oct.	Nov.	Dec.	Year
Mean barometric pressure (700 mm.+)	57·58	56·55	54·02	53·93	53·87	55·27	55·27	54·89	55·10	54·60	55·10	57·09	55·27
" " (29 inches+)	·824	·782	·690	·687	·685	·732	·732	·726	·724	·714	·724	·804	·732
Mean maximum pressure (700 mm.+)	59·57	58·57	55·74	55·55	55·20	56·35	56·24	55·97	56·24	56·29	57·09	58·70	56·79
" " (29 inches+)	·903	·863	·750	·742	·728	·774	·768	·760	·770	·772	·804	·868	·792
Mean minimum pressure (700 mm.+)	55·86	54·98	52·01	52·33	52·52	54·10	54·17	53·89	53·77	53·35	53·69	55·12	53·81
" " (29 inches+)	·754	·729	·610	·623	·632	·694	·696	·686	·681	·664	·678	·724	·682
Mean temperature (C.°)	10°·20	11°·11	12°·49	14°·35	16°·43	19°·20	21°·23	21°·64	19°·94	16°·86	13°·53	10°·61	15°·63
" (Fahr.°)	50°·4	52°·0	54°·5	57°·9	61°·5	66°·6	70°·2	70°·9	67°·8	62°·4	56°·3	51°·1	60°·1
Mean maximum temperature (C.°)	13°·04	14°·29	15°·93	18°·03	20°·38	23°·59	26°·17	26°·77	24°·25	20°·42	16°·58	13°·40	19°·41
" " (Fahr.°)	55°·4	57°·7	60°·6	64°·4	68°·7	74°·5	79°·2	80°·2	75°·7	68°·7	61°·9	56°·1	66°·9
Mean minimum temperature (C.°)	7°·50	8°·25	9°·53	11°·28	13°·13	15°·61	17°·31	17°·74	15°·59	13°·93	10°·88	7°·98	12°·48
" " (Fahr.°)	45°·5	46°·9	49°·1	52°·3	55°·6	60°·1	63°·1	63°·9	60°·1	57°·0	51°·6	46°·4	54°·5
Mean daily range (C.°)	5°·54	6°·04	6°·40	6°·75	7°·25	7°·98	8°·86	9°·03	8°·66	6°·49	5°·70	5°·42	6°·93
" " (Fahr.°)	9°·9	10°·8	11°·5	12°·1	13°·1	14°·4	16°·1	16°·3	15°·6	11°·7	10°·3	9°·7	12°·4
Absolute maximum temperature (C.°)	19°·0	22°·8	28°·3	30°·4	33°·5	37°·5	38°·8	37°·8	35°·1	30°·8	24°·8	19°·0	38°·8
" " (Fahr.°)	66°·2	73°·0	82°·9	86°·7	92°·4	99°·6	101°·9	100°·0	95°·3	87°·5	76°·6	66°·2	101°·9
Absolute minimum temperature (C°.)	-1°·0	-1°·5	1°·1	4°·4	5°·4	9°·5	10°·9	12°·7	10°·4	6°·3	1°·2	-0°·5	-1°·5
" " (Fahr.°)	30°·2	29°·3	34°·0	39°·9	41°·7	49°·1	51°·6	54°·9	50°·7	43°·3	34°·2	31°·1	29°·3
Maximum variation of temperature (C.°)	20°·0	24°·3	27°·2	26°·0	28°·1	28°·0	27°·9	25°·1	24°·7	23°·6	23°·6	19°·5	40°·3
" " (Fahr.°)	36°·0	43°·7	48°·9	46°·8	50°·7	50°·5	50°·3	45°·1	44°·5	44°·2	42°·4	35°·1	72°·6
Mean relative humidity (per cent.)	78·9	75·4	71·2	70·3	68·0	55·4	62·8	61·8	67·2	72·2	77·6	79·1	70·8
Mean amount of cloud (0—10)	5·3	5·2	5·1	5·2	4·5	3·5	2·1	2·1	3·7	4·9	5·3	5·2	4·3
Mean rainfall (in mm.)	92·0	88·7	86·8	67·2	50·5	18·3	3·8	7·5	32·7	81·1	106·2	96·1	730·9
" (in inches)	3·62	3·48	3·41	2·64	1·99	0·72	0·15	0·30	1·28	3·19	4·18	3·78	28·74
Mean number of days with Rain	13·7	12·5	13·3	12·1	9·7	5·3	2·1	2·1	7·1	11·4	13·5	13·4	116·2
Fog	4·3	2·9	1·3	0·5	0·2	0·2	0·3	0·4	1·0	2·2	3·5	4·6	21·3
Thunder	1·1	0·8	0·9	1·1	1·8	1·1	0·4	0·3	1·7	1·4	1·0	1·0	13·4
Hail	0·7	0·8	0·9	0·6	0·3	0·1	0·0	0·0	0·1	0·4	0·4	0·6	4·6

ROME.

The barometric pressure is for 18 years; the temperature observations for 50 years; the rainfall and number of rainy days for 81 years; and the sunshine for 9 years.

	Jan.	Feb.	March	April	May	June	July	Aug.	Sept.	Oct.	Nov.	Dec.	Year
Mean barometric pressure (700 mm.+)	61·65	62·85	63·29	59·48	61·07	61·53	62·17	61·84	61·65	62·09	62·15	61·08	61·74
" " (in inches)	29·99	30·03	30·05	29·90	29·96	29·98	30·01	29·99	29·98	30·00	30·01	29·96	29·91
Mean temperature (C.°)	6·91	8·26	10·42	13·74	17·66	21·65	24·65	24·20	21·15	16·54	11·89	7·80	15·37
" (Fahr.°)	44·42	46·94	50·72	56·66	63·86	71·06	76·46	75·56	71·06	61·70	53·42	46·04	59·72
Absolute maximum temperature (C.°)	18·6	19·9	23·5	26·4	32·2	36·1	40·1	37·2	33·7	32·2	24·5	20·0	—
" " (Fahr.°)	65·48	67·82	74·30	79·52	90·06	97·08	104·28	99·06	92·76	90·06	76·10	68·00	—
Lowest absolute maximum temperature (C.°)	12·9	14·7	16·0	19·9	24·0	27·0	31·5	31·1	27·0	22·8	18·0	13·4	—
Lowest absolute maximum temperature (Fahr.°)	55·22	58·46	60·80	67·82	75·20	80·60	88·80	88·08	80·60	73·04	64·40	56·12	—
Absolute minimum temperature (C.°)	−6·0	−6·2	−4·0	+0·3	3·6	8·8	12·5	11·2	7·2	1·8	−2·5	−5·8	—
" " (Fahr.°)	21·2	20·84	24·8	32·54	38·48	47·84	54·50	52·16	44·96	35·24	27·50	21·56	—
Highest absolute minimum temperature (C.°)	1·8	3·6	4·4	7·6	12·3	16·0	18·0	17·7	14·8	10·6	5·1	8·0	—
Highest absolute minimum temperature (Fahr.°)	35·24	38·48	39·92	45·68	54·14	60·80	64·40	63·86	58·64	51·08	41·18	46·40	—
Mean rainfall (in mm.)	79·4	62·3	68·2	65·6	56·7	40·1	18·0	27·6	71·7	114·5	113·3	90·0	807·4
" " (in inches)	3·13	2·45	2·69	2·58	2·23	1·58	0·32	1·08	2·82	4·51	4·46	3·54	31·39
Mean number of rainy days	10·3	8·9	10·2	9·8	7·9	5·7	2·2	3·3	6·8	10·2	11·4	10·5	97·2
Mean number of hours of sunshine per day	2·72	3·57	3·43	4·22	5·09	7·29	9·21	8·60	5·67	4·02	2·99	2·24	1805·71

BERLIN. Alt. 37 metres.

Compiled from 50 years' observations, 1851—1900.

	Jan.	Feb.	March	April	May	June	July	Aug.	Sept.	Oct.	Nov.	Dec.	Year
Mean barometric pressure (700 mm.+)	62·7	62·3	59·9	60·4	60·7	61·1	60·5	60·9	62·1	61·2	61·6	61·8	61·3
" " (in inches)	30·03	30·01	29·92	29·93	29·95	29·96	29·94	29·96	30·00	29·96	29·98	29·99	29·97
Mean temperature in the City (C.°)	-0·2	+0·8	3·5	8·5	13·3	17·5	19·0	18·3	14·7	9·5	3·9	0·8	9·1
" " (Fahr.°)	31°·64	33°·44	38°·30	47°·30	55°·94	63°·50	66°·20	64°·94	58°·46	49°·10	39°·02	33°·44	48°·38
Mean temperature in suburbs (C.°)	-0·7	+0·3	2·9	7·7	12·7	16·7	18·1	17·4	13·9	9·0	3·6	0·5	8·5
" " (Fahr.°)	30°·74	32°·54	37°·22	45°·86	54°·86	62°·06	64°·58	63°·32	57°·02	48°·20	38°·48	32°·90	47°·30
Mean relative humidity (per cent.)	85	81	76	69	65	65	67	69	73	80	83	85	75
Mean amount of cloud (0—10)	7·4	7·1	6·4	5·8	5·4	5·7	5·7	5·7	5·4	6·7	7·4	7·4	6·4
Mean rainfall (in mm.)	39	37	44	36	49	64	77	57	42	46	43	47	581
" (in inches)	1·54	1·46	1·73	1·42	1·93	2·52	3·03	2·24	1·65	1·81	1·69	1·85	22·87
Mean number of days on which rain was measured	15·7	13·7	15·3	12·5	12·8	13·7	14·5	13·6	12·4	13·8	14·0	16·0	168
Mean number of days on which snow was measured	7·7	7·4	6·7	1·6	0·2	0	0	0	0	0·3	3·1	6·7	33·7
Mean number of days on which thunder was recorded	0·02	0·06	0·22	1·00	2·34	4·00	3·90	2·86	0·88	0·22	0·04	0·06	15·60

The absolute maximum temperature, 37°·0 C. (98°·6 Fahr.), was recorded on the 24th July 1865; the absolute minimum in the city, -24°·9 C. (-12°·82 Fahr.), on the 11th February 1855; the absolute minimum in the suburbs, -31°·0 C. (-23°·80 Fahr.) on the 19th January 1893, while the corresponding figure within the city on that day was -23°·1 C. (-19°·58 Fahr.).

The absolute maximum rainfall on any one day, 166 mm. (6·54 inches), was recorded on the 14th April 1902, and, of this amount, 119 mm. (4·69 inches) fell in 3 hours 22 minutes.

The average yearly amount of sunshine is 1667 hours, out of a possible 4456 hours, or 4·57 hours per day.

PARIS.

The pressure, temperature, and rainfall are for 50 years, the remaining particulars for 30 years; the pressure was recorded at the Paris observatory from 1841—1890, the rainfall both on the terrace and in the court of the observatory for the same period. The temperature observations were recorded at Parc St Maur from 1851—1900: the remainder at Parc St Maur for the period 1874—1903.

	Jan.	Feb.	March	April	May	June	July	Aug.	Sept.	Oct.	Nov.	Dec.	Year
Mean barometric pressure (700 mm.+)	57·12	56·97	55·21	54·12	55·66	56·40	56·44	56·19	56·61	55·02	55·50	57·14	55·99
" " (in inches)	29·81	29·80	29·73	29·69	29·75	29·78	29·78	29·77	29·78	29·73	29·74	29·81	29·76
Mean temperature (C.°)	2°·31	3°·64	5°·91	9°·94	13°·02	16°·52	18°·33	17°·69	14°·74	10°·05	5°·82	2·74	10°·06
" " (Fahr.°)	36°·1	38°·5	42°·6	49°·8	55°·4	61°·7	64°·9	63°·9	58°·5	50°·0	42°·4	36°·9	50°·0
Mean maximum temperature (C.°)	5·14	7·54	11·06	15·85	19·20	22·62	24·55	23·98	20·83	14·94	9·29	5·56	—
" " (Fahr.°)	41°·2	45·5	52·0	60·5	66·6	72·7	76·3	75·2	69·4	58·8	48·7	42·0	—
Mean minimum temperature (C.°)	-0°·06	+0°·65	1·90	5·06	7·67	11·28	13°·02	12·51	10·09	6·44	3·12	0·58	—
" " (Fahr.°)	30°·9	33·2	35·4	41·1	45·8	52·3	55·4	54·5	50·2	43·5	37·6	33·0	—
Absolute maximum temperature (C.°)	15°·7	20·7	23·9	29·4	33·6	35·3	38·4	37·4	35·5	26·7	21·7	17·8	38°·4
" " (Fahr.°)	60°·3	69·3	75·0	84·9	92·5	95·5	101·1	99·3	95·9	80·1	71·1	64·0	101·1
Absolute minimum temperature (C.°)	-17°·0	-15·4	-11·0	-3·5	-1·3	+2·1	5·8	4·9	0·6	-5·3	-15·0	-25·6	-25°·6
" " (Fahr.°)	1·4	4·3	12·2	25·7	29·7	35·8	42·4	40·8	33·1	22·5	5·0	-14·1	-14·1
Mean relative humidity (per cent.)	87·2	82·7	74·8	69·1	69·9	72·7	72·5	74·3	80·5	85·2	86·8	88·6	78·7
Mean amount of cloud (0—10)	7·1	6·6	5·8	5·6	5·2	5·4	5·2	5·0	5·0	5·8	6·9	7·1	5·9
Mean rainfall on Terrace of Observatory (in mm.)	35·2	26·0	32·3	39·1	47·8	53·3	52·7	49·7	47·3	50·1	43·1	35·9	512·5
Mean rainfall on Terrace of Observatory (in inches)	1·38	1·02	1·26	1·54	1·89	2·09	2·07	1·95	1·85	1·97	1·69	1·41	20·18
Mean rainfall in the Court (in mm.)	40·6	29·1	35·9	43·0	51·1	55·3	54·4	52·3	49·9	54·9	48·1	40·8	555·4
" " (in inches)	1·59	1·14	1·41	1·69	2·01	2·18	2·14	2·06	1·96	2·16	1·89	1·60	21·87
Mean number of days of { Rain	13·7	12·7	13·3	13·0	12·9	13·0	12·8	12·3	11·8	15·4	15·3	15·8	162
Snow	3·6	2·8	2·8	0·6	0·1	0	0	0	0	0·1	0·8	2·9	13·7
Hail	0·7	0·9	2·1	1·8	1·7	0·5	0·5	0·3	0·2	0·5	0·2	0·7	10·1
Thunder	0	0·2	0·7	2·3	4·1	5·7	5·6	4·7	2·7	0·9	0·2	0·1	27·2

LONDON (CAMDEN SQUARE). Alt. 118 feet.

Compiled from 40 years' observations.

	Jan.	Feb.	March	April	May	June	July	Aug.	Sept.	Oct.	Nov.	Dec.	Year
Mean barometric pressure (in mm. 700+)	61.53	62.07	59.55	60.24	61.42	61.36	61.00	60.61	61.22	59.60	60.18	60.64	60.83
" " (in inches)	29.98	30.00	29.90	29.93	29.98	29.97	29.96	29.94	29.97	29.90	29.93	29.95	29.95
Mean maximum pressure (in mm. 700+)	75.5	75.2	73.6	71.5	71.6	70.2	69.7	69.2	72.1	72.8	74.3	74.6	—
" " (in inches)	30.53	30.51	30.45	30.37	30.37	30.33	30.31	30.29	30.39	30.42	30.48	30.49	—
Mean minimum pressure (in mm. 700+)	40.7	44.5	41.3	46.0	49.0	50.9	50.5	49.3	47.7	42.9	40.5	41.0	—
" " (in inches)	29.15	29.31	29.18	29.37	27.49	29.57	29.55	29.50	29.44	29.25	29.15	29.17	—
Mean maximum temperature (C.°)	6.1	7.5	10.1	14.5	18.2	21.8	23.4	22.6	19.6	14.2	9.3	6.7	14.5
" " (Fahr.°)	42.98	45.50	50.18	58.10	64.76	71.24	74.12	72.68	67.28	57.56	48.74	44.06	58.10
Mean minimum temperature (C.°)	0.7	1.5	1.9	4.3	6.9	10.5	12.2	11.9	9.8	6.3	3.2	1.3	5.9
" " (Fahr.°)	33.26	34.70	35.42	39.74	44.42	50.90	53.96	53.42	49.64	43.34	37.76	34.34	42.62
Mean temperature (C.°)	3.4	4.5	6.0	9.4	12.5	16.2	17.8	17.2	14.7	10.3	6.2	4.0	10.2
" " (Fahr.°)	38.12	40.10	42.80	48.92	54.50	61.16	64.04	62.96	58.46	50.54	43.16	39.20	50.36
Mean daily range (C.°)	5.4	6.0	8.2	10.2	11.3	11.3	11.2	10.7	9.8	7.9	6.1	5.4	8.6
" " (Fahr.°)	9.72	10.80	14.76	18.36	20.34	20.34	20.16	19.26	17.64	14.22	10.98	9.72	15.48
Mean of absolute maxima (C.°)	11.7	12.9	16.6	21.5	25.6	28.4	29.6	28.9	25.1	20.1	14.7	12.5	—
" " (Fahr.°)	53.06	55.22	61.88	70.70	78.08	83.12	85.26	84.02	77.18	68.18	58.46	54.50	—
Mean of absolute minima (C.°)	-5.7	-4.4	-3.7	-1.2	+1.0	5.4	7.4	6.8	3.9	-0.3	-2.7	-5.1	—
" " (Fahr.°)	21.74	24.08	25.34	29.84	33.80	41.72	45.32	44.24	39.02	31.46	27.14	22.82	—
Mean relative humidity (per cent.)	88	85	83	77	74	73	74	77	82	87	88	84	81
Mean amount of cloud 9 a.m. (0—10)	7.1	7.1	6.4	6.2	5.9	6.0	5.9	6.0	5.9	6.4	6.7	6.8	6.4
" " 9 p.m. "	6.8	6.5	5.8	5.3	5.0	5.2	5.4	5.2	5.3	5.8	6.4	6.7	5.8
Mean number of rainy days	12	10	11	13	15	18	18	17	18	17	15	14	178
Mean amount of rainfall (in mm.)	51	41	43	42	49	59	61	61	61	67	58	54	647
" " (in inches)	2.01	1.61	1.69	1.65	1.93	2.24	2.40	2.40	2.40	2.64	2.28	2.13	25.47

APPENDIX III.

TABLE I.

Conversion of Centigrade Degrees into Fahrenheit Degrees.

C.°	0	1	2	3	4	5	6	7	8	9
	F.°	F.°	F.°	F.°	F.°	F.°	F.°	F.°	F.°	F.°
− 10	+ 14·0	+ 12·2	+ 10·4	+ 8·6	+ 6·8	+ 5·0	+ 3·2	+ 1·4	− 0·4	− 2·2
− 0	32·0	30·2	28·4	26·6	24·8	23·0	21·2	19·4	+ 17·6	+ 15·8
+ 0	32·0	33·8	35·6	37·4	39·2	41·0	42·8	44·6	46·4	48·2
+ 10	50·0	51·8	53·6	55·4	57·2	59·0	60·8	62·6	64·4	66·2
+ 20	68·0	69·8	71·6	73·4	75·2	77·0	78·8	80·6	82·4	84·2
+ 30	86·0	87·8	89·6	91·4	93·2	95·0	96·8	98·6	100·4	102·2
+ 40	104·0	105·8	107·6	109·4	111·2	113·0	114·8	116·6	118·4	120·2

TABLE II.

Values of Centigrade Degrees expressed in Fahrenheit Degrees.

C.°	·0	·1	·2	·3	·4	·5	·6	·7	·8	·9
	F.°	F.°	F.°	F.°	F.°	F.°	F.°	F.°	F.°	F.°
0	0·00	0·18	0·36	0·54	0·72	0·90	1·08	1·26	1·44	1·62
1	1·80	1·98	2·16	2·34	2·52	2·70	2·88	3·06	3·24	3·42
2	3·60	3·78	3·96	4·14	4·32	4·50	4·68	4·86	5·04	5·22
3	5·40	5·58	5·76	5·94	6·12	6·30	6·48	6·66	6·84	7·02
4	7·20	7·38	7·56	7·74	7·92	8·10	8·28	8·46	8·64	8·82

TABLE III.

Conversion of Fahrenheit Degrees into Centigrade Degrees.

F.°	0	1	2	3	4	5	6	7	8	9
	C.°	C.°	C.°	C.°	C.°	C.°	C.°	C.°	C.°	C.°
+ 0	− 17·78	17·22	16·67	16·11	15·56	15·00	14·44	13·89	13·33	12·78
+ 10	− 12·22	11·67	11·11	10·56	10·00	9·44	8·89	8·33	7·78	7·22
+ 20	− 6·67	−6·11	−5·56	−5·00	−4·44	−3·89	−3·33	−2·78	−2·22	−1·67
+ 30	− 1·11	−0·56	0·00	+0·56	+1·11	+1·67	+2·22	+2·78	+3·33	+3·89
+ 40	+ 4·44	+5·00	+5·56	6·11	6·67	7·22	7·78	8·33	8·89	9·44
+ 50	10·00	10·56	11·11	11·67	12·22	12·78	13·33	13·89	14·44	15·00
+ 60	15·56	16·11	16·67	17·22	17·78	18·33	18·89	19·44	20·00	20·56
+ 70	21·11	21·67	22·22	22·78	23·33	23·89	24·44	25·00	25·56	26·11
+ 80	26·67	27·22	27·78	28·33	28·89	29·44	30·00	30·56	31·11	31·67
+ 90	32·22	32·78	33·33	33·89	34·44	35·00	35·56	36·11	36·67	37·22
+100	37·78	38·33	38·89	39·44	40·00	40·56	41·11	41·67	42·22	42·78
+110	43·33	43·89	44·44	45·00	45·56	46·11	46·67	47·22	47·78	48·33
+120	48·89	49·44	50·00	50·56	51·11	51·67	52·22	52·78	53·33	53·89

TABLE IV.

Values of Fahrenheit Degrees expressed in Centigrade Degrees.

F.°	·0	·1	·2	·3	·4	·5	·6	·7	·8	·9
	C.°	C.°	C.°	C.°	C.°	C.°	C.°	C.°	C.°	C.°
0	0·00	0·06	0·11	0·17	0·22	0·28	0·33	0·39	0·44	0·50
1	0·56	0·61	0·67	0·73	0·78	0·84	0·89	0·95	1·00	1·06
2	1·11	1·17	1·22	1·28	1·33	1·39	1·44	1·50	1·56	1·62
3	1·67	1·73	1·78	1·84	1·89	1·95	2·00	2·06	2·11	2·17
4	2·22	2·28	2·33	2·39	2·44	2·50	2·56	2·61	2·67	2·72
5	2·78	2·84	2·89	2·95	3·00	3·06	3·11	3·17	3·22	3·28
6	3·33	3·39	3·44	3·50	3·56	3·62	3·67	3·73	3·78	3·84
7	3·89	3·95	4·00	4·06	4·11	4·17	4·22	4·28	4·33	4·39
8	4·44	4·50	4·56	4·62	4·67	4·73	4·78	4·84	4·89	4·95
9	5·00	5·06	5·11	5·17	5·22	5·28	5·33	5·39	5·44	5·50

TABLE V.

Conversion of Inches to Millimetres.

ins.	·0	·1	·2	·3	·4	·5	·6	·7	·8	·9
	mm.	mm.	mm.	mm.	mm.	mm.	mm.	mm.	mm.	mm.
0	0·00	2·54	5·08	7·62	10·16	12·70	15·24	17·78	20·32	22·86
1	25·40	27·94	30·48	33·02	35·56	38·10	40·64	43·18	45·72	48·26
2	50·80	53·34	55·88	58·42	60·96	63·50	66·04	68·58	71·12	73·66
3	76·20	78·74	81·28	83·82	86·36	88·90	91·44	93·98	96·52	99·06
4	101·60	104·14	106·68	109·22	111·76	114·30	116·84	119·38	121·92	124·46
5	127·00	129·54	132·08	134·62	137·16	139·70	142·24	144·78	147·32	149·86
6	152·40	154·94	157·48	160·02	162·56	165·10	167·64	170·18	172·72	175·26
7	177·80	180·34	182·88	185·42	187·96	190·50	193·04	195·58	198·12	200·66
8	203·20	205·74	208·28	210·82	213·36	215·90	218·44	220·98	223·52	226·06
9	228·60	231·14	233·68	236·22	238·76	241·30	243·84	246·38	248·92	251·46
10	254·00	256·54	259·08	261·62	264·16	266·70	269·24	271·78	274·32	276·86
11	279·40	281·94	284·48	287·02	289·56	292·10	294·64	297·18	299·72	302·26
12	304·80	307·34	309·88	312·42	314·96	317·50	320·04	322·58	325·12	327·66
13	330·20	332·74	335·28	337·82	340·36	342·90	345·44	347·98	350·52	353·06
14	355·60	358·14	360·68	363·22	365·76	368·30	370·84	373·38	375·92	378·46
15	381·00	383·54	386·08	388·62	391·16	393·70	396·24	398·78	401·32	403·86
16	406·40	408·94	411·48	414·02	416·56	419·10	421·64	424·18	426·72	429·26
17	431·80	434·34	436·88	439·42	441·96	444·50	447·04	449·58	452·12	454·66
18	457·20	459·74	462·28	464·82	467·36	469·90	472·44	474·98	477·52	480·06
19	482·60	485·14	487·68	490·22	492·76	495·30	497·84	500·38	502·92	505·46
20	508·00	510·54	513·08	515·62	518·16	520·70	523·24	525·78	528·32	530·86
21	533·40	535·94	538·48	541·02	543·56	546·10	548·64	551·18	553·72	556·26
22	558·80	561·34	563·88	566·42	568·96	571·50	574·04	576·58	579·12	581·66
23	584·20	586·74	589·28	591·82	594·36	596·90	599·44	601·98	604·52	607·06
24	609·60	612·14	614·68	617·22	619·76	622·30	624·84	627·38	629·92	632·46
25	635·00	637·54	640·08	642·62	645·16	647·70	650·24	652·78	655·32	657·86
26	660·40	662·94	665·48	668·02	670·56	673·10	675·64	678·18	680·72	683·26
27	685·80	688·34	690·88	693·42	695·96	698·50	701·04	703·58	706·12	708·66
28	711·20	713·74	716·28	718·82	721·36	723·90	726·44	728·98	731·52	734·06
29	736·60	739·14	741·68	744·22	746·76	749·30	751·84	754·38	756·92	759·46
30	762·00	764·54	767·08	769·62	772·16	774·70	777·24	779·78	782·32	784·86
31	787·40	789·94	792·48	795·02	797·56	800·10	802·64	805·18	807·72	810·26
32	812·80	815·34	817·88	820·42	822·96	825·50	828·04	830·58	833·12	835·66

TABLE VI.

Conversion of Millimetres to Inches.

mm.	0	1	2	3	4	5	6	7	8	9
	ins.	ins.	ins.	ins.	ins.	ins.	ins.	ins.	ins.	ins.
0	0·00	0·04	0·08	0·12	0·16	0·20	0·24	0·28	0·32	0·36
10	0·39	0·43	0·47	0·51	0·55	0·59	0·63	0·67	0·71	0·75
20	0·79	0·83	0·87	0·91	0·95	0·98	1·02	1·06	1·10	1·14
30	1·18	1·22	1·26	1·30	1·34	1·38	1·42	1·46	1·50	1·54
40	1·58	1·61	1·65	1·69	1·73	1·77	1·81	1·85	1·89	1·93
50	1·97	2·01	2·05	2·09	2·13	2·17	2·21	2·24	2·28	2·32
60	2·36	2·40	2·44	2·48	2·52	2·56	2·60	2·64	2·68	2·72
70	2·76	2·80	2·84	2·87	2·91	2·95	2·99	3·03	3·07	3·11
80	3·15	3·19	3·23	3·27	3·31	3·35	3·39	3·43	3·47	3·50
90	3·54	3·58	3·62	3·66	3·70	3·74	3·78	3·82	3·86	3·90
100	3·94	3·98	4·02	4·06	4·10	4·13	4·17	4·21	4·25	4·29
110	4·33	4·37	4·41	4·45	4·49	4·53	4·57	4·61	4·65	4·69
120	4·72	4·76	4·80	4·84	4·88	4·92	4·96	5·00	5·04	5·08
130	5·12	5·16	5·20	5·24	5·28	5·32	5·35	5·39	5·43	5·47
140	5·51	5·55	5·59	5·63	5·67	5·71	5·75	5·79	5·83	5·87
150	5·91	5·95	5·99	6·02	6·06	6·10	6·14	6·18	6·22	6·26
160	6·30	6·34	6·38	6·42	6·46	6·50	6·54	6·58	6·61	6·65
170	6·70	6·73	6·77	6·81	6·85	6·89	6·93	6·97	7·01	7·05
180	7·09	7·13	7·17	7·21	7·25	7·28	7·32	7·36	7·40	7·44
190	7·48	7·52	7·56	7·60	7·64	7·68	7·72	7·76	7·80	7·83
200	7·87	7·91	7·95	7·99	8·03	8·07	8·11	8·15	8·19	8·23
210	8·27	8·31	8·35	8·39	8·43	8·47	8·50	8·54	8·58	8·62
220	8·66	8·70	8·74	8·78	8·82	8·86	8·90	8·94	8·98	9·02
230	9·06	9·09	9·13	9·17	9·21	9·25	9·29	9·33	9·37	9·41
240	9·45	9·49	9·53	9·57	9·61	9·65	9·69	9·73	9·76	9·80
250	9·84	9·88	9·92	9·96	10·00	10·04	10·08	10·12	10·16	10·20
260	10·24	10·28	10·32	10·35	10·39	10·43	10·47	10·51	10·55	10·59
270	10·63	10·67	10·71	10·75	10·79	10·83	10·87	10·91	10·95	10·98
280	11·02	11·06	11·10	11·14	11·18	11·22	11·26	11·30	11·34	11·38
290	11·42	11·46	11·50	11·54	11·58	11·62	11·65	11·69	11·73	11·77
300	11·81	11·85	11·89	11·93	11·97	12·01	12·05	12·09	12·13	12·17
310	12·21	12·24	12·28	12·32	12·36	12·40	12·44	12·48	12·52	12·56
320	12·60	12·64	12·68	12·72	12·76	12·80	12·84	12·87	12·91	12·95
330	12·99	13·03	13·07	13·11	13·15	13·19	13·23	13·27	13·31	13·35
340	13·39	13·43	13·47	13·50	13·54	13·58	13·62	13·66	13·70	13·74
350	13·78	13·82	13·86	13·90	13·94	13·98	14·02	14·06	14·10	14·13
360	14·17	14·21	14·25	14·29	14·33	14·37	14·41	14·45	14·49	14·53
370	14·57	14·61	14·65	14·69	14·72	14·76	14·80	14·84	14·88	14·92
380	14·96	15·00	15·04	15·08	15·12	15·16	15·20	15·24	15·28	15·32
390	15·36	15·39	15·43	15·47	15·51	15·55	15·59	15·63	15·67	15·71

TABLE VI.—(*continued*).

Conversion of Millimetres to Inches.

mm.	0	1	2	3	4	5	6	7	8	9
	ins.	ins.	ins.	ins.	ins.	ins.	ins.	ins.	ins.	ins.
400	15·75	15·79	15·83	15·87	15·91	15·95	15·98	16·02	16·06	16·10
410	16·14	16·18	16·22	16·26	16·30	16·34	16·38	16·42	16·46	16·50
420	16·54	16·58	16·61	16·65	16·69	16·73	16·77	16·81	16·85	16·89
430	16·93	16·97	17·01	17·05	17·09	17·13	17·17	17·21	17·24	17·28
440	17·32	17·36	17·40	17·44	17·48	17·52	17·56	17·60	17·64	17·68
450	17·72	17·76	17·80	17·84	17·87	17·91	17·95	17·99	18·03	18·07
460	18·11	18·15	18·19	18·23	18·27	18·31	18·35	18·39	18·43	18·47
470	18·50	18·54	18·58	18·62	18·66	18·70	18·74	18·78	18·82	18·86
480	18·90	18·94	18·98	19·02	19·06	19·09	19·13	19·17	19·21	19·25
490	19·29	19·33	19·37	19·41	19·45	19·49	19·53	19·57	19·61	19·65
500	19·69	19·72	19·76	19·80	19·84	19·88	19·92	19·96	20·00	20·04
510	20·08	20·12	20·16	20·20	20·24	20·28	20·32	20·35	20·39	20·43
520	20·47	20·51	20·55	20·59	20·63	20·67	20·71	20·75	20·79	20·83
530	20·87	20·91	20·95	20·98	21·02	21·06	21·10	21·14	21·18	21·22
540	21·26	21·30	21·34	21·38	21·42	21·46	21·50	21·54	21·58	21·61
550	21·65	21·69	21·73	21·77	21·81	21·85	21·89	21·93	21·97	22·01
560	22·05	22·09	22·13	22·17	22·21	22·24	22·28	22·32	22·36	22·40
570	22·44	22·48	22·52	22·56	22·60	22·64	22·68	22·72	22·76	22·80
580	22·84	22·87	22·91	22·95	22·99	23·03	23·07	23·11	23·15	23·19
590	23·23	23·27	23·31	23·35	23·39	23·43	23·47	23·50	23·54	23·58
600	23·62	23·66	23·70	23·74	23·78	23·82	23·86	23·90	23·94	23·98
610	24·02	24·06	24·09	24·13	24·17	24·21	24·25	24·29	24·33	24·37
620	24·41	24·45	24·49	24·53	24·57	24·61	24·65	24·69	24·72	24·76
630	24·80	24·84	24·88	24·92	24·96	25·00	25·04	25·08	25·12	25·16
640	25·20	25·24	25·28	25·32	25·35	25·39	25·43	25·47	25·51	25·55
650	25·59	25·63	25·67	25·71	25·75	25·79	25·83	25·87	25·91	25·95
660	25·98	26·02	26·06	26·10	26·14	26·18	26·22	26·26	26·30	26·34
670	26·38	26·42	26·46	26·50	26·54	26·58	26·61	26·65	26·69	26·73
680	26·77	26·81	26·85	26·89	26·93	26·97	27·01	27·05	27·09	27·13
690	27·17	27·21	27·24	27·28	27·32	27·36	27·40	27·44	27·48	27·52
700	27·56	27·60	27·64	27·68	27·72	27·76	27·80	27·84	27·87	27·91
710	27·95	27·99	28·03	28·07	28·11	28·15	28·19	28·23	28·27	28·31
720	28·35	28·39	28·43	28·47	28·50	28·54	28·58	28·62	28·66	28·70
730	28·74	28·78	28·82	28·86	28·90	28·94	28·98	29·02	29·06	29·09
740	29·13	29·17	29·21	29·25	29·29	29·33	29·37	29·41	29·45	29·49
750	29·53	29·57	29·61	29·65	29·69	29·72	29·76	29·80	29·84	29·88
760	29·92	29·96	30·00	30·04	30·08	30·12	30·16	30·20	30·24	30·28
770	30·32	30·35	30·39	30·43	30·47	30·51	30·55	30·59	30·63	30·67
780	30·71	30·75	30·79	30·83	30·87	30·91	30·95	30·98	31·02	31·06
790	31·10	31·14	31·18	31·22	31·26	31·30	31·34	31·38	31·42	31·46
800	31·49.	·31·53	31·57	31·61	31·65	31·69	31·73	31·77	31·81	31·85

K. A.

APPENDIX IV.

EXPLANATION OF TERMS.

Anticyclone or high-pressure system or area, a region of relatively high barometric pressure in which the winds blow spirally outwards from the centre or maximum, in the same sense as the movements of the hands of a watch in the Northern, and in the opposite sense in the Southern Hemisphere.

Cyclone or low-pressure system or area, a region of relatively low barometric pressure in which the winds blow spirally inwards towards the centre or minimum, in the opposite sense to the hands of a watch in the Northern, and in the same sense in the Southern Hemisphere.

Humidity.—*Absolute Humidity*, the total pressure of water vapour in the atmosphere at any place.

Relative Humidity, the ratio of this actual vapour pressure to that of saturated water vapour at the temperature of the air.

Isobars are lines drawn on maps through places with equal atmospheric pressure.

Isohyets are lines drawn on maps through places with the same amount of rainfall.

Isotherms are lines drawn on maps through places with the same temperature.

Mean monthly maximum temperature is used for the monthly averages of the maximum daily temperature of the air registered throughout a number of years.

Mean monthly minimum temperature, similarly, stands for the monthly averages of the minimum daily temperature of the air, registered throughout a number of years.

Absolute maximum temperature. This is used for the highest temperature recorded during a period ; and **absolute minimum temperature** for the lowest temperature recorded during a period.

Mean of absolute maxima is used for the average of the monthly absolute maxima registered throughout a period of years ; similarly **mean of absolute minima** stands for the average of the monthly absolute minima recorded throughout a period of years.

Mean temperature is strictly speaking the mean of the temperature of the air during a day, month, &c. It is only possible, however, to arrive at an absolutely accurate result by the employment of self-recording instruments. Hence, in addition to recording the maximum and minimum temperatures, it is customary, when such self-recording instruments are not available, to take temperature observations at certain specified hours, varying according to circumstances, but generally in the morning, afternoon, and evening ; sometimes these observations are only taken twice a day and sometimes only once. An attempt is then made to calculate the mean temperature of the day from these observations, or, in some cases, from the maximum and minimum temperatures, or, in others, from a combination of these. Various methods are in use at different places, which, it is thought, give, in each case, the closest approximation to the mean temperature. In many cases the arithmetic mean of the maximum and minimum temperatures $\left(i.e.\ \dfrac{\text{max.} + \text{min.}}{2} \right)$ is used for the mean temperature ; in others the arithmetic mean of the temperature recorded at various hours, say at 7 a.m., 2 p.m., and 9 p.m. $\left(i.e.\ \dfrac{7a+2p+9p}{3}\right)$ and so on. In the tables which accompany the text the mean temperature has been calculated, at different places, from the following formulæ, in addition to the two already mentioned :—

$$\frac{8a+8p}{2} \qquad\qquad \frac{7a+2p+8p}{3}$$

$$\frac{9a+9p}{2} \qquad\qquad \frac{7a+3p+10p}{3}$$

$$\frac{6a+1p+10p}{3} \qquad\qquad \frac{8a+1p+6p}{3}$$

$$\frac{6a+2p+9p}{3} \qquad\qquad \frac{8a+2p+10p}{3}$$

$$\frac{6.30a+2p+10p}{3} \qquad\qquad \frac{9a+3p+9p}{3}$$

$$\frac{7a+1p+9p}{3} \qquad\qquad \frac{7a+2p+9p+9p}{4}$$

$$\frac{7a+2p+7p}{3} \qquad\qquad \frac{8a+8p+\text{max.}+\text{min.}}{4}$$

$$\frac{9a+9p+\text{max.}+\text{min.}}{4}$$

From the daily mean temperature is calculated the mean for the month, and from the means for the 12 months the mean for the year ; and the expression *mean temperature* in the tables stands for "the average, for a number of years, of the approximate mean monthly and yearly temperature of the air."

Mean daily range. By subtracting the minimum temperature of the air on any day from the corresponding maximum the extent of the oscillation of the temperature for that day is obtained, and the mean of these results gives the mean diurnal oscillation of the temperature for the month, hence *mean daily range* is used as an abbreviation for the average, extending over a period of years, of the mean diurnal oscillation of the temperature of the air.

Mean relative humidity. As with temperature, so with relative humidity. It is calculated at specified hours, and a mean is struck for the day. From these daily results is calculated a mean for the month, and from the 12 monthly means a mean for the year. Mean relative humidity, in the tables, is thus used for the average, for a number of years, of the mean relative humidity per cent. (See under *Humidity.*)

34—2

INDEX

The figures refer to the pages of the text, the larger figures to matter in the form of lists or tables; the italics refer to productions

Abbasia 2
Abercorn 353, 355
Aberdeen 462
Abu Hammed 21, 28
Aburi 8, 116, 117, 119, **125,** 126, **131**
Abyssinia 14, 28, 274, 284–295 (*see* Somaliland)
  plateau 11, 16, 17, 19, 23, 24, 27, 284, 285, 290
  productions 290, 291
  rain 256, **295**
  S., seasonal changes 287
  uplands 258, 286, 312
  winds 256
Acacia 278, 291, 366, 511, *gloss.*
Acacia arabica 8, 97, 511
Acacia gummifera 33
Accrá 8, 116–119, **122, 131**
Adamawa 150, 196, 197, 206, 213
  health 196
Addis Ábbaba **87,** 284, 286–289, **292, 295**
Addis Alem **293, 295**
Addi Ugri 2, 9, 20, 276–278, **281, 283**
Adghar 80, 81
Adrar 82, **83**
Adwa 285
Afikpo **145**
Afiu 46
Aflu 47
Afmadu 322
Africa, E., rain 85, 86, **87**
  winds and rain 254
  N. 15, 26, 32–83
  productions 32, 33
  N.E. 256
  S. 15–17, 22, 24, 28, 29, 390–510
  ice 392
  mean temperature 391
  rains, régime 390
  snow 392
  temperature, régime 390
  terrace formation, effect of 391
  winds and rain 392, 394

Africa, tropical 84–389
  E. 254–389
  seasons 84–91
  W. 88, 92–253
  N. section 92–206
  S. section 207–253
Agades 80
Agai *see* Hagai
Agave americana 508, 511
Agave sisalana 312, 313, 341, 374, 410, 508, 511
Ain Draham 53, **57**
Ainsworth, Mr 320
Air 25, 80
Ajuba R., region 261
Akaba 76
Akassa 136
Akele Guzai 274, 276
Akobo R. 288
Akwa Town 197
Albert Edward Nyanza *see* Edward L.
Albert Nyanza 17, 308, 309
Alexandria (Egypt) 2, 65, 66, **68, 71**
  (Cape Colony) 469
Algeden 275
Alger 2, 4, 44, 46, **49, 50, 51**
Algeria 2, 4, 5, 15, 16, 18–24, 26, 28, 29, 32, 33, 44–51
  frontier 24
  high plateaux 44, 46
  rains 47, **51**
  rains, compared with Tunis 52
  temperature compared with Tunis 52
Algerian Atlas 22
  Sahara 53
Algiers *see* Alger
Alice 469
Alicedale **487**
Aliwal North 473, **474, 487**
Almonds 33, 410
Aloes 304, 440, 511
Alpha see *Halfa*
Alwa 36
Amalienstein 461, **486**

Amar 147, **155**
Amaramba L. 379
Ambangulu **332**
Ambriz 243
  bay 246
Ambukól 265
Amejove 188, **191, 192**
Amendoim see *Ground nuts*
Amsterdam **425**
Andersson, Mr 490
Anecho **192**
Anglo-Egyptian Sudan 3, 14, 258–273
  northern 265
  productions 265
Anglo-German Boundary (E. Africa) 21, 255, 310, 317, 323, 331
Angola 243–251
  plateau 244
  productions 247
  rainfall 251
  S. coast 245
Angoniland 372, **378**
Angra Pequena *see* Lüderitz Bay
Ankober 284, 286, **294**
Ankole 308–310
  E. 313
Anticyclone, defn. 530
Apples 341, 410
Apricots 291, 410, 426, 432
Arabia 18
Arabian Sea 255, 257
Archer, Mr 99
Arnold, Col. 507
Aro 133
Arochuku 133
Aromatic plants 305
Arussi 287, 290, 291
Aruwimi R. 28, 228
  district 227, 234
Asaba **145**
Ashanti 126–128, **131**
Asmara 276, **283**
Asparagus 278
Assab 275, 277, 278, **282**
Assam rainfall 193
Asses 82
Assinie 179, **181**
Assint 72, **73**
Assobat road, 289
Aswa R. 308, 310
Aswan 23, 72, **74**
Atakora range, 182
Atakpame **192**
Atbai 260, 262
Átbara R. 23, 258
  upper 266
Athi R. 11, 320, **327**
Atlantic coast, N. 25, 34

Atlas Mts. 6, 32, 33
  Algerian 22
  foothills 34
  region 46, 47, 53
  Tunisian 53
Austin, Major 288
Avakubi 235
Axim 86, **89**, 115–119, **120**, 126, **131**
Ayata **51**
Aylmer, Mr 304
  Capt. 320
Ayra, Sig. 58
Ayrshire mine **417**
Azmera rains 285

Badagri **145**
Baga season 285
Bagamoyo 9, **346, 352**
  situation 342
Bagirmi 7, 215
Bahr-el-Ghazal 15, 18
  region, 19, 86, 258–260, 262, 265
Bahr-el-Jebel 15, 16
  highlands 262
  region, lower 264
  region, upper 260
Bai, Gambia, death-rate, 103
Bajianna, Gambia, death-rate 103
Bakel **176**
Balchi 289
Baldrati, Prof. 276
Baliburg (on Cross R.) 193
Baliburg (N.W. Cameroons) 193, 196
Bamboo-palm 198, 237, 247, 291, *gloss.*
Banana 90, 222, 225, 234, **238, 242**
  situation 236
Bananas 98, 237, 248, 278, 291, 313, 341, 410, *gloss.*
Bandama R. 179
Bangala district 227, 228, 233, 234
Bangweulu 16, 28, 353, 354, 355
  district 232
Bani 96
Bania 213
Banken 418, 420, 421, 423
Banzyville 229, 234
Baobab 248
Barberton 418, 420, 421, 423, 425, 426
Baria 275
Baringo L. 10, 11, 319
Barka (Tripoli) 5, 58
Barka R. (Eritrea) 275
  region 275
Barley 33, 82, 265, 278, 291, 324, 410, 426, 432, 440, 484, 496, *gloss.*

Baro R. 9, 286, 288
  region 261
Baroma 380
Barombi 193
Barotse *see* Marotse
Barron, Mr 76, 77
Barter, Maj. 126
Barth 6, 7, 80
Bartholomew, Dr G. J. 1
Barue district 506, 507
Basari **192**
Basoko 28, 86, **89,** 227, 228, 234, **242**
Basutoland 20, 445, **488**
Bathurst 2, 99, 100, 101, **102, 103**
  death-rate 103
  situation 102
Batna 2, 5, 46, **51**
Bauchi 147
  highlands 93, 146, 147, 152
Bayuda Desert 261
Bazen 275
Beans 33, 98, 265, 278, 305, 313, 341, 359, 374
Beaufort West 461
Bechuanaland 390, 445, 483, **488**
  border 420
Beira 14, 390, 506, **509**
Bel, M. Marc 210
Belfast 418
Belgian Congo 86, 90, 222–242
  productions 237
  seasons 233–237
  rainfall **242**
  temperature, winds 224–232
Bell Town (Duala) 197
Bembo 233
Benadir Coast 16, 18, 19, 299
Bende 132, **145**
Benghazi 2, 5, 14, 29, 58, 59, **60,** 61, **63**
Benguela 21, 243, 245, 247
  coast, 244
  current 86, 193, 222, 392, 445, 490
Beni Meskin 36
Benin 20, **145**
Beni Shangul 288
Beni Suef **73**
Benni-seed see *Semsem*
Benue R. 15, 150, 197
  junction 19
  upper 197
Berber 19, 260, 262, **268**
Bérbera 303, 304, **306, 307**
Berbéron 55
Berg winds 458
Beringer, Mr 338
Berlin 522

Bertholon, Dr 53
Bestion, M. 209
Bethanien **503**
Bethel 418, 423
Bethlehem 432
Bethulie 430, **435**
Bet Takue 274
Biafra, Bight of 193
Bibundi 193
Bihe 244
Bingerville 179
Bir Alali 7, 214
Birthday Camp **428**
Bishop's Court (Cape) **485**
Biskra 2, 5, 44, 46, **50, 51**
Bismarckburg 189, **190, 192**
Bissan 203
Bizerta 52, 57
Black-water fever *see* Fever
Black-wattle 440, *gloss.*
Blands R. 420
Blantyre 369, 370, **371, 378**
  district 370, 374
Blida 5, 46
Bloemfontein 391, 429, 430, 432, **433, 435**
Bloemhof (Cape Colony) 462
Bloemhof (Transvaal) 418, **425**
Blondiaux, M. 205
Bogos 274, 275
Boileau, Capt. 355, 372 (note)
Bokula 228, 234
Bolama 203
Bolongo 226
Bolobo 31, 226, **240, 242**
  situation 236
Boma 225
Bomokandi 229, 234
Bonny 132, 133, **141, 145**
  R. 139
  situation 139
Boporu 206
Boraba Kunda 100
Borgu 183
Bornu 146, 197
  N. 147, 150
Borrodale **417**
Boshof 429
Bottego, M. 301
Bougie 2, 4, 46, 47, **50, 51**
Bow-string hemp see *Hemp*
Brava 301
Brazzaville 210, 217, **221, 242**
  death-rate 211
Bremersdorp 420, **428**
Briart, M. 235
Briffet, Gambia, death-rate 103
British *cereals* 32 (*see* European)

British E. Africa 3, 10, 11, 18, 25, 30
  borderlands 302
British South Africa Co. 365
British *vegetables* 291, 313 (*see*
  European)
Broken Hill 366, **368**
Broma 380, 507
Bronkhorstspruit 423, 425
Brussels **519**
Bryntirion 424
Buckley, Mr 318
Buea **200, 202**
Bufarile 46
Bugena 286
Bukoba 30, 331, 339, 341, **350, 351**
  situation 343
Bukumbi gulf 342
Bukuru 147
Bulawayo 390, 394, 395, **396, 397,**
  407, **412, 416, 417**
  Government House **417**
Buloa **332**
Bulpett, Mr 264
Bumban 105
Bungo R. 245
Bunu R. 156
Burao 303
Burghersdorp **487**
Buruli *see* Mruli
Burumba 311
Bu Saada 2, 44
Bussa 94
Butiaba 88, **316**
Buttgenbach, M. 231
Bwigi 302

Cabbages 218
Cabo Delgado Province 384, 386
Cacao see *Cocoa*
Cacheo R. 204
Cairo 2, 21, 65, **70, 71**
Calabar 132-136, **142, 145**
  situation 139
Caledon 450
Calms, belt of, 84, 85, 92
Camels 82, 304
Cameroons 15, 17-20, 29, 86, 132, 150,
  193-202, 206, 207
  boundary 213
  Mt. 18, 20, 22, 133, 194
  N. 98
  productions 198
  rainfall **202,** 330
  R. 24, 198, **199, 202**
Cape Agulhas 254, 391, 445
Cape Blanco 19
Cape Coast (Castle) 8, 116-119, **123,**
  131

Cape Colony 1, 20, 24-26, 29, 390, 391,
  445-488
Cape Colony, Meteorological Dis-
  tricts :
  Basutoland 480
  Bechuanaland 390, 445, 483, **488**
  Cape Peninsula 447-449, **485**
  East-Central Karoo 445, 462, **486**
  Kaffraria 476-479, **488**
  North-East 473-475, **487**
  Northern Border 463, **487**
  Northern Karoo 463, **487**
  South Coast 457, **458, 459, 460,**
    486
  South-East 469-472, **487**
  Southern Karoo 445, 461, 462, **486**
  South-West 450, 451, **453, 454,**
    486
  West-Central Karoo 445, 461, **486**
  West Coast 451, 452, **455, 456,**
    485
Cape Delgado 14, 255, 384
Cape Frio 29
Cape Ghir (Rhir) 34
Cape Juby 2, 21, 35, 36, 37, **41, 43**
Capello, M. 244
Cape Lopez 209
Cape Naze 156
Cape of Good Hope 14
Cape Palmas 18, 20
Cape Peninsula 18, 20-22, 26, 447-449,
  485
Cape Spartel 4, 26, 34, **39, 43**
Cape Three Points 8, 18, 20, 22, 25,
  28, 97, 116, 185, 188
Cape Town 14, 19, 394, 436, 446
  Observatory 436, 445, **485**
  Town House **485**
Cape Verde 22, 26, 158
Caraway 33
Carnot 213
Carob beans 33
Carolina 418
Carrots 218
Casablanca 35-37, **43**
Cassanje 244
Cassava 97, 198, 218, 237, 248, 359
  374, 387, *gloss.*
Castor oil 324, 359
Cataracts district (Congo) 224, 225,
  233
Cattle 291, 304, 305, 313, 324, 359, 366,
  374, 426, 432, 440, 484, 496, 506
Cauliflowers 313
Cavalla R., lower 177
  source, region 205
Cayenne see *Pepper*
Cayor 156, 158

Ceded Mile, Gambia, death-rate 103
Cereals, European *see* European *cereals hardy* 291
Ceres 450, 451, **453, 485**
Chad, L. 6, 7, 15, 19, 20, 21, 22, 28, 29, 97, 146, 150, 152, 197, 214, 216
islands 197
Chad, Military District 208, 213, 214
Chafukuma Mt. 362
"Challenger" Report 82
Chambezi R. 353
Charter **417**
Chekna 215
Chesneau, M. 205
Chevalier, M. 7, 215, 216
Chick-peas 33
Chikwawa 371, **378**
Chillies 312, 341, 374, *gloss.*
Chimpili plateau 353
Chinama's 354
Chinchoxo 252, **253**
Chinde R. 380
Chinde (town) 506
Chiromo 371, **378**, 380
Chishawasha 12, **417**
Choga L. 308, 310, 312
Choma 359
Christansborg 2, 115, 116
Church, Corporal 6
Clanwilliam 19, 451, 452, **455**, 464
Clapperton, Capt. 7
Climatic Province, Kalahari 245
Mediterranean 32, 58, 64
Sahara 58, 72
Tropical 14
Clitandra 217
Clive Town 109
Close, Lt.-Col. 136
Cloud-ring 84, 85
Clozel, M. 213
Cocoa 97, 198, 217, 237, 247, 374, 508, *gloss.*
Cocoa-nut palm 97, 198, 218, 324, 341, 387, *gloss.*
Codrington, Mr Robt. 372
Coffee 97, 217, 247, 248, 291, 304, 312, 313, 341, 359, 374, 387, 426, 440, 508, *gloss.*
Colesberg **487**
Comber **251**
Companha do Nyassa 384
Congo area 17, 27, 29, 222, 224, 237
equatorial 223
Congo, Belgian *see* Belgian Congo
Congo da Lemba **242**
Congo Forest 18
Congo, French *see* French Congo, French Equatorial Africa

Congo R. 13, 15, 22, 24, 26, 86, 211, 236
headwaters 366
lower 14, 223-225, 233
mouth 20, 24, 28, 222, 252
Congo-Zambezi parting 362
Constantine 2, 5, 44, 46, **50, 51,** 55
Conversion Tables 525
Coomassie 116, 117, 126, **127, 131**
Copal 248, 341, *gloss.*
Copra 341, *gloss.*
Coquilhatville 226, 234
Coriander 33
Cork-oak 33, *gloss.*
Cotton 218, 237, 248, 265, 278, 290, 305, 312, 324, 341, 359, 366, 374, 387, 410, 426, 440, 508, *gloss.*
Egyptian 33
Courtet, M. 158
Cradock 461, 463, **487**
Cristal, Monts de, 237
Crocodile R. 423, 425
upper 423
Cross R. 26
Cumin 33, *gloss.*
Cunninghame, Mr 244
Cyclone, defn. 530

Dabus R. 288
Dahomey 20, 116, 182-186, 188
extreme N. 184
hinterland 172, 183
lower 183
productions 97, 98
slope 182
upper 182
Dakar 156, 157, **160, 176**
death-rate 158
situation 161
Damaraland 496
Damerghu 173
Damietta 66
Danakil 275, 299, 303
Danckelmann, Dr von 224
D'Anfreville, Dr 156
Dar-el-Beida 35
Dar-es-Salam 2, 9, 10, 334, 335, **344, 352**
situation 341
Darfur 27, 28, 258, 264
Dar Kuti 216
Darwin, Mt. 12, **417**
Date palm 33, 82, 265, 278, *gloss.*
De Bellay, M. 208
Debunja 2, **202**
Decœur, M. 172
Dega 274, 286, 290, 291
Deki Terfa 274, 275
Delafosse, M. 205

Delagoa Bay 20, 254, 504
Dellys 5, 46
Demagherim 173
Denham, Major 7
Der season 297, 298
Derna 29, 59
De Roquevaire's map 6
Destenave, M. 197
De Wittekranz 425
Dhurra 97, 247, 265, 278, 305, 312, 366, 374, 410, *gloss.* (*see* Kafir corn)
Didessa R. 287, 288
Dingire 171
Diospyros mespiliformis 248
Dire Daua 289
Doba 286
Doleib Hilla 9, **273**
D'Ollone, M. 205
Dominik, M. 197
Dondo 246
Dóngola 258
    province 260, 262
Dongorre 304
Dove, Dr
Drakensberg 418, 419
Drakenstein Mts. 445
Drewin 178, 179
Drought, Indian and Somali 299
Duala 2, 193, 197, **200, 202**
Dueim 9, 260, **269, 273**
Duke Town *see* Calabar
Dumjeri 147, **155**
Dundas, Commander 301, 323
Durban 390, 391, 394, 436, 437, **441, 444**
    wind 438
Duveyrier 45, 46
    M. 79
Duz 53
Dysentery 37, 115, 152, 179, 232, 322, 356, 363

Earth-nuts see *Ground-nuts*
East Africa, winds and rains 254–257
East Africa Protectorate 11, 317–328
    productions 324
    rain 327, 328, 429
Eastern Desert of Egypt 76
Eastern Tropical Africa 254–389
    winds and rains 254
East London 19, 394, 436, 469, **472**
Ebony 210, 217, 265, 278, 291, 324
Ebute Metta 133
Edward L. 28, 30, 308, 310
Egerton, Sir C. 298
Egypt 15, 29, 64–77
    difficulties with instruments 66
    lower 29, 64–71, 256
    upper 72–77

Egyptian Sudan *see* Sudan, Anglo-Egyptian
Ekotoweni 12
Elandsfontein 424
El Arish 34
El Arisha 47
Eldama Ravine 327
El Dueim 9, 260, **269, 273**
Elephant marshes 380
El Golea 46
Elgon, Mt. 308, 317, 321
Elit 275
Elliot, Col. 149
Elmina **131**
El Obeid 9, 86, **89, 271, 273**
Empandeni 396, 408, 409, **414, 416, 417**
Enkeldoorn **417**
Entebbe 2, 9, 11, **87**, 88, 308, 309, 310, 311, 312, **314**, 316
Epe **145**
Equateur district 226, 234
Equateurville 226, 227, 234
Equator, thermic 13, 19, 23, 27, 84, 85
Equatorial Forest 93, 115
Equatorial Nile region 258
Eritrea 23, 28, 30, 274–283
    productions 278
    rainfall **283**
Ermelo 418, 420, 423
Erskine, Mr 506
Eshowe **444**
Estcourt **443**
Ethiope R. 139
Europe, S., *crops* of 32
European *cereals* 98, 291, 305, 324, 341
    fruits 324, 341, 374
    vegetables 98, 218, 237, 278, 324, 341, 359, 374, 387, 410, 426, 432, 440

Fafan, Tug 296
Falabá 110
Faleme R. 156
False Bay 254, 457
Farim 204
Fasher 264
Fauresmith **435**
Fazokli 288
Felkin, Dr 265
Ferlo 156
Fernão Veloso 382
Ferrandi, Sig. 301
Fever 53, 167, 173, 195, 232, 264, 275, 303, 304, 309, 322, 323, 372, 373, 380, 384, 489, 506, 508

Fever—(continued)
  Black-water 37, 104, 106, 152, 355, 356, 372, 373
  Enteric 304, 357
  Intermittent 37, 245, 372
  Malarial *see* Malaria
  Remittent 372
  Yellow, 188
Fez 6, 34
Fezzan 58, 61
Fibre 304, 312, 341, 359, 366, 374, 410, 440
Ficksburg 432
Ficus 312
Field Point 139
Fife 31, **87, 361**
Figig 45
Figs 33, 410, 426
Filabusi **417**
Fischer, Dr 4, 35
Fisher, Rev. A. B. 313
Flamme, M. 228
Flax 33, 278, 359, 366, 410, *gloss.*
Florenkarte 7, 8
Forcados **145**
Fort Anderson 370, **371, 378**
Fort Archambault 8, 28, 214, **219, 221**
Fort Beaufort 469
Fort Crampel 8, 28, 196, **214, 221**
Fort Hall 10, 11, **327**
Fort Hill 31
Fort Hoima 309
Fort Jameson 357–359, **360, 361**
Fort Johnston 31, 370, **371, 377, 378**
Fort Maguire 384
Fort Mlangeni 371, 373, **378**
Fort National 4, 46
Fort Portal 309, **316**
Fort Rixon **417**
Fort Smith 2
Foulon, Rev. P. 355
Foureau, M. 78, 80
Francis Joseph Falls 226
Franzfontein 12
Fras, M. 171
Fraunberger, Dr 3, 6–8
Frazerburg 464
Freetown 2, 28, 105, 106, 109, **112–114**
  death-rate 108
French Congo 29, 207–212
  death-rate 210
  productions 217
  rainfall **221**
French Equatorial Africa 208–221
  productions 217
  rainfall **221**

French Guinea 18, 20, 22, 24, 29, 97, 158, 166–171
  productions 97, 98
  rainfall **176**
French Niger Territory 171–176
  rainfall 176
French W. Africa 4, 156–186
  rainfall **176, 178, 180, 185, 186**
Fruit-trees 33, 34, 426, 432, 484 (*see* European *fruits*)
Fuladon 168
Fuladugu 162
Futa Jallon 168, 171, **176**

Gaba Shamba 9, 90, **273**
Gabes 52, **57**
Gabes Gulf 58
Gabun 209
  R. 86
Gadabursi 304
Gaden 149
Gafsa 53
Gállabat 9, **273**
  district 263
Gambaga 8, 86, **89,** 117, 129, **130, 131,** 188
Gambela 9, **295**
Gambia 17, 26, 99–103, 109
  death-rate 103
  productions 97
  R. 156
Ganale R. 302
Gao 81
Gardulla 291
Garrerès, M. 183
Gautier, M. 8, 81
Gaya 184
Gazaland 17, 379, 390, 504–510
  productions 508
Geba R. 203
Gedáref 9, 20, **273**
  district 261, 263, 266
Gelma 5, 46
Gentil, M. 213
German E. Africa 3, 10, 13, 15, 17, 18, 31, 90, 329–352, 379
  productions 340, 341
  rain 351
  winds 256, 329, 330
German S.W. Africa 3, 12, 14, 26, 31, 390, 391, 394, 489–503
  coast lands 245, 490
  productions 496
  rainfall **503**
  winds and rain 490
German W. Africa 187–202
  rainfall **192, 202**
Géryville 47

Gezira 266
Ghaba Shambi *see* Gaba Shamba
Ghadames 62
Ghardaia 2, 46, **51**
Ghebli 59
Ghee 304, 305
Gibeon **503**
Gibraltar 14, 15
Gidani 288
Ginda 9, **280**
Ginger 98, *gloss.*
Ginir 290
Gleichen, Count 259
Glenconnor 461, **486**
Goats, 82, 291, 304, 313, 366, 374, 484
Gobabis 491, **503**
Gobwein 299
Gojeb R. 289
Gokhas 12
Golbanti 322
Golbo plain 290
Gold Coast 3, 86, 97, 115–131, 188, 206
  death-rate 118
  Northern Territories 86, 97, 98, 128–131, 188
  productions 97
  rainfall **131**
Gomoro R. 308
Gondar **295**
Gondókoro 8, 19, 88, 260, 265, **273**, 308–310, **316**
Gordon, Genl. 265
Goree 2, **176**
Gorongoza Mts. 504
  country 507
Graaf Reinet 390, 461, 462, **465, 486**
Grahamstown 469, **487**
Grand Bassam 29, 179, **180**, 181
Grand Lahu 179
Grape vine 33, 484
Grass-rubber 210, 217, 237
Great Olifants R. 423, 425
Greytown **443**
Gribingi 88, **89, 214, 221**
Griqualand, East 394
  West 464
Grootfontein (Ger. S.W. Af.) 491, 492, **503**
Ground-nuts 97, 198, 237, 313, 324, 341, 359, 366, 374, 387, 508, *gloss.*
Gu season 297, 298
Guardafui 19, 22, 256, 296, 299
Guinea, Upper 86
  hinterland 94
  Lower 86
Guinea corn 247, *gloss.*
Guinea wheat see *Maize*

Guizotia oleifera 278
Gulf of Guinea 14, 17, 100, 182
Gum 33, 265, 278, 291, 305, 366, 497, *gloss.*
Gurma 98, 172, 182
Gutu **417**
Gwanda **417**
Gwelo 395, 396, 406, **411, 416, 417**
Gwynn, Major 287

Habab 274–276
Hadeija 147
Hagai (Agai) 274, 297, 298
Hahn, Dr 4
Haikamchab 493
Halal 274
Halfa 33, 98, 511, *gloss.*
  zone 7, 8, 33
Halhal 274
Hamáda Kessera 55
Hamasen 274, 276
Hanover 463, **467**
Harar 286, 287, 289–291, 296, 304
Harmattan 37, 92, 93, 100, 105, 106, 109, 110, 115, 126, 128, 129, 138, 146, 147, 149, 150, 167, 178, 179, 183, 184, 187, 194, 215
Harrismith **485**
Hartley **417**
Hartmann Hill **417**
Hassi Gernan 78
Hasur 491
Haud 296, 298, 299, 304
Hausberg, Mr 321
Hawash R. 289
Heidelburg 418, 419
Heidke, Dr 341
Hekpoort 424
Hemp 33, *gloss.*
Hemp, bow-string 98, 237, 313, 341, 374, *gloss.*
Henna 82, *gloss.*
"Herald," H.M.S. 380
Herbertson, Dr 3, 7
Hereroland
Herr, Dr 213, 214
Heusis 12
Hex R. 423, 424
High Veld 419–421
Hill Station, Sierra Leone 104, 106 109–111, **112**
Hoakhanas 12, 491, **495**, 496, **503**
Hobley, Mr C. W. 321
Hoggar 80
Hopefontein 394, 396, 408, **413, 416, 417**
Hope Town 464
Hopkins, Mr F. F. 151, 152

Hora plain 373
Horn of Africa 13, 17, 19, 296–307
Horses 374, 440, 484
Hostains, M. 205
Howick **443**
Hübner, M. 206
Huddart, Mr L. H. L. 384
Huib plateau 489
Hume, Mr 76, 77
Humidity, defn. 530
Hupfeld, Dr 188
Hutchins, Mr 320, 321
Hydrographer 4
Hygroscope, difficulties 66

Ibádan **145**
Ibo harbour 382
Ikombe 338
Illig 297
Ilórin 146, 147, **148, 155**
Ilu-Babor Mts. 285
Inagu Mt. 379
Indian corn see *Maize*
Indian Ocean 329
  winds from 254, 397
Indigo 33, 278, 291, 312, 508, *gloss.*
Inhambane 506, 508
In Sala 82, **83**
Inyanga 507
Inyati **417**
In Zize 79
Irangi 341 (*see* Kondoa-Irangi)
Irebu 31, **242**
Iringa 336
Irumu Mts. 362
Isa 304
Isangi 228
Ismailia 2, 66, **68, 71**
Isobar, defn. 530
Isohyet, defn. 530
Isotherm, defn. 530
Itawa 356
Ituri forest 232, 237
Ivens, M. 244
Ivory Coast 13, 20, 21, 177–181, 205, 206
  productions 97, 98
  rain **178, 180**
  Upper 179

Jabbir 229
Jacobsdal 429, 430, **435**
Jacqueville 179
Jagersfontein, 430, **435**
Jalla, M. A. 364
Jamieson R. 139
Jebel Ahmar 54
Jebel Gharian 58

Jebel Marra 264
Jebel Nefus 58
Jedeida 54
Jelinek's Tables 66
Jibuti 22, 26, 85, 302, 303
  death-rate 302, 303
Jilal season 297
Jimma 291
Jinja 9, **316**
Johannesburg 2, 418, 419, 422–424, **427, 428**
Johnston, Sir H. H. 206, 319, 338, 370
Joubert Park 424
Jub R. 24, 299, 301, 305, 317, 323
Jubaland 321, 322
Juniper 291
Jurjura 45
Jute 98, 324, 373, 410, *gloss.*

Kaalfontein 424
Kaapsche Hoop 418
Kabambara 9, 10, 14, 90, **242**
Kabara 94
Kabinda 30, 243, 252, 253
Kabo 198
Kabulwebulwe 362
Kade 168
Kaedi 156
Kaffa 284, 286, 287, 291
  Mts. 285
Kaffraria 20, 436, 476–479
Kafir corn (dhurra) 97, 247, 265, 278, 305, 312, 366, 374, 410, 432, 440 484, 497, *gloss.*
Kafue R. 362
  hook 362, 366, **368**
  tributaries 366
  valley 363
Kagera R. 311, 312, 329, 331
Kairuan 52
Kalahari 390, 419, 483
  Climatic Province 245
Kalomo 366, **368**
Kambamba 246
Kambole 353
Kambove 231, 235, **236**
Kambwire 354
Kanem 7, 215, 216
Kannsala, Gambia, death-rate 103
Kano 146, 147, **148**
Kanoni stream 343
Kansanshi 366, **368**
Kansonso 232
Kaolakh 158
Karagwe 339
Karamojo 308
Karas Mts. 489

Karibib 12
Karimama 184
Karkoj 263
Karonga 370, **371**, 372
Karoo 16, 438, 461–468, 483, **486**, **487**, 490
    E. Central 445, 462, **486**
    N. 463, **487**
    S. 445, 461, 462, **486**
    W. Central 445, 461, **486**
Kasai R. 14, 26
    valley 237
Kasamanse 97, 165, 166
Kasempa 366, **368**
Kasongo 230
Kássala 9, 259, 261, 263, **273**
Kassimbo 246
Katagun 147
Katanga district 231, 235
    highlands 222
Katberg **487**
Katunga 380
Katwe 310
Kavirondo 321
Kawimbe 359
Kayes 156, 161, **163, 176**
Kazembe's 356
Kazungula 362, 364, 365, **367, 368**
Kebale 168
Keetmanshoop 390, 491, 492, **503**
Kef 53
Kenhardt 464
Kenia, Mt. 317, 323
    district 320
    forest region 320
Keremti season 274, 275, 285
Keren 276, **282, 283**
Kero 235
Kete Krachi 2, **192**
Ketosh 321
Keuei 274
Keyser, Consul 303
Khamsin 64, 75, 303
Khan valley 493
Kharif season 264, 303
Khartoum 18, 19, 258, 261, 264–266, **270, 273**
Khoz R. 308
Kibish valley 288
Kiboko R. 308
Kibokolo **251**
Kikuyu 2, 11, 319, **327**
Kilibula **361**
Kilimane 20, 382, 387
Kilimanjaro 256, 329
    district 330, 331
Kilinyatha 320
Kilossa 10

Kilubi 355
Kilwa 352
Kilwa I. (Ger. E. Af.) 334
Kilwa I. (Mweru) 359
Kilwa prov. 334, 337
Kilwa Kisiwani 334
    Kivinje 334
Kimberley 390, 391, 394, 445, 464, **468, 487**
Kimoko 233
Kimuenza 223, **241, 242**
    situation 237
Kindia 168
King William's Town 469, **471, 487**
Kintampo 129
Kioga L. *see* Choga
Kirk range
Kiro 235
Kirui 355
Kisaki 337, **351**
Kismayu 2, 11, 299, 322, 323, **328**
Kissidugu 166
Kisumu 11, **328**
Kita 162
Kitobola 233
Kitui 11
Kivu L. 30, 308, 329
Klein Letaba R. **428**
Klerksdorp 419
Klerkskraal 425
Klipfontein **485**
Knysna 26, **486**
Kodok 9, 262, 264
Kohain 274
Koinadugú 110
Kokomlaka 178
Kokstadt 476, **488**
Kola nuts 97, 198, *gloss.*
Komati Poort **425**
Komati R. 425
    basin 423
Kommbo, British, death-rate 103
Komoe R. 179
Konakri 166, 167, **169**, 171, 172, **176**
Konde, 330, 338, 339
Kondoa Irangi 10, 90, 337, **351**
Kongoa 198
Konta 291
Kontagora 146, **148**
Kordofán 86, 261, 264, 265
    forests 265
    S. 258
Kosha 291
Kowie R. 446
Kpeme 8, 188, **191, 192**
Krabbefontein **425**
Kromdraai 424
Kroonstad 430, **435**

Krugersdorp 423, 424
Kubub 491, 503
Kúkawa 6, **150**
Kunde 215
Kunene R. 14
Kuri archipelago 197
Kurkumes 322
Kuruman 483
Kusu 233
Kutu 233
Kwai **332**, 333, **347, 351**
  situation 342
Kwa-Kwa R. 381
Kwamkoro 330
Kwamouth 226
Kwango district 226, 233
  R. basin 237
Kwania 310
Kwanza R. 246
  lower 248
Kwitta 8, 116–119, **124**
Kwolla 274, 275, 290, 291

Labe 171
La Calle 5, 46
Lac Leopold II district 226, 233
Ladibrand 429, 430, 432, **435**
Ladismith 461
Lado 260, **273**
Lado Enclave 235
Laghuat 2, 5, 46
Lagos 21, 29, 132, 133, 135, **140, 145,** 206
  death-rate 133
  situation 139
Lai 8, 28, 214, **210, 221**
Lake Plateau 15, 24, 28, 317
  district 318
Lamothe 217
Lamu 317
Lancaster, Dr 236, 237
Landolphia 217, 237, *gloss.*
Landolphia owariensis 248, *gloss.*
Landolphia tholloni 217, *gloss.*
Langebergen 445
Lange Kloof Mts. 445
Lapparine, M. 78
Latuka 265, 310
Lauderdale 371, **376, 378**
Lealui 12, 86, **89**, 362, 364–366, **367, 368**
Lebompo Mts. 419, 504
Lefebvre, M. 285, 286
Lelabula season 365
Lemaire, Capt. 227
Lemasle, Dr 179
Lemba 233
Lembu 168

Lemons 33, 410
Lenfant, Capt. 94
Lentils 33, 265, 278
Leopoldville **221,** 225, 226, 233, 237, **242**
Leribe **488**
Liberia 14, 20, 24, 205, 206
  health 206
  productions 97, 98
Libreville 18, 31, 86, **89,** 90, **209, 212,** 221
  death-rate 211
Libyan Desert 75
Lichingo Mt. 379
Lichtemburg 418, 423, 426
Ligonia R. 379
Likungo R. 379
Limpopo R. 24, 395, 504, 507
  bend 19, 26, 29, 507
  valley 422
Lindi 9, 20, 335, **345, 352**
  Creek 342
  situation 342
Linseed 313, *gloss.*
Lisbon **520**
Little Popo 8
Livale 334, 337, **351**
Livingstone 366, **368**
Livingstone, Dr 383
Livingstone Mts. 339
Liwonde 370, **371**
Loanda 2, 14, 15, 17, 245, **249, 251**
  coast 244
Loango 210
Loangwa R. 353
Lobito Bay 247
Logone R. 215
Lolodorf 193
Lokoja 94, **148,** 150, **154, 155**
Lomami R. 228
Lome 8, **192**
London **524**
Lourenço Marques 2, 390, 504, 505, 508, **510**
Lovedale 469, **470, 487**
Lovu R. 353
Low Veld 418, 421
Lualaba district 226, 230, 234
  -Kasai district 233
Luambala R. 384
Luangwa R. 353
  valley 354, 356, 359
Luapula R. 353, 354
Lucerne 33, 410, 440, 484, 496, *gloss.*
Luchiringo R. 379
Lüderitz Bay 491, 495, **503**
Lufila R. 235
Lugh 28, **300, 301,** 302

Lujenda R. 379, 384
Lukanga R. swamps 366
Lukaya R. 237
Lukungu 225
Luli R. 387
Lulua R. 86, 236
Luluabourg 9, 27, 86, **89**, 90, 223, 226, 230, 234, **239, 242**
   situation 236
Lumba 233
Lumbe 364
Lunda 356
Lunga R. 366
Luozi 233
Lurio Bay 383
Lurio R. 379
Luru Hills 232
Lusambo 9, 14, 27, 90, 231, 233, **242**
Lusenfwa R. 362
Luvituku 233
Lydenburg 418, 420, 421, 426
Lyons, Capt. 23, 257, 262

Mabumbu 365
Machabel 366
Machakos 2, 11, 318, 319, 323, **326, 327**
Macheke **417**
Mackinder, Mr 321
Mackinnon Road 11, **327**
Maclear 394
Madagascar 15, 23, 256
Madder 33, *gloss.*
Mafeking 483
Magalies R. 423, 424
Mágdala 284, 286, **294**
Magrotto Mts. 333
Magweru 358, **361**
Mahenge 10, 31, 337, **351**
Mahogany 210, 217, 265, 291, 324
Mai Ambessa 275
Maifoni 146
Maize 32, 97, 98, 237, 248, 305, 313, 319, 324, 341, 359, 374, 387, 432, 440, 484, 496, *gloss.*
Maji 289
Makindani 9
Makindu 11
Makonde plateau 342
Makraka 265
Maktar 2, 5, 53, 55, **56, 57**
Makua country 383
Malagarasi R. 329
Malanje 244
Malaria 104, 137, 147, 152, 153, 167, 206, 207, 263, 275, 299, 309, 320, 334, 353–356, 363, 373, 380, 387, 395, 420, 421, 429, 438, 500, 504, 506, 507

Malemba 252
Malezi 359
Malindi 2, 9, 317, 322, **328**
Malmesbury 451, **485**
Mancarra see *Ground-nuts*
Mandara Mts. 150, 197
Mandioc, manihot see *Cassava, gloss.*
Mandobim see *Ground-nuts*
Manenguba 198
Mangin, M. 205
Mangoes 341
Mangrove bark 508, *gloss.*
Manika 506
   district 507, 508
   plateau 504
   valleys 507
Manioc see *Cassava, gloss.*
Manow 10, 31, 338, 339, **351**
Maples, Bishop 383
Marabastad 421
Maragolia Hills 321
Marandellas **417**
Marangu 331
Mareb R. 275, 285
Maria 274, 275
Maria (season) 365
Marico 418, 420, 421, 426
Maritzburg 436, 437, 439, **442–444**
Marocco 5, 16, 18, 20–22, 24, 26, 28, 30, 32, 33–43
   granary 34
   map 34
   rains 35
Maroccan Atlas 32, 45
Marocco-Algeria frontier 24
Marotseland 362, 364, 365
Márrakesh 6, 34, 35, 37, **42, 43**
Marrows 313
Martino, M. 59
Masailand 331
Masaka **316**
Maseru **488**
Mashonaland 396, 398–405, **416, 417**
Massangano 246
Massaua 9, 30, 31, 275, 277, **279, 283**
Matabeleland 16, 396, 406–**417**
Matadi 225
Matjesfontein **486**
Matoka plateau 366
Matoppo Hills 395
   Mission **417**
Mau escarpment 319
Maud, Capt. 288
Maurer, Dr 331
Mawambi 235
Mayumba **221**, 237
Mazindi **332**
Mbarara 9, 309, **315, 316**

Mbomu R. junction 229
Mbumbi season 365
McCarthy, I., Gambia, death-rate 103
McMillan, Mrs 264
Mealies, 366, 410, 426, 432, 484, *gloss.*
Means, explanation of 530
Medea 5, 44, 46
Medenin 53
Mediterranean Climatic Province 32, 58, 64
 Sea, breadth 47
 Sea, region 25
 Sea, winds from 47
Mejerda 54
Melmoth **444**
Melons 265
Melsetter 395, 399, 400, **404, 416, 417**
Mengo 309
Mensa 274, 275
Mérowe 259, **268**
Metarika 384
Mfumbiro 308
Michel, M. 285
Middleburg 418, 420, 421, 423, 425
Middle Veld 419–421
Mikindani 90, **352**
Millet 32, 82, 98, 237, 248, 291, 312, 319, 359, 374, 387, *gloss.*
Misahöhe 2, **191**
Mizinda 311
Mizon, Capt. 196
Mlangeni *see* Fort Mlangeni
Mlanje 369, 370
 district 370, **378**
 Mt. 379
Mobaye 217, 229, 235
Mocker-Ferryman, Capt. 147
Mogador 33–35, 37, **40, 43**
Mogdishu 20, 296, 299
Mohalie's Hoek 480, **482, 488**
Molo 11, **327**
Mombasa 2, 9, 13, 16, 30, 317, 322, **325, 328**
Mombo **351**
Money, Mr 354, 373
Mongalla 8, **87**, 260, **272, 273**
Monrovia 206
Monsoons 92, 171, 255, 256, 297–299, 303, 318, 322, 329, 330, 370, 385
Montagu 450
Monts de Cristal 237
Monza 364, 366, **368**
Mooi R. 423–425
Mopani tree 248, 359, 497
Mopea 2, 12, 31, 381, **388**
Morabala marshes 380

Morocco *see* Marocco
Morogoro *see* Mrogoro
Moshi 331
Mosi-wathunya 365
Mossamedes 243, 247
 coast 244
Mossel Bay 446, 457, 458, **459, 486**
Mosuril Bay 389
Motsha Mts. 285
Mount Darwin 12, **417**
Mount Edgecombe **444**
Mozambique 20, 379–389, 504
 channel 16, 31, 254, 256, 386
 current 392, 445
 productions 387
 town 381, 382, **389**
Mpwapwa 10, 337, 341, **351**
M'Rewas **417**
Mrogoro 10, 337, 341
Mruli 309
Msalabani-Magila 333, **351**
Msalla **352**
Mtai **382**
Muala district 383
Mualia 384
Muanza 10, 90, 339–341, 349, **351**
 situation 342
Muembe 383
Muhulu Mts. 337
Muidir 79
Muirhead, Mr 109
Mulazzani, Sig. 275
Mulema 311
Mules 304, 484
Mumbwa 366, **368**
Mumia's 2, 309, **328**
Munda season 365
Mundo 334
Muni R. 207
Murzuk 62
Mustard 278
Mweru L. 28, 353, 356, 359
Myara **378**

Nachtigal, M. 7
Naidaus 12
Nairobi 10, 11, 320, 323, **327**
Naivasha 10, 11, 319
Nakuru 11, **327**
Nalisa 364
Nalolo 365
Nama-Dama tableland 489
Namaqualand 46
 Little 19, 445, 451, 464
Namatara R. 308
Namieb desert 490, 492
Namuli Mts. 379, 383
 Pico 379

K. A.

35

Nandi 21, 320, 323
Nasser 9, 90, **273**
Natal 14, 16, 18, 254, 391, 436–444
  productions 439
  rain **444**
Natete 88, **316**
Ndelle 216
Nectarines 426
Nel's Poort **486**
Nemours 4, 46, 47
Nepoko R. 235
Neu Langenberg 31, 338
Newcastle **443**
Newlands 446
Ngaundere 196
Ngombe 336
Ngorongoro 341
Nguru Mts. 329
Nhamonga (Nyamongo) Mts. 504
Niam-Niam 265
Nieuwveld 445
Niger R. 15, 18, 19, 64, 182
  bend 7, 22, 81, 85, 93, 171
  delta 13, 16, 18, 19, 21, 22, 30, 94
  gorge 23
  lower 94, 139
  mouths 24
  upper 96, 97, 205
  valley 94, 146
Nigeria 3, 98, 132–155
  Northern 146–155
    compared with Sierra Leone 109
    death-rate 150
    productions 97, 98
    rainfall **155**
  Southern 132–145, 206
    death-rate 133
    productions 97, 98
    rainfall **145**
Nigeria-Cameroons boundary 136
Nile Province 310
Nile R. 13, 21, 23, 28
  basin 23
  bend 265
  delta 19, 21, 26
  valley 23
  Blue 21, 23, 266, 288
    forests 265
    region 261, 263
    upper 258, 259
  White 16, 21, 265
    districts of Uganda 310
    equatorial region 258
    plain 261
    upper 258, 259, 269
    valley 258, 259
  Victoria 308, 310
Nimule 265, **273**, **316**

Niori R. 211
Niugh 278
Njoro 319, 320
Nkandhla **444**
Nkata Bay 25, 27, 370, **371**, **377**, **378**
No, L. 23, 262, 264
Nogal 298, 304
Nomtsas **503**
Nondweni **444**
Nongoma **444**
Northern Territories, Gold Coast 88,
  97, 98, 128–131, 188
Nouvelle Anvers 20, 31, 227, 228, 233,
  242
Novo Redondo 247
Nubian Desert 260, 262
Nunez R. *see* Rio Nunez
Nyando valley 318
Nyangwe 90, 230
Nyasa 13, 14, 16, 17, 21, 25–28, 31,
  329, 330, 336, 338, 353, 354, 369,
  372, 373, 380
  plateau 379, 385, 387
Nyasaland 86, 363, 369–378
  N. 370, **378**
  productions 373
  rain **378**
  W. 370, **378**
  winds 255, 370, 373
Nyasa-Tanganyika plateau 338, 353,
  355, 356, 372, 373
Nyika plateau 317, 369
Nylstroom 421

Oats 359, 410, 426, 432, 440, 484, 496,
  gloss.
Obbia 298
Obel R. 275
Obok 296
Observatory, Royal (Cape) *see* Cape
  Town
Ógaden 305
Ogowe R. 20, **209**
Oil 278
  seeds 374
Oil-palm 97, 198, 218, 237, 248, 313,
  341, 359, *gloss.*
Oil Rivers 132
Okahanja 2, 390, **503**
Okankulyo 12
Okombahe 12
Old Calabar R. 139
Olifants R. 420
  Great, upper 423, 425
Olives 33, *gloss.*
Olokemeji **145**
Olukonda 2, 12, 494, **503**
Omalako 12

Omaruru 2, 12, 494, **495, 503**
Omo R. 287, 288, 291, 302
  upper 289
Ondangui 12
Ondo **145**
O'Neill, Consul 383
Onicha 132, 133, **144, 145**
  situation 139
Onions 265, 278, 341
O'okiep 391, 394
Opium 33
Oran 2, 4, 33, 44, 46, 47, **50, 51**
  meridian of 47
  S. 45, 46
Orange Free State 18–20, 24, 25, 391,
  425, 429–435
  productions 432
  rain **435**
Orange R. 13, 20, 21
Oranges 33, 291, 410
Orleansville 2, 4, 46, 47, **50, 51**
Oshogbo **145**
Ostriches 484
Otavi 492
Ottawa **444**
Otyikango 12
Otyimbinge **503**
Otyiseva 494, **495**
Oudtshoorn 461, **486**
Outyo 12, 491
Ovampoland 489, 496
Ovington **444**
Oyo (Awyaw) **145**

Paarl 450, **485**
Palm-oil 198, 237
  wine 247
Pandiuli 384
Pangani 9, 334, **352**
Paraku 172
Paramiho 336
Pare Mts. 329
Paris **523**
Parkinson, Mr 304
Pavel, M. 150
Peaches 291, 410, 426, 432
Pears 426
Pearson, Mr 492
Peas 98, 278, 313, 341, 374
Peixo R. 245
Pella 464, **487**
Pemba Bay 384, 386
Pepper 278, 440, *gloss.*
Persian Gulf 23, 256
Peters, Dr 507
Philippeville 5, 46
Philippolis 429–431, **434, 435**
Pico Namuli 379

Pienaars R. 423, 424
Pietermaritzburg *see* Maritzburg
Pietersburg **426**
Piet Retief 420, 421, 426
Pilgrims Rest 418, 420
Pine-apples 218, *gloss.*
Piquetberg 451, **485**
Pitsani 418, 420
Plat, M. 171
Plat R. 423, 424
Plehn, Dr 194, 196
Plums 425, 432
Podor 158, **176**
Poea abyssinica 278
Pomegranates 410
Pondoland 21
Ponthierville 230
Popokabaka 226, 233
Port Alfred 446
Port Elizabeth 14, 19, 22, 394, 436,
  437, 446, 457, **460, 486**
Port Florence 11, **328**
Port Herald 371
Port Nolloth 390, 391, 394, 451, **456,
  485**
Porto Novo 184
  rain **185, 186**
Port Said 2, **71**
Port St John's 436, 476, 477, **478**
Port Shepstone **443**
Portuguese E. Africa, 390 (*see* Gaza-
  land, Mozambique)
Portuguese Guinea 20, 24, 203, 204
  productions 97
Potatoes 98, 218, 313, 324, 341, 426,
  432, *gloss.*
Potchefstroom 418, 423, 426
Powell-Cotton, Major 232, 235
Pretoria 2, 418–421, 423–426, **428**
  Arcadia **425**
Prieska 464
Prince Albert 461, **486**
Progress Farm **417**
Province Orientale 230, 235
Pugu Hills 341
Pumpkins 359, 432
Pungwe R. 504
Putfontein 425

Qudeni 438, **444**
Queen's Town 473, **475, 487**
Quilimane *see* Kilimane
Quinces 410
Quthing **488**

Rainfall maps, latest 3
Raisins 33
Ramie 359, 366, 374, 410, *gloss.*

Rangatan plateaux 317
Raphia vinifera 198, 237, 247, *gloss.*
Red Sea 14, 17, 22–24, 30, 31, 274, 275, 278, 299
Rehoboth 2, 12, 492, 494, **495, 503**
Reid, Mr R. L. 507
Rejaf 265
Resin 305
Rhenosterfontein 425
Rhodesia 3, 12, 21, 232, 353–368, 395–417
 Northern 353–368
 North-Eastern 353–361
  productions 359
  public health 357
  rain **361**
 North-Western 362–368
  productions 366
  rain **368**
  winds 255
 Southern 12, 14, 30, 390, 391, 395–417
  productions 410
  rain **416, 417**
  winds 254, 392–394
Ricardo, M. 207
Rice 33, 97, 198, 218, 237, 312, 313, 324, 341, 373, 374, 387, 508, *gloss.*
Richaud, M. 181
Richmond (Cape Colony) 463
 (Natal) **443**
Rietfontein 419, 424, **428, 486**
Rif 34
Rift Valley 10, 11, 317, 319, 323
Rikatla 505
Rio Nunez 97, 167, 168
Ripon Falls 9, 309
Rohlfs, M. 7
Rome **521**
Root crops 32
Roquevaire's map 6, 34
Roseires **273**
Rosetta 66
Ross, Mr 320, 321
Rouget, M. 7
Ruaha R. 13
Ruanda 339, 341
Rubber 198, 210, 237, 248, 265, 312, 324, 341, 374, 387, 410, 440, 508 (see *Grass-rubber, Sanseviera,* and *gloss.*)
Ruchigga 311
Rudolf, L. 11, 13, 20, 288, 302, 308, 319
Rufigi R. 329
Rukirra 311
Rukusi R. valley 354
Rukwa L. 338

Ru Nzori 235, 308, 310, 313
Rusapi **417**
Rustenburg 418–421, 423, 424, 426
Rutenganio 339
Ruvuma R. 329, 379, 383, 385, 387
Ruwe **236**
Rye 410, 426

Sabaki, R. 322
Sabderat 275
Sabi R. 395, 504, 506
 lower 508
Sable Antelope mine 366, **368**
Saffron 33, *gloss.*
Safi 35
Saganeiti 276
Sahara 6, 17, 19, 25, 29, 44, 46, 78–83, 97, 319
 Algerian 53
  rain percentage 48
 central 79
 N., oases 78
 N.W. 81
 productions 82
 S. 6, 81, 156, 161
 Tripolitan 58, 61
 winds drawn to 61, 92
 winds from 44, 45, 61, 92, 138, 146, 152
Saharan Climatic Province 58, 72
 flora 8, 80
 oases 78, 82
 zone 93–96
Sahel 44, 52, 275
St Louis 2, 6, 26, 28, 98, 156, 157, 158, **159,** 171
 rainfall **176**
 situation 161
St Mary I. **102**
Sakarra **332**
Saki **145**
Salisbury 12, 395, **396, 397,** 398, **402, 416**
 hospital **417**
 public gardens **417**
 Westridge **417**
Salisbury L. 308
Salole 302
Salum R. 156, 158
Samhar 275
Samia Hills 321
Samien 286
Sandfontein 424
Sandwith, Dr 72
Sangoia 197
San Salvador 243, **250, 251**
Sansanne Mangu 8, 188
Sanseviera cylindrica 374, *gloss.*

Sanseveria eherenbergii 304, 313, 341, 374, *gloss.*
Sanseviera longiflora 237, *gloss.*
Sapele 132, 133, **143, 145**
   situation 139
Sassandra 178, 179
Saura, Wad 78
Say 19, 94
Schaaprivier 2, **503**
Schirmer, Dr 82
Schoonspruit 423-425
Sebungwe **417**
Sedhiu 165
Sefula 364, 365
Segu 96
Seif season 264
Sejumi 54
Sékondí 8, 116, 117, 119, **121, 131**
Selit 278
Selukwe **417**
Semliki R. plain 310
Semsem 33, 98, 248, 265, 305, 313, 341, 387, *gloss.*
Senanga 365
Senegal 19, 156-165
   productions 97, 98
   rainfall **176**
   Upper 161-165
      rainfall **176**
      productions 97, 98
   R. 156
      lower 158
Sennar 263
Serae 274, 275
Service Météorologique 4, 156, 167, 181, 184
Sesame see *Semsem*
Sesheke 365
Setif 5, 46
Sfax 5, 52, 53
Shaddocks 33, *gloss.*
Shari R. 213, 215
Sharpe, Sir A. 372
Shawia 35, 36
Shea 98, *gloss.*
Sheep 82, 291, 304, 305, 313, 366, 374, 484
Shehili *see* Sirocco
Shilluk 262, 264
Shiloh **417**
Shimoni 2, 9, **328**
Shirati 10, 330, 331, 340, **351**
Shire R. district 370, **378**
   highlands 369, 372, 374, 385, 386
   valley 369, 380
Shirwa L. 379
Shita season 264
Shoa 284

Siarea 168
Sidamo 284
Sidi-bel-Abbes 4, 46, **49**
Sidra, gulf 58
Sierra Leone 13, 14, 20, 24, 30, 85, 104-114, 205
   compared with other colonies 109
   death-rate 108
   productions 97
Sievers, Dr 7
Sigiri 96
Simon's Town **485**
Simoon 61
Sinai 76, 77
Sinoia **417**
Sipolelos 12, **417**
Sirocco 44-46, 52, 54
Siroko R. 308
Sisal hemp see *Agave sisalana, gloss.*
Sitanda 366
Siwa oases 75
Small-pox 115, 152, 245
Smithfield **435**
Sneeuwberg range 445
Sobat R. 15, 23, 261, 287
   region 258, 260, 261
Sofala 21, 22
Sok-el-Arba 52
Sok-el-Jemaa 53
Sokóde **192**
Sókotó 97, 146, **148, 155**
Somaliland 13, 17, 18, 28, 296-307
   coast, rains 289, 290
      winds 256
   operations 303
   productions 304, 305
   W., winds and rain 256
   Abyssinian 297, 299-302
   British 13, 296, 297, 303, 304
   French 85, 296, 302-304
      death-rate 302, 303
   Italian 297, 299-302
Somerset East 462, **466, 486**
Songea 336
Sorbo 94
Sorghum see *Kafir corn*
South Africa, *see* Africa, S.
South African plateau 418
Southern Europe, *crops* 32
Spanish Guinea 22, 207
Spilsbury, Major 383
Spitzkop 425
Standerton 418, 419, 422-425
Stanger **443**
Stanley, H. M. 310
Stanley Falls 13, 230
   Pool 31, 210, 224
      district 225

Stanleyville 230
Steel, Capt. E. A. 136
Stellenbosch 450
Sterculia 198
Steynsberg 463
Stieler 4
Stock-rearing see *Cattle*
Strawberries 313, 359
Stuhlmann Sound 340
Stutterheim 445, 469
Suakin 2, 30, 31, 258, 260, 261, 263, **273**
Sudan 17, 18, 19, 215, 256, 264
　Anglo-Egyptian 17, 66, 256, 258–273
　difficulties with instruments 66
　productions 265
　rainfall 260, **273**
　temperature 262
　wind 259
　plains 23, 27, 278, 284
　Western 6, 8, 16, 25, 93, 97
Suez 2, 66, **69, 71**
Sugar 33, 98, 218, 237, 248, 265, 291, 341, 387, 440, 508, *gloss.*
Sultan Mamoud 318
Sun, zenithal position 13, 15–17, 19, 21, 23, 25–27, 29, 30
Susa 5, 52, 53
Sutherland 463, **487**
Swakopmund 491, 495, 496, **498, 503**
　situation 497
Swayne, Col. 303
Swaziland 420, 421, 426
Sweet potatoes 98, 198, 218, 237, 248, 313, 359, 366, 374, 387, *gloss.*
Swellendam **486**

Tabarka 4, 46
Table Bay 391, **449**
Table Mountain 392
Tables, conversion 525
Tabóra 10, **87**, 90, 337, 341, **348, 351**
　situation 342
Taf 278
Tafie **192**
Takaungu 2, 9, 317, **328**
Tamarinds 278, *gloss.*
Tamarisks 493
Tana R. 322
　valley 322
Tanga 2, 9, 331, **352**
Tanganbili wind 297
Tanganyika 10, 12, 13, 16, 18–20, 22, 24, 28, 30, 232, 329, 338, 341, 353, 356, 359
　district 331, 338, 359

Tanganyika Concessions, Ltd. 232
Tangier 4, 21, 34–37, **43**
Tapioca see *Cassava, gloss.*
Tarkwa 115
Tea 313, 374, 440, 508, *gloss.*
Tebessa **50**, 55
Tegwani **417**
Tell, 33–35, 44, 46, 52, 53
Temborari 317
Tenda Ba, Gambia, death-rate 103
Tenezruft 80
Teniet-el-Had 5, 46
Terms, explanation of 530
Tesalit 80
Tete 507
Teyateyaneng 480, **481, 488**
Thala 55
Thermal Equator see Equator
Thermometers, spirit, difficulties 66
Thiès 156
Tiaret 5, 46, 47
Tiassale 172
Tibesti 16, 22, 25
Tidikelt 79
Tigre 285
Tilemsi 81
Timbo **170**, 171, **176**
Timbuktu 6, 81, 97, 172, **174, 176**
　situation 173
Timimum 82, **88**
Tirma 288
Tiva R. 320
Tizi-Uzu 5, 46
Tlemsen 5, 44, 46, **50**
Tobacco 33, 82, 97, 218, 237, 248, 265, 278, 313, 324, 341, 359, 366, 374, 387, 410, 426, 432, 440, 484, 508, *gloss.*
Tobruk 5
Togoland 3, 187–192
　health 188
　productions 97, 98
　rainfall **192**, 330
　S. 188
Tokar 260
Tongaland 440
Toro 308–310, 313
Tosamaganga 336
Tosaye 23
Toutée, M. 172
Trade winds 14, 34, 37, 84, 85, 92, 254–256, 297, 329, 330, 338, 353, 394
Transvaal 14, 18, 30, 391, 418–428
　productions 425, 426
　rainfall **428**
Trichardsfontein 425
Tripoli 2, 5, 16, 26. 29, 58–63, **68**

Tripolitan Sahara 58, 61
Tropic of Cancer 16, 21, 23, 30, 85
  Capricorn 19, 21, 22, 24, 30
Tropical Africa *see* Africa, Tropical
Tropical Climatic Province 14
Tropical seasons 84–91
Trotter, Maj.-Gen. J. K. 104, 110
Tseddia 274
Tsoabis 491, 494
Tsomo **488**
Tu R. 308
Tubas R. 492
Tulbagh 450, 461, **485**
Tuli 395, 396, 409, 410, **415, 416, 417**
Tulu Dimtu 286
Tumba 225, 233
Tunis 2, 4, 5, 15, 16, 20, 22, 26, 28, 29, 32, 33, 35, 52–57
  compared with Algeria 52
  health, death-rate 53
  productions 32, 33
  rain 52, **57**
  town **54, 55, 57**
Tunisian Atlas 53
Tuzur 53

Ubangi district 229, 234
Ubangi R. 213, 229, 234
  bend 15, 30
  junction 15, 22, 24, 31
  Upper 216, 217, 229, 235
Ubangi-Shari region 208
  productions 217
Ubangi-Shari-Chad province 213–221
Uganda Protectorate 11, 308–316
  productions 312, 313
  Proper (Buganda) 310, 321
  rainfall **316**
Uganda bark cloth 312, *gloss.*
Ugogo 330, 331
Uha 341
Uhehe 334
  Mts. 329
Uitenhage **486**
Ujiji 10, 12, 90, 338, 341, **351**
Ukamba 10, 320
Ulanga 336
  R. 337
Ulu 320
Uluguru Mts. 337
Umangi 234
Umtali 395, **396**, 399, **403, 416, 417**
  Utopia **417**
Umtata 476, 477, **479**
Umvoti 440
Umzinto **443**
Uniondale 461

Unyamwezi 331
Unyoro 309
Upington 464, **487,** 491
Upper Senegal 161–165
  rainfall **176**
Urema R. 507
Urundi 341
Usagara Mts. 329
Usambara 330, 331, 341
  East 333
  highlands 329
  rainfall **332**
  temperature **332**
  West 333, 342
Usambura 12, 22
Usiba 339
Usoga 321
Uwemba 359

Vaal R. 13
  upper 423, 424
Vandeleur, Capt. 309
Vanga 331
Vanilla 218, 508, *gloss.*
Vasconcellos, Senhor 203
Vasse, M. G. 507
Vereeniging **425**
Verulam **443**
Vetches 33, 278
Vicenti 380
Victoria (Cameroons) 193, **202**
  (Rhodesia) 396, 400, **405, 416, 417**
Victoria Falls 28, **368**
Victoria Nile 308, 310
Victoria Nyanza 10, 16, 18, 22, 25, 28, 30, 308–310, 312, 318, 321, 323, 329, 330, 331, 339, 341–343
Victoria West 463
Villamur, M. 181
Vine see *Grape-vine, gloss.*
Vivi 31, **242**
Vlakfontein 424
Vogel 6
Voi 11
Volksrust **425**
Volta R. 182
  white 129
Von Grünau 75
Vrede 432
Vryburg 390, 483, **488**

Waarkraal 425
Wadai 217
Wádelai 13, 87, 88, 260, **273**, 308, 309, **316**
Wadi Halfa 2, 22, 23, 72, 258–260, 262, **267**

Wad Médani 9, 20, **273**
Wad Rhir (Ghir) 44
Wagadugú **175, 176**
  situation 175
Wagenaar's Kraal (Cape Colony) 463
Wagensdrift 424
Wagger Mts. 298
Wakefield, Mr 322
Wakkerstroom 418, 423
Walamo 284
Walega 288
Walfisch Bay 2, 16, 22, 391, 489, 491, 500, **502, 503**
  wind 501
Wallace, Mr 355, 372 (note)
Walnuts 410
Wangermannshöhe 339
Wargla 46
Warmbad 491, 492, **503**
Wassaw 115
Waterberg (Ger. S.W. Af.) 2, 491, 492, **503**
  (Transvaal) 421, 424, 426
Watherston, Lt.-Col. 129
Watson, Sir C. 263, 265
Wau 8, 86, **89**, 90, 260, 262, **273**
Weatherley, Mr 354
Web R. 302
Webi Shebeli 296, 298, 299, 305
Weenen **443**
Weisgerber, Dr 35
Welle R.
  district 229, 234
Wepener 429, **435**
Weri R. 285
Westacre **417**
West Coast 15, 21, 24, 27, 29, 31, 92–206
  productions 97
Western Tropical Africa, N. Section 92–206
  S. Section 207–253
Wheat 33, 82, 97, 98, 265, 278, 291, 324, 341, 359, 374, 387, 410, 426, 432, 440, 484, 497, 506, 508, *gloss.*
White R. 423, 425
  settlement 425
White, Silva, Mr 75
Whitehouse, Commander 318
Whyte, Mr A. 206
Wida 116, 185
Wilhelmstal 12
Windhoek 2, 390, 394, 489, 491–493, **494**, 495, **499, 503**
  situation 497

Winds and rain (S. Africa) 392–394, 490
Winds of E. Africa 254
Witbank 419
Witklip 425
Witwaters Rand 418–420, 423, **428**
  escarpment 418
Woelffel, M. 205
Woina Dega 274, 276, 278, 286, 290, 291
Wojerate 286
Wolmaranstad 418, 422
Worawora 192
Worcester 450, 451, **454**
Wynberg **485**

Yakoma 229, 234
Yams 98, 168, 218, 341, *gloss.*
Yao country 383
Yaunde 86, **89**, 193, **201, 202**
Yelwa 94
Yola 8, 24, 147, **148, 155**

Zaghuan 52
Zambezi R. 14, 28, 353, 366, 379, 380, 381, 395, 504, 507
  basin 504
  delta 12, 19, 31, 380, 504
  falls, 366
  mouths 19
  source 362
  Upper 86
  valley 362, 363, 380, 387
Zandfontein 425
Zanzibar 22, 25, 26, 29, 30
Zaria 146, 147, **148, 155**
  Hills 93
Zaytoun, Dr 37
Zebe **191**
Zeerust 420
Zeila 13, 303, 304, **306, 307**
Zemio 216
Zesfontein 491, **503**
Zinder 97, 172, 173
Zomba 2, 86, **87**, 370, **371, 375**
  district 370
  Mt. 372
Zonderende plateau 445
Zongo Gorge 229
Zoutpansberg 421, 426
Zululand 22, 438, **444**
  wind 438
Zumbo 381
Zúngerú 8, 147, **148**, 151, **154, 155**
Zusfana, Wad, 78
Zwai L. valley 290
Zwartebergen 445

For EU product safety concerns, contact us at Calle de José Abascal, 56–1°, 28003 Madrid, Spain or eugpsr@cambridge.org.

www.ingramcontent.com/pod-product-compliance
Ingram Content Group UK Ltd.
Pitfield, Milton Keynes, MK11 3LW, UK
UKHW020358210126
466816UK00030B/356